TUBULAR STRING CHARACTERIZATION IN HIGH TEMPERATURE HIGH PRESSURE OIL AND GAS WELLS

Multiphysics Modeling

Series Editors

Jochen Bundschuh
University of Southern Queensland (USQ), Toowoomba, Australia
Royal Institute of Technology (KTH), Stockholm, Sweden

Mario César Suárez Arriaga
Department of Applied Mathematics and Earth Sciences,
School of Physics and Mathematical Sciences, Michoacán University UMSNH,
Morelia, Michoacán, Mexico

ISSN: 1877-0274

Volume 7

Tubular String Characterization in High Temperature High Pressure Oil and Gas Wells

Jiuping Xu
Sichuan University, Chengdu, China

Zezhong Wu
Chengdu University of Information Technology, Chengdu, China

CRC Press
Taylor & Francis Group
Boca Raton London New York Leiden

CRC Press is an imprint of the
Taylor & Francis Group, an **informa** business
A BALKEMA BOOK

CRC Press/Balkema is an imprint of the Taylor & Francis Group, an informa business

Typeset by MPS Limited, Chennai, India
Printed and bound in the United States of America by
Edwards Brothers Malloy on sustainably sourced paper

Published by: CRC Press/Balkema
P.O. Box 11320, 2301 EH Leiden, The Netherlands
e-mail: Pub.NL@taylorandfrancis.com
www.crcpress.com – www.taylorandfrancis.com

Library of Congress Cataloging-in-Publication Data

Xu, Jiuping, 1962–
Tubular string characterization in high temperature high pressure oil and gas wells / Jiuping Xu, Sichuan University, Chengdu, China & Zezhong Wu, Chengdu University of Information Technology, Chengdu, China.
pages cm. – (Multiphysics modeling, ISSN 1877-0274 ; volume 7)
Includes bibliographical references and index.
ISBN 978-1-138-02670-4 (hardcover) – ISBN 978-1-315-74824-5 (ebook)
1. Oil well casing. 2. Oil well drilling. 3. Petroleum engineering. 4. Gas well drilling. 5. Gas drilling (Petroleum engineering) I. Wu, Zezhong, 1970– II. Title. III. Title: Tubular string characterization in high temperature, high pressure oil and gas wells.
TN871.22.W85 2014
622′.338–dc23

2014038628

ISBN: 978-1-138-02670-4 (Hbk)
ISBN: 978-1-315-74824-5 (eBook PDF)

About the book series

Numerical modeling is the process of obtaining approximate solutions to problems of scientific and/or engineering interest. The book series addresses novel mathematical and numerical techniques with an interdisciplinary emphasis that cuts across all fields of science, engineering and technology. It focuses on breakthrough research in a richly varied range of applications in physical, chemical, biological, geoscientific, medical and other fields in response to the explosively growing interest in numerical modeling in general and its expansion to ever more sophisticated physics. The goal of this series is to bridge the knowledge gap among engineers, scientists, and software developers trained in a variety of disciplines and to improve knowledge transfer among these groups involved in research, development and/or education.

This book series offers a unique collection of worked problems in different fields of engineering and applied mathematics and science, with a welcome emphasis on coupling techniques. The book series satisfies the need for up-to-date information on numerical modeling. Faster computers and newly developed or improved numerical methods such as boundary element and meshless methods or genetic codes have made numerical modeling the most efficient state-of-the-art tool for integrating scientific and technological knowledge in the description of phenomena and processes in engineered and natural systems. In general, these challenging problems are fundamentally coupled processes that involve dynamically evolving fluid flow, mass transport, heat transfer, deformation of solids, and chemical and biological reactions.

This series provides an understanding of complicated coupled phenomena and processes, its forecasting, and approaches in problem solving for a diverse group of applications, including natural resources exploration and exploitation (e.g. water resources and geothermal and petroleum reservoirs), natural disaster risk reduction (earthquakes, volcanic eruptions, tsunamis), evaluation and mitigation of human induced phenomena (climate change), and optimization of engineering systems (e.g. construction design, manufacturing processes).

Jochen Bundschuh
Mario César Suárez Arriaga
(Series Editors)

Editorial board of the book series

Table of Contents

Preface

High temperature, high oil pressure, and gas well completion testing have always been technical problem and basic theoretical research is one of the key factors needed to ensure a successful completion test. The completion test basic theory includes: a stress analysis of the completion string, completion string buckling behavior, and temperature and pressure distribution prediction. The completion string is the main bearing and power transmission component for oil and gas well operations and production and it is required to take on a combination of loads, which result in completion string deformation. Because of these complex relationships, completion string stress analysis has become increasingly more complex.

In applied basic theory for deep well testing research, tubular string mechanical analysis is complex as fluid temperature and tubing pressure have a large effect on the force of the tubular string. When well testing, the methods used for determining the temperature and the pressure of the tubular string include: (1) measurements from thermometers and pressure gauges located in the tubing; (2) temperature and pressure measurements at only the well bottom and well head. With these measurements, it is possible to predict the complete distribution using theory analysis technology.

Predicting accurate temperature and pressure profiles in flowing wells can greatly enhance oil and gas production. However, for gas wells with High Temperature High Pressure (HTHP), it is often difficult to operate thermometers and pressure gauges within the tubing itself, and therefore theory analysis can be used as an effective and feasible method.

Over the past decade, the use of numerical simulators with high speed electronic computers has gained wide acceptance throughout the petroleum industry for conducting oil and gas engineering studies. These simulators have been designed for use by engineers who may possess little or no background in the numerical mathematics upon which they are based. Yet in spite of our best efforts to improve these numerical methods to make simulators as reliable, efficient, and automatic as possible, the user of a simulator is constantly faced with a myriad of decisions that have nothing to do with the problem they really want to solve. They need to decide on various numerical questions not germane to the problem at hand. For example, an engineer may need to make a choice between several simulators that use different numerical methods, or may need to choose an iteration method and iteration parameters, a grid spacing, and probably also the time step size. All of these make the process long and complex, adding to production cost and time.

Therefore, with this in mind, we are writing this book for production and testing engineers to provide them with the tools to deal more effectively with the numerical decisions they have to make. To aid in understanding of, and to satisfy curiosity about, the inner workings of the "black box" that is the simulator, we demonstrate the complexities of the oil and gas process by presenting appropriate research studies and developing models the main issues facing the oil and gas exploration industry. Therefore, the first part of the book describes the six conditions in the oil and gas industry. For each condition, the models that are built and algorithms that are designed, implemented and analysed lay the foundation for easy understanding.

But it is not only engineers we had in mind when writing this book. Mathematicians skilled in numerical analysis will, of course, find much of the material already familiar to them. For differential equations and partial differential equation established models, we give proofs for their existence and uniqueness, thus, providing research ideas for mathematicians.

It is our hope that this book will provide food for thought leading to further progress in numerical simulation. All the equations in this book are free from numerical constants (which are dependent on the units being used) and are valid for any consistent set of units. The use of dimension-free equations should become more common as the industry moves to the adoption of the SI (System International) standard of units, as is now being proposed. Accordingly, in the nomenclature following each chapter, we have specified various quantities in the basic SI units of kilograms, metres, and seconds, together with the derived Newton units for force (which equal kg * m/s^2) and the Pascal for pressure (which equals N/mZ). These form a consistent set of units. If the reader prefers, any other consistent set of units can be used, and the equations will still be correct.

Following the oil and gas testing and production process sequence, this book has ten chapters. In the first chapter, a brief background introduction is given for the six conditions: Placing test string, siting, perforating, injection, production, shut-in, and re-opening. In the second chapter, we present the basic theory to be used in the following chapters including the differential geometry, optimization and variational methods. In Chapter 3, HTHP super-deep deviated gas wells are studied which includes a comprehensive consideration of the axial tubular string load, the internal and external fluid pressure, the normal pressure between the tubular and the well wall, the friction and viscous friction of the fluid flow, and the presentation of a new nonlinear differential equations model. In this chapter, also is proved the nonlinear differential equations that are equivalent to a functional extremum problem using a variational method. In Chapter 4, we provide a mechanical analysis of the test string placement. In Chapter 5, we provide a mechanical analysis of setting. When the string is played out from underground, for a string with a packer, the corresponding string deformation needs to be calculated with the packer re-opened. Thus, in Chapter 6, then, we provide a mechanical analysis for the re-opened condition. In Chapter 7, we consider changes to the pressure and temperature with well depth change in HTHP injection wells, and discuss the different heat transfer states from the second interface to the stratum, which contains both steady and unsteady heat transfers. In Chapter 8, we investigate the changes in the pressure and temperature with well depth change in HTHP production wells, and also discuss the changes in the pressure, temperature, density and velocity in the production process. Well control problems have always been difficult, inattention of which could result in serious consequences. In shut-in procedures, ascertaining the downhole status of the gas is essential for effective well control measures, as is knowing the pressure and temperature distribution. In particular, it is very important to determine the maximum wellhead pressure to enable the selection of a proper wellhead assembly to ensure shut-in processes are safe. Thus, in Chapter 9, we also discuss the change in the pressure and temperature with a well depth change in a shut-in condition. In Chapter 10, we discuss software design and development which includes calculation programs and databases. The calculation process completes the basis data input, calculation, and the output of the results. The database selection discuss data input, data saving and delete functions for the tubing and casing.

This monograph has been supported by the National Science Foundation for Distinguished Young Scholars, P. R. China (Grant No. 70425005); the Key Program of National Science Foundation, P. R. China (Grant No. 70831005); the Key Project of China Petroleum and Chemical Corporation (Grant No. GJ-73-0706) and the Key Project of China Petroleum Corporation (Grant No. 2008-89). The authors take this opportunity to thank senior engineers at the Research School of Engineering Technology, The Southwest Petroleum and Gas Corp, China Petroleum and Chemical Corp, for contributing valuable insights and information, S. Wang, B. Qi and Z. Qiao. We would also like to thank the senior engineers at the Chuanqing Drilling Engineering Company and the China National Petroleum Corporation, X. Wang, J. Song and C. Yang, from whom the authors got significant assistance in petroleum and gas theory. For discussions and advice, the authors also thank researchers from the Uncertainty Decision-Making Laboratory of Sichuan University, in particular, L. Yao, M. Tao, Y. Liu, J. Hu, M. Luo, C. Ding, X. Li, K. Chen, X. Zhao, and J. Yang, who have done much work in this field and have made a number of corrections.

Sichuan University, *Jiuping Xu*
Chengdu University of Information Technology, *Zezhong Wu*
November, 2014

Acronyms

AASDT-SITP	Analyzing axial stress and deformation of tubular for steam injection process in HTHP wells based on the varied (T, P) fields
APDTU-VTPF	Analyzing packer's deformation of tubular for unsetting process in HTHP wells based on the varied (T, P) fields
DFA-SIPVF	Dryness fraction analysis for steam injection process of HTHP wells in the varied (T, P) fields
HTHP	High temperature high pressure
NMSOGW-TTBF	Numerical modeling study of oil-gas-water three-phase transient bubbly flow in HTHP wells
NMSQ-DWV	Numerical modelling of steam quality in deviated wells with variable (T, P) fields
OPPV-BCSF	Optimization of perforation parameters for perforated vertical HTHP wells based on comprehensive skin factor
PDPT-IW	Predicting dryness fraction of gas, pressure and temperature in HTHP injection wells
PDPT-SIBUHT	Predicting dryness fraction of gas, pressure and temperature for steam injection based on unsteady heat transmission
PDPTVD-TBF	Predicting on distribution of pressure, temperature, velocity, density of three-phase bubbly flow in HTHP wells
PDTPVD-GLTPTF	Predicting on distribution of temperature, pressure,velocity and density of gas liquid two-phase transient flow in HTHP wells
PPTHVD-STF	Prediction of pressure, temperature, hold-up, velocity and density distribution of steady-state three-phase flow in HTHP wells
PPT-SPDW	Prediction of pressure and temperature in shut-in procedures for HTHP deviated wells
PPTVD-TFSP	Prediction of pressure, temperature, velocity and density of two-phase flow in shut-in procedures for the HTHP gas wells
PTPD-IGWTE	Prediction of temperature and pressure distribution in HTHP injection gas wells considering thermal effect of wellbore
PTP-GW	Predicting temperature and pressure in HTHP gas wells
PTPTF-IWLFM	Pressure and temperature prediction of transient flow in HTHP injection wells by Lax-Friedrichs method
PTPTV-GW	Prediction of temperature, pressure, density, velocity distribution in HTHP gas wells
TSCOGW	Tubular string characterization in HTHP oil and gas wells

Symbols

a	Perforation length [m]
A	Cross-sectional area flow conduit [m^2]
C_f	Specific heat of formation [$J/(kg \cdot K)$]
C_p	Heat capacity of fluids [$J/(kg \cdot K)$]
C_p	Specific heat of flowing [$J/(kg \cdot K)$]
d	Hydraulic diameter [m]
D	Well-bore diameter [m]
f	Friction factor, dimensionless
f_{t_D}	Dimensionless transient heat conduction time function for formation
g	Acceleration constant of gravity [m/s^2]
Gr	Grashof number, dimensionless
h	Enthalpy [J^2/s^2]
H	Vertical depth of the borehole [m]
h_c	Heat transfer coefficient for natural convection based on the outside tubing surface and the temperature difference between the outside tubing and inside casing surface [$J/m \cdot s \cdot K$]
h'_c	Heat transfer coefficient for natural convection based on the outside insulation surface and the temperature difference between the outside tubing and the inside casing surface [$J/m \cdot s \cdot K$]
h_r	Heat transfer coefficient for radiation based on the outside tubing surface and the temperature difference between the outside tubing and inside casing surface [$J/m \cdot s \cdot K$]
h'_r	Heat transfer coefficient for radiation based on the outside insulation surface and the temperature difference between the outside tubing and inside casing surface [$J/m \cdot s \cdot K$]
k	Reynolds number
k_{cas}	Thermal conductivity of the casing material at the average casing temperature [$J/m{\cdot}K$]
k_{cem}	Thermal conductivity of the cement at the average cement temperature and pressure [J/m K]
K_e	Formation conductivity [$J/m \cdot K$]
L_p	Length of perforated vertical well [m]
M	Molecular mass of gas [kg/mol]
$\dot{m}$	Mass flow rate [kg/s]
N	Total perforation number
P	Pressure of flowing liquid [Pa]
P_{pc}	Critical pressure [Pa]
P_{pr}	Comparative pressure, dimensionless
Pr	Prandtl number, dimensionless
q	Heat flux [$J/m \cdot s$]
q_D	Dimensionless heat flux
q_i	Flow-rate in through i perforation [m^3/d]
Q_i	Flow-rate into ith perforation

R	Gas constant
$r_{c_{cem}}$	External radius of cement mantle [m]
r_{c_o}	External radius of casing [m]
r_{ci}	Internal radius of casing [m]
r_D	Dimensionless radius
r_{t_i}	Internal radius of conduit [m]
r_{t_o}	External radius of conduit [m]
Re	Renold number
S	Skin factor constant
t	Production time of gas [s]
T	Temperature of flowing liquid [K]
T_{grad}	Thermal gradient of formation [K]
T_{pc}	Critical temperature [K]
T_{pr}	Comparative temperature, dimensionless
T_c	Temperature of formation/earth surface [K]
t_D	Dimensionless time
T_e	Initial temperature of formation [K]
T_k	Second interface temperature [K]
T_r	Temperature of formation/earth [K]
T_w	Wellbore temperature [K]
U	Average flow velocity [m/s]
U_{t_o}	Overall-heat-transfer coefficient [W/m · K]
v	Velocity of flowing liquid [m/s]
w	Mass flow rate of gas [kg/s]
x_i	Position of the ith perforation [m]
z	The distance co-ordinate in the flow direction along the conduit [m]
Z	Total length of the flow conduit [m]
z_D	Dimensionless depth
Z_g	Gas deviation factor
θ	Inclination angle flow conduit [°]
ρ	Density of flowing liquid [kg/m^3]
λ_f	Thermal conduction coefficient of formation [J/m · K]
κ	Permeability [m^2]
δ	Perforation radius [m]
ϵ	Roughness height [m]
τ_w	Wall shear stress [N/m^2]
μ	Flowing viscosity [MPa · s]
Δ	Pressure drop [MPa]
ϕ	Porosity of formation/earth

About the authors

Jiuping Xu was born on September 11, 1962 in Chongqing, China. He obtained his Ph.D. of applied mathematics from Tsinghua University, Beijing, China; and Ph.D. of physical chemistry from Sichuan University, Chengdu, China, in 1995 and 1999, respectively. He has been appointed as Lifetime Academician of the International Academy for Systems and Cybernetic Sciences and appointed as the Lifetime Academician of Lotfi Zadeh International Academy of Sciences in recognition of his contribution to the development of science and organization of scientific forums. He has been recognized by a number of distinctions including the Distinguished Professor of "Cheung Kong Scholars Program" by the MOE of China at Sichuan University, the winner of National Natural Science Foundation for Outstanding Young Scholars of China, the State Class Persons of National New Century Excellent Talents Program of China, and the China Youth Prize of Science and Technology. Prof. Xu is also the Editor-in-Chief of International Journal of Management Science and Engineering Management and World Journal of Modelling and Simulation since 2006, the president of International Society for Management Science and Engineering Management since 2007, the vice-president of the Systems Engineering Society of China between 2006 and 2014, the vice-president of the Chinese Society of Optimization & Overall Planning and Economical Mathematics since 2009, and the vice-president of the Management Science and Engineering of China since 2013. He has devoted his long term research at the forefront of engineering practice to establish decision-making and technology innovation paradigms for the large scale of complex systems. His research has brought both theoretical and practical implications into the fields of energy development, circular economy, water resources management, low-carbon and environmental issues, and civil engineering and management. He has published more than 40 books in well recognized presses such as Springer, Taylor & Francis and over 600 academic papers in peer-reviewed journals such as Computer-aided Civil and Infrastructure Engineering, IEEE Transaction on Fuzzy Systems, IEEE Transactions on Systems, Man and Cybernetics, Chemical Engineering Science, Industrial & Engineering Chemistry Research, Quality of Life Research, The International Journal of Management Sciences, Renewable and Sustainable Energy Reviews, Renewable Energy, Energy Policy, Energy Conversion and Management, Energy, Journal of Construction Engineering and Management-ASCE, Journal of Management in Engineering-ASCE, Journal of Infrastructure Systems-ASCE, Journal of Water Resources Planning and Management-ASCE, International Journal of Project Management, Applied Mathematical Modelling, Mathematical and Computer Modelling, Journal of Computational and Applied Mathematics, Computers and Mathematics with Applications, Applied Mathematics and Computation, Mathematics and Computers in Simulation, Journal of Computational Analysis and Applications, Fuzzy Sets and Systems, International Journal of Approximate Reasoning, Journal of Scheduling, Annals of Operations Research, Computers & Industrial Engineering, The Canadian Journal of Chemical Engineering, Petroleum Science and Technology, The Journal of Petroleum Science and Engineering and so on. His outstanding career has been highlighted by numerous awards. The International Federation of Operational Research Societies awarded him the IFORS Prize for OR in recognition of his work in 1996 and 2014, respectively. Prof. Jiuping Xu has been the first awardee in 9 ministry and provincial first-class prizes including the Chinese Ministry of Education Prize for Progress in Science and Technology, Chinese Ministry of Education Prize for Natural Science, Chinese Sichuan Provincial Prize for Progress in Science and Technology, Chinese Sichuan Provincial Prize for Philosophy Social Science, and Chinese Sichuan Provincial Teaching Achievement Prize.

Zezhong Wu was born on December 7, 1970 in Chongqing, China. He received his Ph.D. of Management Science and Engineering in Sichuan University, China in 2010. Dr. Wu is currently Professor at Chengdu University of Information Technology, the director of The Mathematical Model and The Applied Research Institute of Chengdu University of Information Technology. His current research interests include the areas of applied mathematics, petroleum and gas, mathematical modelling and simulation. He has published 30 journal papers in Fuzzy Sets and Systems, The Journal Fuzzy Mathematics, Applied and Computational Mathematics, The Canadian Journal of Chemical Engineering, Petroleum Science and Technology.

The corresponding author:

Prof. Dr. Jiuping Xu
Uncertainty Decision-Making Laboratory, Sichuan University, Chengdu 610064, P. R. China
Tel: 86-28-85418522; Fax: 86-28-85415143
Email: xujiuping@scu.edu.cn

CHAPTER 1

Background

The oil and gas process can be divided into seven conditions: placing the test string, seating, perforating, injection, production, shut-in, and re-opening. In this chapter, a brief background is given for each condition.

1.1 PLACING THE TEST STRING

Placing the test string is the first part of the oil and gas process, as shown in Figure 1.1:

1. The (a group) tubing is lifted using mechanical equipment so that the suspended tubing is perpendicular to the ground plane, and the base of the tubing is above the ground just above the well-bore;
2. The aligned hole of the tubing string is slowly decentralized. When the top of the tubing reaches the ground, the tubing is fixed to the wellhead using fixtures;

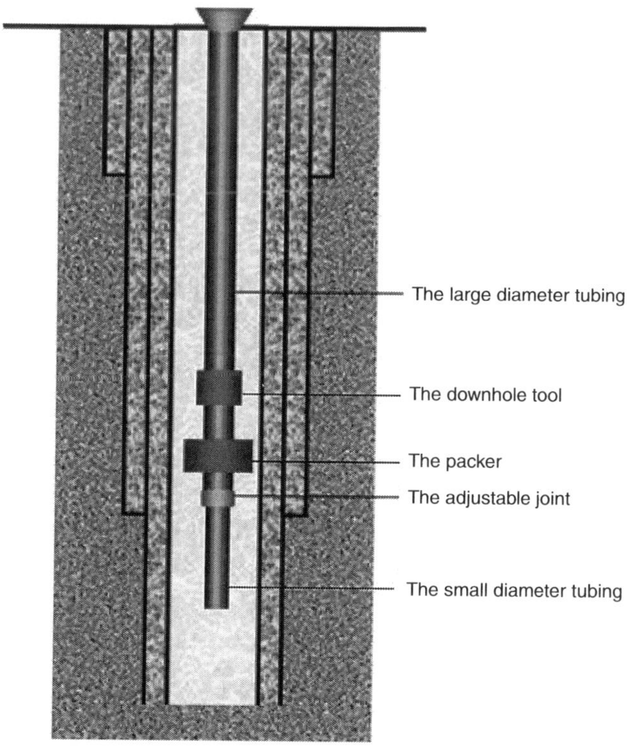

Figure 1.1 Figure of the test string.

3. Another pipeline is lifted using mechanical equipment, and the suspended tubing string is slowed until it is connected to the fixed tubing at the wellhead;
4. After the tubing is connected, the wellhead fixtures are released, and the tubing is placed slowly into the well;
5. The action is repeated until the last of the tubing is in the well;
6. When the last of the tubing is placed into the well, a wimble-pipe is connected, and this is placed into the well. Because the diameter of the wimble-pipe is greater than that of the wellhead equipment, the wimble-pipe remains at the well-head and is fixed, thus bearing the weight of the lower tubing.

We can see from the placement process that of all the tubing in the well, the string tension near the wellhead is the largest and has the greatest deformation, so stress analysis is the main consideration for the test string in the well.

1.2 SEATING CONDITION

There are three methods for the seating of the packer: mechanical seating, hydraulic seating and intubation seating. When the packer is seated, there are three states between the packer and the tubing: free movement, limited mobility and immobility, as shown in Figure 1.2:

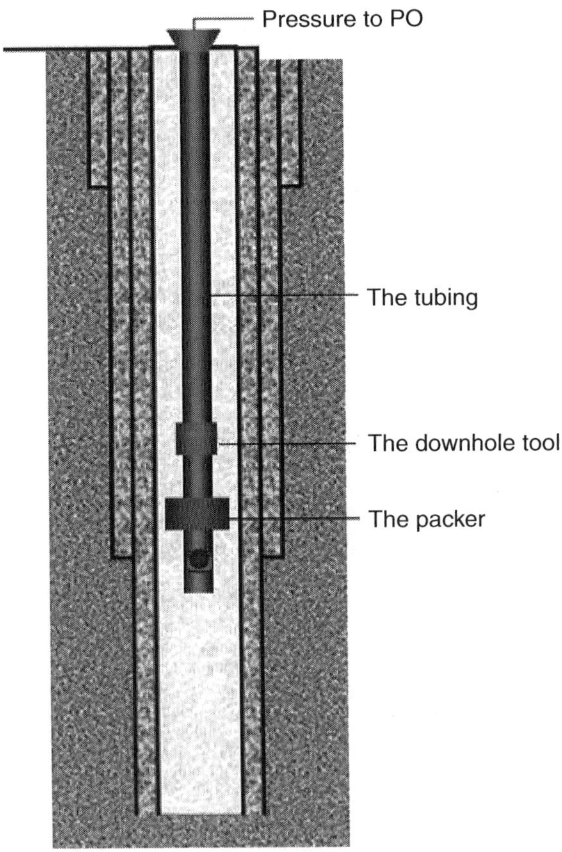

Figure 1.2 Figure of the seated packer.

(1) Mechanical seating process

Mechanical seating refers to the seating of the packer using the string weight to ensure that the packer is sealed. There are two seating processes, with the first having the following steps:

(i) When the tubing is placed in the desired location, the tube is lifted so as to put the transposition structure in the active position;
(ii) The tubing is loaded, and the packer centre mandrel of the packer is loosened so that the anchor is pushed and fixed in the casing wall. Then the tubing is decentralized and pressure added to compress the packer's rubber tube, so that make the expansion of the rubber tube casing greases the annular seal;
(iii) When the plastic tube is expanded, the centre device of the packer tube protrudes to support the inner plastic tube surface;
(iv) After the tubing's top pressure is released, the packer's internal support mechanisms ensure that the plastic tube is in an expansion state to ensure it remains sealed.

(2) The hydraulic packer seating process

(i) When the tubing is placed into the desired location, the wellhead equipment is fixed;
(ii) A steel ball is cast from the wellhead, which falls to the tee to seal the tubing channel;
(iii) Liquid is injected from the wellhead, and the steel ball seals the lower tubing so as to increase the tubing's internal pressure;
(iv) With the increase in fluid pressure within the tubing, the liquid flows from the centre inlet hole of the packer into the centre tubing. When the tubing fluid pressure reaches a certain value, the pressure pushes the pin head and cuts the packer. From the pressure difference between the internal and external liquids, the plastic tube is compressed inflating the rubber which then seals the ring space between the oil and the tubing;
(v) With a packer expansion plastic tube, the packer centre tube is moved and eventually locked in position to support the packer plastic tube;
(vi) When the tubing pressure is released, the tube in the centre is under strong pressure, and the plastic tube locking the clip reed cannot be released, meaning that the plastic tube is in the oil and tubing ring space at all times.

(3) The intubation seating process

(i) The packer and the lower packer are loaded into the desired position using the seating tool, then the packer is seated;
(ii) Remove the seating tool;
(iii) The upper tubing of the packer (the lower tubing of the packer is intubation) is placed on the top of the packer, and the intubation is aligned with the packer sealing tube, and then inserted into the packer sealing tube using the weight of the upper tubing;
(iv) The initial intubation position is adjusted according to need to complete the sealing operation.

When the packer is seated, two situations may occur: one is that fluid may remain inside and outside the tubing or, because there are two different systems inside and outside the tubing, there may be two different densities.

After the packer is seated, the packer and the string may have three different conditions: free movement, limited mobility and immobility:

(1) Free movement

Free movement is when the sealed tube at the bottom can move freely up and down in the packer seal cavity. In this case, the main consideration is where the sealed tube is in the packer chamber and how long the short section of the sealed tube is (the tube with a sealed ring). As the sealed segment cannot be removed from the sealed chamber, there could be a channelling up and down movement through the packer with a resultant shortening or elongation of the tubing. When the pressure is high and the packer seal chamber aperture is large, a permanent spiral deformation or damage due to excessive tension may occur.

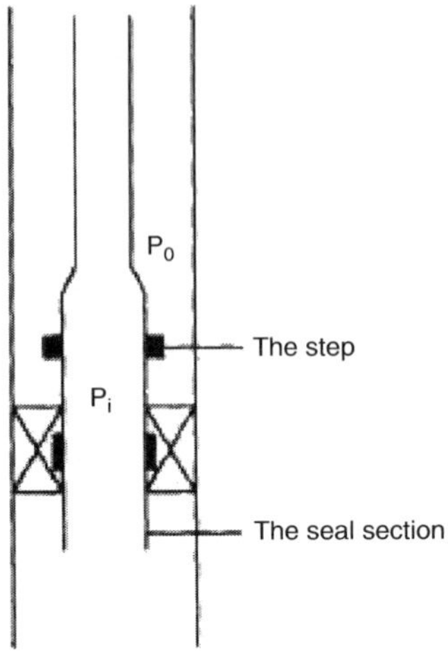

Figure 1.3 Limited movement of the seated packer.

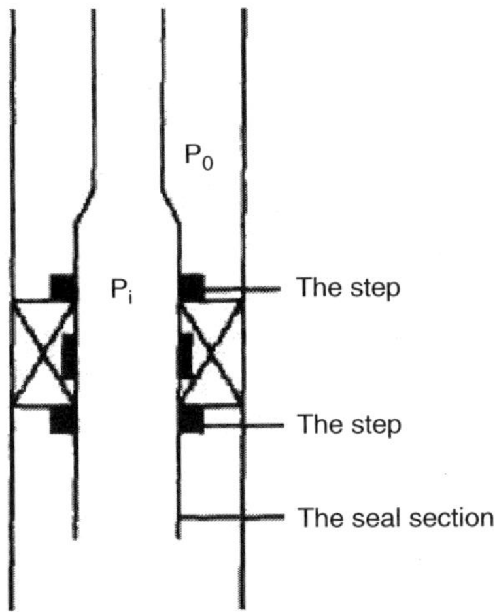

Figure 1.4 Not movement of the seated packer.

(2) Limited movement

As shown in Figure 1.3, limited mobility is when the sealed tube at the bottom of the tubing can only move in one direction. If the tubing becomes shorter because of pressure or temperature, it has no restrictions. On the contrary, if the tubing becomes longer, it places a downward force on the packer. In practical engineering, this can have a number of effects on the packer string structure, such as a slip in the hanging packer string, or a short positioned drilling packer or a dual packer. As a result, the following possible problems could result.

(i) When the tubing is contracted, the sealed segment may move out of the packer sealing chamber resulting in channelling. (ii) The force of the string may lead to difficulties with the rope operation and lift the packer, or cause permanent spiral bending.

(3) Immobility

As shown in Figure 1.4, immobility refers to the string section at the bottom of the seal being completely confined in the packer, and being unable to move freely up and down. As the tubing cannot move, changes in temperature and pressure in the tubing lead to elongation or shortening, with the packer exerting force up or down the tubing.

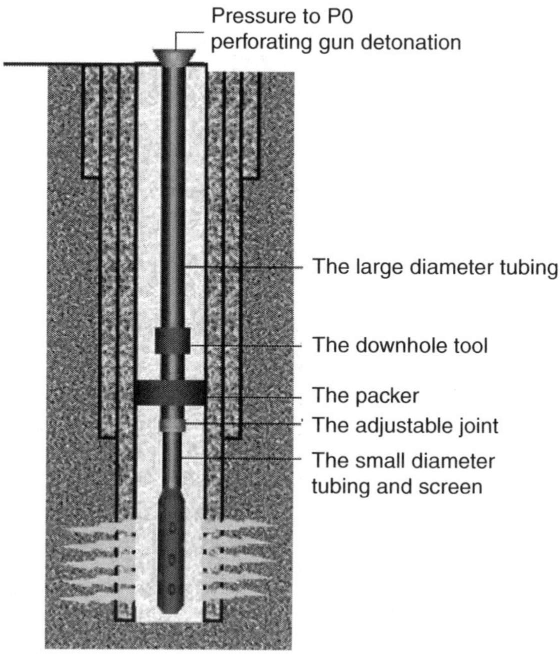

Figure 1.5 Figure of the perforation.

In practical engineering, this type of tubing is common, for example packer tubing with side slips, and drilling packer tubing with a sealed locking bolt. In these cases, the following problems could occur:

(i) Because the tubular contraction leads to excessive tension, this will result in a rupturing of the tubing or the central packer tube.
(ii) The helical bending caused by the tubular elongation can have a harmful effect on rope operations and pumping production.

1.3 PERFORATION CONDITION

As can be seen in Figure 1.5, before perforating, the tubing has to be divided into two separate internal and external systems by the packer. The tubing and the lower annulus of the packer is system A, above the annulus of the packer is system B.

Hydraulic detonation perforating process:

(1) After the packer is seated, the liquid is injected from the wellhead to the tubing, and as system A is a confined space, system A's pressure will rise;
(2) When the wellhead pressure reaches a certain value, the fluid pressure in the tubing increases to cut detonator pins and detonate the perforating charges. In this, perforating process, the possibility of tubular deformation should be considered.

1.4 INJECTION CONDITION

As can be seen in Figure 1.6, injection operations include fracturing, acidizing, snubbing, and the wash well, with the common feature being the injection of fluid into the well-bore. To simplify

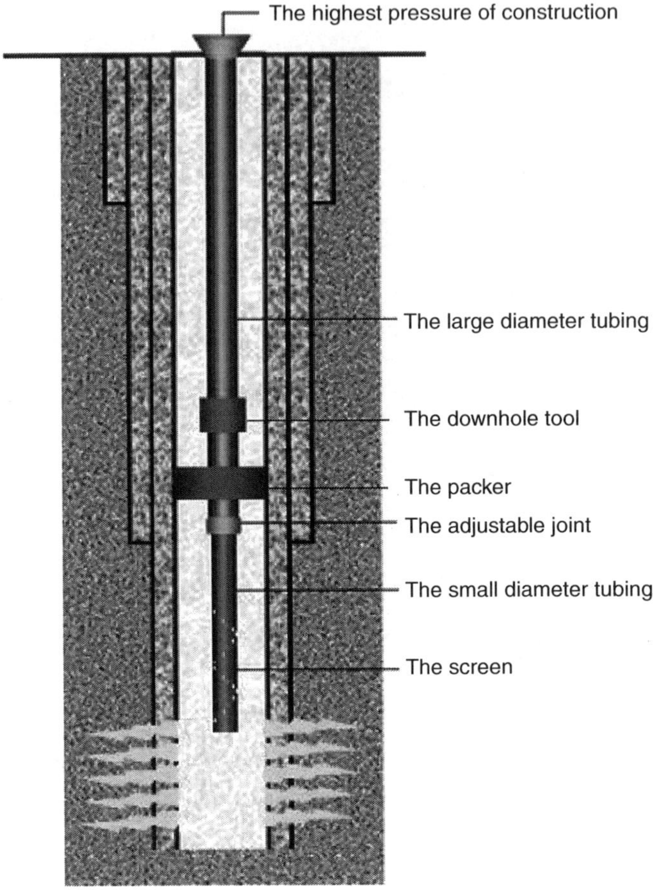

Figure 1.6 Figure of the injection.

the analysis, fracturing and acidizing construction are only considered in the case of a single-layer construction, whereas snubbing construction is only considered in demanding construction jobs when lighter techniques are not suitable.

Injection operations need to calculate the tubing deformation under a maximum pump pressure, and determine the sealing effect of the packer.

In steam injection, as steam is injected either down the well-bore or down the tubing/casing annulus to the formation being flooded, heat is transferred from or to the surrounding earth as a result of the difference in geothermal and injected fluid temperature. To evaluate the feasibility of an injection project, a reasonable estimate for the effective amount of heat carried by the fluid, its temperature, pressure and gas dryness fraction is need. Therefore, we have considered not only temperature and pressure but also the gas dryness fraction in the computation.

1.5 PRODUCTION CONDITION

As can be seen in Figure 1.7, before opening the wells for production, the production string has a certain deformation degree; for normal production, the bottomhole pressure, well-bore temperature and pressure distribution, and well-bore fluid properties will change. In the adjustment

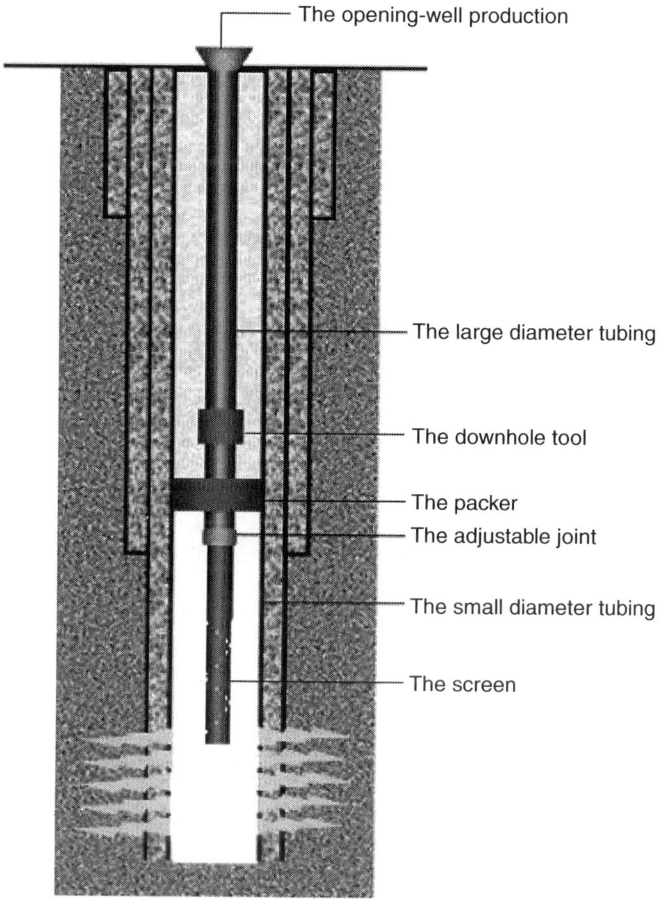

Figure 1.7 Figure of the production.

system, because of the greater production pressure, the tubing has a large deformation, resulting in a damaged tubular, therefore the maximum allowable deformation for the corresponding production pressure needs to be calculated.

1.6 SHUT-IN CONDITION

As shown in Figure 1.8, when the wellhead is shut-in because the formation pressure is greater than the bottom pressure of the well-bore, the formation fluid continues to flow into the well-bore, and as the well-bore is a confined space, the well-bore pressure gradually increases until the bottom fluid pressure of the well-bore equals that of the formation pressure, at which point the formation fluid stops flowing into the well-bore.

1.7 RE-OPENED CONDITION

As shown in Figure 1.9, for a string with a packer, when the downhole string needs to be uploaded, the corresponding tubular deformation needs to be calculated for the re-opened

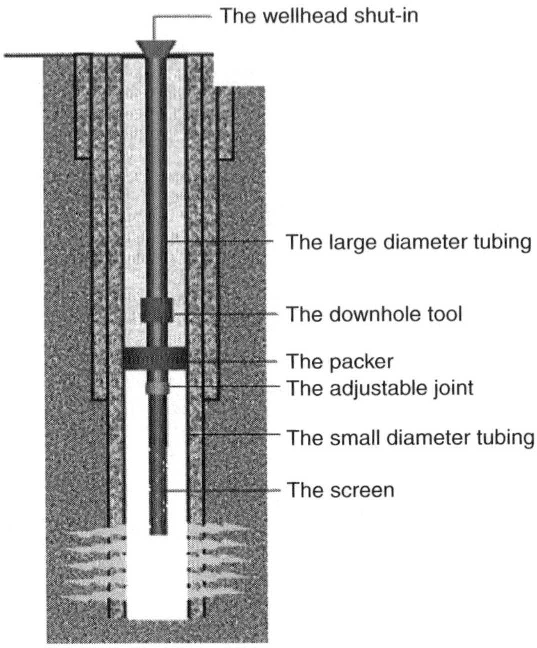

Figure 1.8 Figure of the shut-in.

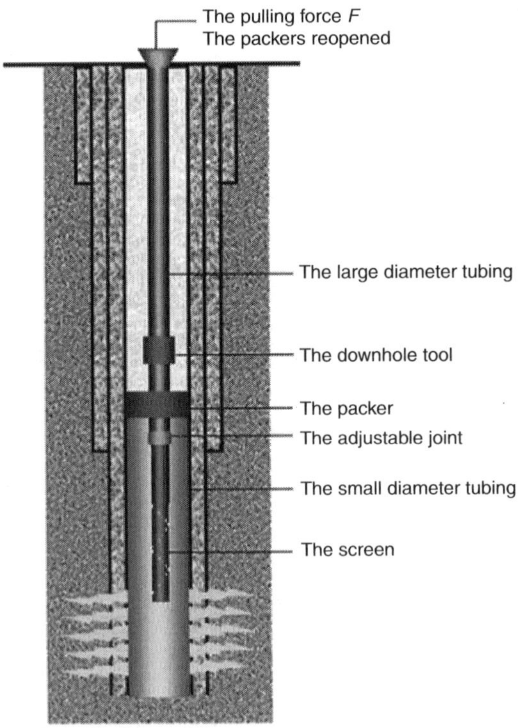

Figure 1.9 Figure of the re-opened.

packer process. As there are three kinds of packers, each has a corresponding re-opening operations process:

(1) Mechanical re-opening process
   (i) When the tubing is put on to reduce the weight of the upper packer the compressive force is reduced;
   (ii) As the centre tube moves up, the centre tube support body returns to a contracted state;
   (iii) The packer rubber tube automatically shrinks without compression or internal support, so the packer is re-opened;
   (iv) Continuous pressure is put on the string, so the packer's slip automatically shrinks due to the string tension, and thus the tubing is re-opened.

(2) Hydraulic packer re-opening process
   (i) When the tubing is put on, the centre tube body moves up, and the support agencies return to a contracted state;
   (ii) The packer rubber tube automatically shrinks without compression or internal support, so the packer is re-opened;
   (iii) Continuous pressure is put on the string, so the packer's slip automatically shrinks due to the spring tension, thus the tubing is re-opened.

(3) Intubation packer re-opening process
   (i) When the tubing is put on, the intubation tube is pulled out from the sealed tube of the packer;
   (ii) Using specific tools, the packer is re-opened, and the lower tubing is pulled out.

CHAPTER 2

Element theory

In this chapter, we write down the basic theory which is used in the following chapters including differential geometry, optimization and variational methods.

2.1 DIFFERENTIAL GEOMETRY

This section is devoted to the fundamentals of differential geometry.

2.1.1 Frenet frame

What distinguishes a circle or a helix from a line is their curvature, i.e., the tendency of the curve to change direction. We shall now see that we can associate to each smooth (e^3) arclength-parametrized curve α, a natural "moving frame" (an orthonormal basis for $\Re^3$ chosen at each point on the curve, adapted to the geometry of the curve as much as possible).

We begin with a fact from vector calculus that will appear throughout this section.

Lemma 2.1. (Alekseevskići *et al.*, 1991) Suppose $f, g : (a, b) \to \Re^3$ are differentiable and satisfy $f(t) \cdot g(t) = c$, for all t (c represents constant).

Then:

$$f'(t) \cdot g(t) + f(t) \cdot g'(t) = 0.$$

In particular,

$$\|f(t)\| = c, \quad \text{if and only if} \quad f(t) \cdot f'(t) = 0, \quad \text{for all } t.$$

Using Lemma 2.1 repeatedly, we now construct the Frenet frame of suitable regular curves. We assume throughout that the curve α is parametrized by arclength. Then, for starters, α's is the unit tangent vector to the curve, which we denote by $\tau(s)$. Since τ has constant length, $\tau'(s)$ will be orthogonal to $\tau(s)$. Assuming $\tau'(s) \neq 0$, define the principal normal vector $n(s) = \frac{\tau(s)}{\|\tau'(s)\|}$ and the curvature $K(s) = \|\tau'(s)\|$. So far, we have:

$$\tau'(s) = K(s) \cdot n(s).$$

If $K(s) = 0$, the principal normal vector is not defined. Assuming $K(s) \neq 0$, we continue. Define the binormal vector as,

$$b(s) = \tau(s) \times n(s).$$

Then $\{\tau(s), n(s), b(s)\}$ form a right-handed orthonormal basis for $\Re^3$. Now, $n'(s)$ must be a linear combination of $\tau(s)$, $n(s)$, $b(s)$. But we know that:

$$n'(s) \cdot n(s) = 0$$

and that:

$$n'(s) \cdot \tau(s) = -\tau'(s) \cdot n(s) = -K(s).$$

We define the torsion $T(s) = n'(s) \cdot b'(s)$. This gives us:

$$n'(s) = -K(s)\tau(s) + T(s)B(s).$$

Finally, $b'(s)$ must be a linear combination of $n(s)$, $\tau(s)$ and $b(s)$. Lemma 2.1 tells us that $b'(s)\cdot b(s)=0$, $b'(s)\cdot\tau(s)=-\tau'(s)\cdot b(s)=0$ and $b'(s)\cdot n(s)=-b(s)\cdot n'(s)=0$.

Thus:

$$b'(s)=-T(s)n(s).$$

In summary, we have:

$$\begin{aligned} \tau'(s) &= \qquad\qquad K(s)n(s) \\ n'(s) &= -K(s)\tau(s) \qquad\qquad + T(s)b(s)\,. \\ b'(s) &= \qquad\qquad -T(s)n(s) \end{aligned}$$

The skew-symmetry of these equations is made clearest when we state the Frenet formulas in matrix form:

$$[\tau'(s)\quad n'(s)\quad b'(s)]=[\tau(s)\quad n(s)\quad b(s)]\begin{bmatrix} 0 & -K(s) & 0 \\ K(s) & 0 & -T(s) \\ 0 & T(s) & 0 \end{bmatrix}.$$

Indeed, note that the coefficient matrix appearing on the right is skew-symmetric. This is the case whenever we differentiate an orthogonal matrix depending on a parameter.

2.1.2 Geometric description of the 3D curved borehole

As in Figure 2.1, letting the wellhead P be the origin point, we upbuild a $PXYZ$ right-handed coordinate system. $\vec{i},\vec{j},\vec{k}$ represented the unit vectors of X-axis, Y-axis, Z-axis separately. On the axis line of the well-bore, the geometry position of arbitrary point $O(x,y,z)$ can be represented by vector $\vec{r}_0(s_o)$:

$$\vec{r}_0(s_o)=x_0(s_0)\vec{i}+y_0(s_0)\vec{j}+z_0(s_0)\vec{k} \tag{2.1}$$

the corresponding increment can be represented as:

$$d\vec{r}_0(s_o)=dx_0(s_0)\vec{i}+dy_0(s_0)\vec{j}+dz_0(s_0)\vec{k}.$$

In Figure 2.1, $\vec{\tau}_0$ is the unit vector of the tangential direction of the well-bore axis line and passing point o. The included angle φ between $\vec{\tau}_0$ and k is called the deviation angle, the included angle ψ between projection $\vec{\tau}_0'$ of $\vec{\tau}_0$ on the $PXYZ$ plane and $\vec{i}$ is called the azimuth angle.

From Figure 2.1, we have:

$$\frac{dx_0}{ds_0}=\sin\varphi\cos\psi,\qquad \frac{dy_0}{ds_0}=\sin\varphi\sin\psi,\qquad \frac{dz_0}{ds_0}=\cos\varphi, \tag{2.2}$$

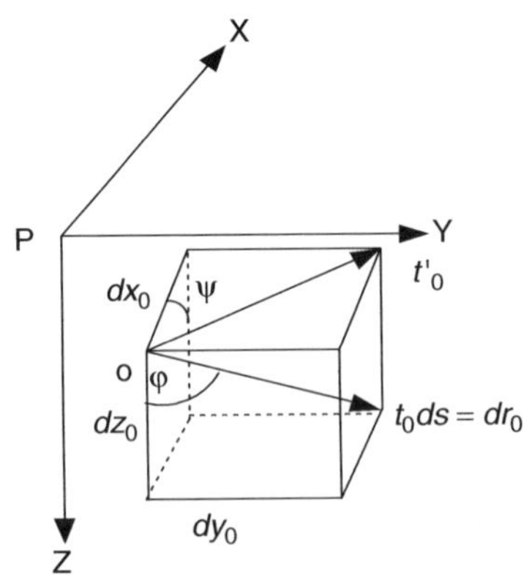

Figure 2.1 The 3D figure of an inclined well-bore.

$$\vec{\tau}_0 = \sin\varphi\cos\psi\,\vec{i} + \sin\varphi\sin\psi\,\vec{j} + \cos\varphi\,\vec{k}. \tag{2.3}$$

The curvature K_0 and torsion T_0 of the corresponding trail can be written as:

$$K_0 = \left|\frac{d\vec{\tau}_0}{ds_0}\right|, \quad T_0 = \frac{1}{K_0^2}\left(\vec{\tau}_0, \frac{d\vec{\tau}_0}{ds_0}, \frac{d^2\vec{\tau}_0}{ds_0^2}\right), \tag{2.4}$$

therefore, we get:

$$K_0^2 = \left(\frac{d\varphi}{ds_0}\right)^2 + \sin^2\varphi\left(\frac{d\psi}{ds_0}\right)^2 \tag{2.5}$$

$$K_0^2 T_0 = \sin\varphi\left(\frac{d\varphi}{ds_0}\frac{d^2\psi}{ds_0^2} - \frac{d\psi}{ds_0}\frac{d^2\varphi}{ds_0^2}\right) + \cos\varphi\left[2\left(\frac{d\varphi}{ds_0}\right)^2\frac{d\psi}{ds_0} + \sin^2\varphi\left(\frac{d\psi}{ds_0}\right)^3\right]. \tag{2.6}$$

The unit vectors $\vec{n}_0$ and $\vec{b}_0$ of principal normal and subnormal of point O in the well-bore can be represented as:

$$\vec{n}_0 = \frac{1}{K_0}\frac{d\vec{\tau}_0}{ds_0} = \frac{1}{K_0}\left[\frac{d}{ds_0}(\sin\varphi\cos\psi)\vec{i} + \frac{d}{ds_0}(\sin\varphi\sin\psi)\vec{j} + \frac{d\cos\varphi}{ds_0}\vec{k}\right], \tag{2.7}$$

$$\begin{aligned}\vec{b}_0 &= \left(\frac{d\varphi}{ds_0}\right)^2 \times \vec{n}_0 \\ &= -\frac{1}{K_0}\left(\sin\varphi\frac{d\varphi}{ds_0} + \frac{\sin 2\varphi}{2}\frac{d\sin\psi}{ds_0}\right)\vec{i} + \frac{1}{K_0}\left(\cos\varphi\frac{d\varphi}{ds_0} + \frac{\sin 2\varphi}{2}\frac{d\cos\psi}{ds_0}\right)\vec{j} + \frac{\sin^2\varphi}{K_0}\frac{d\psi}{ds_0}\vec{k}.\end{aligned} \tag{2.8}$$

Therefore, the transformation matrix between the natural coordinates $(\vec{\tau}_0, \vec{n}_0, \vec{b}_0)$ and cartesian coordinate system $(\vec{i}, \vec{j}, \vec{k})$ is as follows:

$$C = \frac{1}{K_0}\begin{bmatrix} K_0\sin\varphi\cos\psi & K_0\sin\varphi\sin\psi & K_0\cos\varphi \\ \dfrac{d}{ds_0}\sin\varphi\cos\psi & \dfrac{d}{ds_0}\sin\varphi\sin\psi & \dfrac{d\cos\varphi}{ds_0} \\ -\left(\sin\psi\dfrac{d\varphi}{ds_0} + \dfrac{\sin 2\varphi}{2}\dfrac{d\sin\psi}{ds_0}\right) & \cos\psi\dfrac{d\varphi}{ds_0} + \dfrac{\sin 2\varphi}{2}\dfrac{d\cos\psi}{ds_0} & \sin^2\varphi\dfrac{d\psi}{ds_0}\end{bmatrix}.$$

When $\left(\frac{d\varphi}{ds_0}\right)^2 + \left(\frac{d\psi}{ds_0}\right)^2 \neq 0$, C^{-1} exists and $C^{-1} = C^T$, then:

$$(\vec{\tau}_0, \vec{n}_0, \vec{b}_0)^T = C(\vec{i}, \vec{j}, \vec{k})^T, \quad (\vec{i}, \vec{j}, \vec{k})^T = C^T(\vec{\tau}_0, \vec{n}_0, \vec{b}_0)^T. \tag{2.9}$$

If $\varphi(s_0)$, $\psi(s_0)$ is known, then we can get accordingly curvature K_0, torsion T_0 and unit vector $\vec{\tau}_0, \vec{n}_0, \vec{b}_0$ of tangent direction, principal normal direction and subnormal direction separately on well-bore axis line trail. According to differential geometry theory (Alekseevskići *et al.*, 1991), there is a relation between $\vec{\tau}_0, \vec{n}_0, \vec{b}_0$ as follows:

$$\begin{pmatrix} \dfrac{d\vec{\tau}_0}{ds_0} \\ \dfrac{d\vec{n}_0}{ds_0} \\ \dfrac{d\vec{b}_0}{ds_0} \end{pmatrix} = \begin{pmatrix} 0 & K_0 & 0 \\ -K_0 & 0 & T_0 \\ 0 & -T_0 & 0 \end{pmatrix}\begin{pmatrix} \vec{\tau}_0 \\ \vec{n}_0 \\ \vec{b}_0 \end{pmatrix}. \tag{2.10}$$

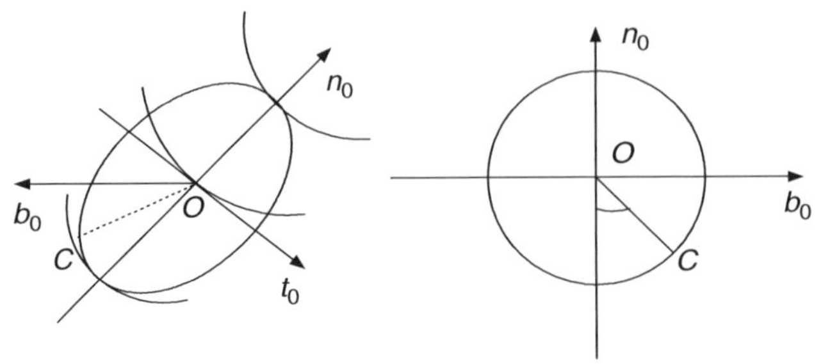

Figure 2.2 The 3D figure of tubular string.

2.1.3 **Geometry description of tubular string in 3D inclined well-bore**

As Figure 2.2, at point O, we suppose that normal plane $O\vec{n}_0\vec{b}_0$ of the axis trail cuts the axis of the tubular string at C point, and the tubular string keeps continuous contact with the well wall all the time, then C is bound to be on the circular torus at O as centre of a circle and $r = r_w - t_{to}$ as radius of a circle. Where r_w is the radius of the well-bore, r_{to} is the external radius of the tubular string, and r is the clearance. Thus, the location of C can be fixed by the deflection angle θ of the deviating principal normal $\vec{n}_0$. Location of C in 3D space can be represented as follows:

$$\vec{r} = \vec{r}_o + \vec{r}_{oc} = \vec{r}_o - r\cos\theta\,\vec{n}_0 + r\sin\theta\,\vec{b}_0 \tag{2.11}$$

$$\vec{\tau} = \frac{ds_0}{ds}\left[(1 + rK_0\cos\theta)\vec{\tau}_0 + r\sin\theta\left(\frac{d\theta}{ds_0} - T_0\right)\vec{n}_0 + r\cos\theta\left(\frac{d\theta}{ds_0} - T_0\right)\vec{b}_0\right], \tag{2.12}$$

where ds_0 is the differential arc element length of the well-bore axis, ds is the differential arc length which the subnormal of the two terminals of differential arc element ds_0 cuts the tubular string. $\vec{\tau}, \vec{n}, \vec{b}$ are the unit tangent vector, unit principal normal vector and unit subnormal vector of the tubular string axis, separately. Therefore,

$$\begin{aligned}\|\tau\|^2 &= \left[(1 + rK_0\cos\theta)^2 + r^2\sin^2\theta\left(\frac{d\theta}{ds_0} - T_0\right)^2 + r^2\cos^2\theta\left(\frac{d\theta}{ds_0} - T_0\right)^2\right]\left(\frac{ds_0}{ds}\right)^2 \\ &= \left[(1 + rK_0\cos\theta)^2 + r^2\left(\frac{d\theta}{ds_0} - T_0\right)^2\right]\left(\frac{ds_0}{ds}\right)^2 \\ &= 1\end{aligned}$$

i.e.,

$$\frac{ds}{ds_0} = \sqrt{(1 + rK_0\cos\theta)^2 + r^2\left(\frac{d\theta}{ds_0} - T_0\right)^2}. \tag{2.13}$$

In fact, for the axis trail of the well-bore and tubular string, curvature K_0, torsion T_0 and efficiency clearance r are very small ($\ll 1$). Thus, as a matter of convenience, in procession of discussion and analysis, we will ignore higher order term $K_0^{n_1}T_0^{n_2}r^{n_3}$ $(n_1 + n_2 + n_3 > 1)$ and every derivative term $\frac{d^nK_0}{ds^n}, \frac{d^nT_0}{ds^n}$ $(n \geq 1)$. Thus, $ds = ds_0$, we can look upon ds and ds_0 as identical.

Ignoring the higher order term of K_0, T_0, r and the derivative term, we can get:

$$\vec{\tau} = \vec{\tau}_0 - r\frac{d\cos\theta}{ds}\vec{n}_0 + r\frac{d\sin\theta}{ds}\vec{b}_0.$$

From (2.10), we have

$$K\vec{n} = \frac{d\vec{\tau}}{ds} = \frac{d\vec{\tau}_0}{ds} - r\frac{d^2\cos\theta}{ds^2}\vec{n}_0 - r\frac{d\cos\theta}{ds}\frac{d\vec{n}_0}{ds} + r\frac{d^2\sin\theta}{ds^2}\vec{b}_0 + r\frac{d\sin\theta}{ds}\frac{d\vec{b}_0}{ds}$$
$$= K_0\vec{n}_0 - r\frac{d^2\cos\theta}{ds^2}\vec{n}_0 - r\frac{d\cos\theta}{ds}(-K_0\vec{\tau}_0 + T_0\vec{b}_0) + r\frac{d^2\sin\theta}{ds^2}\vec{b}_0 + r\frac{d\sin\theta}{ds}(-T_0\vec{n}_0),$$

ignoring $K_0^{n_1}T_0^{n_2}r^{n_3}$ $(n_1+n_2+n_3>1)$, then:

$$K\vec{n} = \left(K_0 - r\frac{d^2\cos\theta}{ds^2}\right)\vec{n}_0 + r\frac{d^2\sin\theta}{ds^2}\vec{b}_0,$$

where K is the curvature of the curved tubular string.

From $K\vec{b} = K\vec{\tau}\times\vec{n} = \vec{\tau}\times K\vec{n}$, we have:

$$K\vec{b} = \begin{vmatrix} \vec{\tau}_0 & \vec{n}_0 & \vec{b}_0 \\ 1 & -\dfrac{rd\cos\theta}{ds} & \dfrac{rd\sin\theta}{ds} \\ 0 & K_0 - r\dfrac{d^2\cos\theta}{ds^2} & r\dfrac{d^2\sin\theta}{ds^2} \end{vmatrix}.$$

Thus,

$$K\vec{b} = K\vec{\tau}\times\vec{n} = -r\frac{d^2\sin\theta}{ds^2}\vec{n}_0 + \left(K_0 - r\frac{d^2\cos\theta}{ds^2}\right)\vec{b}_0 - \left[r^2\left(\frac{d\theta}{ds}\right)^2 + rK_0\cos\theta\frac{d\theta}{ds}\right]\vec{\tau}_0. \tag{2.14}$$

2.2 VARIATIONAL METHODS

In this section, we first introduce the basic concepts of a norm, functional and variation of functional, state the fundamental theorem of the calculus of variations, and then give the Euler-Lagrange equation.

Definition 2.1. (Buttazzo *et al.*, 1999) A norm on a vector space X is a function:

$$\|\cdot\| : X \to R^+ := [0, +\infty) : X \to \|X\|$$

that satisfies, for all $x, y \in X$ and $\alpha \in F$:

(i) $\|x\| = 0$, if and only if $x = 0$ (faithfulness),
(ii) $\|\alpha x\| = |\alpha|\,\|x\|$ (homogeneity),
(iii) $\|x+y\| \le \|x\| + \|y\|$ (subadditivity).

A seminorm on X is a function $p : X \to R^+$ that satisfies (ii) and (iii) above.

Such a function is called a norm on X and $(X, \|\cdot\|)$ is called a normed linear space. Most often we will be working with only one specific $\|\cdot\|$ on any given vector space X thus we omit writing $\|\cdot\|$ and simply say that X is a normed linear space.

Definition 2.2. (Buttazzo *et al.*, 1999) Let $\mathbb{F} = \{y(x)\}$ be a functional set satisfying certain conditions, R be real set $\{J\}$ (J is variable of R). If for every function $y(x)$ of $\mathbb{F}$, there exists only number J that is correspondence of functional $y(x)$, then J is called function of $y(x)$ and is denoted by:

$$J = J[y(x)], \quad J = J[y],$$

the functional set $\mathbb{F}$ is called domain.

When x represents a point $(x_1,x_2,...,x_n)$ of R^n, $\mathbb{F}$ is a multivariate function satisfying some conditions. Then the function is denoted by:

$$J=J[y(x_1,x_2,...,x_n)].$$

Similarly,

$$J=J[y_1(x),y_2(x),...,y_m(x)],$$

where, $y_1(x),y_2(x),...,y_m(x)$ are independent, every function $y_i(x)$ belongs to a specific functional set.

Definition 2.3. (Buttazzo *et al.*, 1999) Let the function $J[y]$ be:

$$J[y]=\int_{x_0}^{x_n} f(x,y,y')dx,$$

where, $y\in C^1[x_0,x_n], f\in C^1$. Then integral:

$$\int_{x_0}^{x_n} (f_y\delta y+f_{y'}\delta y')dx,$$

is called the variation of function $J[y]$ at $y=y(x)$ and is denoted by δJ, i.e.,

$$\delta J=\int_{x_0}^{x_n} \delta f\ dx=\int_{x_0}^{x_n} (f_y\delta y+f_{y'}\delta y')dx,$$

where, $C^n[x_0,x_1], n=0,1,2,...$, represents the function n order continuous differential at interval $[x_0,x_1]$.

Lemma 2.1. (Buttazzo *et al.*, 1999) Let function $f(x)$ be continuous on interval $[0,l]$, and arbitrary function $\delta\theta(x)$ satisfy the following conditions:

(i) $\delta\theta(x)$ is continuously differential at the interval $(0,l)$,
(ii) at two endpoints of the interval $[0,l]$, $|\delta\theta(x)|<\epsilon$.

In case that the integral $\int_0^l f(x)\delta\theta(x)dx=0$ always holds, then on $[0,l]$,

$$f(x)\equiv 0.$$

Proof. By contradiction, suppose that $f(x)$ is not identically equal to zero at the interval $[0,l]$. By means of continuity of $f(x)$, there at least is a point $\bar{x}$, as Figure 2.3 (where, $(x_1,x_2)=(\bar{x}-\delta,\bar{x}+\delta)$), at the interval $(0,l)$, such that $f(\bar{x})\neq 0$.

Without loss of generality, letting $f(\bar{x})>0$, then there is the small neighborhood of $\bar{x}:(\bar{x}-\delta, \bar{x}+\delta)$ $(\delta>0)$, when $0<\bar{x}-\delta<\bar{x}<\bar{x}+\delta, f(\bar{x})>0$. Where, we select:

$$\delta\theta(x)=\begin{cases} e^{\frac{1}{(x-\bar{x})^2-\delta^2}}, & s\in(\bar{x}-\delta,\bar{x}+\delta),\\ 0, & x\in[0,\bar{x}-\delta]\cup[\bar{x}+\delta,l]. \end{cases}$$

Therefore, we have:

$$\int_0^l f(x)\delta\theta(x)dx=\int_0^l f(x)e^{\frac{1}{(x-\bar{x})^2-\delta^2}}ds>0.$$

This contradicts condition $\int_0^l f(x)\delta\theta(x)dx=0$. Thus, $f(x)\equiv 0$.

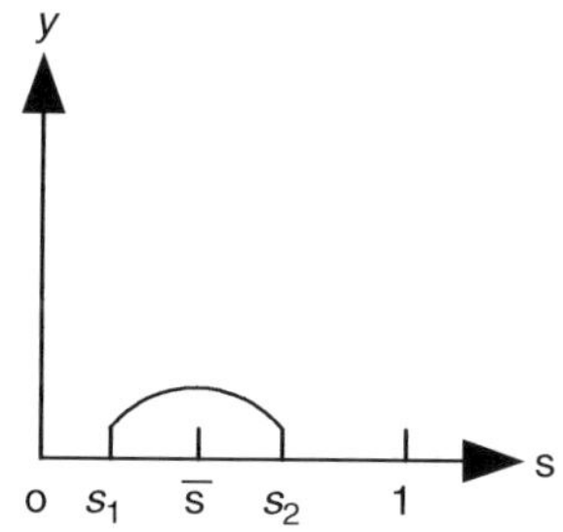

Figure 2.3 The selection point.

For the multi-variable problem, there is a similar preparation theorem.

Lemma 2.2. (Buttazzo *et al.*, 1999) Let $F(x,y)$ is continuous in the plane domain S, and $\delta z(x,y)$ is 0 on the boundary of the domain S, $\|\delta z\| < \varepsilon$, $\|\delta z'_x\| < \varepsilon$, $\|\delta z'_y\| < \varepsilon$, and satisfy the continuity and the first-order or the higher-order differentiable. If for the selected $\delta z(x,y)$,

$$\iint_S F(x,y)\delta z(x,y)dx\,dy = 0$$

always holds, then in the domain S,

$$F(x,y) \equiv 0.$$

Theorem 2.1. (Buttazzo *et al.*, 1999) Such that the simplest functional:

$$J[y(x)] = \int_{x_0}^{x_1} F(x,y,y')dx$$

to obtain the extreme and satisfy fixed boundary conditions:

$$y(x_0) = y_0, \quad y(x_1) = y_1,$$

this extremal curve $y = y(x)$ should satisfy the solution of the following necessary condition:

$$F_y - \frac{d}{dx}F_{y'} = 0, \tag{2.15}$$

where, F are the known functions of x, y, y' and the second-order continuously differentiable function.

(2.15) also is called the Euler-Lagrange equation, it can be written as:

$$F_y - F_{xy'} - F_{xy'}y' - F_{y'y'}y'' = 0. \tag{2.16}$$

Theorem 2.2. (Buttazzo *et al.*, 1999) Such that the functional:

$$J[y(x), z(x)] = \int_{x_0}^{x_1} F(x,y,y',z,z')dx$$

to obtain the extreme and satisfy the fixed boundary conditions:

$$y(x_0) = y_0, \quad y(x_1) = y_1, \quad z(x_0) = z_1, \quad z(x_1) = z_1,$$

this extremal curve $y = y(x)$, $z = z(x)$ must satisfy Euler equations:

$$\begin{cases} F_y - \dfrac{d}{dx}F_{y'} = 0 \\ F_z - \dfrac{d}{dx}F_{z'} = 0. \end{cases} \tag{2.17}$$

For the variational problem of the functional with higher-order derivatives, we only give the Euler-Poisson Equation of the functional with the second derivative:

$$J[y(x)] = \int_{x_0}^{x_1} F(x, y, y', y'')dx,$$

where, F is the third-order continuously differential function, and y is the fourth-order continuously differential function.

Theorem 2.3. (Buttazzo *et al.*, 1999) Such that the functional:

$$J[y(x)] = \int_{x_0}^{x_1} F(x, y, y', y'')dx,$$

to obtain the extreme and satisfy the fixed boundary conditions:

$$y(x_0) = y_0, \quad y(x_1) = y_1, \quad y'(x_0) = y_0', \quad y'(x_1) = y_1',$$

this extremal curve $y = y(x)$ must satisfy differential equation:

$$F_y - \frac{d}{dx}F_{y'} + \frac{d^2}{dx^2}F_{y''} = 0. \tag{2.18}$$

For the variational problem of the function of several variables, we only give the function with two variables.

Theorem 2.4. (Buttazzo *et al.*, 1999) Let D be the plane area, $(x, y) \in D$, $u(x, y) \in C^2(D)$. Such that the functional:

$$J[u(x, y)] = \iint_D F(x, y, u, u_x, u_y)dx\,dy,$$

to obtain the extreme and the known extreme function $u = u(x, y)$ on boundary curve L, this extremal function must satisfy partial differential equation:

$$F_u - \frac{\partial}{\partial x}F_{u_x} - \frac{\partial}{\partial y}F_{u_y} = 0. \tag{2.19}$$

CHAPTER 3

Tubular string buckling theoretical analysis

3.1 INTRODUCTION

The buckling behavior of the tubular string has an important impact on well design and production operations. For example, the tubular movement caused by the buckling influences seal length design, and the bending stresses caused by buckling may dictate tubular weight and grade. The problem of tube sinusoidal buckling was first studied by Paslay and Bogy (1964). In their analysis, the ends of the tube were supported by hinges and the critical force for sinusoidal buckling at the bottom of the tube was found to be a function of the length of the tube because of the compression in the tube increases with depth, the amplitude of the sinusoidal buckling also increases with depth. Paslay and Bogy found that the number of buckling sinusoids increased with the length of the tube. An asymptotic solution for the sinusoidal buckling of an extremely long tube was analysed by Dawson (1984) based on a sinusoidal buckling of constant amplitude. In their work, simple expressions for the buckling force and wave numbers were derived.

In the post-buckling stage, the shape of the buckled tube can change from the sinusoidal mode to a helical mode. The vertical tube helical buckling problem was first analysed by Lubinski and Althouse (1962) based on the principle of minimum potential energy. Since then, the helical buckling of tubes in vertical, horizontal or inclined well-bores have been studied by looking at applied forces by Cheatham (1984), He and Kyllingstad (1995), Miska and Cunha (1995) and Wu and Juvkam-Wold (1995). An experimental study of the helical buckling of a horizontal rod in a tube was conducted by McCann and Suryanarayana (1994). Their observation confirmed the idea that the sinusoidally buckling rod can change its shape to the helical mode when the load increases.

The most generally accepted method for the analysis of buckling, tubular movement, and packer selection is the method developed by Lubinski and Althouse (1962), which considered only the vertical well with no friction. The analyses conducted following the basic approach of Lubinski and Althouse focused on developing methods for more complicated tubing configurations, such as tapered strings (Hammerlindl, 1977), (Hammerlindl, 1980). Henry Woods, in the appendix to Lubinski and Althouse (1962), developed a mechanical model for well buckling behavior that was able to predict the buckled configuration as a function of the well loads. Mitchell developed a more general approach that replaced the virtual work relations with a full set of beam-column equations constrained to be in contact with the casing (Mitchell, 1988). In this formation, helical buckling in a deviated well can be described by a fourth order non-linear differential equation. For a vertical well, the solution to this equation can be accurately approximated using a simple algebraic equation proposed by Lubinski and Woods. However, this solution is not valid for deviated or horizontal wells because of the lateral gravity forces. Numerical solutions were also sought by Mitchell (1997) using the Galerkin technique, and these solutions confirmed the thought that under a general loading the deformed shape of the tube is a combination of helices and sinusoids while helical deformation occurs only under special values for the applied load. For further research on tubular string buckling deformation refer to (Chen and Adnan, 1993; Huang and Pattillo, 2000; Mitchell, 1996; 2006, Qui *et al.*, 1998).

In the analysis of string buckling behavior, many research papers have only considered a frictionless well. However, in some research, the importance of friction has been recognized. Hammerlindl (1980) showed an experimentally measured 50% error in frictionless buckling length change, which was attributed to friction. In this work, analytical solutions to two basic problems were developed: (1) the slacking off from the surface and (2) the upward loading from the base of the tubular. Analytical solutions were possible because the load application was only in a single direction. So, while limited in application, this work showed the considerable importance of friction forces on buckling and on tubular signs. For instance the effect of friction can greatly reduce set down force regardless of the surface slack-off. More research about the influence of friction can be found in He and Kyllingstad (1995) and Mitchell (1997; 2006).

3.2 DEFORMATION DIFFERENTIAL EQUATIONS MODELLING

The buckling behavior of the tubular string has an important impact on well design and production operations. For example, the tubular movement caused by the buckling influences seal length design, and the bending stresses caused by the buckling may dictate tubular weight and grade.

3.2.1 Tubular string differential element force analysis

We can intercept a differential element for the tubular string between s and $s+ds$ for the loads on the differential element including:

1. Internal force vector $\vec{F}(s)$, $-\vec{F}(s+ds)$ and the external moment of the force vector $\vec{M}(s)$, $-\vec{M}(s+ds)$ on the tubular string interface:

$$\vec{F}(s) = \vec{F}_\tau(s)\vec{\tau}_0 + \vec{F}_n(s)\vec{n}_0 + \vec{F}_b(s)\vec{b}_0,$$

$$\vec{M}(s) = \vec{M}_\tau(s)\vec{\tau}_0 + \vec{M}_n(s)\vec{n}_0 + \vec{M}_b(s)\vec{b}_0.$$

2. Distribution external force vector $\vec{f}(s)ds$ for the tubular string differential element:
   (1) Tubular string deadweight $q\vec{k}$
   (2) Normal contact pressure of the curved tubular string and well wall $\vec{N} = N(\cos\theta\,\vec{n}_0, -\sin\theta\,\vec{b}_0)$
   (3) Tubular string internal and external flowing fluid viscous friction force $(\vec{f}_{uo} + \vec{f}_{ui})\vec{\tau}_0$
   (4) Axial friction force $\vec{f}_{f_1} = -f_1 N\vec{\tau}_0$
   (5) Circumferential hoop friction force $\vec{f}_{f_2} = -f_2 N(\sin\theta\,\vec{n}_0 + \cos\theta\,\vec{b}_0)$

   Thus, the whole external force vector can be represented as follows:

$$\vec{f}(s) = (f_{ui} + f_{uo} - f_1 N)\vec{\tau}_0 + N\cos\theta\,\vec{n}_0 - N\sin\theta\,\vec{b}_0 - f_2 N(\sin\theta\,\vec{n}_0 + \cos\theta\,\vec{b}_0) + q\vec{k},$$

   where, f_1 represents the axial friction coefficient between the tubular string and well wall; f_2 represents the circumferential hoop friction coefficient between the tubular string and well wall; f_{ui},f_{uo} represents the viscous friction coefficient of the external and internal fluid acting on the tubular string.
3. Fluid pressure acting on the internal and external side wall of the tubular string's infinitesimal.

The equivalent static force of the internal pressure acting on the tubular string's infinitesimal is as shown in Figure 3.1. From Figure 3.1, we can determine that the equivalent static force of the internal pressure acting on the infinitesimal of the tubular string is equal to enacting a coupled axial constringent load $p_i(s)A_i, p_i(s+ds)A_i$ and a distribution load $d\vec{F}_i(s)$ on the section between s and $s+ds$, respectively:

$$\begin{aligned} d\vec{F}_i(s) &= -S_i\left(p_i - \frac{f_{ui}}{A_i}\right)d\vec{A} = \iiint_{V_i}\left(\frac{\partial p_i}{\partial z}\vec{K} - \frac{f_{ui}}{A_i}\vec{\tau}_0\right)dV \\ &= \iiint_{V_i}\left(\frac{\partial p_i}{\partial S_{n0}}\vec{n}_0 + \frac{\partial p_i}{\partial S_{b0}}\vec{b}_0 + \frac{\partial p_i}{\partial S_{\tau 0}}\vec{\tau}_0\right)dV \end{aligned}$$

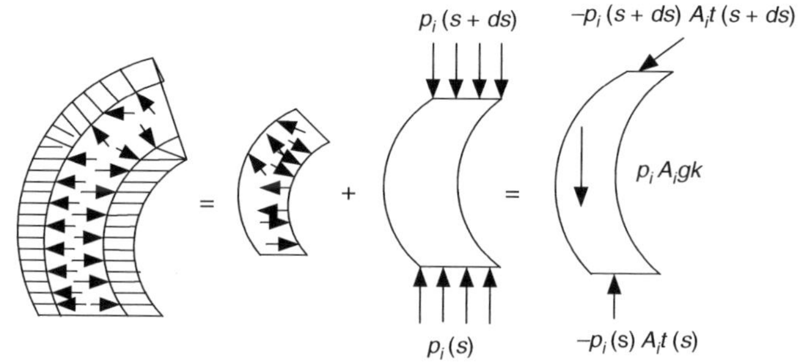

Figure 3.1 The effective action of internal pressure.

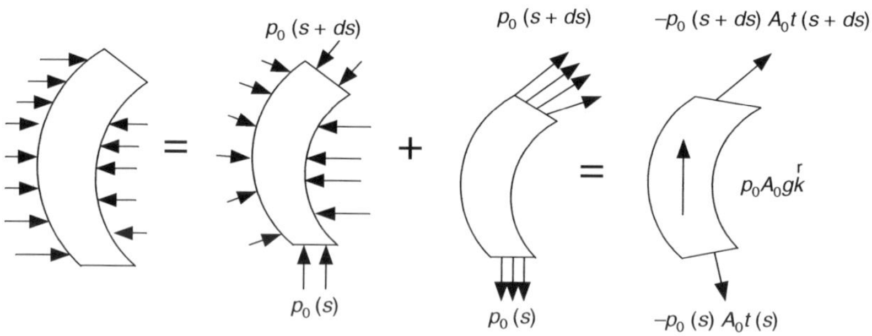

Figure 3.2 The effective action of external pressure.

where,

$$\frac{\partial p_i}{\partial S_{n_0}} = -\rho_i g \frac{\sin\varphi}{K_0}\frac{d\varphi}{ds_0}, \qquad \frac{\partial p_i}{\partial S_{b_0}} = \rho_i g \frac{\sin^2\varphi}{K_0}\frac{d\psi}{ds_0}, \qquad \frac{\partial p_i}{\partial S_{\vec{\tau}_0}} = \rho_i g\cos\varphi - \frac{f_{ui}}{A_i}.$$

Thus,

$$\frac{d\vec{F}_i(s)}{ds} = \rho_i g \frac{\sin\varphi}{K_0}\frac{d\varphi}{ds_0}\vec{n}_0 + \rho_i g \frac{\sin^2\varphi}{K_0}\frac{d\psi}{ds_0}\vec{b}_0 + \left(\rho_i g\cos\varphi - \frac{f_{ui}}{A_i}\right)A_i\vec{\tau}_0$$

or,

$$\frac{d\vec{F}_i(s)}{ds} = \rho_i g A_i \vec{K} - f_{ui}\vec{\tau}_0.$$

As shown in Figure 3.2, the equivalent static force of the external pressure acting on the infinitesimal of the tubular string is equal to enacting a coupled axial tensile load $p_0(s)A_0$, $p_0(s+ds)A_0$ and a distribution load $d\vec{F}_o(s)$ on the section between s and $s+ds$, respectively:

$$d\vec{F}_0(s) = S_0\left(p_0 - \frac{f_{uo}+f_{uw}}{A_h}\right)d\vec{A} = -\iiint_{V_o}\left(\frac{\partial p_0}{\partial S_{n_0}}\vec{n}_0 + \frac{\partial p_0}{\partial S_{b_0}}\vec{b}_0 + \frac{\partial p_0}{\partial S_{\tau_0}}\vec{\tau}_0\right)dV$$

i.e.,

$$\frac{d\vec{F}_0(s)}{ds} = -\rho_0 g A_0 \vec{K} + \frac{f_{uo}+f_{uw}}{A_h}A_0\vec{\tau}_0,$$

where, A_i, A_0 is the internal and external cross-sectional area of the tubular string, f_{uw} is the friction force of well wall, and A_h is annular area.

3.2.2 Static force equilibrium equation for the tubular string infinitesimal

Theorem 3.1. Let the curvature and torsion of the well-bore trail be K_0 and T_0 separately, the deflection angle be θ, and the deviation angle and azimuth angle be φ and ψ. Then the static force equilibrium equation for the tubular string infinitesimal is:

$$\begin{cases} \dfrac{dF_{\tau e}}{ds} = K_0F_n + f_{uo} - f_1N + q_e\cos\varphi + \dfrac{(f_{ui}+f_{uw})A_0}{A_h} \\ \dfrac{dF_n}{ds} = T_0F_b - F_{\tau e}K_0 + N\cos\theta - f_2N\sin\theta - \dfrac{q_e\sin\varphi}{K_0}\dfrac{d\varphi}{ds} \\ \dfrac{dF_b}{ds} = -T_0F_n - N\sin\theta - f_2N\cos\theta + \dfrac{q_e\sin^2\varphi}{K_0}\dfrac{d\psi}{ds} \\ \dfrac{d\vec{M}_\tau}{ds} = K_0M_n + F_n\dfrac{dr\sin\theta}{ds} + F_b\dfrac{dr\cos\theta}{ds} \\ \dfrac{d\vec{M}_n}{ds} = -K_0M\vec{\tau} + T_0M_b + F_b - F_{\tau e}\dfrac{dr\sin\theta}{ds} \\ \dfrac{d\vec{M}_b}{ds} = -T_0M_n + \left(F_{\tau e}\dfrac{dr\cos\theta}{ds} + F_n\right). \end{cases}$$

Proof. Since the tubular string is in a state of equilibrium under the actions of the internal and external force, and the internal and external fluid flow pressure, thus we can determine the following equilibrium equation:

$$\sum\vec{F}_i(s)=0,\quad \sum\vec{M}_0(\vec{F}_i)=0.$$

From $\sum\vec{F}_i(s)=0$, we have:

$$\vec{F}(s)-\vec{F}(s+ds)+[p_i(s)A_i-p_o(s)A_o]\vec{\tau}_0(s)+d\vec{F}_i(s)+d\vec{F}_o(s)+\vec{f}(s)ds$$
$$+[p_o(s+ds)A_0-p_i(s+ds)A_i]\vec{\tau}_0(s+ds)=0$$

i.e.,

$$\frac{\vec{F}(s+ds)-\vec{F}(s)}{ds}+\frac{[p_i(s+ds)\vec{\tau}_0(s+ds)-p_i(s)\vec{\tau}_0(s)]A_i}{ds}$$
$$-\frac{[p_o(s+ds)\vec{\tau}_0(s+ds)-p_o(s)\vec{\tau}_0(s)]A_o}{ds}=\frac{d\vec{F}_i(s)}{ds}+\frac{d\vec{F}_o(s)}{ds}+\vec{f}(s).$$

Thus,

$$\frac{d\vec{F}(s)+dp_i(s)\vec{\tau}_0(s)A_i-dp_o(s)\vec{\tau}_0(s)A_o}{ds}=\frac{d\vec{F}_i(s)}{ds}+\frac{d\vec{F}_o(s)}{ds}+\vec{f}(s).$$

Let:

$$F_{\tau e}(s)=F_\tau(s)+p_i(s)A_i-p_o(s)A_o,$$

then,

$$\vec{F}_e(s)=\vec{F}(s)+[p_i(s)A_i-p_o(s)A_o]\vec{\tau}_0(s)=F_{\tau e}(s)\vec{\tau}_0+F_n(s)\vec{n}_0+F_b(s)\vec{b}_0.$$

Therefore,

$$\vec{f}_e(s)=\vec{f}(s)+\frac{d\vec{F}_i}{ds}+\frac{d\vec{F}_0}{ds}=f_{\tau e}(s)\vec{\tau}_0+f_{ne}(s)\vec{n}_0+f_{be}(s)\vec{b}_0.$$

Let $q_e = q + \rho_i g A_i - \rho_0 g A_0$, then from (2.9), $\vec{k}$, $\frac{d\vec{F}_i}{ds}$ and $\frac{d\vec{F}_0}{ds}$, we get:

$$\begin{aligned} f_{\tau e}(s) &= f_{uo} - f_1 N + q_e \cos\varphi + \frac{(f_{ui} + f_{uw})A_0}{A_h} \\ f_{ne}(s) &= N\cos\theta - f_2 N \sin\theta - \frac{q_e \sin\varphi}{K_0}\frac{d\varphi}{ds} \\ f_{be}(s) &= -N\sin\theta - f_2 N\cos\theta + \frac{q_e \sin^2\varphi}{K_0}\frac{d\psi}{ds}. \end{aligned}$$

Thus,

$$\frac{d\vec{F}_e}{ds} = \vec{f}_e(s),$$

where, $F_{\tau e}$ is the equivalent axial load under the internal and external pressure equivalent action, F_τ is actual axial load, $f_{\tau e}$ is the equivalent axial external force, f_{ne} is the equivalent external force for the principal normal, f_{be} is the equivalent external force for the subnormal. q_e is the equivalent dead weight after considering the internal and external fluid action on the tubular string, and q is the actual dead weight.

From $\sum \vec{M}_0(\vec{F}_i) = 0$, we have:

$$\vec{M}(s) - \vec{M}(s+ds) - \vec{F}_e(s) \times \vec{r}(s) + \vec{F}_e(s+ds) \times \vec{r}(s+ds) - \vec{f}_e(s) \times \vec{r}(s)ds = 0,$$

i.e.,

$$\frac{d\vec{M}}{ds} = \vec{F}_e(s) \times \vec{\tau}(s).$$

Because of $\vec{F}_e(s) = F_{\tau e}(s)\vec{\tau}_0 + F_n(s)\vec{n}_0 + F_b(s)\vec{b}_0$, then:

$$\frac{d\vec{F}_e}{ds} = \frac{dF_{\tau e}}{ds}\vec{\tau}_0 + F_{\tau e}\frac{d\vec{\tau}_0}{ds} + \frac{dF_n}{ds}\vec{n}_0 + F_n\frac{d\vec{n}_0}{ds} + \frac{dF_b}{ds}\vec{b}_0 + F_b\frac{d\vec{b}_0}{ds},$$

from (2.10), we can get:

$$\begin{aligned} \frac{d\vec{F}_e}{ds} &= \frac{dF_{\tau e}}{ds}\vec{\tau}_0 + F_{\tau e}K_0\vec{n}_0 + \frac{dF_n}{ds}\vec{n}_0 + F_n(-K_0\vec{\tau}_0 + T_0\vec{b}_0) + \frac{dF_b}{ds}\vec{b}_0 - F_b T_0 \vec{n}_0 \\ &= \left(\frac{dF_{\tau e}}{ds} - K_0 F_n\right)\vec{\tau}_0 + \left(F_{\tau e}K_0 + \frac{dF_n}{ds} - T_0 F_b\right)\vec{n}_0 + \left(T_0 F_n + \frac{dF_b}{ds}\right)\vec{b}_0. \end{aligned}$$

Again using $\frac{d\vec{F}_e}{ds} = \vec{f}_e(s)$, we have:

$$\frac{dF_{\tau e}}{ds} - K_0 F_n = f_{uo} - f_1 N + q_e \cos\varphi + \frac{(f_{ui} + f_{uw})A_o}{A_h} \tag{3.1}$$

$$F_{\tau e}K_0 + \frac{dF_n}{ds} - T_0 F_b = N\cos\theta - f_2 N\sin\theta - \frac{q_e \sin\varphi}{K_0}\frac{d\varphi}{ds} \tag{3.2}$$

$$T_0 F_n + \frac{dF_b}{ds} = -N\sin\theta - f_2 N\cos\theta + \frac{q_e \sin^2\varphi}{K_0}\frac{d\psi}{ds}. \tag{3.3}$$

Similarly,

$$\frac{d\vec{M}}{ds} = \vec{F}_e(s) \times \vec{\tau}(s) = \begin{vmatrix} \vec{\tau}_0 & \vec{n}_0 & \vec{b}_0 \\ F_{\tau e} & F_n & F_b \\ 1 & -r\frac{d\cos\theta}{ds} & r\frac{d\sin\theta}{ds} \end{vmatrix}$$

i.e.,

$$\frac{d\vec{M}}{ds} = \left(F_n\frac{dr\sin\theta}{ds} + F_b\frac{dr\cos\theta}{ds}\right)\vec{\tau}_0 + \left(F_b - F_{\tau e}\frac{dr\sin\theta}{ds}\right)\vec{n}_0 - \left(F_{\tau e}\frac{dr\cos\theta}{ds} + F_n\right)\vec{b}_0.$$

From $\vec{M} = \vec{M}_\tau\vec{\tau}_0 + \vec{M}_n\vec{n}_0 + \vec{M}_b\vec{b}_0$, we have:

$$\frac{d\vec{M}}{ds} = \frac{d\vec{M}_\tau}{ds}\vec{\tau}_0 + \vec{M}_\tau\frac{d\vec{\tau}_0}{ds} + \frac{d\vec{M}_n}{ds}\vec{n}_0 + \vec{M}_n\frac{d\vec{n}_0}{ds} + \frac{d\vec{M}_b}{ds}\vec{b}_0 + \vec{M}_b\frac{d\vec{b}_0}{ds}$$

again using (2.10), we can get:

$$\frac{d\vec{M}}{ds} = \left(\frac{d\vec{M}_\tau}{ds} - K_0M_n\right)\vec{\tau}_0 + \left(\frac{d\vec{M}_n}{ds} + K_0M\vec{\tau} - T_0M_b\right)\vec{n}_0 + \left(\frac{d\vec{M}_b}{ds} + T_0M_n\right)\vec{b}_0,$$

therefore,

$$\frac{d\vec{M}_\tau}{ds} - K_0M_n = F_n\frac{dr\sin\theta}{ds} + F_b\frac{dr\cos\theta}{ds} \tag{3.4}$$

$$\frac{d\vec{M}_n}{ds} + K_0M\vec{\tau} - T_0M_b = F_b - F_{\tau e}\frac{dr\sin\theta}{ds} \tag{3.5}$$

$$\frac{d\vec{M}_b}{ds} + T_0M_n = -\left(F_{\tau e}\frac{dr\cos\theta}{ds} + F_n\right). \tag{3.6}$$

Thus, the static force equilibrium equation of differential element of tubular string is:

$$\begin{cases} \dfrac{dF_{\tau e}}{ds} = K_0F_n + f_{uo} - f_1N + q_e\cos\varphi + \dfrac{(f_{ui}+f_{uw})A_0}{A_h} \\ \dfrac{dF_n}{ds} = T_0F_b - F_{\tau e}K_0 + N\cos\theta - f_2N\sin\theta - \dfrac{q_e\sin\varphi}{K_0}\dfrac{d\varphi}{ds} \\ \dfrac{dF_b}{ds} = -T_0F_n - N\sin\theta - f_2N\cos\theta + \dfrac{q_e\sin^2\varphi}{K_0}\dfrac{d\psi}{ds} \\ \dfrac{d\vec{M}_\tau}{ds} = K_0M_n + F_n\dfrac{dr\sin\theta}{ds} + F_b\dfrac{dr\cos\theta}{ds} \\ \dfrac{d\vec{M}_n}{ds} = -K_0M\vec{\tau} + T_0M_b + F_b - F_{\tau e}\dfrac{dr\sin\theta}{ds} \\ \dfrac{d\vec{M}_b}{ds} = -T_0M_n + \left(F_{\tau e}\dfrac{dr\cos\theta}{ds} + F_n\right). \end{cases}$$

3.2.3 The buckling differential equation for the tubular string

Theorem 3.2. Let the curvature and torsion of the well-bore trail be K_0 and T_0 separately, the deflection angle be θ, the deviation angle and azimuth angle be φ and ψ, and the length of the tubular string be l. Suppose that the well-bore tubular string experiences a buckling deformation. Then we can determine the buckling tubular string's deformation system:

(1) The main differential equations:

$$\begin{cases} \dfrac{dF_{\tau e}}{ds} = f_{uo} - f_1N + q_e\cos\varphi + \dfrac{(f_{ui}+f_{uw})A_0}{A_h} \\ \dfrac{d^4\theta}{ds^4} - 6\left(\dfrac{d\theta}{ds}\right)^2\dfrac{d^2\theta}{ds^2} + \dfrac{3M_T}{EI}\dfrac{d\theta}{ds}\dfrac{d^2\theta}{ds^2} + \dfrac{F_{\tau e}}{EI}\dfrac{d^2\theta}{ds^2} + \dfrac{f_n}{EIr}\sin(\theta-\gamma) + \dfrac{f_2N}{EIr} = 0. \end{cases}$$

(2) The differential equations' equilibrium condition:

$$N = -EIr\left[\left(\frac{d\theta}{ds}\right)^4 - 4\frac{d\theta}{ds}\frac{d^3\theta}{ds^3} - 3\left(\frac{d^2\theta}{ds^2}\right)^2\right] + 3M_T r\left[\left(\frac{d\theta}{ds}\right)^3 - \frac{d^3\theta}{ds^3}\right]$$

$$+ F_{\tau e} r\left(\frac{d\theta}{ds}\right)^2 + f_n\cos(\theta - \gamma),$$

(3) The boundary condition:

$$\left.\frac{d\theta}{ds}\right|_{s=0} = 0, \quad \left.\frac{d\theta}{ds}\right|_{s=l} = \theta_1,$$

where,

$$f_{n1} = \frac{q_e \sin^2\varphi}{K_0}\frac{d\psi}{ds}, \quad f_{n2} = K_0 F_{\tau e} + \frac{q_e \sin\varphi}{K_0}\frac{d\varphi}{ds}, \quad \tan\gamma = \frac{f_{n1}}{f_{n2}},$$

$$f_n = (f_{n1}^2 + f_{n2}^2)^{1/2} = \left[\left(F_{\tau e}\frac{d\varphi}{ds} + q_e\sin\varphi\right)^2 + \left(F_{\tau e}\sin\varphi\frac{d\psi}{ds}\right)^2\right]^{1/2}.$$

Proof. Because of the well wall limits, the tubular string deformation has only a small elasticity deformation. So, in this case, the higher terms can be ignored, and the relationship can be considered as one of a landscape orientation deformation and an internal moment in time,

$$\vec{M}(s) = -EI(K\vec{b}) + \vec{M}_T\vec{\tau} = \left(EIr\frac{d^2\sin\theta}{ds^2}\right)\vec{n}_0 + \left(EIr\frac{d^2\cos\theta}{ds^2} - EIK_0\right)\vec{b}_0 + M_T\vec{\tau},$$

i.e.,

$$M_\tau(s) = M_T \tag{3.7}$$

$$M_n(s) = EIr\frac{d^2\sin\theta}{ds^2} - M_T r\frac{d\cos\theta}{ds} \tag{3.8}$$

$$M_b(s) = EIr\frac{d^2\cos\theta}{ds^2} - EIK_0 + M_T r\frac{d\sin\theta}{ds}. \tag{3.9}$$

From (3.8) and (3.9), we have:

$$\frac{dM_n(s)}{ds} = EIr\frac{d^3\sin\theta}{ds^3} - M_T r\frac{d^2\cos\theta}{ds^2} \tag{3.10}$$

$$\frac{dM_b(s)}{ds} = EIr\frac{d^3\cos\theta}{ds^3} + M_T r\frac{d^2\sin\theta}{ds^2}. \tag{3.11}$$

Combining (3.5), (3.8), (3.10) and (3.11), then:

$$EIr\frac{d^3\sin\theta}{ds^3} - M_T r\frac{d^2\cos\theta}{ds^2} + K_0 M_T - T_0\left(EIr\frac{d^2\cos\theta}{ds^2} - EIK_0 + M_T r\frac{d\sin\theta}{ds}\right)$$

$$= F_b - F_{\tau e}\frac{dr\sin\theta}{ds}. \tag{3.12}$$

Considering $K_0^{n_1} T_0^{n_2} r^{n_3}$ $(n_1 + n_2 + n_3 > 1)$, (3.12) can be approximated as follows:

$$F_b = EIr\frac{d^3\sin\theta}{ds^3} - M_T r\frac{d^2\cos\theta}{ds^2} + K_0 M_T + F_{\tau e}\frac{dr\sin\theta}{ds}. \tag{3.13}$$

Similarly,

$$F_n = -EIr\frac{d^3\sin\theta}{ds^3} - M_T r\frac{d^2\sin\theta}{ds^2} - F_{\tau e}\frac{dr\cos\theta}{ds}. \tag{3.14}$$

From (3.14) and (3.1), we have:

$$\begin{aligned}&\frac{dF_{\tau e}}{ds} - K_0\left(-EIr\frac{d^3\sin\theta}{ds^3} - M_T r\frac{d^2\sin\theta}{ds^2} - F_{\tau e}\frac{dr\cos\theta}{ds}\right)\\ &= f_{uo} - f_1 N + q_e\cos\varphi + \frac{(f_{ui}+f_{uw})A_o}{A_h}.\end{aligned} \tag{3.15}$$

Ignoring combination of K_0, T_0, r, then (3.15) can be approximated:

$$\frac{dF_{\tau e}}{ds} = f_{uo} - f_1 N + q_e\cos\varphi + \frac{(f_{ui}+f_{uw})A_o}{A_h}. \tag{3.16}$$

Let (3.2) × cos θ – (3.3) × sin θ, we have:

$$N = \left(K_0F_{\tau e} + \frac{dF_n}{ds} - T_0F_b + \frac{q_e\sin\varphi}{K_0}\frac{d\varphi}{ds}\right)\cos\theta - \left(T_0F_n + \frac{dF_b}{ds} - \frac{q_e\sin^2\varphi}{K_0}\frac{d\psi}{ds}\right)\sin\theta. \tag{3.17}$$

Let (3.2) × sin θ + (3.3) × cos θ,

$$\left(K_0F_{\tau e} + \frac{dF_n}{ds} - T_0F_b + \frac{q_e\sin\varphi}{K_0}\frac{d\varphi}{ds}\right)\sin\theta + \left(T_0F_n + \frac{dF_b}{ds} - \frac{q_e\sin^2\varphi}{K_0}\frac{d\psi}{ds}\right)\cos\theta + f_2N = 0. \tag{3.18}$$

Since:

$$\frac{dF_n}{ds} = -EIr\frac{d^4\sin\theta}{ds^4} - M_T r\frac{d^3\sin\theta}{ds^3} - F_{\tau e}r\frac{d^2\cos\theta}{ds^2} \tag{3.19}$$

$$\frac{dF_b}{ds} = EIr\frac{d^4\sin\theta}{ds^4} - M_T r\frac{d^3\cos\theta}{ds^3} + K_0M_T + F_{\tau e}r\frac{d^2\sin\theta}{ds^2}. \tag{3.20}$$

Combining (3.1), (3.2), (3.17), (3.19) and (3.20), we have:

$$\begin{aligned}N = &\left(K_0F_{\tau e} + \frac{dF_n}{ds} - T_0F_b + \frac{q_e\sin\varphi}{K_0}\frac{d\varphi}{ds} - EIr\frac{d^4\sin\theta}{ds^4} - M_T r\frac{d^3\sin\theta}{ds^3} - F_{\tau e}r\frac{d^2\cos\theta}{ds^2}\right)\cos\theta\\ &- \left(T_0F_n + \frac{dF_b}{ds} - \frac{q_e\sin^2\varphi}{K_0}\frac{d\psi}{ds} + EIr\frac{d^4\sin\theta}{ds^4} - M_T r\frac{d^3\cos\theta}{ds^3} + K_0M_T + F_{\tau e}r\frac{d^2\sin\theta}{ds^2}\right)\sin\theta.\end{aligned} \tag{3.21}$$

Again,

$$T_0F_n = -EIrT_0\frac{d^3\sin\theta}{ds^3} - M_T rT_0\frac{d^2\sin\theta}{ds^2} - F_{\tau e}rT_0\frac{d\cos\theta}{ds}$$

$$T_0F_b = EIrT_0\frac{d^3\sin\theta}{ds^3} - M_T rT_0\frac{d^2\cos\theta}{ds^2} + K_0T_0M_T + F_{\tau e}rT_0\frac{d\sin\theta}{ds}.$$

Considering K_0T_0, T_0r are very small geometrical products, therefore ignoring T_0F_b and T_0F_n, (3.21), (3.18) can be approximated separately,

$$N = \left(K_0F_{\tau e} + \frac{dF_n}{ds} + \frac{q_e \sin\varphi}{K_0}\frac{d\varphi}{ds} - EIr\frac{d^4\sin\theta}{ds^4} - M_T r\frac{d^3\sin\theta}{ds^3} - F_{\tau e}r\frac{d^2\cos\theta}{ds^2}\right)\cos\theta - \left(\frac{dF_b}{ds} - \frac{q_e \sin^2\varphi}{K_0}\frac{d\psi}{ds} + EIr\frac{d^4\sin\theta}{ds^4} - M_T r\frac{d^3\cos\theta}{ds^3} + K_0M_T + F_{\tau e}r\frac{d^2\sin\theta}{ds^2}\right)\sin\theta \tag{3.22}$$

$$\left(K_0F_{\tau e} + \frac{dF_n}{ds} + \frac{q_e \sin\varphi}{K_0}\frac{d\varphi}{ds} - EIr\frac{d^4\sin\theta}{ds^4} - M_T r\frac{d^3\sin\theta}{ds^3} - F_{\tau e}r\frac{d^2\cos\theta}{ds^2}\right)\sin\theta + \left(\frac{dF_b}{ds} - \frac{q_e \sin^2\varphi}{K_0}\frac{d\psi}{ds} + EIr\frac{d^4\sin\theta}{ds^4} - M_T r\frac{d^3\cos\theta}{ds^3} + K_0M_T + F_{\tau e}r\frac{d^2\sin\theta}{ds^2}\right)\cos\theta + f_2N = 0. \tag{3.23}$$

Because of,

$$\begin{aligned}
\frac{d\cos\theta}{ds} &= -\sin\theta\frac{d\theta}{ds},\\
\frac{d^2\cos\theta}{ds^2} &= -\sin\theta\frac{d^2\theta}{ds^2} - \cos\theta\left(\frac{d\theta}{ds}\right)^2,\\
\frac{d^3\cos\theta}{ds^3} &= \sin\theta\left(\frac{d\theta}{ds}\right)^3 - 3\cos\theta\frac{d^2\theta}{ds^2}\frac{d\theta}{ds} - \sin\theta\frac{d^3\theta}{ds^3},\\
\frac{d^4\cos\theta}{ds^4} &= \cos\theta\left(\frac{d\theta}{ds}\right)^4 + 6\sin\theta\left(\frac{d\theta}{ds}\right)^2\frac{d^2\theta}{ds^2} - 3\cos\theta\left(\frac{d^2\theta}{ds^2}\right)^2\\
&\quad - 4\cos\theta\frac{d\theta}{ds}\frac{d^3\theta}{ds^3} - \sin\theta\frac{d^4\theta}{ds^4}.\\
\frac{d\sin\theta}{ds} &= \cos\theta\frac{d\theta}{ds},\\
\frac{d^2\sin\theta}{ds^2} &= \cos\theta\frac{d^2\theta}{ds^2} - \sin\theta\frac{d\theta}{ds},\\
\frac{d^3\sin\theta}{ds^3} &= -\cos\theta\left(\frac{d\theta}{ds}\right)^3 - 3\sin\theta\frac{d\theta}{ds}\frac{d^2\theta}{ds^2} + \cos\theta\frac{d^3\theta}{ds^3},\\
\frac{d^4\sin\theta}{ds^4} &= \sin\theta\left(\frac{d\theta}{ds}\right)^4 - 6\cos\theta\left(\frac{d\theta}{ds}\right)^2\frac{d^2\theta}{ds^2} - 3\sin\theta\left(\frac{d^2\theta}{ds^2}\right)^2\\
&\quad - 4\sin\theta\frac{d\theta}{ds}\frac{d^3\theta}{ds^3} + \cos\theta\frac{d^4\theta}{ds^4}.
\end{aligned}$$

Then,

$$\begin{aligned}
\cos\frac{d^4\cos\theta}{ds^4} &= \cos^2\theta\left(\frac{d\theta}{ds}\right)^4 + 6\sin\theta\cos\theta\left(\frac{d\theta}{ds}\right)^2\frac{d^2\theta}{ds^2} - 3\cos^2\theta\left(\frac{d^2\theta}{ds^2}\right)^2\\
&\quad - 4\cos^2\theta\frac{d\theta}{ds}\frac{d^3\theta}{ds^3} - \sin\theta\cos\theta\frac{d^4\theta}{ds^4},
\end{aligned}$$

$$\cos\theta\frac{d^3\sin\theta}{ds^3} = -\cos^2\theta\left(\frac{d\theta}{ds}\right)^3 - 3\sin\theta\cos\theta\frac{d\theta}{ds}\frac{d^2\theta}{ds^2} + \cos^2\theta\frac{d^3\theta}{ds^3},$$

$$\cos\theta\frac{d^2\cos\theta}{ds^2} = -\sin\theta\cos\theta\frac{d^2\theta}{ds^2} - \cos^2\theta\left(\frac{d\theta}{ds}\right)^2,$$

and

$$\begin{aligned}\sin\theta\frac{d^4\sin\theta}{ds^4} &= \sin^2\theta\left(\frac{d\theta}{ds}\right)^4 - 6\sin\theta\cos\theta\left(\frac{d\theta}{ds}\right)^2\frac{d^2\theta}{ds^2} - 3\sin^2\theta\left(\frac{d^2\theta}{ds^2}\right)^2 \\ &\quad - 4\sin^2\theta\frac{d\theta}{ds}\frac{d^3\theta}{ds^3} + \sin\theta\cos\theta\frac{d^4\theta}{ds^4},\end{aligned}$$

$$\sin\theta\frac{d^3\cos\theta}{ds^3} = \sin^2\theta\left(\frac{d\theta}{ds}\right)^3 - 3\sin\theta\cos\theta\frac{d^2\theta}{ds^2}\frac{d\theta}{ds} - \sin^2\theta\frac{d^3\theta}{ds^3},$$

$$\sin\theta\frac{d^2\sin\theta}{ds^2} = \sin\theta\cos\theta\frac{d^2\theta}{ds^2} - \sin^2\theta\frac{d\theta}{ds}.$$

Therefore,

$$\begin{aligned}N = &-EIr\left[\left(\frac{d\theta}{ds}\right)^4 - 4\frac{d\theta}{ds}\frac{d^3\theta}{ds^3} - 3\left(\frac{d^2\theta}{ds^2}\right)^2\right] + 3M_T r\left[\left(\frac{d\theta}{ds}\right)^3 - \frac{d^3\theta}{ds^3}\right] + F_{\tau e}r\left(\frac{d\theta}{ds}\right)^2 \\ &+ \left(K_0F_{\tau e} + \frac{q_e\sin\varphi}{K_0}\frac{d\varphi}{ds}\right)\cos\theta + \frac{q_e\sin^2\varphi}{K_0}\frac{d\psi}{ds}\sin\theta. \end{aligned} \tag{3.24}$$

Similarly,

$$\begin{aligned}&EIr\left[\left(\frac{d^4\theta}{ds^4}\right) - 6\left(\frac{d\theta}{ds}\right)^2\frac{d^2\theta}{ds^2}\right] + 3M_T r\left[\frac{d\theta}{ds}\frac{d^2\theta}{ds^2}\right] + F_{\tau e}r\frac{d^2\theta}{ds^2} \\ &+ \left(K_0F_{\tau e} + \frac{q_e\sin\varphi}{K_0}\frac{d\varphi}{ds}\right)\sin\theta - \frac{q_e\sin^2\varphi}{K_0}\frac{d\psi}{ds}\cos\theta + f_2N = 0,\end{aligned}$$

i.e.,

$$\begin{aligned}&\frac{d^4\theta}{ds^4} - 6\left(\frac{d\theta}{ds}\right)^2\frac{d^2\theta}{ds^2} + \frac{3M_T}{EI}\frac{d\theta}{ds}\frac{d^2\theta}{ds^2} + \frac{F_{\tau e}}{EI}\frac{d^2\theta}{ds^2} + \frac{K_0F_{\tau e} + \frac{q_e\sin\varphi}{K_0}\frac{d\varphi}{ds}}{EIr}\sin\theta \\ &- \frac{1}{EIr}\frac{q_e\sin^2\varphi}{K_0}\frac{d\psi}{ds}\cos\theta + \frac{f_2N}{EIr} = 0.\end{aligned} \tag{3.25}$$

Let,

$$f_{n1} = \frac{q_e\sin^2\varphi}{K_0}\frac{d\psi}{ds}, \quad f_{n2} = K_0F_{\tau e} + \frac{q_e\sin\varphi}{K_0}\frac{d\varphi}{ds}, \quad \tan\gamma = \frac{f_{n1}}{f_{n2}}, \tag{3.26}$$

then,

$$\left(K_0F_{\tau e} + \frac{q_e\sin\varphi}{K_0}\frac{d\varphi}{ds}\right)\cos\theta + \frac{q_e\sin^2\varphi}{K_0}\frac{d\psi}{ds}\sin\theta = f_n\cos(\theta-\gamma),$$

$$\left(K_0F_{\tau e} + \frac{q_e\sin\varphi}{K_0}\frac{d\varphi}{ds}\right)\sin\theta - \frac{q_e\sin^2\varphi}{K_0}\frac{d\psi}{ds}\cos\theta = f_n\sin(\theta-\gamma),$$

where,

$$f_n=\left(f_{n1}^2+f_{n2}^2\right)^{1/2}=\left[\left(\frac{q_e\sin^2\varphi}{K_0}\frac{d\psi}{ds}\right)^2+\left(K_0F_{\tau e}+\frac{q_e\sin\varphi}{K_0}\frac{d\varphi}{ds}\right)^2\right]^{1/2}$$

$$=\left[\left(F_{\tau e}\frac{d\varphi}{ds}+q_e\sin\varphi\right)^2+\left(F_{\tau e}\sin\varphi\frac{d\psi}{ds}\right)^2\right]^{1/2}. \tag{3.27}$$

Thus, the buckling deformation system can be got as the following:

(1) the main differential equations:

$$\begin{cases}\dfrac{dF_{\tau e}}{ds}=f_{uo}-f_1N+q_e\cos\varphi+\dfrac{(f_{ui}+f_{uw})A_o}{A_h}\\[2ex]\dfrac{d^4\theta}{ds^4}-6\left(\dfrac{d\theta}{ds}\right)^2\dfrac{d^2\theta}{ds^2}+\dfrac{3M_T}{EI}\dfrac{d\theta}{ds}\dfrac{d^2\theta}{ds^2}+\dfrac{F_{\tau e}}{EI}\dfrac{d^2\theta}{ds^2}+\dfrac{f_n}{EIr}\sin(\theta-\gamma)+\dfrac{f_2N}{EIr}=0\end{cases} \tag{3.28}$$

(2) the equilibrium condition of differential equations:

$$N=-EIr\left[\left(\frac{d\theta}{ds}\right)^4-4\frac{d\theta}{ds}\frac{d^3\theta}{ds^3}-3\left(\frac{d^2\theta}{ds^2}\right)^2\right]+3M_Tr\left[\left(\frac{d\theta}{ds}\right)^3-\frac{d^3\theta}{ds^3}\right]$$

$$+F_{\tau e}r\left(\frac{d\theta}{ds}\right)^2+f_n\cos(\theta-\gamma),$$

(3) the boundary condition:

$$\left.\frac{d\theta}{ds}\right|_{s=0}=0,\quad \left.\frac{d\theta}{ds}\right|_{s=l}=\theta_1.$$

Let $\theta'=\frac{d\theta}{ds}$, $\theta''=\frac{d^2\theta}{ds^2}$, $\theta'''=\frac{d^3\theta}{ds^3}$, $\theta^{(4)}=\frac{d^4\theta}{ds^4}$, then the above equation can be written as:

(1) the main differential equations:

$$\begin{cases}\dfrac{dF_{\tau e}}{ds}=f_{uo}-f_1N+q_e\cos\varphi+\dfrac{(f_{ui}+f_{uw})A_o}{A_h}\\[1ex]\theta^{(4)}-6(\theta')^2\theta''+\dfrac{3M_T}{EI}\theta'\theta''+\dfrac{F_{\tau e}}{EI}\theta''+\dfrac{f_n}{EIr}\sin(\theta-\gamma)+\dfrac{f_2N}{EIr}=0\end{cases}$$

(2) the equilibrium condition of differential equations:

$$N=-EIr[(\theta')^4-4\theta'\theta'''-3(\theta'')^2]+3M_Tr[(\theta')^3-\theta''']+F_{\tau e}r(\theta')^2+f_n\cos(\theta-\gamma)$$

(3) the boundary condition:

$$\theta'(0)=\theta_0,\quad \theta'(l)=\theta_1.$$

Which completes Theorem 3.2.

3.3 THE EQUIVALENT VARIATIONAL PROBLEM

From the above, we can determine the tubular string's deformation differential equations using statistics, but it is very difficult to obtain an analytical solution to the non-linear differential equations. Therefore, in the following section, we will discuss an equivalent variational problem with non-linear differential equations by variational method.

3.3.1 Tubular displacement analysis

Assume that the tubular displacement caused by the axial force is $u_a(s)$, then, corresponding to the axial force, the axial strain is:

$$\varepsilon_a(s) = \frac{du_a(s)}{ds} = \frac{F_{\tau e}}{EA},$$

therefore, the tubular string displacement is:

$$u_a(s) = \int_0^l \frac{F_{\tau e}}{EA} ds. \tag{3.29}$$

Assume that the axial displacement caused by the bending and buckling deformation is $u_b(s)$, so then:

$$ds - ds_0 = \left[rK_0 \cos\theta + \frac{1}{2} r^2 \left(\frac{d\theta}{ds_0} \right)^2 \right] ds_0. \tag{3.30}$$

Let ds_0 be the arc infinitesimal of the well-bore axis, ds be the arc infinitesimal length for the normal plane of the two endpoints of ds_0 cutting the tubular string, then corresponding to ds_0, the length ds_0' without bending is:

$$ds_0' = (1 + rk_0) ds_0. \tag{3.31}$$

From (3.31) and (3.30), we can determine the displacement of pre and post bending,

$$du_\tau(s) = ds - ds_0' = (ds - ds_0) + ds_0 - ds_0' = \left[\frac{1}{2} r^2 \left(\frac{d\theta}{ds_0} \right)^2 - rK_0(1 - \cos\theta) \right] ds_0. \tag{3.32}$$

From $ds = ds_0$, (3.32) can be written as:

$$du_\tau(s) = \left[\frac{1}{2} r^2 \left(\frac{d\theta}{ds} \right)^2 - rK_0(1 - \cos\theta) \right] ds. \tag{3.33}$$

Let the torsion angle displacement caused by the bending deformation and buckling be $\gamma_b(s)$, from (2.14), so we get:

$$\frac{dr_b(s)}{ds} = k\vec{b} \cdot \vec{\tau} = r^2 \left(\frac{d\theta}{ds} \right)^3 + rk_0 \cos\theta \frac{d\theta}{ds},$$

therefore,

$$r_b(s) = \int_0^l \left[r^2 \left(\frac{d\theta}{ds} \right)^3 + rk_0 \cos\theta \frac{d\theta}{ds} \right] ds. \tag{3.34}$$

Assume that the vertical displacement is $u_z(s)$, and the displacement s along the principal normal $\vec{n}_0$ and subnormal $\vec{b}_0$ is separate,

$$u_{n_0} = r(1 - \cos\theta), \quad u_{b_0} = r \sin\theta. \tag{3.35}$$

From the relationship between $(\vec{\tau}_0, \vec{n}_0, \vec{b}_0)$ and $(\vec{i}, \vec{j}, \vec{k})$, we get the vertical displacement,

$$u_z(s) = -[u_{n_0}(s)\vec{n}_0 \cdot \vec{k} + u_{b_0}\vec{b}_0 \cdot \vec{k}],$$

i.e.,

$$u_z(s) = r(1 - \cos\theta) \frac{\sin\varphi}{k_0} \frac{d\varphi}{ds} - r\sin\theta \frac{\sin^2\varphi}{k_0} \frac{d\psi}{ds}. \tag{3.36}$$

3.3.2 External force and deformation energy analysis

The bending deformation energy of the tubular string is $\vec{M}(s) = -EIk\vec{b} + M_t\vec{\tau}$. From (2.14), then we get:

$$\vec{M}(s) = \left(EIr\frac{d^2\sin\theta}{ds^2}\right)\vec{n}_0 + \left(EIr\frac{d^2\cos\theta}{ds^2} - EIk_0\right)\vec{b}_0 + M_t\vec{\tau}.$$

Thus,

$$M^2(s) = \left(EIr\frac{d^2\sin\theta}{ds^2}\right)^2 + \left(EIr\frac{d^2\cos\theta}{ds^2} - EIk_0\right)^2 + M_t^2,$$

i.e.,

$$M^2(s) = \left(EIr\frac{d^2\sin\theta}{ds^2}\right)^2 + \left(EIr\frac{d^2\cos\theta}{ds^2}\right)^2 - 2(EI)^2k_0r\frac{d^2\cos\theta}{ds^2} + (EIk_0)^2 + M_t^2.$$

Since:

$$\frac{d^2\sin\theta}{ds^2} = -\sin\theta\cdot\left(\frac{d\theta}{ds}\right)^2 + \cos\theta\cdot\frac{d^2\theta}{ds^2}, \quad \frac{d^2\cos\theta}{ds^2} = -\cos\theta\cdot\left(\frac{d\theta}{ds}\right)^2 - \sin\theta\cdot\frac{d^2\theta}{ds^2},$$

it follows that:

$$\begin{aligned} M^2(s) &= M_t^2 + (EIr)^2\left[\left(\frac{d^2\theta}{ds^2}\right)^2 + \left(\frac{d\theta}{ds}\right)^4\right] + (EIk_0)^2 \\ &\quad + 2E^2I^2k_0r\left[\cos\theta\cdot\left(\frac{d\theta}{ds}\right)^2 + \sin\theta\cdot\frac{d^2\theta}{ds^2}\right]. \end{aligned} \tag{3.37}$$

Therefore, the tubular string deformation energy is $U = \int_0^l \frac{M^2(s)}{2EI}ds$, i.e.,

$$\begin{aligned} U &= \int_0^l \left\{\frac{M_t^2}{2EI} + \frac{1}{2}EIr^2\left[\left(\frac{d^2\theta}{ds^2}\right)^2 + \left(\frac{d\theta}{ds}\right)^4\right] + \frac{1}{2}EIk_0^2 \right. \\ &\quad \left. + EIk_0r\left[\cos\theta\cdot\left(\frac{d\theta}{ds}\right)^2 + \sin\theta\cdot\frac{d^2\theta}{ds^2}\right]\right\} ds. \end{aligned} \tag{3.38}$$

The doing work of axial compress load is $W_F = W_{Fa} + W_{Fb}$,

$$W_{Fa} = \int_0^l \frac{F_{\tau e}^2}{EA}ds = \frac{F_{\tau e}^2 l}{EA}, \quad W_{Fb} = \int_0^l F_{\tau e}\left[\frac{1}{2}\left(r\frac{d\theta}{ds}\right)^2 - rk_0(1-\cos\theta)\right]ds.$$

Therefore,

$$W_F = \frac{F_{\tau e}^2 l}{EA} + \int_0^l F_{\tau e}\left[\frac{1}{2}\left(r\frac{d\theta}{ds}\right)^2 - rk_0(1-\cos\theta)\right]ds. \tag{3.39}$$

The doing work of equivalent gravity is $W_q = W_{qa} + W_{qb}$, where,

$$W_{qa} = \int_0^l \frac{q_e^2}{EA}ds = \frac{q_e^2 l}{EA}, \quad W_{qb} = -\int_0^l q_e\left[r(1-\cos\theta)\frac{\sin\varphi}{k_0}\frac{d\varphi}{ds} - r\sin\theta\frac{\sin^2\varphi}{k_0}\frac{d\psi}{ds}\right]ds,$$

whence,

$$W_q = \frac{q_e^2 l}{EA} - \int_0^l q_e \left[r(1-\cos\theta)\frac{\sin\varphi}{k_0}\frac{d\varphi}{ds} - r\sin\theta\frac{\sin^2\varphi}{k_0}\frac{d\psi}{ds} \right] ds. \tag{3.40}$$

The doing work of torque is $W_T = W_{Ta} + W_{Tb}$, where:

$$W_{Ta} = \int_0^l \frac{M_T^2}{GJ} ds = \frac{M_T^2 l}{GJ},$$

$$W_{Tb} = \frac{1}{2}\int_0^l M_T(s)\frac{d\gamma_b}{ds} ds = \frac{1}{2}\int_0^l M_T \left[r^2\left(\frac{d\theta}{ds}\right)^3 + rk_0\cos\theta\frac{d\theta}{ds} \right] ds,$$

i.e.,

$$W_T = \frac{M_T^2 l}{GJ} + \frac{1}{2}\int_0^l M_T \left[r^2\left(\frac{d\theta}{ds}\right)^3 + rk_0\cos\theta\frac{d\theta}{ds} \right] ds. \tag{3.41}$$

The doing work of annular friction is:

$$W_{f_2} = -\int_0^l f_2 N r\theta\, ds.$$

Therefore, the total potential energy can be represented as: $\prod = U - W_F - W_q - W_T - W_{f_2}$, i.e.,

$$\begin{aligned}
\prod = &\int_0^l \left\{ \frac{M_T^2}{2EI} + \frac{1}{2}EIr^2\left[\left(\frac{d^2\theta}{ds^2}\right)^2 + \left(\frac{d\theta}{ds}\right)^4 \right] \right.\\
&\left. + \frac{1}{2}EIk_0^2 + EIk_0 r\left[\cos\theta\cdot\left(\frac{d\theta}{ds}\right)^2 + \sin\theta\cdot\frac{d^2\theta}{ds^2} \right] \right\} ds\\
&- \frac{F_{\tau e}^2 l}{EA} - \int_0^l F_{\tau e}\left[\frac{1}{2}\left(r\frac{d\theta}{ds}\right)^2 - rk_0(1-\cos\theta) \right] ds - \frac{q_e^2 l}{EA}\\
&+ \int_0^l q_e\left[r(1-\cos\theta)\frac{\sin\varphi}{k_0}\frac{d\varphi}{ds} - r\sin\theta\frac{\sin^2\varphi}{k_0}\frac{d\psi}{ds} \right] ds\\
&- \frac{M_T^2 l}{GJ} - \frac{1}{2}\int_0^l M_T\left[r^2\left(\frac{d\theta}{ds}\right)^3 + rk_0\cos\theta\frac{d\theta}{ds} \right] ds + \int_0^l f_2 N r\theta\, ds.
\end{aligned}$$

Thus,

$$\begin{aligned}
\prod = &\frac{M_T^2 l}{2EI} + \frac{1}{2}EIlk_0^2 - \frac{F_{\tau e}^2 l}{EA} - \frac{q_e^2 l}{EA} - \frac{M_T^2 l}{GJ}\\
&+ \int_0^l \left\{ \frac{1}{2}EIr^2\left[\left(\frac{d^2\theta}{ds^2}\right)^2 + \left(\frac{d\theta}{ds}\right)^4 \right] + EIk_0 r\left[\cos\theta\cdot\left(\frac{d\theta}{ds}\right)^2 + \sin\theta\cdot\frac{d^2\theta}{ds^2} \right] \right.\\
&- F_{\tau e}\left[\frac{1}{2}\left(r\frac{d\theta}{ds}\right)^2 - rk_0(1-\cos\theta) \right] + q_e\left[r(1-\cos\theta)\frac{\sin\varphi}{k_0}\frac{d\varphi}{ds} - r\sin\theta\frac{\sin^2\varphi}{k_0}\frac{d\psi}{ds} \right]\\
&\left. - \frac{1}{2}M_T\left[r^2\left(\frac{d\theta}{ds}\right)^3 + rk_0\cos\theta\frac{d\theta}{ds} \right] + f_2 N r\theta \right\} ds.
\end{aligned}$$

Let $\theta' = \frac{d\theta}{ds}$, $\theta'' = \frac{d^2\theta}{ds^2}$, $\theta''' = \frac{d^3\theta}{ds^3}$, $\theta^{(4)} = \frac{d^4\theta}{ds^4}$, then the above equation can be written as

$$\begin{aligned}\prod = {} & \frac{M_T^2 l}{2EI} + \frac{1}{2}EIlk_0^2 - \frac{F_{\tau e}^2 l}{EA} - \frac{q_e^2 l}{EA} - \frac{M_T^2 l}{GJ} \\ & + \int_0^l \left\{ \frac{1}{2}EIr^2[(\theta'')^2 + (\theta')^4] + EIk_0 r[\cos\theta \cdot (\theta')^2 + \sin\theta \cdot \theta''] \right. \\ & -F_{\tau e}\left[\frac{1}{2}(r\theta')^2 - rk_0(1-\cos\theta)\right] + q_e\left[r(1-\cos\theta)\frac{\sin\varphi}{k_0}\frac{d\varphi}{ds} - r\sin\theta\frac{\sin^2\varphi}{k_0}\frac{d\psi}{ds}\right] \\ & \left. -\frac{1}{2}M_T[r^2(\theta')^3 + rk_0\cos\theta\theta'] + f_2 Nr\theta \right\} ds.\end{aligned}$$

3.3.3 The equivalent variational problem

Let:

$$\begin{aligned}\Omega = {} & \frac{M_T^2}{2EI} + \frac{1}{2}EIk_0^2 - \frac{F_{\tau e}^2}{EA} - \frac{q_e^2}{EA} - \frac{M_T^2}{GJ} \\ & + \frac{1}{2}EIr^2[(\theta'')^2 + (\theta')^4] + EIk_0 r[\cos\theta \cdot (\theta')^2 + \sin\theta \cdot \theta''] \\ & - F_{\tau e}\left[\frac{1}{2}(r\theta')^2 - rk_0(1-\cos\theta)\right] + q_e\left[r(1-\cos\theta)\frac{\sin\varphi}{k_0}\frac{d\varphi}{ds} - r\sin\theta\frac{\sin^2\varphi}{k_0}\frac{d\psi}{ds}\right] \\ & - \frac{1}{2}M_T[r^2(\theta')^3 + rk_0\cos\theta\theta'] + f_2 Nr\theta \end{aligned} \tag{3.42}$$

then the function corresponding to the total potential energy is:

$$\prod = \int_0^l \Omega\, ds.$$

Theorem 3.1. Suppose that the function $\theta(s)$ is n continuously differentiable on $[0, l]$, and $\theta(0) = 0, \theta(l) = \theta_1$, $\theta'(0) = 0, \theta'(l) = \theta_1'$ (θ_1, θ_1' are fixed constants). Then the variational problem ($VP1$) of the function $\prod = \int_0^l \Omega\, ds$ is equivalent to the boundary problem ($DP1$) of the differentiable equation, where, ($VP1$) is:

$$\begin{cases} \min \prod(s) \\ \theta(0) = 0, \quad \theta(l) = \theta_1, \\ \theta'(0) = 0, \quad \theta'(l) = \theta_1'. \end{cases}$$

($DP1$) is:

$$\begin{cases} EIr\left[\left(\frac{d^4\theta}{ds^4}\right) - 6\left(\frac{d\theta}{ds}\right)^2\frac{d^2\theta}{ds^2}\right] + 3M_T r\left[\frac{d\theta}{ds}\frac{d^2\theta}{ds^2}\right] + F_{\tau e} r\frac{d^2\theta}{ds^2} + \left(K_0 F_{\tau e} + \frac{q_e \sin\varphi}{K_0}\frac{d\varphi}{ds}\right)\sin\theta \\ -\frac{q_e\sin^2\varphi}{K_0}\frac{d\psi}{ds}\cos\theta + f_2 N = 0 \\ \theta(0) = 0, \quad \theta(l) = \theta_1, \\ \theta'(0) = 0, \quad \theta'(l) = \theta_1'. \end{cases}$$

Proof. According to the variational principle, we have:

$$\delta \prod = \int_0^l \delta\Omega\, ds = \int_0^l \Omega_\theta \delta_\theta\, ds + \int_0^l \Omega_{\theta'} \delta_{\theta'}\, ds + \int_0^l \Omega_{\theta''} \delta_{\theta''}\, ds.$$

From (3.42), we get:

$$\begin{aligned}
\Omega_\theta &= EIrk_0[-\sin\theta(\theta')^2 + \cos\theta\theta''] + F_{\tau e} rk_0 \sin\theta \\
&\quad + q_e r\left[\sin\theta \frac{\sin\varphi}{k_0}\frac{d\varphi}{ds} - r\cos\theta \frac{\sin^2\varphi}{k_0}\frac{d\psi}{ds}\right] + \frac{1}{2}M_T rk_0 \sin\theta\theta' + f_2 Nr, \\
\Omega_{\theta'} &= 2EIr^2(\theta')^3 + 2EIrk_0\cos\theta\theta' - F_{\tau e} r^2\theta' - \frac{3}{2}M_T r^2(\theta')^2 - \frac{1}{2}M_T rk_0 \cos\theta, \\
\Omega_{\theta''} &= EIr^2\theta'' + EIrk_0 \sin\theta.
\end{aligned}$$

Therefore,

$$\begin{aligned}
\int_0^l \Omega_\theta \delta_\theta ds &= \int_0^l \Big\{ EIrk_0[-\sin\theta(\theta')^2 + \cos\theta\theta''] + F_{\tau e} rk_0 \sin\theta \\
&\quad + q_e r\left[\sin\theta \frac{\sin\varphi}{k_0}\frac{d\varphi}{ds} - r\cos\theta \frac{\sin^2\varphi}{k_0}\frac{d\psi}{ds}\right] + \frac{1}{2}M_T rk_0 \sin\theta\theta' + f_2 Nr \Big\} \delta\theta\, ds.
\end{aligned} \tag{3.43}$$

By the formula of integration by parts for $\int_0^l \Omega_{\theta'} \delta_{\theta'} ds$, we have:

$$\int_0^l \Omega_{\theta'} \delta_{\theta'}\, ds = \int_0^l \Omega_{\theta'}\, d\delta_\theta = [\Omega_{\theta'} \delta_\theta]_0^l - \int_0^l \frac{d\Omega_{\theta'}}{ds}\delta\theta\, ds,$$

and

$$\begin{aligned}
\frac{d\Omega_{\theta'}}{ds} &= 6EIr^2(\theta')^2\theta'' + 2EIrk_0[-\sin\theta(\theta')^2 + \cos\theta\theta''] - F_{\tau e} r^2\theta'' \\
&\quad - 3M_T r^2\theta'\theta'' + \frac{1}{2}M_T rk_0 \sin\theta\theta'.
\end{aligned}$$

Therefore,

$$\begin{aligned}
\int_0^l \Omega_{\theta'} \delta_{\theta'}\, ds &= \left\{\left[2EIr^2(\theta')^3 + 2EIrk_0\cos\theta\theta' - F_{\tau e} r^2\theta' - \frac{3}{2}M_T r^2(\theta')^2 - \frac{1}{2}M_T rk_0 \cos\theta\right]\delta\theta\right\}_0^l \\
&\quad - \int_0^l \Big\{6EIr^2(\theta')^2\theta'' + 2EIrk_0[-\sin\theta(\theta')^2 + \cos\theta\theta''] - F_{\tau e} r^2\theta'' - 3M_T r^2\theta'\theta'' \\
&\quad + \frac{1}{2}M_T rk_0 \sin\theta\theta' \Big\} \delta\theta\, ds.
\end{aligned} \tag{3.44}$$

Similarly,

$$\begin{aligned}
\int_0^l \Omega_{\theta''} \delta_{\theta''}\, ds &= \int_0^l \Omega_{\theta''}\, d\delta_{\theta'} = [\Omega_{\theta''} \delta_{\theta'}]_0^l - \int_0^l \frac{d\Omega_{\theta''}}{ds}\delta\theta' ds \\
&= [\Omega_{\theta''} \delta_{\theta'}]_0^l - \int_0^l \frac{d\Omega_{\theta''}}{ds} d\delta\theta = [\Omega_{\theta''} \delta_{\theta'}]_0^l - \left[\frac{d\Omega_{\theta''}}{ds}\delta\theta\right]_0^l + \int_0^l \frac{d^2\Omega_{\theta''}}{ds^2}\delta\theta\, ds,
\end{aligned}$$

and

$$\frac{d\Omega_{\theta''}}{ds} = EIr^2\theta''' + EIrk_0\cos\theta\theta',$$
$$\frac{d^2\Omega_{\theta''}}{ds^2} = EIr^2\theta^{(4)} + EIrk_0[-\sin\theta(\theta')^2 + \cos\theta\theta'].$$

Therefore,

$$\int_0^l \Omega_{\theta''}\delta_{\theta''}ds = \{[EIr^2\theta'' + EIrk_0\sin\theta]\delta_{\theta'}\}_0^l - \{[EIr^2\theta''' + EIrk_0\cos\theta\theta']\delta\theta\}_0^l + \int_0^l \{EIr^2\theta^{(4)} + EIrk_0[-\sin\theta(\theta')^2 + \cos\theta\theta']\}\delta\theta\, ds. \tag{3.45}$$

Combining (3.43), (3.44) and (3.45), we have:

$$\begin{aligned}\delta\prod &= \int_0^l \Omega_\theta\delta_\theta ds + \int_0^l \Omega_{\theta'}\delta_{\theta'}ds + \int_0^l \Omega_{\theta''}\delta_{\theta''}ds\\ &= \int_0^l \Bigg\{\Bigg\{EIrk_0[-\sin\theta(\theta')^2 + \cos\theta\theta''] + F_{\tau e}rk_0\sin\theta\\ &+q_er\left[\sin\theta\frac{\sin\varphi}{k_0}\frac{d\varphi}{ds} - r\cos\theta\frac{\sin^2\varphi}{k_0}\frac{d\psi}{ds}\right] + \frac{1}{2}M_Trk_0\sin\theta\theta' + f_2Nr\Bigg\}\\ &-\Bigg\{6EIr^2(\theta')^2\theta'' + 2EIrk_0[-\sin\theta(\theta')^2 + \cos\theta\theta''] - F_{\tau e}r^2\theta''\\ &- 3M_Tr^2\theta'\theta'' + \frac{1}{2}M_Trk_0\sin\theta\theta'\Bigg\} + \{EIr^2\theta^{(4)} + EIrk_0[-\sin\theta(\theta')^2 + \cos\theta\theta']\}\Bigg\}\delta\theta\, ds\\ &+\{[EIr^2\theta'' + EIrk_0\sin\theta]\delta_{\theta'}\}_0^l + \Bigg\{\Bigg\{\Bigg[2EIr^2(\theta')^3 + 2EIrk_0\cos\theta\theta' - F_{\tau e}r^2\theta'\\ &- \frac{3}{2}M_Tr^2(\theta')^2 - \frac{1}{2}M_Trk_0\cos\theta\Bigg] - [EIr^2\theta''' + EIrk_0\cos\theta\theta']\Bigg\}\delta\theta\Bigg\}_0^l.\end{aligned}$$

Thus,

$$\begin{aligned}\delta\prod &= \int_0^l \Bigg\{\Bigg\{F_{\tau e}rk_0\sin\theta + q_er\left[\sin\theta\frac{\sin\varphi}{k_0}\frac{d\varphi}{ds} - r\cos\theta\frac{\sin^2\varphi}{k_0}\frac{d\psi}{ds}\right] + f_2Nr\Bigg\}\\ &-\{6EIr^2(\theta')^2\theta'' - F_{\tau e}r^2\theta'' - 3M_Tr^2\theta'\theta''\} + \{EIr^2\theta^{(4)}\}\Bigg\}\delta\theta\, ds\\ &+\{[EIr^2\theta'' + EIrk_0\sin\theta]\delta_{\theta'}\}_0^l + \Bigg\{\Bigg\{\Bigg[2EIr^2(\theta')^3 + 2EIrk_0\cos\theta\theta' - F_{\tau e}r^2\theta'\\ &- \frac{3}{2}M_Tr^2(\theta')^2 - \frac{1}{2}M_Trk_0\cos\theta\Bigg] - [EIr^2\theta''' + EIrk_0\cos\theta\theta']\Bigg\}\delta\theta\Bigg\}_0^l.\end{aligned}$$

By the principle of minimum potential energy, we have $\delta\prod = 0$, thus:

$$\int_0^l \left\{\left\{F_{\tau e}rk_0\sin\theta + q_e r\left[\sin\theta\frac{\sin\varphi}{k_0}\frac{d\varphi}{ds} - r\cos\theta\frac{\sin^2\varphi}{k_0}\frac{d\psi}{ds}\right] + f_2 Nr\right\} - \{6EIr^2(\theta')^2\theta''\right.$$

$$- F_{\tau e}r^2\theta'' - 3M_T r^2\theta'\theta''\} + \left\{EIr^2\theta^{(4)}\right\}\delta\theta\, ds + \{[EIr^2\theta'' + EIrk_0\sin\theta]\delta_{\theta'}\}_0^l$$

$$+ \left\{\left\{\left[2EIr^2(\theta')^3 + 2EIrk_0\cos\theta\theta' - F_{\tau e}r^2\theta' - \frac{3}{2}M_T r^2(\theta')^2\right.\right.\right.$$

$$\left.\left.\left. - \frac{1}{2}M_T rk_0\cos\theta\right] - [EIr^2\theta''' + EIrk_0\cos\theta\theta']\right\}\delta\theta\right\}_0^l = 0.$$

From $\theta(0) = 0, \theta(l) = \theta_1, \theta'(0) = 0, \theta'(l) = \theta_1'$, we get $[\delta\theta]_0^l = 0, [\delta\theta']_0^l = 0$. Thus we have:

$$\int_0^l \left\{F_{\tau e}rk_0\sin\theta + q_e r\left[\sin\theta\frac{\sin\varphi}{k_0}\frac{d\varphi}{ds} - r\cos\theta\frac{\sin^2\varphi}{k_0}\frac{d\psi}{ds}\right] + f_2 Nr - 6EIr^2(\theta')^2\theta''\right.$$

$$\left. + F_{\tau e}r^2\theta'' - 3M_T r^2\theta'\theta'' + EIr^2\theta^{(4)}\right\}\delta\theta\, ds = 0.$$

Let:

$$J(s) = F_{\tau e}rk_0\sin\theta + q_e r\left[\sin\theta\frac{\sin\varphi}{k_0}\frac{d\varphi}{ds} - r\cos\theta\frac{\sin^2\varphi}{k_0}\frac{d\psi}{ds}\right] + f_2 Nr - 6EIr^2(\theta')^2\theta''$$

$$+ F_{\tau e}r^2\theta'' - 3M_T r^2\theta'\theta'' + EIr^2\theta^{(4)}.$$

Since $\theta(s)$ is n order continuously differentiable on $[0, l]$, thus $J(s)$ is continuous on $[0, l]$. By Lemma 3.1, we have $J(s) = 0$. i.e.,

$$F_{\tau e}rk_0\sin\theta + q_e r\left[\sin\theta\frac{\sin\varphi}{k_0}\frac{d\varphi}{ds} - r\cos\theta\frac{\sin^2\varphi}{k_0}\frac{d\psi}{ds}\right] + f_2 Nr - 6EIr^2(\theta')^2\theta'' + F_{\tau e}r^2\theta''$$

$$-3M_T r^2\theta'\theta'' + EIr^2\theta^{(4)} = 0.$$

Considering the primal conditions, we have:

$$\begin{cases} EIr\left[\left(\dfrac{d^4\theta}{ds^4}\right) - 6\left(\dfrac{d\theta}{ds}\right)^2\dfrac{d^2\theta}{ds^2}\right] + 3M_T r\left[\dfrac{d\theta}{ds}\dfrac{d^2\theta}{ds^2}\right] + F_{\tau e}r\dfrac{d^2\theta}{ds^2} \\ + \left(K_0 F_{\tau e} + \dfrac{q_e\sin\varphi}{K_0}\dfrac{d\varphi}{ds}\right)\sin\theta - \dfrac{q_e\sin^2\varphi}{K_0}\dfrac{d\psi}{ds}\cos\theta + f_2 N = 0 \\ \theta(0) = 0, \quad \theta(l) = \theta_1, \\ \theta'(0) = 0, \quad \theta'(l) = \theta_1'. \end{cases}$$

Corollary 3.1. Suppose that the function $\theta(s)$ is n order continuously differentiable on $[0, l]$, $\theta(0) = 0, \theta(l) = \theta_1, \theta'(0) = 0, \theta'(l) = \theta_{11}$ (the functional value of one endpoint is fixed), $\theta(l), \delta\theta(l), \delta\theta'(l)$ are independent. Then variational problem $(VP2)$ of $\prod = \int_0^l \Omega\, ds$ is equivalent to the differential equation boundary problem including the natural boundary conditions $(DP2)$ where, $(VP2)$ is:

$$\begin{cases} \min \prod(s) \\ \theta(0) = 0, \quad \theta(l) = \theta_1, \\ \theta'(0) = 0, \quad \theta'(l) = \theta_{11}. \end{cases}$$

$(DP2)$ is:

$$\begin{cases} q_e r\left[\left(\dfrac{F_{\tau e}k_0}{q_e}+\dfrac{\sin\varphi}{k_0}\dfrac{d\varphi}{ds}\right)\sin\theta - r\cos\theta\dfrac{\sin^2\varphi}{k_0}\dfrac{d\psi}{ds}\right]+f_2Nr-6EIr^2(\theta')^2\theta'' \\ +F_{\tau e}r^2\theta''+3M_T r^2\theta'\theta''+EIr^2\theta^{(4)}=0, \\ \theta(0)=0,\quad \theta(l)=\theta_1, \\ \theta'(0)=0,\quad \theta'(l)=\theta_{11}, \\ \left\{\theta'''+\dfrac{F_{\tau e}}{EI}\theta'+\dfrac{3M_T}{2EI}(\theta')^2-2(\theta')^3\right\}_{s=l}=0, \\ \{\theta''\}_{s=l}=0. \end{cases}$$

Proof. According to the proof of Theorem 3.1, we have:

$$\int_0^l\left\{\left\{F_{\tau e}rk_0\sin\theta+q_e r\left[\sin\theta\frac{\sin\varphi}{k_0}\frac{d\varphi}{ds}-r\cos\theta\frac{\sin^2\varphi}{k_0}\frac{d\psi}{ds}\right]+f_2Nr\right\}-\{6EIr^2(\theta')^2\theta''\right.$$
$$\left.-F_{\tau e}r^2\theta''-3M_T r^2\theta'\theta''\}+\{EIr^2\theta^{(4)}\}\right\}\delta\theta\,ds+\{[EIr^2\theta''+EIrk_0\sin\theta]\delta_{\theta'}\}_0^l$$
$$+\left\{\left\{\left[2EIr^2(\theta')^3+2EIrk_0\cos\theta\theta'-F_{\tau e}r^2\theta'-\frac{3}{2}M_T r^2(\theta')^2-\frac{1}{2}M_T rk_0\cos\theta\right]\right.\right.$$
$$\left.\left.-[EIr^2\theta'''+EIrk_0\cos\theta\theta']\right\}\delta\theta\right\}_0^l=0.$$

From $\theta(0)=0, \theta'=0$, we get $\delta\theta(0)=0, \delta\theta'(0)=0$. Thus, the above equation can be written as:

$$\int_0^l\left\{\left\{F_{\tau e}rk_0\sin\theta+q_e r\left[\sin\theta\frac{\sin\varphi}{k_0}\frac{d\varphi}{ds}-r\cos\theta\frac{\sin^2\varphi}{k_0}\frac{d\psi}{ds}\right]+f_2Nr\right\}-\{6EIr^2(\theta')^2\theta''\right.$$
$$\left.-F_\tau r^2\theta''-3M_T r^2\theta'\theta''\}+\{EIr^2\theta^{(4)}\}\right\}\delta\theta\,ds+\{[EIr^2\theta''+EIrk_0\sin\theta]\delta\theta'\}_{s=l}$$
$$+\left\{\left\{\left[2EIr^2(\theta')^3+2EIrk_0\cos\theta\theta'-F_{\tau e}r^2\theta'-\frac{3}{2}M_T r^2(\theta')^2-\frac{1}{2}M_T rk_0\cos\theta\right]\right.\right.$$
$$\left.\left.-[EIr^2\theta'''+EIrk_0\cos\theta\theta']\right\}\delta\theta\right\}_{s=l}=0.$$

i.e.,

$$\int_0^l\left\{\left\{F_{\tau e}rk_0\sin\theta+q_e r\left[\sin\theta\frac{\sin\varphi}{k_0}\frac{d\varphi}{ds}-r\cos\theta\frac{\sin^2\varphi}{k_0}\frac{d\psi}{ds}\right]+f_2Nr\right\}-\{6EIr^2(\theta')^2\theta''\right]$$
$$\left.-F_{\tau e}r^2\theta''-3M_T r^2\theta'\theta''\}+\{EIr^2\theta^{(4)}\}\right\}\delta\theta ds+\{[EIr^2\theta''+EIrk_0\sin\theta]\}_{s=l}\delta\theta'(l)$$
$$+\left\{\left[2EIr^2(\theta')^3+2EIrk_0\cos\theta\theta'-F_{\tau e}r^2\theta'-\frac{3}{2}M_T r^2(\theta')^2-\frac{1}{2}M_T rk_0\cos\theta\right]\right.$$
$$\left.-[EIr^2\theta'''+EIrk_0\cos\theta\theta']\right\}_{s=l}\delta\theta(l)=0.$$

Since $\theta(l), \delta\theta(l), \delta\theta'(l)$ are independent, therefore we obtain:

$$q_e r\left[\left(\frac{F_{\tau e}k_0}{q_e}+\frac{\sin\varphi}{k_0}\frac{d\varphi}{ds}\right)\sin\theta - r\cos\theta\frac{\sin^2\varphi}{k_0}\frac{d\psi}{ds}\right]+f_2Nr-6EIr^2(\theta')^2\theta''$$
$$+F_{\tau e}r^2\theta''+3M_Tr^2\theta'\theta''+EIr^2\theta^{(4)}=0, \tag{3.46}$$

$$\{EIr^2\theta''+EIrk_0\sin\theta\}_{s=l}=0, \tag{3.47}$$

$$\{2EIr^2(\theta')^3+2EIrk_0\cos\theta\theta'-F_{\tau e}r^2\theta'-\frac{3}{2}M_Tr^2(\theta')^2-\frac{1}{2}M_Trk_0\cos\theta$$
$$-EIr^2\theta'''-EIrk_0\cos\theta\theta'\}_{s=l}=0. \tag{3.48}$$

We call (3.47), (3.48) as the partial natural boundary conditions of (3.47). If ignoring the term rk_0, then we obtain the partial natural boundary conditions as:

$$\begin{cases}\left\{EIr^2\theta'''+F_{\tau e}r^2\theta'+\frac{3}{2}M_Tr^2(\theta')^2-2EIr^2(\theta')^3\right\}_{s=l}=0\\ \{EIr^2\theta''\}_{s=l}=0.\end{cases}$$

i.e.,

$$\begin{cases}\left\{\theta'''+\frac{F_{\tau e}}{EI}\theta'+\frac{3M_T}{2EI}(\theta')^2-2(\theta')^3\right\}_{s=l}=0,\\ \{\theta''\}_{s=l}=0.\end{cases}$$

Therefore, the variational problem of function $\prod(s)$ is equivalent to the following differential equation including the primal conditions and the partial natural boundary conditions:

$$\begin{cases}q_e r\left[\left(\frac{F_{\tau e}k_0}{q_e}+\frac{\sin\varphi}{k_0}\frac{d\varphi}{ds}\right)\sin\theta-r\cos\theta\frac{\sin^2\varphi}{k_0}\frac{d\psi}{ds}\right]+f_2Nr-6EIr^2(\theta')^2\theta''\\ +F_{\tau e}r^2\theta''+3M_Tr^2\theta'\theta''+EIr^2\theta^{(4)}=0,\\ \theta(0)=0,\quad \theta(l)=\theta_1,\\ \theta'(0)=0,\quad \theta'(l)=\theta_{11},\\ \left\{\theta'''+\frac{F_{\tau e}}{EI}\theta'+\frac{3M_T}{2EI}(\theta')^2-2(\theta')^3\right\}_{s=l}=0,\\ \{\theta''\}_{s=l}=0.\end{cases}$$

Corollary 3.2. Suppose that the function $\theta(s)$ is n continuously differentiable on $[0, l]$, $\theta(0)=\theta_0, \theta(l)=\theta_1, \theta'(0)=\theta_{00}, \theta'(l)=\theta_{11}$ (the functional value of two endpoints are not fixed), $\theta, \delta\theta, \delta\theta'$ are independent. Then variational problem $(VP3)$ of $\prod=\int_0^l \Omega\, ds$ is equivalent to the differential equation boundary problem including the natural boundary conditions $(DP3)$ where, $(VP3)$ is:

$$\begin{cases}\min\ \prod(s)\\ \theta(0)=\theta_0,\quad \theta(l)=\theta_1,\\ \theta'(0)=\theta_{00},\quad \theta'(l)=\theta_{11}.\end{cases}$$

$(DP3)$ is:

$$\begin{cases} q_e r\left[\left(\frac{F_{\tau e}k_0}{q_e}+\frac{\sin\varphi}{k_0}\frac{d\varphi}{ds}\right)\sin\theta - r\cos\theta\frac{\sin^2\varphi}{k_0}\frac{d\psi}{ds}\right]+f_2Nr-6EIr^2(\theta')^2\theta'' \\ +F_{\tau e}r^2\theta''+3M_Tr^2\theta'\theta''+EIr^2\theta^{(4)}=0, \\ \theta(0)=\theta_0, \quad \theta(l)=\theta_1, \\ \theta'(0)=\theta_{00}, \quad \theta'(l)=\theta_{11}, \\ \left\{\theta'''+\frac{F_{\tau e}}{EI}\theta'+\frac{3M_T}{2EI}(\theta')^2-2(\theta')^3\right\}_0^l=0, \\ \{\theta''\}_0^l=0. \end{cases}$$

Proof. According to the proof of Theorem 3.1, we have:

$$\int_0^l \left\{\left\{F_{\tau e}rk_0\sin\theta+q_e r\left[\sin\theta\frac{\sin\varphi}{k_0}\frac{d\varphi}{ds}-r\cos\theta\frac{\sin^2\varphi}{k_0}\frac{d\psi}{ds}\right]+f_2Nr\right\}-\{6EIr^2(\theta')^2\theta''\right.$$

$$\left. -F_{\tau e}r^2\theta''-3M_Tr^2\theta'\theta''\}+\{EIr^2\theta^{(4)}\}\right\}\delta\theta\, ds+\{[EIr^2\theta''+EIrk_0\sin\theta]\delta_{\theta'}\}_0^l$$

$$+\left\{\left\{\left[2EIr^2(\theta')^3+2EIrk_0\cos\theta\theta'-F_{\tau e}r^2\theta'-\frac{3}{2}M_Tr^2(\theta')^2-\frac{1}{2}M_Trk_0\cos\theta\right]\right.\right.$$

$$\left.\left. -[EIr^2\theta'''+EIrk_0\cos\theta\theta']\right\}\delta\theta\right\}_0^l=0.$$

From $\delta\theta$, $\delta\theta'$ and θ are independent, we can get:

$$q_e r\left[\left(\frac{F_{\tau e}k_0}{q_e}+\frac{\sin\varphi}{k_0}\frac{d\varphi}{ds}\right)\sin\theta-r\cos\theta\frac{\sin^2\varphi}{k_0}\frac{d\psi}{ds}\right]+f_2Nr-6EIr^2(\theta')^2\theta''$$
$$+F_{\tau e}r^2\theta''+3M_Tr^2\theta'\theta''+EIr^2\theta^{(4)}=0, \tag{3.49}$$

$$\{EIr^2\theta''+EIrk_0\sin\theta\}_0^l=0, \tag{3.50}$$

$$\left\{2EIr^2(\theta')^3+2EIrk_0\cos\theta\theta'-F_{\tau e}r^2\theta'-\frac{3}{2}M_Tr^2(\theta')^2-\frac{1}{2}M_Trk_0\cos\theta\right.$$
$$\left. -EIr^2\theta'''-EIrk_0\cos\theta\theta'\right\}_0^l=0. \tag{3.51}$$

We call (3.50) and (3.51) as the natural boundary conditions of (4.21). Ignoring the term rk_0, then we obtain the natural boundary conditions as:

$$\begin{cases} \left\{EIr^2\theta'''+F_{\tau e}r^2\theta'+\frac{3}{2}M_Tr^2(\theta')^2-2EIr^2(\theta')^3\right\}_0^l=0 \\ \{EIr^2\theta''\}_0^l=0. \end{cases}$$

i.e.,

$$\begin{cases} \left\{\theta''' + \dfrac{F_{\tau e}}{EI}\theta' + \dfrac{3M_T}{2EI}(\theta')^2 - 2(\theta')^3\right\}_0^l = 0, \\ \{\theta''\}_0^l = 0. \end{cases}$$

Therefore, the variational problem of function $\prod(s)$ is equivalent to the following differential equation including the primal conditions and the natural boundary conditions:

$$\begin{cases} q_e r\left[\left(\dfrac{F_{\tau e}k_0}{q_e} + \dfrac{\sin\varphi}{k_0}\dfrac{d\varphi}{ds}\right)\sin\theta - r\cos\theta\dfrac{\sin^2\varphi}{k_0}\dfrac{d\psi}{ds}\right] + f_2 Nr - 6EIr^2(\theta')^2\theta'' \\ +F_{\tau e}r^2\theta'' + 3M_T r^2\theta'\theta'' + EIr^2\theta^{(4)} = 0, \\ \theta(0) = \theta_0, \quad \theta(l) = \theta_1, \\ \theta'(0) = \theta_{00}, \quad \theta'(l) = \theta_{11}, \\ \left\{\theta''' + \dfrac{F_{\tau e}}{EI}\theta' + \dfrac{3M_T}{2EI}(\theta')^2 - 2(\theta')^3\right\}_0^l = 0, \\ \{\theta''\}_0^l = 0. \end{cases}$$

From the above theorem, we can obtain the differential equation and the natural boundary conditions using the variational method. But we unable to obtain the positive force differential equation. It is easier to solve functional extremal problems than to solve differential equations.

3.4 SIMPLIFIED ANALYSIS OF THE MODEL

The tubular string of lost stability is $\frac{d_{\tau e}}{ds} << \frac{d\theta}{ds}$. In other words, $F_{\tau e}(s)$ is a slowly varying function relative to $\theta(s)$. Thus, when analysing the solution for (3.28), we can assume $F_{\tau e}(s)$ is a constant (the slowly varying parameter). Similarly, f_n, γ are slowly varying parameters. Thus, replacing θ with $\theta - \gamma$ a reasonable assumption. When $f_n \neq 0$, (3.27) can determine the differential equations using the following translation.

Let:

$$\mu = \left(\frac{f_n}{EIr}\right)^{1/4}, \quad \alpha = \mu s, \quad \beta = \frac{F_{\tau e}}{2EI\mu^2} = \frac{F_{\tau e}}{2}\left(\frac{r}{EIf_n}\right)^{1/2}, \quad m = \frac{M_T}{4EI\mu},$$

$$\gamma_1 = -\frac{\mu f_1 \gamma}{2}, \quad \gamma_2 = \frac{q_e\cos\varphi + f_{uo} + \frac{(f_{ui}+f_{uw})A_o}{A_h}}{2EI\mu^3}, \quad \gamma_3 = \frac{f_2}{f_n}, \quad n = \frac{N}{f_n},$$

where, α represents dimensionless length, n represents dimensionless positive pressure, γ_1 represents dimensionless axial friction coefficient, γ_3 represents dimensionless annular friction coefficient, γ_2 represents dimensionless distribution load coefficient, β represents dimensionless axial load, m represents dimensionless torsion moment.

From dimensionless parameters and (3.27), we can get:

$$\frac{dF_{\tau e}}{ds} = 2EI\mu^3\frac{d\beta}{d\alpha} = f_{uo} - f_1 N + q_e\cos\varphi + \frac{(f_{ui} + f_{uw})A_o}{A_h},$$

i.e.,

$$\frac{d\beta}{d\alpha} = \gamma_1 n + \gamma_2.$$

Similarly,

$$N = -f_n\left[\left(\frac{d\theta}{d\alpha}\right)^4 - 4\frac{d\theta}{d\alpha}\frac{d^3\theta}{d\alpha^3} - 3\left(\frac{d^2\theta}{d\alpha^2}\right)^2\right] + \frac{M_T f_n}{EI\mu}\left[\left(\frac{d\theta}{d\alpha}\right)^3 - \frac{d^3\theta}{d\alpha^3}\right]$$
$$+\frac{F_{\tau e}f_n}{EI\mu^2}\left(\frac{d\theta}{d\alpha}\right)^2 + f_n\cos\theta$$

i.e.,

$$\frac{N}{f_n} = -\left(\frac{d\theta}{d\alpha}\right)^4 + 4\frac{d\theta}{d\alpha}\frac{d^3\theta}{d\alpha^3} + 3\left(\frac{d^2\theta}{d\alpha^2}\right)^2 + 4m\left[\left(\frac{d\theta}{d\alpha}\right)^3 - \frac{d^3\theta}{d\alpha^3}\right] + 2\beta\left(\frac{d\theta}{d\alpha}\right)^2 + \cos\theta.$$

Thus,

$$n = -\left(\frac{d\theta}{d\alpha}\right)^4 + 4\frac{d\theta}{d\alpha}\frac{d^3\theta}{d\alpha^3} + 3\left(\frac{d^2\theta}{d\alpha^2}\right)^2 + 4m\left[\left(\frac{d\theta}{d\alpha}\right)^3 - \frac{d^3\theta}{d\alpha^3}\right] + 2\beta\left(\frac{d\theta}{d\alpha}\right)^2 + \cos\theta,$$

and

$$\frac{d^4\theta}{d\alpha^4} - 6\left(\frac{d\theta}{d\alpha}\right)^2\frac{d^2\theta}{d\alpha^2} + 12m\frac{d\theta}{d\alpha}\frac{d^2\theta}{d\alpha^2} + 3\beta\frac{d^2\theta}{d\alpha^2} + \sin\theta + f_2 n = 0.$$

Therefore, the differential equation dimensionless system of the buckling deformation tubular string can be reduced to the following system:

$$\begin{cases} \dfrac{d\beta}{d\alpha} = \gamma_1 n + \gamma_2 \\ \dfrac{d^4\theta}{d\alpha^4} - 6\left(\dfrac{d\theta}{d\alpha}\right)^2\dfrac{d^2\theta}{d\alpha^2} + 12m\dfrac{d\theta}{d\alpha}\dfrac{d^2\theta}{d\alpha^2} + 3\beta\dfrac{d^2\theta}{d\alpha^2} + \sin\theta + f_2 n = 0 \end{cases} \tag{3.52}$$

the differential equations equilibrium condition:

$$n = -\left(\frac{d\theta}{d\alpha}\right)^4 + 4\frac{d\theta}{d\alpha}\frac{d^3\theta}{d\alpha^3} + 3\left(\frac{d^2\theta}{d\alpha^2}\right)^2 + 4m\left[\left(\frac{d\theta}{d\alpha}\right)^3 - \frac{d^3\theta}{d\alpha^3}\right] + 2\beta\left(\frac{d\theta}{d\alpha}\right)^2 + \cos\theta$$

the boundary condition:

$$\theta(0) = \theta_0, \quad \theta(\mu l) = \theta_1.$$

(3.52) comprehensively considers the following cases:

(1) External load: seat force, internal and external force, viscous resistance, piston force, bending moment, torsional moment, normal force and axial friction resistance of the tubular string and well wall.
(2) The geometry and physical characteristics of the tubular string: length, external diameter, internal diameter, bore area, outside area, cross-sectional area, inertia moment, bending stiffness, modulus of elasticity, Poisson ratio, frictional coefficient.
(3) Configuration parameter of well: internal diameter of well-bore, length, well deviation angle, azimuth.
(4) Operation parameter: internal and external fluid density, viscosity, wellhead pressure, mass flow rate, and so on.

Therefore, (3.52) is a general mechanics equation which describes the buckling deformation and load distribution rule for the tubular strings.

On the basis of these dimensionless deformation differential equations, we can discuss the critical dimensionless load under sinusoidal buckling and helical buckling states.

3.4.1 The buckling critical load and tubular string deformation solution

According to function theory, the functional of the deformation differential equations is:

$$\prod = \int_0^L \Omega \, ds,$$

where,

$$\Omega = \frac{M_T^2}{2EI} + \frac{1}{2}EIk_0^2 - \frac{F_{\tau e}^2}{EA} - \frac{q_e^2}{EA} - \frac{M_T^2}{GJ} + \frac{1}{2}EIr^2[(\theta'')^2 + (\theta')^4]$$
$$+EIk_0 r[\cos\theta \cdot (\theta')^2 + \sin\theta \cdot \theta''] - F_{\tau e}\left[\frac{1}{2}(r\theta')^2 - rk_0(1 - \cos\theta)\right].$$

Property 3.1. Supposed that the sinusoidal buckling of tubular string took place, and it has n sinusoidal wave on tubular string of the length l. Let: $\omega = \frac{2n\pi}{l}, \theta = a_1 \sin \omega s$; then the approximate critical load of the sinusoidal buckling is:

$$\beta_{crs} = 1, \quad F_{crs} = \sqrt{\frac{4EIf_n}{r}}.$$

and the deformation differential equations has approximate cycle solution:

$$\theta(\alpha) = \sqrt{\frac{4}{3}(\beta - 1)} \sin \alpha.$$

Proof. From $\omega = \frac{2n\pi}{l}, \theta = a_1 \sin \omega s$, we have,

$$\theta' = \frac{d\theta}{ds} = a_1 \omega \cos \omega s, \quad \theta'' = \frac{d^2\theta}{ds^2} = -a_1 \omega^2 \sin \omega s.$$

By Taylor expanded form, we have:

$$\cos\theta = \cos(a_1 \sin \omega s) = 1 - \frac{a_1^2 \sin^2 \omega s}{2!} + \frac{a_1^4 \sin^4 \omega s}{4!} + \circ(a_1^4),$$

thus,

$$1 - \cos\theta = \frac{a_1^2 \sin^2 \omega s}{2} - \frac{a_1^4 \sin^4 \omega s}{24} + \circ(a_1^4).$$

Similarly,

$$\sin\theta = \sin(a_1 \sin \omega s) = a_1 \sin \omega s - \frac{a_1^3 \sin^3 \omega s}{6} + \circ(a_1^3).$$

From

$$\int_0^l [\cos\theta \cdot (\theta')^2 + \sin\theta \cdot \theta'']ds = \int_0^l \cos\theta \cdot \theta' d\theta + \int_0^l \sin\theta \cdot \theta' ds$$
$$= \sin\theta\theta' \Big|_0^l - \int_0^l \sin\theta\theta' ds + \int_0^l \sin\theta \cdot \theta' ds$$
$$= a_1 \cos \omega s \cdot \left(a_1 \sin \omega s - \frac{a_1^3 \sin^3 \omega s}{6}\right)\Bigg|_0^l$$
$$= 0,$$

$$\begin{aligned}\int_0^l (\theta'')^2 ds &= a_1^2\omega^4 \int_0^l \sin^2\omega s\, ds \\ &= a_1^2\omega^4 \int_0^l \frac{1-\cos 2\omega s}{2} ds \\ &= \frac{1}{2}a_1^2\omega^4 l - \frac{1}{4}a_1^2\omega^3 \int_0^l \cos 2\omega s\, d2\omega s \\ &= \frac{1}{2}a_1^2\omega^4 l,\end{aligned}$$

$$\begin{aligned}\int_0^l (\theta')^4 ds &= a_1^4\omega^4 \int_0^l \cos^4\omega s\, ds \\ &= a_1^4\omega^4 \int_0^l \left(\frac{1+\cos 2\omega s}{2}\right)^2 ds \\ &= a_1^4\omega^4 \int_0^l \frac{1+2\cos 2\omega s+\cos^2 2\omega s}{4} ds \\ &= \frac{3}{8}a_1^4\omega^4 l,\end{aligned}$$

$$\begin{aligned}\int_0^l (1-\cos\theta) ds &= \frac{a_1^2}{2}\int_0^l \sin^2\omega s\, ds - \frac{a_1^4}{24}\int_0^l \sin^4\omega s\, ds \\ &= \frac{a_1^2}{2}\int_0^l \frac{1-\cos 2\omega s}{2} ds - \frac{a_1^4}{24}\int_0^l \left(\frac{1-\cos 2\omega s}{2}\right)^2 ds \\ &= \frac{a_1^2}{4} l - \frac{a_1^4}{64} l,\end{aligned}$$

$$\begin{aligned}\int_0^l \sin\theta\, ds &= \int_0^l \left(a_1 \sin\omega s - \frac{a_1^3 \sin^3\omega s}{6}\right) ds \\ &= \frac{a_1}{\omega}\int_0^l \sin\omega s\, d\omega s - \frac{a_1^3}{6}\int_0^l \sin^3\omega s\, ds \\ &= \frac{a_1}{\omega}\int_0^l \sin\omega s\, d\omega s + \frac{a_1^3}{6\omega}\int_0^l \sin^2\omega s\, d\cos\omega s \\ &= 0,\end{aligned}$$

$$\begin{aligned}\int_0^l (\theta')^2 ds &= a_1^2\omega^2 \int_0^l \cos^2\omega s\, ds \\ &= a_1^2\omega^2 \int_0^l \frac{1+\cos 2\omega s}{2} ds \\ &= \frac{1}{2}a_1^2\omega^2 l,\end{aligned}$$

$$\begin{aligned}\int_0^l (\theta')^3 ds &= a_1^3\omega^3 \int_0^l \cos^3 \omega s\, ds \\ &= a_1^3\omega^2 \int_0^l \cos^3 \omega s\, d\omega s \\ &= a_1^3\omega^2 \int_0^l \cos^2 \omega s\, d\sin \omega s \\ &= a_1^3\omega^2 \int_0^l (1-\sin^2 \omega s) d\sin \omega s \\ &= 0,\end{aligned}$$

$$\begin{aligned}\int_0^l \cos\theta\theta' ds &= \int_0^l \cos\theta\, d\theta \\ &= 0,\end{aligned}$$

$$\begin{aligned}\int_0^l \theta\, ds &= \int_0^l a_1 \sin \omega s\, ds \\ &= 0,\end{aligned}$$

we have:

$$\begin{aligned}\prod_s &= \int_0^l \Omega\, ds \\ &= \frac{M_T^2 l}{2EI} + \frac{1}{2}EIk_0^2 l - \frac{F_{\tau e}^2 l}{EA} - \frac{q_e^2 l}{EA} - \frac{M_T^2 l}{GJ} + \frac{1}{4}EIr^2\left[a_1^2\omega^4 l + \frac{3}{4}a_1^4\omega^4 l\right] \\ &\quad - \frac{1}{4}F_{\tau e}r^2 a_1^2\omega^2 l + f_n \cos\gamma r\left[\frac{a_1^2}{4}l - \frac{a_1^4}{64}l\right].\end{aligned}$$

Considering γ is a slowly varying parameter, we have:

$$\begin{aligned}\prod_s &= \int_0^l \Omega\, ds \\ &= \frac{M_T^2 l}{2EI} + \frac{1}{2}EIk_0^2 l - \frac{F_{\tau e}^2 l}{EA} - \frac{q_e^2 l}{EA} - \frac{M_T^2 l}{GJ} + \frac{1}{4}EIr^2\left[a_1^2\omega^4 l + \frac{3}{4}a_1^4\omega^4 l\right] \\ &\quad - \frac{1}{4}F_{\tau e}r^2 a_1^2\omega^2 l + f_n r\left[\frac{a_1^2}{4}l - \frac{a_1^4}{64}l\right].\end{aligned}$$

Therefore,

$$\begin{aligned}\prod_s &= \int_0^l \Omega\, ds \\ &= \frac{M_T^2 l}{2EI} + \frac{1}{2}EIk_0^2 l - \frac{F_{\tau e}^2 l}{EA} - \frac{q_e^2 l}{EA} - \frac{M_T^2 l}{GJ} \\ &\quad + \frac{1}{4}EIr^2 l\left[a_1^2\omega^4 + \frac{3}{4}a_1^4\omega^4 - \frac{F_{\tau e}}{EI}a_1^2\omega^2 + \frac{f_n}{EIr}\left(a_1^2 - \frac{a_1^4}{16}\right)\right].\end{aligned} \tag{3.53}$$

By the principle of stationary potential energy, we get:

$$\begin{cases} 2a_1\omega^4 + 3a_1^3\omega^4 - \dfrac{2F_{\tau e}}{EI}a_1\omega^2 + \dfrac{2f_n}{EIr}\left(a_1 - \dfrac{a_1^3}{8}\right) = 0 \\ 4a_1^2\omega^3 + 3a_1^4\omega^3 - \dfrac{2F_{\tau e}}{EI}a_1^2\omega = 0 \end{cases}$$

i.e.,

$$\begin{cases} 2\omega^4 + 3a_1^2\omega^4 - \dfrac{2F_{\tau e}}{EI}\omega^2 + \dfrac{2f_n}{EIr}\left(1 - \dfrac{a_1^2}{8}\right) = 0 \\ 4\omega^2 + 3a_1^2\omega^2 - \dfrac{2F_{\tau e}}{EI} = 0. \end{cases} \tag{3.54}$$

From $\mu = \left(\frac{f_n}{EIr}\right)^{1/4}$, $\beta = \frac{F_{\tau e}}{2EI\mu^2} = \frac{F_{\tau e}}{2}\left(\frac{r}{EIf_n}\right)^{1/2}$, we have:

$$\mu^4 = \frac{f_n}{EIr}, \quad 2\beta\mu^2 = \frac{F_{\tau e}}{EI}.$$

Thus, (3.54) can be written:

$$\begin{cases} 2\omega^4 + 3a_1^2\omega^4 - 4\beta\mu^2\omega^2 + 2\mu^4\left(1 - \dfrac{a_1^2}{8}\right) = 0 \\ 4\omega^2 + 3a_1^3\omega^2 - 4\beta\mu^2 = 0. \end{cases} \tag{3.55}$$

Ignoring $\frac{a_1^2}{8}$ of (3.30), we can get new equations:

$$\begin{cases} 2\omega^4 + 3a_1^2\omega^4 - 4\beta\mu^2\omega^2 + 2\mu^4 = 0 \\ 4\omega^2 + 3a_1^3\omega^2 - 4\beta\mu^2 = 0. \end{cases} \tag{3.56}$$

Thus,

$$\omega = \mu, \quad a_1 = \sqrt{\frac{4(\beta - 1)}{3}}. \tag{3.57}$$

Combining (3.57) and (3.53), we have:

$$\begin{aligned} \prod_s &= \frac{M_T^2 l}{2EI} + \frac{1}{2}EIk_0^2 l - \frac{F_{\tau e}^2 l}{EA} - \frac{q_e^2 l}{EA} - \frac{M_T^2 l}{GJ} + \frac{1}{4}EIr^2 l\left[\frac{4(\beta-1)}{3}\mu^4 + \frac{3}{4}\left[\frac{4(\beta-1)}{3}\right]^2\mu^4\right. \\ &\quad \left. - 2\beta\frac{4(\beta-1)}{3}\mu^4 + \mu^4\left(\frac{4(\beta-1)}{3} - \frac{1}{16}\left[\frac{4(\beta-1)}{3}\right]^2\right)\right] \\ &= \frac{M_T^2 l}{2EI} + \frac{1}{2}EIk_0^2 l - \frac{F_{\tau e}^2 l}{EA} - \frac{q_e^2 l}{EA} - \frac{M_T^2 l}{GJ} \\ &\quad + \frac{1}{4}EIr^2 l\left[\frac{4(\beta-1)}{3}\mu^4 + \frac{4(\beta-1)^2}{3}\mu^4 - \frac{8\beta(\beta-1)}{3}\mu^4 + \frac{4(\beta-1)}{3}\mu^4 - \frac{(\beta-1)^2}{9}\mu^4\right] \\ &= \frac{M_T^2 l}{2EI} + \frac{1}{2}EIk_0^2 l - \frac{F_{\tau e}^2 l}{EA} - \frac{q_e^2 l}{EA} - \frac{M_T^2 l}{GJ} - \frac{13}{36}EIr^2 l(\beta-1)^2\mu^4. \end{aligned}$$

Initial potential energy of buckling well-bore is:

$$\prod_0 = \frac{M_T^2 l}{2EI} + \frac{1}{2}EIk_0^2 l - \frac{F_{\tau e}^2 l}{EA} - \frac{q_e^2 l}{EA} - \frac{M_T^2 l}{GJ}. \tag{3.58}$$

Therefore, when $\prod_s = \prod_0$, we can get the approximate critical load of the sinusoidal buckling:

$$\beta_{crs} = 1, \quad F_{crs} = \sqrt{\frac{4EIf_n}{r}}.$$

At the same time, the deformation differential equations has cycle solution:

$$\theta(\alpha) = \sqrt{\frac{4}{3}(\beta - 1)} \sin \alpha.$$

The above theorem discusses the approximate critical load and cycling solution for the sinusoidal buckling string. Similarly, when helical buckling of the string occurs, we have the following result.

Property 3.2. Supposed that the tubular string occurs helical bucking. Letting $\theta = a_0 s, a_0 = \frac{2n\pi}{L}$. Then, the approximate critical load of the helical buckling is:

$$\beta_{crh} = 1.469, \quad F_{crh} = 1.469\sqrt{\frac{4EIf_n}{r}}$$

and the helical solution of deformation differential equations is:

$$\theta(\alpha) = \sqrt{\beta}\alpha.$$

Proof. From $\theta = a_0 s,\ a_0 = \frac{2n\pi}{L}$, we have:

$$\theta' = \frac{d\theta}{ds} = a_0, \quad \theta'' = \frac{d^2\theta}{ds^2} = 0, \quad \sin\theta = \sin(a_0 s),$$

$$1 - \cos\theta = 1 - \cos(a_0 s), \quad \frac{d^2 \cos\theta}{ds^2} = -a_0^2 \cos(a_0 s).$$

Ignoring small variable of higher order, then:

$$\prod_h = \frac{M_T^2 l}{2EI} + \frac{1}{2}EIk_0^2 l - \frac{F_{\tau e}^2 l}{EA} - \frac{q_e^2 l}{EA} - \frac{M_T^2 l}{GJ} + \frac{1}{2}EIr^2 l\left[a_0^4 - \frac{F_{\tau e}}{EI}a_0^2 - \frac{M_T}{EI}a_0^3 - \frac{f_2 N}{EIr}a_0 l + \frac{2f_n}{EIr}\right].$$

Considering friction force, i.e., $f_2 \approx 0$, then above equation can be reduced to:

$$\prod_h = \frac{M_T^2 l}{2EI} + \frac{1}{2}EIk_0^2 l - \frac{F_{\tau e}^2 l}{EA} - \frac{q_e^2 l}{EA} - \frac{M_T^2 l}{GJ} + \frac{1}{2}EIr^2 ld\left[a_0^4 - \frac{F_{\tau e}}{EI}a_0^2 - \frac{M_T}{EI}a_0^3 + \frac{2f_n}{EIr}\right]. \tag{3.59}$$

According to principal of stationary potential energy, we have:

$$4a_0^2 - 3\frac{M_T}{EI}a_0 - \frac{F_{\tau e}}{EI} = 0.$$

Therefore,

$$a_0 = \frac{3M_T}{8EI} \pm \sqrt{\frac{9}{64}\left(\frac{M_T}{EI}\right)^2 + \frac{F_{\tau e}}{2EI}}.$$

From $m=\frac{M_T}{4EI\mu}$, $\mu=\left(\frac{f_n}{EIr}\right)^{1/4}$, $\beta=\frac{F_{\tau e}}{2EI\mu^2}=\frac{F_{\tau e}}{2}\left(\frac{r}{EIf_n}\right)^{1/2}$, we can determine:

$$a_0=\frac{3}{2}m\mu\pm\sqrt{\frac{9}{4}m^2\mu^2+\beta\mu^2},$$
$$=\frac{3}{2}m\mu\pm\mu\sqrt{\frac{9}{4}m^2+\beta}. \tag{3.60}$$

Combining (3.59) and (3.60), we have:

$$\prod_h=\frac{M_T^2l}{2EI}+\frac{1}{2}EIk_0^2l-\frac{F_{\tau e}^2l}{EA}-\frac{q_e^2l}{EA}-\frac{M_T^2l}{GJ}$$
$$+\frac{1}{4}EIr^2l\frac{f_n}{EIr}\left[27m^4+(18m^3+8\beta m)\sqrt{\frac{9}{4}m^2\mu^2+\beta}+18\beta m^2+2\beta^2-4\right].$$

When $\prod_h=\prod_s$, the relation equation of the critical helical buckling load F_{crh} and the dimensionless torsional moment m is:

$$27m^4+(18m^3+8\beta m)\sqrt{\frac{9}{4}m^2\mu^2+\beta}+18\beta m^2+5\beta^2+26\beta-49=0.$$

Again when $m=0$, we can get the dimensionless critical load of helical buckling: $\beta_{crh}=1.469$. The critical load of helical buckling:

$$F_{crh}=1.469\sqrt{\frac{4EIf_n}{r}}.$$

Thus, when downhole tubular string occurs helical bucking, the helical solution of deformation differential equations is:

$$\theta(\alpha)=\sqrt{\beta}\alpha.$$

From $F_{crh}=1.469\sqrt{\frac{4EIf_n}{r}}$ and $F_{crs}=\sqrt{\frac{4EIf_n}{r}}$, we have:

$$F_{cr}=2\beta_{cr}\sqrt{\frac{EIf_n}{r}},$$

where, β_{cr} represents the dimensionless critical load. When tubular string occurs the sinusoidal buckling, then $\beta_{cr}=1$; When tubular string occurs the helical buckling, then $\beta_{cr}=1.469$. From $F_{cr}=2\beta_{cr}\sqrt{\frac{EIf_n}{r}}$, we have:

$$F_{cr}^4=\left(\frac{4\beta_{cr}^2EI}{r}\right)^2\left[\left(F_{cr}\frac{d\varphi}{ds}+q_e\sin\varphi\right)^2+\left(F_{cr}\sin\varphi\frac{d\psi}{ds}\right)^2\right].$$

When azimuth varies very small, i.e., $\frac{d\varphi}{ds}>>\frac{d\psi}{ds}$, we can ignore the influence of $\frac{d\psi}{ds}$, then:

$$F_{cr}^2=\left(\frac{4\beta_{cr}^2EI}{r}\right)\left(F_{cr}\frac{d\varphi}{ds}+q_e\sin\varphi\right),$$

therefore,

$$F_{cr1}=\left(\frac{2\beta_{cr}EIK_0}{r}\right)\left(1+\sqrt{1+\frac{q_er\sin\varphi}{EIK_0^2\beta_{cr}^2}}\right).$$

When $\frac{q_e r \sin\varphi}{EIK_0^2\beta_{cr}^2}$ is vary small, we ignore its influence, then:

$$F_{cr2}=\frac{4\beta_{cr}EIK_0}{r}.$$

When $\frac{q_e r \sin\varphi}{EIK_0^2\beta_{cr}^2}$ is vary large, then:

$$F_{cr3}=2\beta_{cr}\sqrt{\frac{EIq_e\sin\varphi}{r}}.$$

Therefore, the downhole tubular string has three equilibrium states with varying of the dimensionless axial load β:

1. When $\beta<1$, the tubular string is in the stable state;
2. When $1\leq\beta<1.469$, the tubular string is in the sinusoidal buckling state;
3. When $\beta\geq1.469$, the tubular string is in the helical buckling state.

At varied equilibrium states, solution of θ for the deformation system of differential equations is:

$$\theta(\alpha)=\begin{cases}0, & \beta<1\\ \sqrt{\frac{4(\beta-1)}{3}}\sin\alpha, & 1\leq\beta<1.469\\ \sqrt{\beta}\alpha, & \beta\geq1.469\end{cases}$$

the dimensionless positive force of varied equilibrium states is:

$$n=\begin{cases}1, & \beta<1\\ \frac{8}{9}\left(\beta+\frac{1}{2}\right)^2-1, & 1\leq\beta<1.469\\ \beta^2, & \beta\geq1.469.\end{cases}$$

3.4.2 The axial buckling deformation analysis of the downhole string

The tubular string deformation includes transverse deformation and longitudinal deformation. Because the transverse length (order of magnitude 10^{-3} m) is smaller than the longitudinal length (order of magnitude 10^3 m), the transverse buckling deformation length is smaller than the longitudinal buckling deformation length. Therefore, we mainly consider the axial (longitudinal) buckling deformation for the tubular string deformation analysis. Up to now, general research has divided the buckling deformation into four deformations: temperature deformation, ballooning deformation, axial load deformation, and helical buckling deformation. Thus we can determine the axial deformation formation for temperature variation, internal and external fluid pressure variation, and the axial load which results in the tubular string buckling:

1. **The temperature variation axial deformation**

   Suppose that the temperature at point s of the tubular string is $T(s)$, the initial temperature is $T_0(s)$, and the warm balloon coefficient of the tubular string is α, then:

$$\frac{du_T(s)}{ds}=\alpha[T(s)-T(s_0)],$$

i.e.,

$$u_T(s)=u_T(s_0)+\int_{s_0}^{s}\alpha[T(s)-T(s_0)]ds,$$

where, $u_T(s)$ represents the axial deformation of temperature variation.

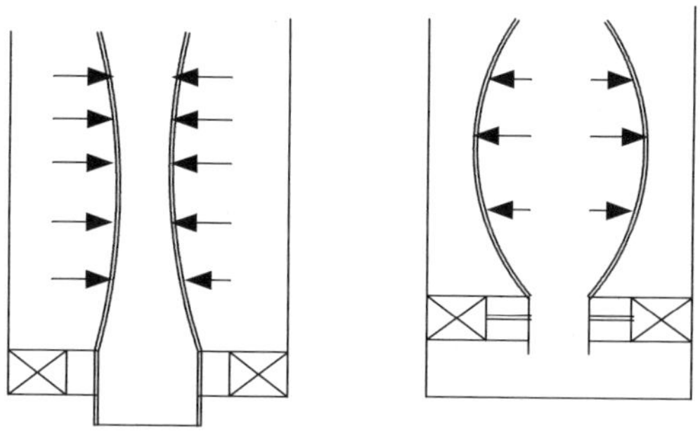

Figure 3.3 The ballooning effect.

2. **The internal and external fluid pressure variation axial deformation**

   As shown in Figure 3.3, the tubular string has a "the ballooning effect" because of the internal and external fluid pressure. According to the HuKe rule, the axial stress variation ε_p is:

$$\varepsilon_p(s) = \frac{2\mu(p_o r_0^2 - p_i r_i^2)}{E(r_o^2 - r_i^2)},$$

   i.e.,

$$\frac{du_p}{ds} = \varepsilon_p(s) = \frac{2\mu(p_o r_o^2 - p_i r_i^2)}{E(r_o^2 - r_i^2)}.$$

   Thus,

$$u_p(s) = u_{po}(s) + \frac{2\mu}{E(r_o^2 - r_i^2)}\left[r_o^2 \int_{s0}^{s} p_o(s)ds - r_i^2 \int_{s0}^{s} p_i(s)ds\right].$$

3. **The axial deformation of the axial load**

   According to the HuKe rule, the axial stress variation $\varepsilon_F(s)$ relative to the axial stress $\sigma_F(s)$ is:

$$\varepsilon_F(s) = \frac{\sigma_F(s)}{E} = -\frac{F_\tau(s)}{E(A_o - A_i)},$$

   i.e.,

$$\frac{du_F}{ds} = \varepsilon_F(s) = \frac{\sigma_F(s)}{E} = -\frac{F_\tau(s)}{E(A_o - A_i)},$$

$$u_F(s) = u_F(s_0) - \frac{1}{E(A_o - A_i)} \int_{s0}^{s} F_\tau(s)ds.$$

4. **The axial deformation $u_B(s)$ of the tubular string buckling**

$$\frac{du_B(s)}{ds} \approx \frac{1}{2} r \left(\frac{d\theta}{ds}\right)^2 - rK_0(1 - \cos\theta),$$

   thus,

$$u_B(s) = u_B(s_0) + \int_{s0}^{s} \left[\frac{1}{2} r \left(\frac{d\theta}{ds}\right)^2 - rK_0(1 - \cos\theta)\right] ds.$$

5. **The whole axial deformation $u_B(s)$ of the downhole string**

   The whole axial deformation $u(s)$ is an algebraic sum:

$$u(s) = u_p(s) + u_T(s) + u_F(s) + u_B(s),$$

i.e.,

$$u(s) = u(s_0) + \frac{2\mu}{E(r_0^2 - r_i^2)}\left[r_0^2\int_{s_0}^{s} p_0(s)ds - r_i^2\int_{s_0}^{s} p_i(s)ds + \alpha\int_{s_0}^{s}[T(s) - T(s_0)]\,ds\right]$$
$$-\frac{1}{E(A_o - A_i)}\int_{s_0}^{s} F_\tau(s)ds - \int_{s_0}^{s}\left[\frac{1}{2}r\left(\frac{d\theta}{ds}\right)^2 - rK_0(1-\cos\theta)\right]ds.$$

Because of $F_{\tau e} = F_\tau(s) + p_i(s)A_i - p_o(s)A_o$, then the above equation can be written as:

$$u(s) = u(s_0) + \frac{1+2\mu}{E(r_0^2 - r_i^2)}\left[r_0^2\int_{s_0}^{s} p_0(s)ds - r_i^2\int_{s_0}^{s} p_i(s)ds + \alpha\int_{s_0}^{s}[T(s) - T(s_0)]ds\right]$$
$$-\frac{1}{E(A_o - A_i)}\int_{s_0}^{s} F_{\tau e}(s)ds - \int_{s_0}^{s}\left[\frac{1}{2}r\left(\frac{d\theta}{ds}\right)^2 - rK_0(1-\cos\theta)\right]ds.$$

The above equation includes three sections: the axial deformation of the internal and external pressure $u_{p'}(s)$, the axial deformation from the temperature variation $u_T(s)$, the axial deformation of from the equivalent axial force and the buckling effect $u_{F,B}(s)$:

1. The axial deformation from the internal and external pressure
   If the influence of the friction force is ignored, then the pressure distribution of the internal and external fluid is:

   $$\frac{dp_i}{ds} = \rho_i g\cos\varphi, \quad p_i = p_{i0} + \rho_i gs\cos\varphi, \quad \frac{dp_o}{ds} = \rho_o g\cos\varphi, \quad p_o = p_{o0} + \rho_i gs\cos\varphi.$$

   Thus,

   $$u_p{}'(s) = \frac{1+2\mu}{E(r_o^2 - r_i^2)}[p_{o0} + \rho_o g\cos\varphi]r_o^2 - \frac{1+2\mu}{E(r_o^2 - r_i^2)}[p_{i0} + \rho_i gs\cos\varphi]r_i^2.$$

2. The axial deformation from temperature variation is:

   $$u_T(s) = \alpha\Delta Ts,$$

   where, $\Delta T = T(s) - T(s_0)$.
3. The equivalent axial force and buckling axial deformation:

   $$u_{F,B}(s) = u_{F,B}(s_0) - \frac{1}{E(A_o - A_i)}\int_{s_0}^{s} F_{\tau e}(s)ds - \int_{s_0}^{s}\left[\frac{1}{2}r\left(\frac{d\theta}{ds}\right)^2 - rK_0(1-\cos\theta)\right]ds,$$

   because rK_0 is very small, thus:

   $$u_{F,B}(s) = u_{F,B}(s_0) - \frac{1}{E(A_o - A_i)}\int_{s_0}^{s} F_{\tau e}(s)ds - \int_{s_0}^{s}\left[\frac{1}{2}r\left(\frac{d\theta}{ds}\right)^2\right]ds.$$

When the tubular string is in different equilibrium states (the stable state, the sinusoidal buckling state, the helical bucking state), the distribution of $\beta(\alpha)$, $\frac{d\theta}{ds}$ is different, thus the deformation length is different:

(1) The stable state axial deformation:
   at the stable state, $\frac{d\theta}{d\alpha} = 0$, $\frac{d\beta_w}{d\alpha} = \gamma_1 + \gamma_2$, we have

   $$\beta_\omega(a) = \beta_\omega(a_{\omega 0}) + (\gamma_1 + \gamma_2)(a - a_{\omega 0}),$$

thus,

$$u_\omega(a)=u_\omega(a_{\omega 0})+\beta_\omega(a_{\omega 0})(a-a_{\omega 0})+\frac{1}{2}(\gamma_1+\gamma_2)(a-a_{\omega 0})^2.$$

Assume that at the stable state, $C_\omega=\frac{u_\omega(a)-u_\omega(a_{\omega 0})}{a-a_{\omega 0}}$ represents the relative deformation length of the tubular string, $a_\omega=a-a_{\omega 0}$ is the tubular string length, then:

$$C_\omega=\beta_\omega(a_{\omega 0})+\frac{1}{2}(\gamma_1+\gamma_2)(a-a_{\omega 0}).$$

(2) The axial deformation of the sinusoidal buckling state:
at the sinusoidal buckling state: $\frac{d\theta}{d\alpha}=\sqrt{\frac{4}{3}(\beta_s-1)}\cos\alpha$, $\frac{d\beta_s}{d\alpha}=\left[\frac{8}{9}\left(\beta_s+\frac{1}{2}\right)^2-1\right]\gamma_1+\gamma_2$.
When $\gamma_1=0$, $\beta_s(\alpha)=\beta_s(\alpha_{s0})+\gamma_2(a-a_{s0})$, we have

$$u_s(a)=u_s(a_{s0})+\left[\left(1+\frac{2}{3}a\right)\beta_s(a_{s0})-\frac{2}{3}a\right](a-a_{s0})+\frac{1}{2}\left(1+\frac{2}{3}a\right)\gamma_2(a-a_{s0})^2.$$

Assume that at the sinusoidal buckling state, $C_s=\frac{u_s(a)-u_s(a_{s0})}{a-a_{s0}}$ represents the relative deformation length of the tubular string, $a_s=a-a_{s0}$ is the length of the tubular string, then:

$$C_s=\left[\left(1+\frac{2}{3}a\right)\beta_s(a_{s0})-\frac{2}{3}a\right]+\frac{1}{2}\left(1+\frac{2}{3}a\right)\gamma_2(a-a_{s0}).$$

When $\gamma_1\neq 0$,

$$d\alpha=\frac{1}{\frac{16}{9}\left(\beta_s+\frac{1}{2}\right)\gamma_1}d\ln\left|\left[\frac{8}{9}\left(\beta_s+\frac{1}{2}\right)^2-1\right]\gamma_1+\gamma_2\right|,$$

thus, the tubular string deformation length can be represented,

$$\begin{aligned}u_s(a)&=u_s(a_{s0})+\frac{9+6a}{16\gamma_1}\left(\ln\left|\left[\frac{8}{9}\left(\beta_s(a)+\frac{1}{2}\right)^2-1\right]\gamma_1+\gamma_2\right|\right.\\&\left.-\ln\left|\left[\frac{8}{9}\left(\beta_s(a_{s0})+\frac{1}{2}\right)^2-1\right]\gamma_1+\gamma_2\right|\right)-\frac{2}{3}a(a-a_{s0}).\end{aligned}$$

Let $C_s=\frac{u_s(a)-u_s(a_{s0})}{a-a_{s0}}$, $a_s=a-a_{s0}$, then:

$$\begin{aligned}C_s&=\frac{9+6a}{16\gamma_1 a_s}\left(\ln\left|\left[\frac{8}{9}\left(\beta_s(a)+\frac{1}{2}\right)^2-1\right]\gamma_1+\gamma_2\right|\right.\\&\left.-\ln\left|\left[\frac{8}{9}\left(\beta_s(a_{s0})+\frac{1}{2}\right)^2-1\right]\gamma_1+\gamma_2\right|\right)-\frac{2}{3}a.\end{aligned}$$

(3) The axial deformation of the helical buckling state:
at the helical buckling state: $\frac{d\theta}{d\alpha}=\sqrt{\beta_h}$, $\frac{d\beta_h}{d\alpha}=(\beta_h)^2\gamma_1+\gamma_2$.

When $\gamma_1=0$, $\beta_h(a)=\beta_h(a_{h0})+\gamma_2(a-a_{h0})$, then:

$$u_h(a)=u_h(a_{h0})+(1+a)\left[\beta_h(a_{h0})(a-a_{h0})+\frac{1}{2}(a-a_{h0})^2\gamma_2\right].$$

Letting $C_h=\frac{u_h(a)-u_h(a_{h0})}{a-a_{h0}}$, $a_h=a-a_{s0}$, then:

$$C_h=(1+a)\left[\beta_h(a_{h0})+\frac{1}{2}a_h\gamma_2\right].$$

When $\gamma_1 \neq 0$, $d\alpha = \frac{1}{2\beta_h \gamma_1} d \ln|\beta_h^2 \gamma_1 + \gamma_2|$, then the relative deformation length of the tubular string is:

$$u_h(a) = u_h(a_{s0}) + \frac{1+a}{2\gamma_1}[\ln|\beta_h^2(a)\gamma_1 + \gamma_2| - \ln|\beta_h^2(a_{h0})\gamma_1 + \gamma_2|].$$

Letting $C_h = \frac{u_h(a) - u_h(a_{h0})}{a - a_{h0}}$, $a_h = a - a_{s0}$, then:

$$C_h = \frac{1+a}{2\gamma_1 a_h}[\ln|\beta_h^2(a)\gamma_1 + \gamma_2| - \ln|\beta_h^2(a_{h0})\gamma_1 + \gamma_2|].$$

If there are three states, then the whole deformation length be represented as:

$$u = C_\omega a_\omega + C_s a_s + C_h a_h.$$

In this chapter, the deformation differential equations for the tubular string were improved using the classical infinitesimal analysis method, and the dimensionless differential equations were obtained using variable replacement. The deformation differential equations for the tubular string were solved using functional theory, and three different states were obtained: a stable state, a sinusoidal buckling state, and a helical buckling state: through solving the critical load. At the same time, for the three different states, the axial load, the normal pressure, the friction force, and four deformation lengths for the tubular string were obtained using difference equations.

CHAPTER 4

Mechanical analysis for the placement of the test string

We can see from the pipeline placement process that when all the pipelines enter the well, the tubing string nearest the wellhead bears the greatest tension and undergoes the largest deformation. Therefore, the main force analysis is conducted after all test strings have been placed in the well.

4.1 MECHANICAL ANALYSIS

When the test string is stable after all the test strings have been placed in the well, the string has the following forces:

String weight: $q\vec{k}$;
The positive pressure between a bent string and the borehole wall: $\vec{N} = N(\cos\theta\,\vec{n_0} - \sin\theta\,\vec{b_0})$;
Axial friction: $\vec{f_{f1}} = -f_1 N_{\tau 0}$;
Centre friction: $\vec{f_{f2}} = -f_2 N(\cos\theta\,\vec{n_0} + \sin\theta\,\vec{b_0})$;
Fluid pressure acting on the lateral walls of the string: $p_i(s)$, $p_o(s)$.

4.2 TEMPERATURE DISTRIBUTION

The test string phase is not related to heat exchange, so the inside string temperature distribution is the same as the formation temperature distribution:

$$T = aH + T_0,$$

where a is geothermal gradient, H is vertical depth of the borehole, T_0 is surface temperature.

4.3 PRESSURE DISTRIBUTION

When the test string is stable, the internal and external fluids do not flow, so the inside and outside pressure distribution of the string has a hydrostatic pressure distribution.

4.4 MODEL CALCULATION

From Chapter 3, we know that the model for the placement of the test tubular is similar to the deformation differential equations model, only with some changed conditions.

To calculate the deformation length, the tubular string of total length L is divided into m sections, including $m+1$ joints with an order of $0, 1, \ldots, m$ from the bottom to the top. Using a dimensionless method, we can calculate the dimensionless axial load of each joint, the real axial load, various deformation lengths and the total deformation length.

4.4.1 The internal and external pressure calculation

$$P_{i,j+1} = P_{i,j} + \rho_{i,j} g \left(\cos \varphi_j - \frac{\lambda_{i,j} u_{i,j}^2}{4 g r_{i,j}} \right) \Delta s_j, \tag{4.1}$$

$$P_{o,j+1} = P_{o,j} + \rho_{o,j} g \left(\cos \varphi_j - \frac{\lambda_{o,j} u_{o,j}^2}{4 g r_{o,j}} \right) \Delta s_j, \tag{4.2}$$

where, $P_{i,j}$, $P_{o,j}$ represent the internal and external pressure of the tubular string at j joint; $\rho_{i,j}$, $\rho_{o,j}$ are the internal and external fluid density at j joint; φ_j is the deviation angle of j joint; $\lambda_{i,j}$, $\lambda_{o,j}$ represent the internal and external fluid dimensionless viscous friction coefficient at j point; $u_{i,j}$, $u_{o,j}$ represent the internal and external fluid velocity; $r_{i,j}$, $r_{o,j}$ are the internal and external radius of the j unit tubular string; $\Delta s_j = s_j - s_{j-1}$ is the j unit length.

When the wellhead pressure $P_{i,0}$ and $P_{o,0}$ is known, the internal and external pressure of every joint is calculated by means of the above difference equations.

4.4.2 The axial force distribution, the normal pressure and the moment calculation

The tubing axial force varies with hole depth, which results in tension near the wellhead and compression near the bottom, so that the tubing's upper section remains straight but the lower part may become helically buckled. The axial force may distribute linearly if there is no buckling. However, when buckling occurs, the buckled segment is in contact with the casing and thus friction occurs, which can cause the axial force distribution to change significantly. The temperature, pressure, liquid density and fluid velocity within the tubing may change with hole depth, time and operation, so the axial force is changing constantly. It is very important for the safety evaluation to predict the axial forces in high pressure-high temperature wells. A large compression load at the lower end can induce tubular plastic deformation and consequently damage the packer. A large tension load at the top end may unpack the packer or cause the tubing to break. If the tubing fails, the borehole cannot maintain its integrity resulting in significant safety concerns.

Initial conditions:

$$F_{\tau e0} = (\rho_{i0} g A_{i0} - \rho_{o0} g A_{o0}) \Delta s_0$$

when the tubular density of the fluid inside and outside is the same, $\Delta s_0 = 1m$, then:

$$F_{\tau e0} = \frac{1}{4} \rho g \pi (d_i^2 - d_o^2), \quad d_i = d_0 - 2r.$$

Tubular equivalent weight:

$$q_{ej} = q_j + (\rho_{ij} g A_{ij} - \rho_{oj} g A_{oj}) \Delta s_j, \quad q_j = \rho_{lj} s_j, \quad q_0 = 0.$$

Assume that the equivalent axial force of the joint j is known, then according to (3.27), we get:

$$f_{nj} = \left[\left(F_{\tau e,j} \frac{d\varphi_j}{ds} + q_{ej} \sin \varphi_j \right)^2 + \left(F_{\tau e,j} \sin \varphi_j \frac{d\psi_j}{ds} \right)^2 \right]^{1/2}, \tag{4.3}$$

and the dimensionless axial force β_j:

$$\beta_j = \frac{F_{\tau ej}}{2} \left(\frac{r_j}{E_j I_j f_{nj}} \right)^{1/2}. \tag{4.4}$$

According to the dimensionless axial force β_j, we get the dimensionless normal pressure n_j and the real normal pressure N_j:

$$n_j = \begin{cases} 1, & \beta_j < 1 \\ \frac{8}{9}\left(\beta_j + \frac{1}{2}\right)^2 - 1, & 1 \leq \beta_j < 1.469, \\ \beta_j^2, & \beta_j \geq 1.469 \end{cases} \tag{4.5a}$$

$$N_j = n_j \beta_j. \tag{4.5b}$$

Thus, we can get the axial force of the joint $j+1$:

$$F_{\tau e,j+1} = F_{\tau e,j} + (q_{ej}\cos\varphi_j + f_{vej} - f_j N_j)\Delta s_j. \tag{4.6}$$

Therefore,

$$F_{\tau,j} = F_{\tau e,j} - P_{ij}A_{ij} + P_{oj}A_{oj}, \tag{4.7}$$

the resultant bending moment M_j of the joint j:

$$M_j = \begin{cases} (EIK_0)_j, & \beta_j < 1 \\ (EIK_0)_j + \frac{4}{3}(EIrf_n)_j^{1/2}(\beta_j - 1), & 1 \leq \beta_j < 1.469, \\ (EIK_0)_j + (EIrf_n)_j^{1/2}(\beta_j - 1), & \beta_j \geq 1.469 \end{cases} \tag{4.8}$$

where, E_j is material elastic modulus of the j unit tubular string; $I_j = \frac{\pi}{4}(r_{oj}^4 - r_{ij}^4)$ is the main moment of inertia of the j unit tubular string cross section (m^4); $A_{i,j}$, $A_{o,j}$ are the internal and external cross sectional area of the j unit tubular string.

4.4.3 Calculation procedures

The steps of calculating are as follows,

Step 1: Obtain each point's inclination angle and azimuth:

$$\varphi_j = \varphi_{j-1} + \frac{\varphi_k - \varphi_{k-1}}{\Delta s_k}\Delta s_j,$$

$$\psi_j = \psi_{j-1} + \frac{\psi_k - \psi_{k-1}}{\Delta s_k}\Delta s_j,$$

where j represents a selected segment point, Δs_k represents the measurement of the depth of the inclination angle φ_k and φ_{k-1}, or the measurement of the depth of the azimuth ψ_k and ψ_{k-1}, Δs_j is step length of calculation.

Step 2: Calculating the distribution of the internal and external fluid pressure $P_{i,j+1}, P_{o,j+1}$.

Step 3: Calculating the equivalent pipe weight q_{ej},

$$q_{ej} = q_j + (\rho_{ij}gA_{ij} - \rho_{oj}gA_{oj})\Delta s_j.$$

Step 4: When the equivalent axial force $F_{\tau e,j}$ of joint j is known, calculating f_{nj} and β_j.

Step 5: Calculating the dimensionless axial pressure n_j and axial pressure N_j.

Step 6: Calculating the equivalent axial force at joint $j+1$,

$$F_{\tau e,j+1} = F_{\tau e,j} + (q_{ej}\cos\varphi_j + f_{vej} - f_j N_j)\Delta s_j.$$

Step 7: Calculating the real axial force $F_{\tau,j}$,

$$F_{\tau,j} = F_{\tau e,j} - P_{ij}A_{ij} + P_{oj}A_{oj}.$$

Step 8: Assume that various kinds of deformation length of the tubular string are known at joint j, then we can obtain various kinds of deformation length of the joint $j+1$, $U_{F,j+1}, U_{M,j+1}, U_{P,j+1}, U_{T,j+1}$:

1. $U_{P,j+1} = U_{P,j} + \frac{2\mu_j}{E_j(A_{oj}-A_{ij})}(A_{oj}P_{oj} - A_{ij}P_{ij})\Delta s_j$.
2. $U_{T,j+1} = U_{T,j} + \alpha_j(T_j - T_0)\Delta s_j$.
3. For $U_{F,j+1}$:
When $\beta_j \leq 1$,

$$U_{F,j+1} = U_{F,j} + \beta_j\Delta s_j + \frac{1}{2}(\gamma_{2j} + \gamma_{1j})\Delta s_j^2.$$

When $1 < \beta_j \leq 1.469$,
(1) $\gamma_{1j} = 0$,

$$U_{F,j+1} = U_{F,j} + \left[\left(1 + \frac{2}{3}a\right)\beta_j - \frac{2}{3}a\right]\Delta s_j + \frac{1}{2}\left(1 + \frac{2}{3}a\right)\gamma_{2j}\Delta s_j^2.$$

(2) $\gamma_{1j} \neq 0$,

$$\begin{aligned} U_{F,j+1} &= U_{F,j} + \frac{9+6a}{16\gamma_{1j}}\left(\ln\left|\left[\frac{8}{9}\left(\beta_{j+1} + \frac{1}{2}\right)^2 - 1\right]\gamma_{1j} + \gamma_{2j}\right|\right. \\ &\quad \left. - \ln\left|\left[\frac{8}{9}\left(\beta_j + \frac{1}{2}\right)^2 - 1\right]\gamma_{1j} + \gamma_{2j}\right|\right) - \frac{2}{3}a\Delta s_j. \end{aligned}$$

When $\beta_j > 1.469$,
(1) $\gamma_{1j} = 0$,

$$U_{F,j+1} = U_{F,j} + \left[(1+a)\beta_j\Delta s_j + \frac{1}{2}(1+a)\gamma_{2j}\Delta s_j^2.\right.$$

(2) $\gamma_{1j} \neq 0$,

$$U_{F,j+1} = U_{F,j} + \frac{1+a}{2\gamma_{1j}}(\ln|\beta_{j+1}^2\gamma_{1j} + \gamma_{2j}| - \ln|\beta_j\gamma_{1j} + \gamma_{2j}|).$$

4. For $U_{M,j+1}$:
When $\beta_j \leq 1$,

$$U_{M,j+1} = U_{M,j}.$$

When $1 < \beta_j \leq 1.469$,

$$U_{M,j+1} = U_{M,j} - \frac{2}{3}\left(\frac{f_{nj}}{E_jI_jr_j}\right)^{1/2} r_j^2(\beta_j - 1).$$

When $\beta_j > 1.469$,

$$U_{M,j+1} = U_{M,j} - \frac{1}{2}\left(\frac{f_{nj}}{E_jI_jr_j}\right)^{1/2} r_j^2\beta_j.$$

Step 9: Calculating the total deformation length of the joint $j+1$,

$$U_{j+1} = U_{P,j+1} + U_{T,j+1} + U_{F,j+1} + U_{M,j+1}.$$

4.5 EXAMPLE CALCULATION

As was described previously, the algorithm begins with a calculation for the tubular string at the bottom. The deformation calculations are performed for each successive "segment" from the bottom to the top.

Table 4.1 Pipe parameters.

Diameter [mm]	88.9	88.9	88.9	73	73
Thickness [mm]	9.53	7.34	6.45	7.82	5.51
Weight [kg]	18.9	15.18	13.69	12.8	9.52
Expansion coefficient [m/m·K]	1.15×10^{-5}	1.15×10^{-5}	1.15×10^{-5}	1.15×10^{-5}	1.15×10^{-5}
Young's modulus [GPa]	215	215	215	215	215
Poisson's ratio [–]	0.3	0.3	0.3	0.3	0.3
Using length [m]	1400	750	4200	600	150

Table 4.2 Pipe parameters.

Measured depth [m]	Internal diameter [mm]	External diameter [mm]
4325.69	168.56	193.7
6301.7	168.3	193.7
7100	121.42	146.1

Table 4.3 Azimuth, inclination and vertical depth.

Number	Measured depth [m]	Azimuth [°]	Inclination [°]	Vertical depth [m]
1	0	0	120.33	0
2	303	1.97	121.2	302.87
3	600	1.93	120.28	599.73
4	899	0.75	126.57	898.59
5	1206	1.25	124.9	1205.45
6	1505	1.04	124.62	1504.32
7	1800	0.49	123.75	1799.18
8	2105	2.49	125.27	2104.04
9	2401	1.27	123.13	2399.91
10	2669	2.44	120.12	2667.79
11	3021	0.14	127.39	3019.63
12	3299	1.18	122.6	3297.5
13	3605	2.05	123.25	3603.36
14	3901	0.16	121.45	3899.22
15	4183	2.92	121.24	4181.09
16	4492	2.73	129.22	4489.95
17	4816.07	1.98	121.61	4813.87
18	5099.07	2.74	129.93	5096.74
19	5394.07	0.13	120.46	5391.61
20	5706.07	0.63	129.59	5703.47
21	5983.07	2.09	120.14	5980.34
22	6302.07	2.69	122.91	6299.19
23	6597.07	2.45	129.41	6594.06
24	6911.12	0.15	124.88	6907.96
25	7100	1.15	123.2	7085.88

4.5.1 Simulation parameters

In this simulation, we study a pipe in X well, which is in Sichuan Province, China. All the needed parameters are as follows:

Internal fluid density = 1000 kg/m^3
External fluid density = 1000 kg/m^3

Table 4.4 Axial force, normal pressure and friction force.

Number	Depth [m]	Axial force [N]	Normal pressure [N]	Friction [N]
1	1	896824.686	24006.786	−14404.229
2	200	864615.152	24273.330	−29127.996
3	400	832242.283	37578.845	−45094.615
4	600	799868.877	22489.576	−26987.491
5	800	767493.223	33083.491	−39700.190
6	1000	735119.086	25484.629	−30581.555
7	1200	702748.575	23329.635	−27995.562
8	1400	670372.392	21196.733	−25436.079
9	1600	644313.862	13115.568	−5870.521
10	1800	618256.154	12013.854	−6678.965
11	2000	592195.533	11512.675	−10949.109
12	2200	566766.256	14600.976	−17521.171
13	2400	543218.953	16351.727	−19622.073
14	2600	519673.889	22555.647	−27066.777
15	2800	496127.403	10996.424	−13195.709
16	3000	472576.411	8530.028	−852.853
17	3200	449031.288	15227.762	−18273.315
18	3400	425482.484	9258.117	−11109.740
19	3600	401929.406	6771.052	−4280.916
20	3800	378387.666	9203.803	−11044.564
21	4000	354836.402	8307.531	−9969.037
22	4200	331288.923	13785.527	−16542.632
23	4400	307737.851	12651.684	−15182.021
24	4600	284192.958	11758.267	−14109.920
25	4800	260649.291	9456.659	−11347.991
26	5000	237100.543	4014.502	−4817.403
27	5200	213557.963	3541.713	−4250.056
28	5400	190014.301	1340.633	−1109.288
29	5800	142924.840	4148.071	−4977.686
30	6200	95839.602	349.462	−419.355
31	6600	50754.240	2481.551	−2977.861
32	6800	28814.446	1406.575	−1687.890
33	7000	8266.029	271.122	−325.347

Depth of the well = 7100 m
Friction coefficient = 1.2
Ground surface temperature = 16°C
Ground thermal conductivity parameter = 2.06 W/m · K
Geothermal gradient = 2.18°C/100 m
Gas outputs = 500000 m^3/day
Length of one segment = 1 m

The parameters of the pipes, the inclined well, the inclination, the azimuth and vertical depth are given in Table 4.3.

4.5.2 Main results

After performing the calculations, we obtain a series of results for this well. The detailed source code can be seen in **Appendix 1**.

At the same time, we obtain a comparative result with a measurement value.

From the above comparative result, the relative error is less than 10%, proving the algorithm is feasible and effective.

Table 4.5 Various kinds of deformation length.

Number	Depth [m]	$U_{F,j}$ [m]	$U_{P,j}$ [m]	$U_{T,j}$ [m]	$U_{M,j}$ [m]	U_j [m]
1	1	0	0	0	0	0
2	200	0.139	0.01141	0.028	0	0.1786
3	400	0.275	0.02255	0.060	0	0.3579
4	600	0.407	0.03336	0.097	0	0.5378
5	800	0.535	0.04384	0.138	0	0.7172
6	1000	0.659	0.05399	0.181	−0.00600	0.8873
7	1200	0.779	0.06382	0.228	−0.00600	1.0640
8	1400	0.895	0.07331	0.280	−0.00600	1.2415
9	1600	1.007	0.08248	0.337	−0.00700	1.4199
10	1800	1.115	0.09133	0.400	−0.00700	1.5982
11	2000	1.218	0.09984	0.467	−0.00700	1.7783
12	2200	1.318	0.10802	0.540	−0.00700	1.9589
13	2400	1.414	0.11588	0.618	−0.00800	2.1402
14	2600	1.506	0.12341	0.701	−0.01000	2.3197
15	2800	1.594	0.13061	0.789	−0.01200	2.5012
16	3000	1.678	0.13749	0.883	−0.02200	2.6757
17	3200	1.758	0.14403	0.981	−0.02400	2.8580
18	3400	1.833	0.15025	1.085	−0.02700	3.0407
19	3600	1.905	0.15614	1.193	−0.03600	3.2178
20	3800	1.973	0.16170	1.307	−0.04000	3.4021
21	4000	2.037	0.16693	1.426	−0.04400	3.5861
22	4200	2.096	0.17184	1.550	−0.07100	3.7469
23	4400	2.152	0.17642	1.680	−0.08600	3.9221
24	4600	2.204	0.18067	1.814	−0.09000	4.1078
25	4800	2.252	0.18459	1.952	−0.09400	4.2950
26	5000	2.295	0.18818	2.091	−0.10500	4.4697
27	5200	2.335	0.19145	2.219	−0.11900	4.6263
28	5400	2.371	0.19438	2.352	−0.14600	4.7705
29	5800	2.430	0.19927	2.616	−0.19600	5.0489
30	6200	2.474	0.20285	2.855	−0.22900	5.3023
31	6600	2.501	0.20511	3.113	−0.27500	5.5439
32	6800	2.509	0.20575	3.249	−0.28600	5.6779
33	7000	2.512	0.20607	3.391	−0.32800	5.7808

Table 4.6 The axial force measurement results.

Location	Axial force [N]	Relative error [–]
Well head	869918.788	0.0309
7000 m	8079.036	0.0228

Table 4.7 The deformation length.

Location depth [m]	The total deformation length [m]	Relative error [–]
2000	1.7239	0.0315
7000	5.4738	0.0561

In this chapter, the deformation differential equations for the tubular string were improved using the classical differential element analysis method, and dimensionless differential equations were obtained using variable replacement. The deformation differential equations were developed using functional theory, and three different states were obtained: a stable state, a sinusoidal buckling state; a helical buckling state: by solving the critical load. At the same time, for the three different states, the axial load, normal pressure, friction force, and four deformation tubular string lengths were determined using the numerical solution from the difference equations. Basic data from *X* Well (high temperature-high pressure gas well), 7100 m deep in China, were used for case history calculations. When comparing with the measurements, an error rate of less than 10% was achieved, proving that our method is feasible.

CHAPTER 5

Setting the mechanical analysis

The constrained buckling of a tube in a slanted well under gravity and the compressive force has been analysed by several authors over the last three decades. This study of force analysis in the setting stage was motivated by an interest in the packer's initial conditions and boundary conditions. The buckling behavior of the pipe string influences the design of the well and the production operations. For example, an axial displacement influences the setting length design, and the bending stress may influence weight and grade. The problem with a sinusoidal buckling of the pipe string was first studied by Paslay and Bogy (1964). In their work, the end of the pipe string was supported by hinges. The critical force at the bottom of the tube for sinusoidal buckling was found to be a function of the length of the pipe. Since the pressure in the pipe increases with depth, the amplitude of the sinusoidal buckling also increases with depth so Paslay and Bogy found that the number of sinusoids in the buckling multiplied with the length of the pipe. An asymptotic solution to the sinusoidal buckling of an extremely long pipe was analysed by Dawson (1984) based on a sinusoidal buckling with a constant amplitude. In their work, simple descriptions for the buckling force and wave numbers were derived.

The most generally accepted method for the analysis of buckling, tubular movement, and packer selection was developed by Lubinski and Althouse (1962), which considered only a vertical well with no friction. Some further analysis using Lubinski's approach has been done for more complicated tubular configurations such as tapered pipe strings (Cheatham, 1984). Henry Woods, in the appendix to Lubinski and Althouse (1962), developed a mechanical model to predict the buckling configuration for tubular buckling behavior. Mitchell developed a more general approach that replaced the virtual work relations with a full set of beam-column equations constrained to be in contact with the casing (He and Kyllingstad, 1995). In this formation, helical buckling in a deviated well, was described using a fourth order non-linear differential equation. For a vertical well, the solution to this equation can be accurately approximated using the simple algebraic equation discovered by Lubinski and Woods. This solution, however, is not valid for deviated or horizontal wells because of the lateral gravitational forces. Using the Galerkin technique Miska and Cunha (1995) sought numerical solutions, which confirmed the thought that under a general load the deformed shape of the pipe is a combination of helices and sinusoids, while helical deformation occurs only under special values for the applied load. For more research on the pipe string buckling deformation, readers can refer to McCann and Suryanarayana (1994), Mitchell (1988; 1997), (Xu *et al.*, 2012b) and Wu and Juvkam-Wold (1995).

Much of the previous research focusing on the analysis of buckling behavior has not considered friction. However, in some studies the importance of friction has been recognized. For example, Hammerlindl (1977) showed an experimentally measured 50% error in the frictionless buckling length change. In this work, analytical solutions for two basic problems were developed: (1) slacking from the surface and (2) loading upward from the base of the tubular. Analytical solutions were possible because the load application was in a single direction. While limited in application, this work showed the considerable importance of friction forces on buckling and tubular signs. For instance, the effect of friction was shown to greatly reduce the set down force regardless of the surface slack-off. More research on the influence of friction can be found in (Mitchell, 1997), (Chen and Xu, 2008), (Yao *et al.*, 2011) and (Liu *et al.*, 2013).

5.1 HYDRAULIC PACKER FORCE ANALYSIS IN DEVIATED HPHT WELLS

The tubular string used for testing or production is commonly very long. The top end hangs out of the wellhead and the lower end has a packer and many other accessories. The packer is firmly fixed with a casing near the bottom-hole.

The axial force of the tubing varies with hole depth, as there is tension near the wellhead and compression near the bottom-hole, so that the upper section of the tubing remains straight but the lower part may helically buckle. The axial force is distributed linearly if there is no buckling but if buckling occurs, the buckled segment comes into contact with the casing and friction occurs, which can result in a significant change to the axial force distribution.

Temperature, pressure, liquid density and fluid velocity within the tubing may change with hole depth, time and operation, so that the axial force may be constantly changing. For a comprehensive safety evaluation of HPHT wells it is very important to be able to predict the axial forces as a large compression load at the lower end can result in plastic tubing deformation and packer damage. A large tension load at the top end, on the other hand, may unpack the packer or cause the tubing to break. If the tubing fails, the borehole is unable to maintain its integrity leading to significant safety concerns.

5.1.1 **Model building**

In this section, some basic assumptions are given, then a force model is developed using force balance, and the initial condition and boundary condition are determined:

(i) The curvature of the wellhead in the considered modular section is constant.
(ii) On the upper side or underside of the section where the pipe touches the well wall, the curvature is the same as that of the wellhead.
(iii) The shearing force on the transverse section in the pipe and the impact from the stiffness of the pipe is not considered, but axial stress is considered.
(iv) The modular section is on an inclined plane in one space.

5.1.1.1 *Force analysis of the pipe*

Before considering the setting of the packer, we analyse the underground tube's three-dimensional track. As shown in Figure 5.1, since the trace of the tube is skewed, a space rectangular coordinate is used to describe it.

Let wellhead P be the origin, we upbuild $Pxyz$ a right-handed coordinate system. $\vec{i}, \vec{j}, \vec{k}$ denote the unit vector of x-axis, y-axis, z-axis, respectively. On the axis line of the well-bore, the geometry position of arbitrary point $O(x, y, z)$ can be represented by vector $\vec{r}_0(s_0)$:

$$\vec{r}_0(s_0) = x_0(s_0)\vec{i} + y_0(s_0)\vec{j} + z_0(s_0)\vec{k} \tag{5.1}$$

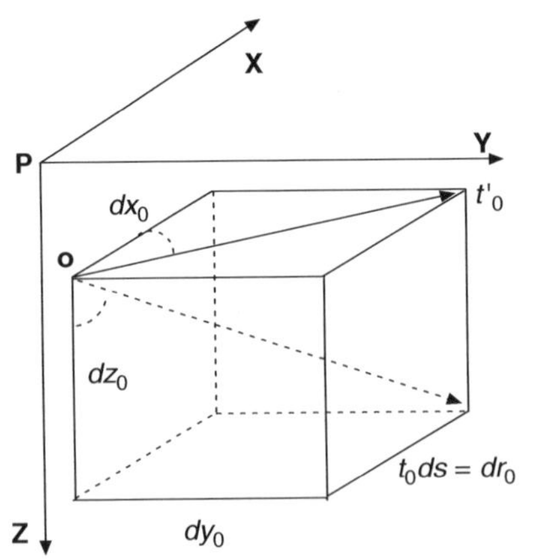

Figure 5.1 The 3D figure of inclined well-bore.

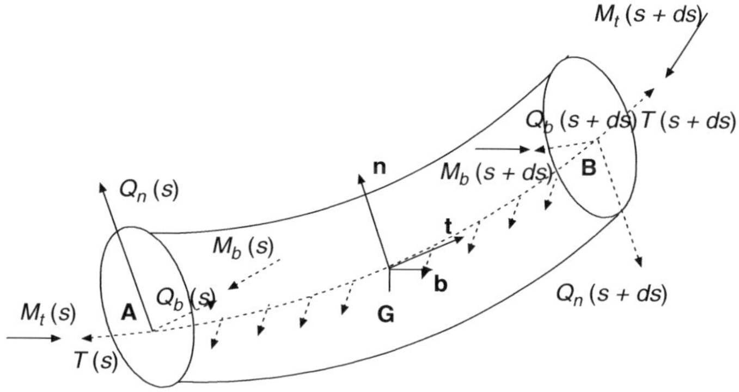

Figure 5.2 The tubular force analysis.

and the corresponding increment is represented as follows:

$$d\vec{r}_0(s_0) = dx_0(s_0)\vec{i} + dy_0(s_0)\vec{j} + dz_0(s_0)\vec{k}. \tag{5.2}$$

Let $\vec{\tau}_0$ be the unit vector of the tangential direction of the well-bore axis line and pass point O. The inclined angle φ between $\vec{\tau}_0$ and $\vec{k}_0$ is called deviation angle, the included angle ψ between projection $\vec{\tau}_0'$ of $\vec{\tau}_0$ on the $PXYZ$ plane and $\vec{i}$ is called azimuth angle. From Figure 5.1, we have:

$$\frac{dx_0}{ds_0} = \sin\varphi\cos\psi, \quad \frac{dy_0}{ds_0} = \sin\varphi\sin\psi, \quad \frac{dz_0}{ds_0} = \cos\varphi. \tag{5.3}$$

$$\vec{\tau}_0 = \sin\varphi\cos\psi\vec{i} + \sin\varphi\sin\psi\vec{j} + \cos\varphi\vec{k}. \tag{5.4}$$

The curvature K_b and torsion K_n of the corresponding trail can be obtained as follows:

$$K_b = \sqrt{K_\varphi^2 + \sin^2\varphi K_\psi^2} \tag{5.5}$$

$$K_n = \frac{K_\varphi K_\psi' - K_\varphi' K_\psi}{K_b^2} + K_\psi\left(1 + \frac{K_\varphi^2}{K_b^2}\right)\cos\varphi, \tag{5.6}$$

where K_φ and K_ψ are respectively the change rate of the deviation angle and the azimuth angle, K_φ' and K_ψ' are the derivatives of K_φ and K_ψ.

Next, it comes to the force analysis of the infinitesimal between s and $s + ds$ (see Figure 5.2):

(1) The concentrated force at A is:

$$\vec{F}(s) = (-T(s) + M_t(s) \quad Q_n(s) \quad Q_b(s) - M_b(s)) \begin{bmatrix} \vec{\tau}(s) \\ \vec{n}(s) \\ \vec{b}(s) \end{bmatrix},$$

where $T(s)$ is the axis force, Q_n and Q_b are the shear forces, M_b is the bending moment and M_t is the torsional moment. According to the realistic situation and assumptions, the torsional moment is usually considered to be 0.

(2) The concentrated force at B is:

$$\vec{F}(s+ds) = [T - M_t + dT - dM_t \quad -(Q_n + dQ_n) \quad -(Q_b + dQ_b) + M_b + dM_b] \cdot \begin{bmatrix} \vec{\tau}(s) + d\vec{\tau} \\ \vec{n}(s) + d\vec{n} \\ \vec{b}(s) + d\vec{b} \end{bmatrix}. \tag{5.7}$$

(3) The contact force of the infinitesimal is:

$$\vec{q}(s) = [\pm\mu N \quad N_n \quad N_b]\begin{bmatrix}\vec{\tau}(s)\\ \vec{n}(s)\\ \vec{b}(s)\end{bmatrix}.$$

(4) The submerged weight of the infinitesimal:

$$\vec{W}_p = G\vec{g}\cdot\begin{bmatrix}\vec{\tau}(s)\\ \vec{n}(s)\\ \vec{b}(s)\end{bmatrix} = q_m[\vec{g}\cdot\vec{\tau}(s) \quad \vec{g}\cdot\vec{n}(s) \quad \vec{g}\cdot\vec{b}(s)].$$

According to the equilibrium condition of these forces, we have:

$$\vec{F}(s) + \vec{F}(s+ds) + \vec{q}ds + \vec{W}_p ds = 0. \tag{5.8}$$

Then it follows that:

$$\begin{cases} T\dfrac{d\vec{\tau}}{ds} + \dfrac{dT}{ds} \pm \mu_\varphi N + q_m\vec{g}\cdot\vec{\tau} = 0 \\ -Q_n\dfrac{d\vec{n}}{ds} - \dfrac{dQ_n}{ds} + N_n + q_m\vec{g}\cdot\vec{n} = 0 \\ -Q_b\dfrac{d\vec{b}}{ds} - \dfrac{dQ_b}{ds} + N_b + q_m\vec{g}\cdot\vec{b} = 0 \end{cases}. \tag{5.9}$$

Here exists the relationship among the three directions as follows:

$$\begin{cases} T\dfrac{d\vec{\tau}}{ds} = K_b Q_n \\ Q_n\dfrac{d\vec{n}}{ds} = -K_b T + \tau Q_b \\ Q_b\dfrac{d\vec{b}}{ds} = -K_n Q_n \end{cases}. \tag{5.10}$$

In addition, it follows from Figure 5.1 that:

$$\begin{cases} \vec{g}\cdot\vec{\tau} = \vec{K}\cdot\vec{\tau} = -\cos\varphi \\ \vec{g}\cdot\vec{n} = -\dfrac{1}{K_b}\dfrac{d\varphi}{ds}\sin\varphi \\ \vec{g}\cdot\vec{b} = -\dfrac{1}{K_b}\dfrac{d\psi}{ds}\sin^2\varphi \end{cases}. \tag{5.11}$$

Above all, we obtain the coupled system model of differential equations on the force analysis of the infinitesimal as follows:

$$\begin{cases} \dfrac{d}{ds}\left[T + M_b^2/(2EI)\right] = -q\cos\varphi \pm \mu N \\ N_n = \dfrac{d^2M_b}{ds^2} - K_b T - K_n^2 M_b - q\dfrac{K_\varphi}{K_b}\sin\varphi \\ N_b = \dfrac{d}{ds}(K_n M_b) + K_n\dfrac{dM_b}{ds} + q\dfrac{K_\psi}{K_b}\sin^2\varphi \\ N^2 = N_b^2 + N_n^2 \\ M_b = EIK_b \end{cases}, \tag{5.12}$$

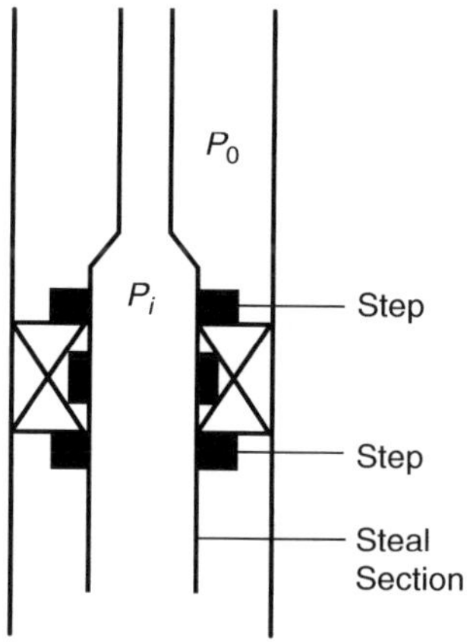

Figure 5.3 The hydraulic packer.

where E is the elastic modulus of the tabular and I is the tabular inertia moment.

5.1.1.2 *Hydraulic packer force analysis*

The setting process for the hydraulic packer is described as follows (see Figure 5.3):

(1) After the tubular is laid in the predetermined location, the wellhead assembly is fixed.
(2) A steel ball is dropped from the wellhead and falls to the ball seat sealing the pipe's passageway is sealed.
(3) The liquid is injected through the wellhead and, as the tubular bottom is sealed by the steel ball, the pressure in the tubular increases.
(4) As the pressure in the pipe increases, the liquid in the pipe flows into the central tube from the packer inlet hole. When the pressure reaches a certain level, the pushing of the pressure head bearing cuts the packing pin. The differing pressures inside and outside the pipe compress the packer rubber so that it balloons and seals the interspace between the tubular and the radial collar.
(5) As the packer rubber balloons, the inside of the central packer tube moves and is finally locked at a certain location to support the packer rubber.
(6) After the pressure on the pipe is released, the packer rubber is unable to return as it is locked in the snap spring with support from inside the central packer tube. The packer rubber, therefore, continues to seal the interspace between the tubular and the radial collar.

When packed using a hydraulic packer, a hydraulic force is added to the wellhead, which leads to an elongation of the well. After the hydraulic force is released, the pipe shrinks, and there is a resulting pulling force at the bottom of the packer. This force can be computed as follows:

(1) When the hydraulic packer is packed using hydraulic force, there is a concentrated force F_h on the bridge plug below the packer:

$$F_h = A \cdot \triangle P_h, \tag{5.13}$$

where A denotes the area and $\triangle P_h$ denotes the pressure by considering the difference between the initial liquid and the liquid in the hydraulic well. The elongation of the pipe is caused by F_h and the bulging deformation is caused by the differing pressure on the inside and outside of the well-bore. The elongation indicator $\triangle L_h$ can be computed by:

$$\triangle L_h = \triangle L_{1h} + \triangle L_{3h}, \tag{5.14}$$

where $\triangle L_{1h}$ and $\triangle L_{3h}$ then be computed by the above section.

(2) After the hydraulic force is released, the packer can freely move and the string shrinks. The length of the sealed section is $\triangle L_h$. After releasing the hydraulic force, there is a resultant pulling force F_h^q:

$$F_h^q = \triangle L_h \frac{EA_S}{L}. \tag{5.15}$$

If the packer is unable to move, F_h^q is considered to be the packing force when computing the relationship between the length of the sealed section and the pulling force or computing force in each section of the packer. Hence, the real pulling force is:

$$F_S' = F_S - F_h^q. \tag{5.16}$$

When computing the force at each point of the tabular when using a hydraulic packer, the complete process must satisfy model (5.12). The axial force initial condition is as follows:

$$T_0 = T_P + F_S', \tag{5.17}$$

where T_P denotes the axial force at the packer after putting down the test pipe.

5.1.2 Computing parameters

Other parameters which need to be computed are as follows:

(1) Computing the axial force T_P after the test pipes are put down or before packing. Because of gravity the pipe is deformed, and consequently the initial axial force changes. When the test pipe is initially placed, the initial condition (the axial force at the bottom of the pipe) is described as follows:

$$T_0 = P_o^0 A_o^0 - P_i^0 A_i^0, \tag{5.18}$$

where P_o^0 and P_i^0 are the external and internal pressure at the bottom of the pipe, and A_o^0 and A_i^0 are the external and internal areas at the bottom of the string. By including the initial condition in the coupled differential equations model for force analysis, we are able to determine the axial force T_P after placing down the test pipe or before packing.

(2) Computing the piston change $\triangle L_{1h}$. According to the Hooke's law, we have:

$$\triangle L_{1h} = \frac{L_h}{E_h}(\triangle F_h), \tag{5.19}$$

where L_h denotes the length of the pipe at the packer, E_h denotes the axial rigidity of the pipe at the packer, and $\triangle F_h$ denotes the centralized Hooke's force and it can be computed using:

$$\triangle F_h = (A_P - A_i)\triangle P_i - (A_P - A_o)\triangle P_o, \tag{5.20}$$

where A_P, A_i, A_o respectively denote the sealing area, the internal area and external area of the pipe, $\triangle P_i$ and $\triangle P_o$ express the internal and external pressure difference.

(3) Computing the ballooning change $\triangle L_{3h}$:

$$\triangle L_{3h} = \frac{\mu}{E}\left[\frac{\rho_C - R^2\rho_K - (1+2\mu)\tau/(2\mu)}{R^2 - 1} \cdot L^2 + 2\frac{P_C - R^2 P_K}{R^2 - 1} \cdot L\right], \tag{5.21}$$

where μ denotes Poisson steel ratio, E denotes the elastic modulus of the steel, ρ_C denotes the mass density of the internal liquid, ρ_K denotes the mass density of the hohlraum liquid, P_C and P_K respectively denote the pressure inside the packer and in the hohlraum, τ denotes the friction resistance coefficient movement of liquid in the pipe, and R denotes the ratio of the internal and external radius of the pipe.

5.1.3 Algorithm

Since the coupled differential equations model has ordinary differential equations and partial differential equations, the four-order Runge-Kutta method and the finite difference method are suggested solutions to this problem. The detailed algorithm is summarized as follows.

Step 1. Give the initial values θ_0, L_h, respectively, where L_h denotes the length of the pipe at the packer.
Step 2. Compute all the coefficients in the coupled system.
Step 3. Compute the pressure P_i, P_o, P_K for the inside and outside of the pipe and the hohlraum.
Step 4. Compute T_P, M_t, M_b, N_n and N_b at the packer.
Step 5. Depending on the geometric shape and structure of the pipe, the whole pipe is divided into n segments. If T_i and M_{bi} are determined obtained at the ith point, the axial force T_{i+1} and the bending moment M_{bi} at the $(i+1)$th point can be computed using the difference equation. The difference scheme for the Equation (5.12) is written as follows:

$$\begin{cases} \dfrac{T_i - T_{i-1}}{\Delta s_i} + K_{bi}\dfrac{M_{bi} - M_{bi-1}}{\Delta s_i} = -q_i \cos\varphi_i \pm \mu N_i \\ N_{ni} = \dfrac{M_{bi+1} + M_{bi} - 2M_{bi-1}}{\Delta s_i^2} - K_{bi}T_i - K_{ni}^2 M_{bi} + q_i \dfrac{K_{\varphi i}}{K_{bi}} \sin\varphi_i \\ N_{bi} = M_{bi}\dfrac{K_{ni} - K_{ni-1}}{\Delta s_i} + 2K_{ni}\dfrac{M_{bi} - M_{bi-1}}{\Delta s_i} - q_i \dfrac{K_{\psi i}}{K_{bi}} \sin^2\varphi_i \\ N_i^2 = N_{bi}^2 + N_{ni}^2 \\ M_b = EI_i K_{bi} \end{cases} . \quad (5.22)$$

Considering the initial condition $T_0 = P_o^0 A_o^0 - P_i^0 A_i^0$, we can obtain the axial force T_P after putting down the test pipe or before packing.
Step 6. Compute F_h, ΔL_h, F_h^q, F'_S and T_0.
Step 7. Take T_0 as the initial condition for the Equation (5.22) and from which the axial force distribution is determined.

5.1.4 Numerical simulation

As described previously, the algorithm begins with a calculation at the bottom of the pipe. The calculations are performed for each successive "segment" of the pipe from the bottom to the top.

5.1.4.1 *Parameters*

The parameters for an X well in Sichuan province, China are presented as follows:

Internal fluid density = 1000 kg/m^3
External fluid density = 1000 kg/m^3
Depth of the well = 7100 m
Ground temperature = 20°C
Ground thermal conductivity parameter = 2.06 W/m · K
Geothermal gradient = 2.18°C/100 m
Length of one segment = 1 m

The parameters for the pipes including the inclination, the azimuth and the vertical depth are given in Tables 5.1, 5.2 and 5.3, respectively.

Table 5.1 Pipe parameters.

Diameter [mm]	88.9	88.9	88.9	73	73
Thickness [mm]	9.53	7.34	6.45	7.82	5.51
Weight [kg]	18.9	15.18	13.69	12.8	9.52
Expansion coefficient [m/m · K]	1.15×10^{-5}	1.15×10^{-5}	1.15×10^{-5}	1.15×10^{-5}	1.15×10^{-5}
Young's modulus [GPa]	215	215	215	215	215
Poisson's ratio [–]	0.3	0.3	0.3	0.3	0.3
Using length [m]	1400	750	4200	600	150

Table 5.2 Well parameters.

Measured depth [m]	Internal diameter [mm]	External diameter [mm]
4325.69	168.56	193.7
6301.7	168.3	193.7
7100	121.42	146.1

Table 5.3 Azimuth, inclination and vertical depth.

Number	Measured depth [m]	Azimuth [°]	Inclination [°]	Vertical depth [m]
1	0	0	120.33	0
2	303	1.97	121.2	302.87
3	600	1.93	120.28	599.73
4	899	0.75	126.57	898.59
5	1206	1.25	124.9	1205.45
6	1505	1.04	124.62	1504.32
7	1800	0.49	123.75	1799.18
8	2105	2.49	125.27	2104.04
9	2401	1.27	123.13	2399.91
10	2669	2.44	120.12	2667.79
11	3021	0.14	127.39	3019.63
12	3299	1.18	122.6	3297.5
13	3605	2.05	123.25	3603.36
14	3901	0.16	121.45	3899.22
15	4183	2.92	121.24	4181.09
16	4492	2.73	129.22	4489.95
17	4816.07	1.98	121.61	4813.87
18	5099.07	2.74	129.93	5096.74
19	5394.07	0.13	120.46	5391.61
20	5706.07	0.63	129.59	5703.47
21	5983.07	2.09	120.14	5980.34
22	6302.07	2.69	122.91	6299.19
23	6597.07	2.45	129.41	6594.06
24	6911.12	0.15	124.88	6907.96
25	7100	1.15	123.2	7085.88

5.1.4.2 *Main results*

Based on the Runge-Kutta method and the finite difference method, the algorithm was implemented using *C#* language, which allows for a dynamic management of the data base for the addition, deletion, and modification of the casing tube. We obtained a series of results from this well. The axial force, normal pressure and friction are shown in Table 5.4. The detailed source code can be seen in **Appendix 2**.

5.1.5 **Discussion**

Figures 5.4–5.7 show the force distribution and axial displacement caused by the force.

Table 5.4 The axial force, normal pressure and friction.

Depth [m]	Axial force [N]	Normal pressure [N]	Friction [N]
1	896824.686	24006.786	−14404.229
200	864615.152	24273.330	−29127.996
400	832242.283	37578.845	−45094.615
600	799868.877	22489.576	−26987.491
800	767493.223	33083.491	−39700.190
1000	735119.086	25484.629	−30581.555
1200	702748.575	23329.635	−27995.562
1400	670372.392	21196.733	−25436.079
1600	644313.862	13115.568	−5870.521
1800	618256.154	12013.854	−6678.965
2000	592195.533	11512.675	−10949.109
2200	566766.256	14600.976	−17521.171
2400	543218.953	16351.727	−19622.073
2600	519673.889	22555.647	−27066.777
2800	496127.403	10996.424	−13195.709
3000	472576.411	8530.028	−852.853
3200	449031.288	15227.762	−18273.315
3400	425482.484	9258.117	−11109.740
3600	401929.406	6771.052	−4280.916
3800	378387.666	9203.803	−11044.564
4000	354836.402	8307.531	−9969.037
4200	331288.923	13785.527	−16542.632
4400	307737.851	12651.684	−15182.021
4600	284192.958	11758.267	−14109.920
5000	237100.543	4014.502	−4817.403
5200	213557.963	3541.713	−4250.056
5400	190014.301	1340.633	−1109.288
5800	142924.840	4148.071	−4977.686
6200	95839.602	349.462	−419.355
6600	50754.240	2481.551	−2977.861
6800	28814.446	1406.575	−1687.890
7000	8266.029	271.122	−325.347

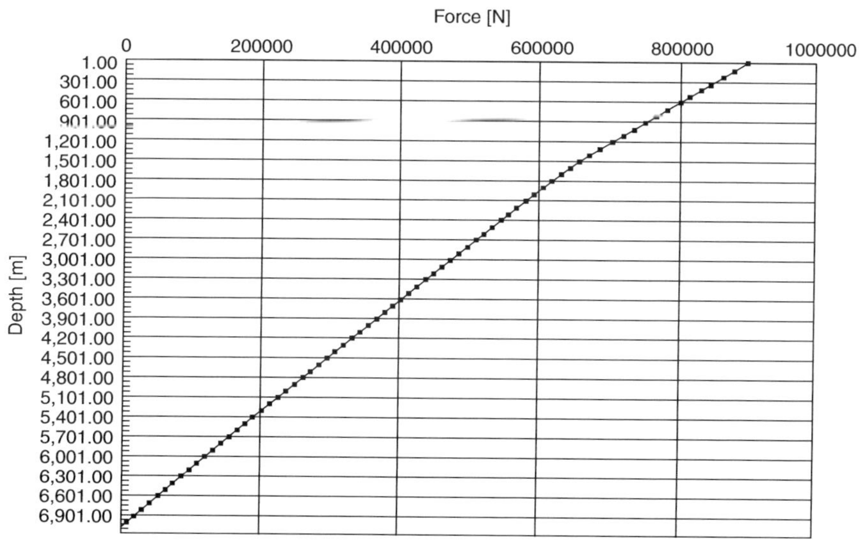

Figure 5.4 The force distribution.

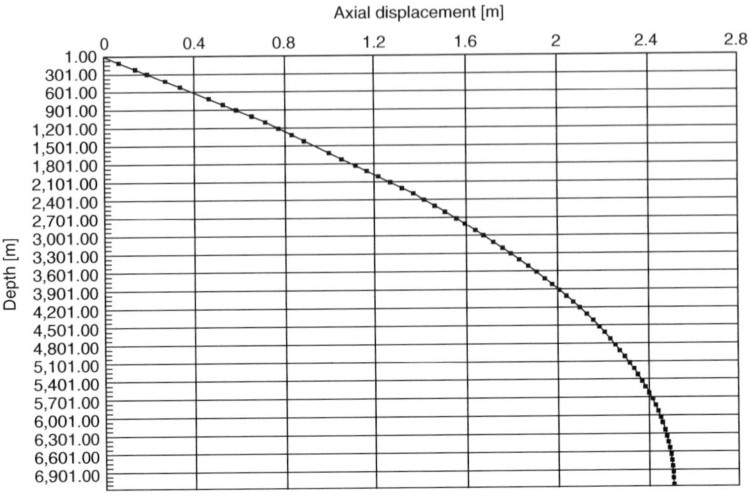

Figure 5.5 Axial displacement curve.

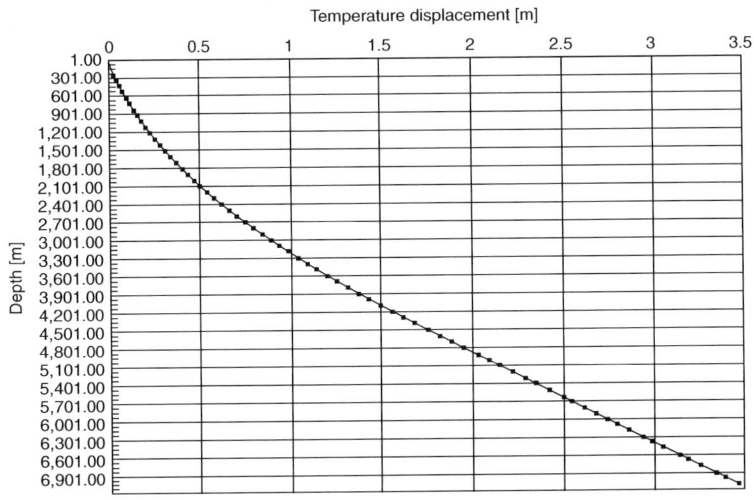

Figure 5.6 Temperature displacement curve.

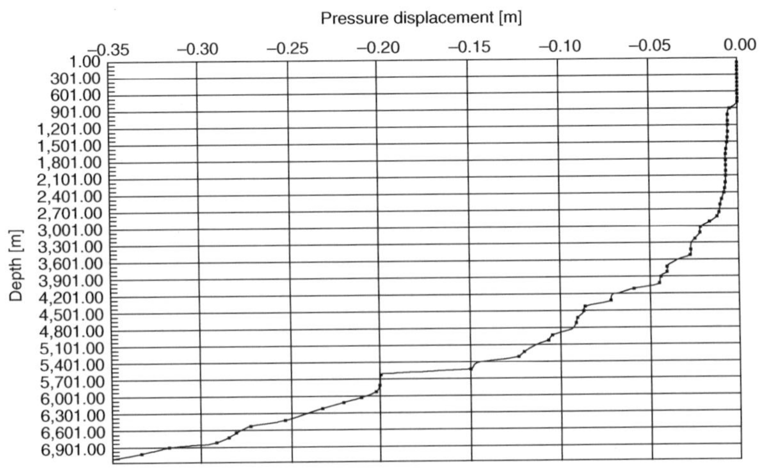

Figure 5.7 Pressure displacement curve.

By comprehensively considering the axial load of the pipe, the internal and external pressure, the normal pressure between the pipe and the well wall, the fluid flow friction and the viscous friction, a highly dimensional nonlinear differential equations model was derived using the classical differential element analysis method. This demonstrated a significant improvement compared with other differential models that ignore the effect of friction or provide analyses without considering the external normal force, or those that only consider straight wells. The axial force, normal pressure, friction, and pressure can be obtained using a dimensionless iterative interpolation algorithm consisting of dimensionless differential equations and the backward difference method. This effectively overcomes the calculation difficulties.

CHAPTER 6

Re-opened mechanical analysis

6.1 INTRODUCTION

For a string with a packer, when the string is played out underground, the corresponding string deformation needs to be calculated with the packer re-opened. Corresponding to the three ways the packer can be seated, there are three corresponding ways to re-open it. Here, we only consider the mechanical and hydraulic re-opening processes. When the string is pulled upward by force F, the action on the packer is reduced, thus the reaction of the packer on the string is also reduced. The calculation begins at the packer, with the initial value being at the sealing stage with an addition of tension F (the downward direction is positive, so F is considered negative). The pressure and temperature distribution is not affected, so these remain the same as in the shut-in distribution condition. Similarly, for the hydraulic reopening process, pressure is only added in the pipe or the annulus, so for the shut-in case only the pressure distribution changes from the initial values.

Because of peak-modulation or maintenance demands, the well needs to be frequently turned on and off during the gas well production process. In applied basic theory for deep well testing, tubular string mechanical analysis is very complex, as fluid temperature and tubing pressure significantly affect the force of the tubular string. Some conditions, such as the failure of the packer, abnormal formation pressure and pipe leakage have a great impact on testing production and are key technical problems affecting testing success. Packers are an important well tool, which have complicated and volatile working conditions and are therefore more easily damaged than other tools. As is known, both gas well temperature and pressure change during production and closing. For HTHP wells, excessive pressure results in a large pressure difference on the packer, which not only damages the packer rubber but also causes the slips to move upward resulting in packer failure. Further, the fluid flow, temperature change, tubular deformation and the increase in the axial forces result in a reduction in the packer bearing capacity. At the same time, because of the pressure difference, the sealing arrangement is crushed, also resulting in packer failure. The rubber tube intensity is obviously decreased if the bottom-hole temperature exceeds its rated working temperature. In this case, the rubber tube is easily crowded out and causes the packer to lose its effectiveness. With the varying temperatures and pressures, the deformation and load exerted on the tubing strings as well as the pressure and fluids from the gas reservoir changes result in safety concerns. If the tubing fails, the borehole is unable to maintain its integrity or safety (Gao and Gao, 2004). Therefore, it is very important to predict the axial forces for HTHP wells.

A significant contribution to tubular mechanics was made by Hammerlindl (1977). He proposed four effects on the packer forces and tubing length change: the temperature effect, the ballooning effect, the axial load effect and the helical buckling effect. There has been significant research on the buckling behavior effect. This research showed that an axial force inflexion is caused under certain conditions, which results in parts of the drill string colliding with the well bore. When the tubular buckles beyond the well-hole's control, the load increases the buckling configuration and transforms it from a state of sinusoidal buckling stability to helical buckling. The tube buckling problem was first studied and the theory put into practice by Lubinski and Althouse (1962). He emulated and experimented on the buckling behavior of a tube in deviated wells and found a computer formula for the critical buckling load. The discovery that the number of sinusoids in the buckling mode increases with the length of the tube was made by Paslay and Bogy (1964). The buckling behavior on the inner and outer fluid pressure of the tubing was analysed and the mathematical relationship between the pitch and the axial pressure was deduced based on the principle of minimum potential energy (Gao and Gao, 2004). The mptotic solution for the sinusoidal buckling of an extremely long tube was analysed by Dawson (1984) using a sinusoidal buckling mode of constant amplitude. Numerical solutions were also sought by Mitchell (1997) using basic mechanics equations. His solutions confirmed the thought that, under a general load, the deformed shape of the tube is a combination of helices and sinusoids while helical deformation occurs only under special values of the applied load. This formula for tubing forces, however, was too simple to accommodate the complicated states in deep wells. Up to now, much research has been centred on water injection tubulars but not well shut-in conditions, and the temperature and pressure values are considered as constant or lineal functions, which causes large errors in tubular deformation computations (Ding and Yan, 2005). For further study, see Liu *et al.* (2012).

6.2 APDTU-VTPF

The material properties of the tubular, such as the elastic modulus and the linear expansion coefficient, all change with temperature and pressure, so the stress on the tubing is very difficult to calculate. With this in mind, the following assumptions are considered in this section:

1. The hole curvature of the considered modular section is constant.
2. On the upperside or underside of the section which is the point of contact between the pipe and the tube wall, the curvature is the same as the hole curvature.
3. The string radius, in contrast to the curvature of the borehole, is insignificant.
4. The string is in a linear elastic relationship state.

6.2.1 HTHP wells characteristics

Generally, the working conditions and technological features of HTHP wells can be reduced to two simple points. The bottomhole temperature is high at around 160°C, and there are multiple tubing measurements. The tool combinations, such as the bottomhole test valve, the safety valve and the packer, are complex. The details for the tubular mechanics analysis is as follows:

(1) The temperature and pressure distribution on the tubing are significantly different under variable outputs (flow velocity), but this is not a simple linear relationship and the fluid density is not constant.
(2) The force and deformation in the tubular as a result of temperature, pressure, fluid density, viscous friction loss and coulomb friction between the tubing and the well increases with the depth of well.
(3) The stretching force and creeping displacement of the downhole strings impact the sealed state of the packer and may even cause the packer to unset.

6.2.2 The packer principle

Setting: After the hydraulic pressure pressurises the tubing, this hydrostatic pressure impacts the fluid cylinder of the packer through the centre bore and pushes up the control piston. When the pressure reaches a certain level, the shear stud connectors shear to control the upward movement of the piston. The connecting block is released, and the packer setting process starts. If there is sufficient hydrostatic pressure when the connecting block is freed, the setting organization gives downhole support. Then, the outer central tube moves down to hold open the anchor and compress the packer. The setting of the packer is then completed.

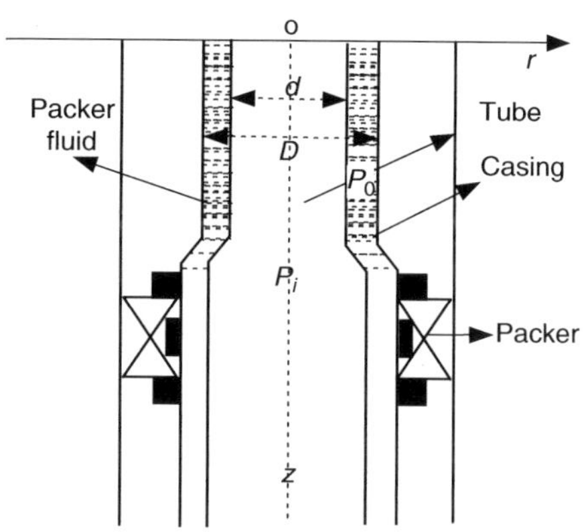

Figure 6.1 The physical figure of forces analysis on tube.

Unsetting: The tubing string is lifted while the outer central tube and the anchor organization hold their positions relative to the casing. Then, the inner central tube moves upward to cut the unset shear ring. The tube continues going up, and the packer is then released. The inner tube snap spring drives the outer tube up and the anchor becomes ineffective. The packer is then unset.

6.2.3 **Theoretical model**

Tubing deformation because of a pressure or temperature changes, which causes damage to the shear ring, is the main cause of packer failure in wells. However, if a larger sized shear ring is selected, the unset force increases, which leads to added inventory costs. Therefore, it is necessary to research the stress distribution of the packer before the unset process. The hydraulic packer structural diagram is shown in Figure 6.1.

6.2.3.1 *Factors affecting the unset force*

Generally, the factors affecting the unset force of a hydraulic packer can be summarized are as follows:

1. A ballooning effect and piston effect from the pressure variation.
2. A temperature effect from the tubing temperature variation.
3. A sucker-rod pumping effect.
4. A piston force effect from the supporting packer's pressure.

6.2.3.2 *Basic equations*

As shown in Figure 6.1, there is a constant cross-sectional flow area A, an inner diameter d, an outer diameter D, material density ρ_1, packer fluid density ρ_2 and a total length Z. Gas flows through this tubing from the bottom to the top with a mass flow rate W. The distance co-ordinate in the flow direction along the tubing is denoted z. The cylindrical coordinate system $r\theta z$, the origin of which is in the wellhead, and the Z axis as the borehole axis, are used.

6.2.3.3 *Initial axial load*

The distance of the section from the wellhead is z (m). The axial static load is calculated as the tubular deadweight:

$$N_{qz} = \int_z^L q \cos\alpha \, dz = \frac{\pi}{4}\rho_1 g(D^2 - d^2)\int_z^L \cos\alpha \, dz, \tag{6.1}$$

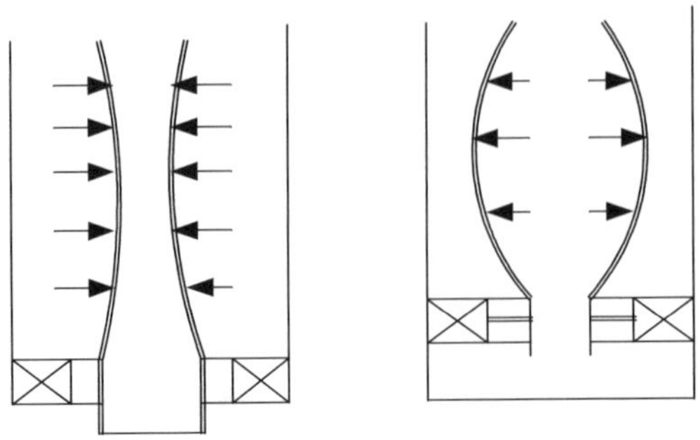

Figure 6.2 The figure of ballooning effect.

where, N_{qz} is the deadweight of tubular, L is the length of tubular, ρ_1 is the density of tubular, α is the inclination angle. The axial static load by the buoyant weight:

$$N_{bz} = -\rho_2 g z A_2 = -\rho_2 g z \left(\frac{D}{2}\right)^2, \tag{6.2}$$

where, N_{bz} is the tubular buoyant weight, ρ_2 is the packer fluid density.

Therefore, summing the (6.1) and (6.2), the axial forces in the section are determined as follows:

$$F_i = N_{qz} + N_{bz}. \tag{6.3}$$

6.2.3.4 *Pressure effect on the packer setting*

When the packer is setting, the tube's internal pressure is higher than the external pressure, which produces an axial load piston effect, the force of which can be determined using the follow equation:

$$F_z = \frac{\pi}{4} d^2 \Delta P_{start}, \tag{6.4}$$

where, F_z represents the axial tensile strength, and ΔP_{start} is the differential pressure at startup. After the slips anchor the packer, the tube's internal pressure continues to increase, but the length of the tube does not change. Therefore, the differential pressure in (6.3) is at startup.

The differential pressure between the casing and the tube produces a ballooning effect which forces the tube length to decrease as shown in Figure 6.2. Depending on the thickness of the wall cylinder and the generalized Hooke's law, the length change can is calculated using the following formula (the equation was obtained from Lubinski and Althouse (1962)):

$$\Delta L_{bi} = -\frac{\mu \Delta Z_i^2}{E} \frac{d^2 \Delta \rho_{ii} g - D^2 \Delta \rho_{oi} g - \delta(1+2\mu)/(2\mu)}{D^2 - d^2} - \frac{2\mu \Delta Z_i}{E} \frac{d^2 \Delta P_{ii} - D^2 \Delta P_{oi}}{D^2 - d^2}, \tag{6.5}$$

where, ΔP_{ii} represents a change in the tubing pressure at i length, ΔP_{oi} represents a change in the annular pressure at i length, E is the steel elastic modulus of the tubular, δ is the drop in pressure in the tubing due to flow per unit length, $\Delta \rho_{ii}$ is the change in the density of the liquid in the tubing at i length, μ is the Poisson' ratio for the material, and $\Delta \rho_{oi}$ is the change in the density of the liquid in the casing at i length. From the assumption ($\Delta P_{oi} = 0$, $\Delta \rho_i = 0$, $\Delta \rho_o = 0$, $\delta = 0$), the equation can be reduced as follows:

$$\Delta L_{bi} = -\frac{2\mu \Delta Z_i}{E} \frac{d^2 \Delta P_{ii}}{D^2 - d^2}. \tag{6.6}$$

The total axial deformation of the varied pressure fields can be determined by accumulating each element:

$$\Delta L_b = \sum_{i=1}^{N} \Delta L_{bi}. \tag{6.7}$$

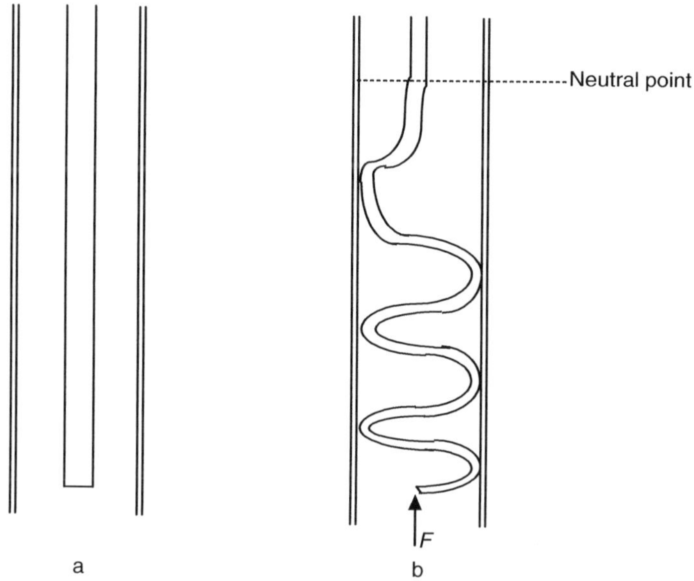

Figure 6.3 Buckling of tubular.

The force, which is affected by the temperature, can be calculated using the following formula according to Hooke's law:

$$F_b = \frac{\Delta L_b}{L} EA. \tag{6.8}$$

6.2.3.5 *The contact force and friction between the casing and the tubing*
Researchers generally call the buckling a bending effect. The tubular is freely suspended in the absence of fluid inside, as shown in Figure 6.3. A force F is applied at the lower end of the tubular and if the force is large enough, the tubular buckles as shown in Figure 6.3.

Lubinski and Althouse (1962) did significant research in this area, from which we are able to determine the buckling effect. First the virtual axial force of the tubing needs to be defined as follows:

$$F_f = A_p(P_1 - P_0), \tag{6.9}$$

where, P_1 is the pressure inside the tubular at the packer length, P_0 is the pressure outside the tubular at the packer length and A_p is the area corresponding to the packer bore.

Whether the tubular is buckling or not can be determined using Equation (6.13). The string buckles if F_f is positive, and remains straight if F_f is negative or zero. The axial deformation for tubular string buckling is:

$$\Delta L_3 = -\frac{r^2 A_p^2 (\Delta P_1 - \Delta P_0)^2}{8EIW}. \tag{6.10}$$

The contact force between the helically buckled tubing and casing within an axial unit length can be expressed as follows (Mitchell, 1997):

$$N = -\frac{r^2 F_f^2}{8EI}. \tag{6.11}$$

From the Coulomb's friction principle, the friction force is derived by multiplying the contact force by the friction factor:

$$F_f = \mu N. \tag{6.12}$$

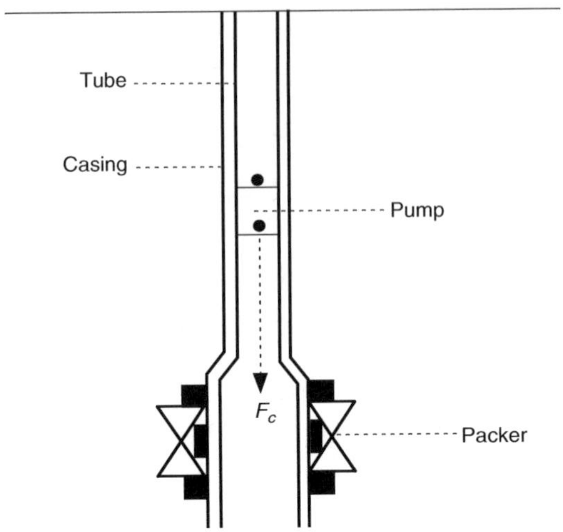

Figure 6.4 The physical figure of forces analysis on pumping effect.

6.2.3.6 *Temperature effect*

Steam injection, oil production and well shut-in after the packer is set cause a rise in the tubing temperature, which in turn causes changes in the tubular length. The temperature in the injection or shut-in processes, in particular, changes significantly compared with other operations. In this section, we have selected the temperature change on well shut-in as the temperature effect for the unset force.

For the tubular microelement dz, the temperature change unit deformation is as follows (Li, 2008; Dou and Zhang, 2007; Meriam and Kraige, 2005):

$$\Delta L_{ti} = \int_{Z_{i-1}}^{Z_i} \frac{\sigma_{zt}}{E} dz = \beta \Delta T_i \Delta L_i, \tag{6.13}$$

where, σ_{zt} represents the axial thermal stress, E is the steel elastic modulus of the tubular, β is the warm balloon coefficient of the tubular string, and ΔT is the temperature change before and after well shut-in. The principle is that the total axial deformation of the various temperature fields can be determined by accumulating each element:

$$\Delta L_t = \sum_{i=1}^{N} \Delta L_{ti}. \tag{6.14}$$

The force which is led by the temperature effect can be calculated using follow formula according to the Hooke's law:

$$F_t = \frac{\Delta L_t}{L} EA. \tag{6.15}$$

6.2.3.7 *Sucker-rod pumping effect*

As the sucker-rod pumps, the up stroke of the pump decreases the load on the tubular and the down stroke of pump increases the load, both of which force a change to the top tubular located at the packer. These expansion or contraction forces may lead to packer unset.expansion or contraction forces As shown in Figure 6.4, this pumping force can be determined using the following:

$$F_s = \frac{\pi}{4t} d^2 h \rho_i g v_s, \tag{6.16}$$

where, F_s represents the pumping force, h is the depth of the top tubular located at the packer, ρ_i is the density of the fluid in the tubing, g is the gravity acceleration, v_s is the velocity of the down stroke, and t is the timing of the down stroke. Because there are only small changes in the fluid density, ρ_i here is regarded as a constant.

6.2.3.8 *Piston effect to support the packer's pressure*
According to the work principium mentioned in 6.2, the structure of the hydrostatic packer uses the inner central tube connected with the down joint to support the deferential pressure from the top to the bottom. The hydrostatic pressure is transmitted to the casing through the slips. While supporting the deferential pressure from the bottom to the top, the packer pushes up the outer central tube and the hydrostatic pressure is transmitted to the casing through the hydraulic anchor, the value of which can be obtained using following equation:

$$F_c = \Delta P_c A_c, \tag{6.17}$$

where, F_c represents the piston force for the supporting packer pressure, ΔP_c is the differential pressure from the top to the bottom, and A_c is the effective area.

6.2.3.9 *Analysis of pressure and temperature fields*
In the course of modelling the above, the factors which affect the packer's unset force are functions of temperature and pressure. In fact, these two parameters vary according to the depth and the change in time. The variation in the pressure and temperature have a significant effect on the unset force so the varied (T, P) fields need to be researched. As we know, the variation maxims are different for steam injection and the well shut-in process. Here, we have chosen the varied (T, P) fields for well shut-in as the study objectives.

In previous studies for HTHP wells, the varied (T, P) fields were deduced based strictly on the mass, momentum and energy balances in the shut-in procedures. The model focused on the heat transmission in the stratum and then calculated the temperature through a differential equation based on the Cullender-Smith method:

$$\begin{cases} \dfrac{\partial \upsilon}{\partial z} = -\dfrac{1}{\rho}\left(\dfrac{\partial \rho}{\partial t} + \upsilon\dfrac{\partial \rho}{\partial z}\right) \\ \dfrac{\partial P_i}{\partial z} = -\rho\dfrac{\partial \upsilon}{\partial t} + \upsilon\dfrac{\partial \rho}{\partial t} + \upsilon^2\dfrac{\partial \rho}{\partial z} - \rho g\cos\theta - \dfrac{f\rho\upsilon^2}{2d} \\ \dfrac{\partial T}{\partial z} = -\dfrac{1}{C_p}\left[\dfrac{\partial}{\partial t}(\rho C_p T)A - C_J C_p\dfrac{\partial P}{\partial z} + \dfrac{\partial}{\partial z}(\partial \upsilon^3 A/2 + \rho g\upsilon Az\cos\theta) + a(T - T_e) + \dfrac{f\rho\upsilon^3 A}{2d}\right] \\ \dfrac{\partial T_e}{\partial t_D} = \dfrac{\lambda_e}{C_{p_e}\rho_e}\left(\dfrac{\partial^2 T_e}{\partial r_D^2} + \dfrac{1}{r_D}\dfrac{\partial T_e}{\partial r_D}\right) \\ \rho = \dfrac{MP_i}{RZT} \\ \left.\dfrac{\partial T_e}{\partial r_D}\right|_{r_D=1} = -\dfrac{dQ}{dz}(2\pi\lambda_e)^{-1}, \qquad \left.\dfrac{\partial T_e}{\partial r_D}\right|_{r_D\to\infty} = 0 \\ \dfrac{d(\rho_g\alpha_g\upsilon_g^2)}{dz} = -\dfrac{d(\alpha_g P_g)}{dz} - \rho_g\alpha_g g\cos\theta \\ P(z,0) = P_0(z),\ \ T(z,0) = T_0(z),\ \ \upsilon(z,0) = \upsilon_0(z),\ \ T_e(r_D,0) = T_e^0(r_D) \\ P(0,t) = \bar{P}_0(t),\ \ T(0,t) = \bar{T}_0(z),\ \ \upsilon(0,t) = \bar{\upsilon}_0(z),\ \ T_e(0,t) = T_e^0(t), \end{cases} \tag{6.18}$$

where P_i is the pressure in the tubing, T_i is the temperature in the tubing, υ_i is the velocity of the fluid in the tubing, ρ_i is the density of the fluid in the tubing, C_J is the Joule-Thompson (J-T) coefficient, C_P is the heat capacity of the fluids, T_e is the initial formation temperature, t_D is dimensionless time.

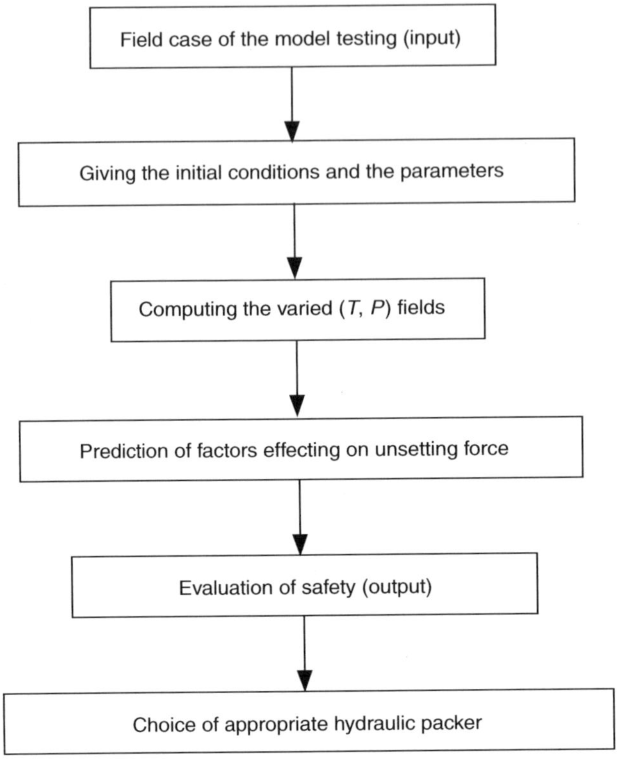

Figure 6.5 Architecture of the proposed methodology.

6.2.4 Solution methodology

To simplify the calculation, we divided the well into several short segments of the same length. The length of a segment varies depending on variations in wall thickness, hole diameter, the fluid density inside and outside the pipe and well geometry. The model begins with a calculation at the top of the pipe.

Setting the depth step length we denote the relative tolerance error as ε. The smaller h, and ε are, the more accurate the results. However, this leads to a rapid increase in calculation time. Here, we set $h = 1$ (m), and $\varepsilon = 5\%$.

With reference to Figure 6.5, the proposed methodology involves the following steps:

Step 1. Generate a sample of the model test. This sampling data can be either experimental or field measured.

Step 2. Conduct numerical simulations using the sample (input) from the previous step and obtain the relative parameters.

Step 3. Use the parameters obtained in the previous steps and compute the varied (T, P) fields for the shut-in process.

Step 4. Calculate the factors affecting the unset force using the given model with the varied (T, P) fields in step 3.

Step 5. Conduct numerical simulations using the values obtained in the previous step to confirm their performance level.

Step 6. The designer now selects the solution that satisfies the most preferred structure among the best confirmed design values.

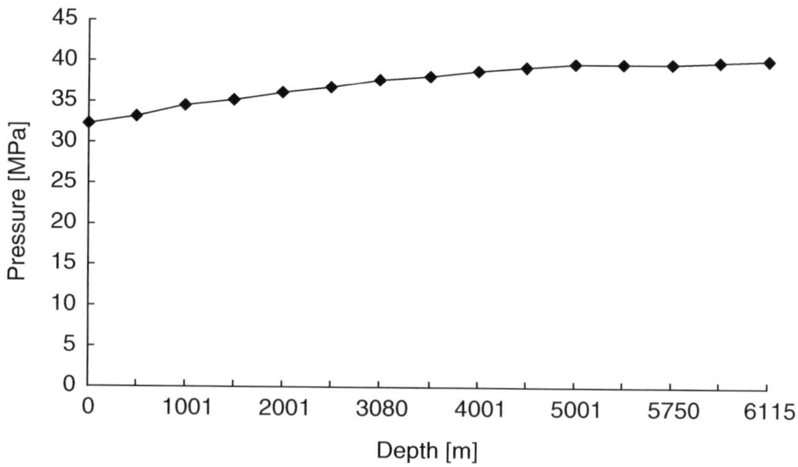

Figure 6.6 Pressure distribution.

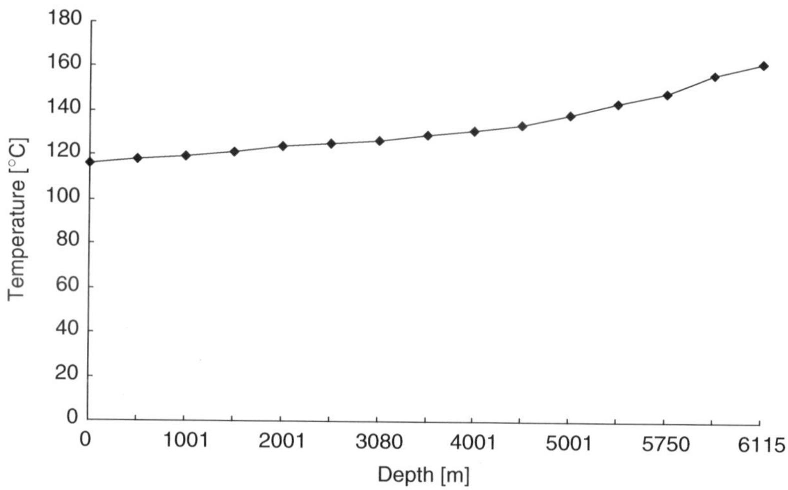

Figure 6.7 Temperature distribution.

6.2.5 Analysis of field case

6.2.5.1 *Parameters*

For investigation of the model validity and its performance, case field data was studied. Data for this case were taken from *X* well, in China. All the needed parameters are given as follows:

Depth of the well = 7100 m
Well bottom pressure = 40 MPa
Critical pressure = 4.968 MPa
Gas specific weight = 0.6434 kg/m^3
Ground thermal conductivity parameter = 2.06 W/m · K
Ground temperature = 16°C
Geothermal gradient = 2.18°C/100 m
Roughness of the inner surface of the well = 0.000015

Table 6.1 Parameters of pipes.

Diameter [mm]	Thickness [mm]	Weight [kg]	Expansion coefficient [m/m · K]	Young's modulus [GPa]	Poisson's ratio	Using length [m]
88.9	12.95	23.791	0.0000115	215	0.3	700
88.9	9.53	18.28	0.0000115	215	0.3	2850
88.9	7.34	15.034	0.0000115	215	0.3	1430
88.9	6.45	13.582	0.0000115	215	0.3	950
73	5.51	9.493	0.0000115	215	0.3	185

Table 6.2 Well parameters.

Measured depth [m]	Internal diameter [mm]	External diameter [mm]
3301.7	154.78	193.7
5936.83	152.5	177.8
6115	108.62	127

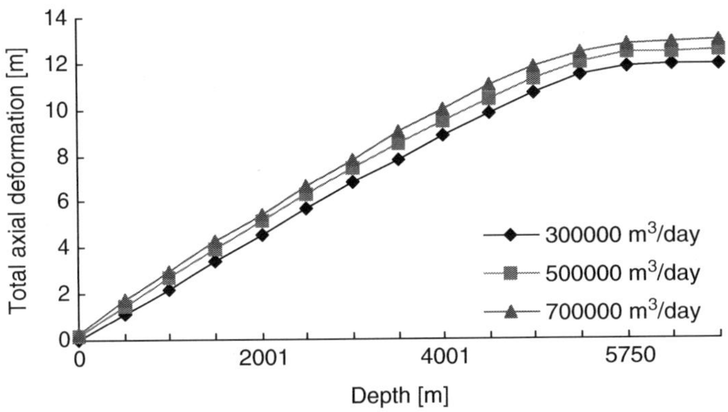

Figure 6.8 The total axial deformation under varied outputs.

The parameters of the pipes, the inclined well, the inclination, the azimuth and the vertical depth are given in Tables (6.1), (6.2) and (6.3). The varied (T, P) fields are shown in Figures (6.7) and (6.6) (Xu and Wu, 2012).

6.2.5.2 *Main results and results analysis*

After calculation, we obtained a series of results for this well as shown in Table 6.4. The influence of the outputs on the axial deformation of the tubing was investigated as shown in Figure 6.8. From the results as shown in Table 6.4, some useful analyses can be made:

1. The upward force on the packer design should be reduced as much as possible to improve the force-bearing conditions of the packer and the tubular, extend the effective life and increase total economic efficiency.
2. The unset force analysis of the hydrostatic packer should take into consideration the structural principles and all associated processes such as the effects of the setting and the bearing pressure. The main factors that affect the unset force of the hydraulic packer are the ballooning and

Table 6.3 Azimuth, inclination and vertical depth.

Number	Measured depth [m]	Inclination [°]	Azimuth [°]	Vertical depth [m]
1	1000	2.82	240.84	999.88
2	1200	2.28	237.69	1199.53
3	1300	1.13	213.69	1299.49
4	2800	1.19	26.21	2799.41
5	3000	1.74	44.39	2999.25
6	3400	1.92	190.95	3399.21
7	3900	1.98	268.9	3899.14
8	4000	2.00	297.38	3999.11
9	4100	4.68	324.34	4098.96
10	4200	1.97	302.88	4198.74
11	4300	1.03	204.57	4298.72
12	4400	1.54	164.16	4398.68
14	4500	2.37	195.11	4498.61
15	4600	2.12	214.67	4598.54
16	4700	1.96	216.31	4698.47
17	4800	3.04	229.14	4798.38
18	4900	3.59	243.86	4898.23
19	5000	5.79	366.45	4997.87
20	5100	8.14	258.61	5097.01
21	5200	7.01	236.71	5196.12
22	5300	5.78	239.1	5295.51
23	5400	5.05	244.42	5395.04
24	5500	3.92	228.03	5494.72
25	5600	4.44	233.71	5594.49
26	5700	5.03	234.87	5694.17
27	5800	5.13	233.21	5793.77
28	5900	4.53	234.82	5893.44
29	6000	3.67	232.4	5993.21
30	6115	4.94	233.11	6107.88

Table 6.4 The results of the axial force and various kinds of deformation length.

Number	Depth [m]	Displacement by temperature changed [m]	Displacement by pressure changed [m]	Axial deformation [m]	Buckling deformation [m]	Total deformation [m]
1	500	0.1139	0.0209	0.033	0	0.168
2	1000	0.9425	0.1709	0.334	0	1.447
3	1500	1.711	0.3209	0.702	−0.005	2.729
4	2000	2.4177	0.4512	1.114	−0.006	3.977
5	2500	3.0559	0.5512	1.565	−0.006	5.167
6	3000	3.634	0.635	2.058	−0.007	6.319
7	3500	4.152	0.685	2.593	−0.016	7.414
8	4000	4.6077	0.7208	3.168	−0.022	8.474
9	4500	4.9955	0.7208	3.782	−0.048	9.450
10	5000	5.3232	0.7075	4.435	−0.057	10.409
11	5500	5.5908	0.6575	5.075	−0.067	11.255
12	6000	5.7957	0.6075	5.704	−0.103	12.003
13	6115	5.8857	0.5775	6.029	−0.109	12.383

piston effect, the temperature effect, the sucker-rod pumping effect and the piston force effect supporting the packer's pressure. The simulation results show that the piston force at the packer setting is a fixed value because of the fixed value of the packer's differential pressure at startup.

3. The length of the tubular deformation increases with an increase in outputs, but more slowly.
4. Thermal stress is the main factor influencing the tubular deformation.
5. The packer can greatly improve the stress above the cement surface of the casing under high-pressure conditions. To reduce the force or the tubular deformations, a retractable compensation device should be added.

CHAPTER 7

Predicting pressure and temperature in HTHP injection wells

7.1 INTRODUCTION

Completion testing of deep wells is a new problem. A well depth of 4570 m is considered a deep well, and a well depth of 6100 m is considered a super-deep well. In deep well testing applied basic theoretical research, the tubular string mechanical analysis is very complex as the fluid temperature and tubing pressure heavily affect the force of the tubular string. Therefore, it is necessary to ensure an accurate data calculation for the temperature and inflation effects and for the pressure on the tubular string.

When conducting well testing, the temperature and pressure of the tubular string need to be considered. (1) Thermometer and pressure gauges are located in the tubing; (2) By applying theoretical analysis technology, it is possible to predict the overall distribution by measuring the temperature and the pressure at both the well bottom and the well head.

However, for high temperature high pressure gas wells, it is often difficult to operate the thermometer and pressure gauges. Thus, an effective and feasible method is required to give a prediction of the overall distribution and this can be done by applying theoretical analysis technology.

The earliest work in this area was based on analytical solutions with a line source concept-that is, the oil well was considered to be a cylinder of infinite length in an infinite medium. Moss and White (1959) derived an expression for the calculation of the temperature of water during hot water injection as a function of time, *t*. They assumed the following:

1. The physical properties of the fluid and the formation were independent of the depth and temperature.
2. The heat transfer factors for the completion test were ignored.
3. The frictional losses and the kinetic energy effects were negligible.
4. The heat transfer in the well-bore was considered to be in a steady state.

Assumption 4 is inherent in all the subsequent models developed. If we compare the heat absorbed by the well-bore with the heat transferred through the well-bore, it is possible that, except for very short times, this assumption is valid. Holst and Flock (1966) discuss this point in more detail.

Ramey, (1962) extended Moss and White's algorithm to incorporate an overall heat transfer coefficient. The overall heat transfer calculations were based on the work of McAdams (1954). He assumed (1) that heat flows radially away from the well-bore, (2) there is steady-state loss in the well-bore, and (3) there are constant thermal and physical properties. Ramey's analysis assumed that a perfect gas was used and the fluid did not undergo any phase changes. Satter (1965) extended Ramey's model to steam injection cases, where he calculated the quality of a condensing fluid as a function while the pressure and temperature of the injected fluid was constant.

All the models discussed so far have considered the physical system as a whole. Holst and Flock (1966), however, divided the physical system into the fluid, the well-bore, and the formation. These parts were then linked by the heat flow. Subsequent models developed by Pacheco and Farouq Ali (1972) were designed in a similar manner. Holst and Flock improved Ramey's and Satter's model by extending the analysis to include steam pressure calculations and included frictional and kinetic energy effects. Pacheco and Farouq Ali's model improved the accuracy of the heat loss calculations by the inclusion of simultaneous calculations for steam pressure and quality. However, the model did not account for either the slip concept or the flow regimes that might occur during a two-phase flow in a vertical pipeline. The overall heat transfer coefficient calculation was based on Willhite's method (Willhite, 1967; Willhite and Dietrich, 1967). Faroug-Ali (1981) proposed a comprehensive model by combining Pacheco and Farouq Ali's (1972) model with standard two-phase-flow correlations to account for the slip concept and flow regimes. Furthermore, in this model, the heat loss to the surrounding areas was treated rigorously. The assumption that the heat transfer in the well-bore is in a steady state, however, is still inherent in the model. The resulting mathematical model consists of a set of algebraic ordinary and partial differential equations. Farouq solved the system of equations using a numerical scheme. Apart from these studies, Farouq-Ali (1981) and (Ah, 1981) show the effects of the temperature on the casing. Durrant and Thambynayagam (1986) presented a straightforward iterative procedure for the well-bore heat transmission problem during an upward or downward flow of a steam/water mixture that includes vertical heat conduction. Hagoort (2004) assessed Ramey's classic method for the calculation of temperatures in injection and production wells, and demonstrated that Ramey's method was an excellent approximation, except for an early transient period in which the calculated temperatures were significantly overestimated. They presented a simple graphical correlation to estimate the length of this early transient period. In recent years, the calculation of heat loss and pressure drop has been focused on in actual applications, such as those in Akin (2006), Akin and Bagci (2001), Cazarez-Candia and Vásquez-Cruz (2005), Wang and Dusseault (2003), Wu *et al.* (2011) and Xu *et al.* (2011).

The early methods which computed the pressure drop in the steam or water injection, such as the Beggs-Brill (1987) method and the Orkiszewski (1967) method all derived the pressure drop equation by considering the mechanical energy conservation of the infinitesimal and omitting the change in the internal energy and the heat exchange within the external environment. The usual method was to assume that the change in the internal steam energy was equal to the external and so the pressure and temperature were computed in the well-bore which resulted in a major error. In later research, although many factors which impacted the temperature and pressure were considered when computing the dryness, the steam friction loss was usually ignored. Pacheco and Farouq Ali (1972) improved the assumption that ignored the loss of steam pressure in the model of Satter (1965), and computed the pressure by only considering a single medium in the well-bore. Farouq Ali (1981) further improved this model and computed the pressure drop by considering the impact of the two-phase flow pattern. Further, the heat loss to the surroundings was also treated rigorously, but the assumption that the heat transfer in the well-bore is in a steady state is still inherent in this model. The resulting mathematical model consists of a set of algebraic, ordinary and partial differential equations.

In fact, in the injection process, not only does the internal energy of the steam change over time, but a heat transfer between the steam and the surroundings also regularly occurs, which means that the physical parameters themselves are a function of pressure, temperature and dryness. Wu *et al.* (2011) presented a coupled differential equations system model covering the dryness fraction of the gas, the pressure and the temperature in high temperature-high pressure steam injection wells according to mass, momentum and energy balances. However, the construction of the energy balance equation in this model ignores the loss of energy caused by the friction and assumes that an temperature of the stratum does not change and is only impacted by depth and the geothermal gradient. Some similarly handled processes can be found in associated research such as Deberne *et al.* (1999), Lin *et al.* (2008), Tiyao *et al.* (2010), Wolcott *et al.* (1995), Babadagli *et al.* (2009), Liu *et al.* (2009; 2013), and Xu *et al.* (2012b; 2013a; 2013b).

The borehole temperature and pressure distribution is a necessary parameter for the design and performance analysis of gas wells, and is obtained using a direct measurement. Rzasa [115] presented an average temperature and compressibility method, which resulted in large calculation errors. Cullender and Smith (1956) established a solution method based on numerical integration which considered pressure as a variable with variations in well depths. Later Ramey (1962) presented an approximate method for predicting the temperature of either a single-phase incompressible liquid or a single-phase ideal gas flow in injection and production wells. Shiu and Beggs (1980) simplified Ramey's method by correlating for a specific coefficient in Ramey's equation. Since the late 1980s, the mechanistic approach has been used and begins with local instantaneous conservation equations and is then systematically developed for the variables of interest. Grolman and Fortuin (1997), Ouyang and Aziz (2000), and Hurlburt and Hanratty (2002) all provided several mechanistic models, assuming the flow is under steady-state conditions. However, these models predict only the pressure profiles. Hagoort (2002) presented a simple and transparent analytical solution for the prediction of well-bore temperatures in gas production wells. Most research in this area calculates the temperature and pressure distribution predictions separately, and their interdependence is ignored. In fact, it is well known that the temperature and pressure are interdependent in injection and production wells, and particularly for HTHP wells. Liao and Feng (2005) integrated the coupling influence between the pressure and the temperature in the well bore and presented a coupling calculation model in which heat-transmission is considered but this model did not consider the pressure and temperature variance at different times in the injection and production process. Wu *et al.* (2011) presented a coupled difference equations system model concerning pressure and temperature in HTHP wells according to mass, momentum and energy balances and presented an algorithm, along with the fourth order Runge-Kutta method but this model also failed to consider the variations in the pressure and temperature at different times in the injection and production process.

The gas flow through oil wells and pipes is a complex phenomenon. In general, flow states depend on well diameter, the thermo-physical properties of the fluids, the inclination angle, and the flow rate of each phase. Both the gas temperature and pressure in the reservoir are of interest, yet the determination of well and earth temperature is a complex task. There are two parts to the well-bore thermal motion: the flowing stream and the formation. Many empirical or semi-empirical predictive techniques have been developed but these may not be suitable for predicting pressure during transient periods for predicting the pressure.

Ramazanov and Nagimov, 2007 earlier researched the dependence between changes in the time and space reservoir for temperature and pressure during a steady gas inflow. Lauwerier (1955) presented a widely known mathematical model, which considered a constant injection rate with a constant fluid temperature into a semi-infinite non-permeable formation. These models assumed uniform temperatures in the reservoir at any depth, but the direction of the gas flow in the reservoir and the formation thermal conductivities were neglected. Based on Lauwerier's assumption, Marx and Langenheim (1959) developed a radial flow system, which assumed the total heated reservoir area was at a constant temperature. Ramey extended their work to examine the variable heat injection rate, and observed that the solution was independent of flow geometry.

Fayers (1962) demonstrated that heat and mass transfer were only weakly coupled in the one-dimensional non-isothermal two-phase displacement in porous media. Thus a calculation of temperatures in a one-dimensional single-phase non-isothermal liquid flow in porous media has value when studying hot water injection. Lesem *et al.* (1957) and Moss and White (1959) were the first to determine a method for the calculation of the temperature distribution in flowing gas wells, demonstrating that linking a flowing temperature calculation to soil temperatures is an important part of the formulation of a well-bore thermal simulator. Kirkpatrick (1959) earlier presented a model on predicting temperature profiles in a flowing gas well. He presented a simple flowing temperature gradient chart that could be used to predict the gas-lift valve temperatures at injection depths. Much of the classic work in this area was developed by Ramey (1962), who presented approximate methods for predicting the temperature of either a single-phase incompressible liquid or a single-phase ideal gas flow in injection and gas production wells. Ramey's solution consisted of an apt approximation of heat losses to the formation through the consideration of a steady-state heat flow through the gas well-bore heat resistances and a transient heat flow using conduction in the formation.

Satter (1965) later improved Ramey's method by discussing the phase changes that occur within steam-injection projects. Willhite (1967) presented a detailed analysis of the overall heat-transfer mechanism in an injection gas well. Sagar *et al.* (1991) presented a simple model suitable for calculation by hand to predict temperature profiles in two-phase flowing wells. Alves *et al.*

(1992) extended the Ramey solution to real gases and presented a more general model to predict the temperature of a gas well-bore. The Alves *et al.*'s solution included considering the pressure gradient in the well-bore as a constant parameter, obtained using a separate calculation. Recent articles have begun to investigate the heat transfer that exists between the flowing gas and the earth. Heat must pass through several materials to travel from the fluid gas to the formation.

Wu *et al.* (2011) presented a coupled differential equations system model concerning pressure and temperature in high-temperature-high-pressure gas wells according to mass, momentum and energy balances and presented an algorithmic solution model, along with the four order Runge-Kutta method, but this model failed to consider the thermal effect. Hagoort (2004) assessed Ramey's classic method for the calculation of temperatures in gas injection and production wells. He showed that Ramey's method gave an excellent approximation, except for an early transient period in which the calculated temperatures are significantly overestimated. For further research on the prediction of temperature and pressure in gas wells, refer to Chaudhry (2004), Hagoort (2007), Wolcott *et al.* (1995) and Wang and Dusseault (2003).

Most thermal recovery methods have been applied to high viscosity oil reservoirs with the objective of increasing oil production by reducing oil viscosity. Heat can be injected into the reservoir as hot water or steam, or can be generated in-situ by burning part of the reservoir crude oil. Of all these processes, steam injection is the most reliable (and has enjoyed by far the most commercial success). In the steam injection process, the pressure and temperature of the steam should be high enough so that the amount of fuel carried by a unit mass of steam is greater, which is beneficial to further energy consumption reductions in injection processes. However, in an actual engineering situation, and under reservoir regulations, an extravagant injection pressure could cause a rupture of the reservoir. The purpose of steam injection is to enhance oil recovery (EOR). Significant energy is consumed in the steam production and injection. So it is necessary to be aware of the increasing gas oil ratio during thermal recovery. Therefore, when using steam stimulation to recover heavy oil, the calculation of the on-the-way fluid parameters in the well-bores can not only analyse the steam utilization efficiency and save high field temperature testing costs, but can also define the bottom-hole steam state and provide important information for the analysis of the steam injection recovery mechanisms and production performance evaluations.

There have been various research papers on the modelling of both injection and production wells. One of the first papers goes back to Woods *et al.* (1998), which has been referred to by many subsequent works modelling well-bore heat loss and pressure drop. Ramey extended Moss and White's algorithm to incorporate an overall heat transfer coefficient. The overall heat transfer calculations were based on the work of McAdams (1954). In that paper, the author simplified the heat balance equation to solve it analytically. Satter (1965) improved Ramey's analytical model by considering a depth-dependent overall heat transfer coefficient and temperature-dependent fluid properties. Holst and Flock (1966) divided the physical system into the fluid, the well-bore, and the formation. These parts were then linked together by the heat flow. The subsequent models of Pacheco and Farouq Ali (1972) and Farouq-Ali (1981) were designed in a similar manner. Flock and Holst improved Ramey's and Satter's model by extending the analysis to include steam pressure calculations-that is, they included the frictional and kinetic energy effects. Pacheco and Farouq Ali's model improved the accuracy of the heat loss calculations by the inclusion of simultaneous calculations for steam pressure and quality, but the model did not account for either the slip concept or the flow regimes that might occur during a two-phase flow in a vertical pipeline. The overall heat transfer coefficient calculation was based on Willhite's method. Farouq-Ali (1981) proposed a comprehensive model by combining Pacheco and Farouq Ali's previous model with standard two-phase-flow correlations to account for the slip concept and flow regimes. Furthermore, in Farouq Ali's model, the heat loss to the surroundings was treated rigorously. Note however, the assumption that the heat transfer in the well-bore is in a steady state is still inherent in the model. The resulting mathematical model consisted of a set of algebraic and ordinary and partial differential equations. Farouq Ali solved the system of equations with a numerical scheme. Hasan and Kabir (1991) developed an analytical model to determine the flowing fluid temperature inside the well. They started with a steady-state energy balance equation and combined it with the definition for fluid enthalpy in terms of heat capacity and the Joule-Thompson coefficient.

Using some simplifications, they then converted the original partial differential equation to an ordinary differential equation and solved it with appropriate boundary conditions. Livescu *et al.* (2010) developed a comprehensive numerical non-isothermal multiphase well-bore model.

7.2 PDPT-IW[1]

Considering the differential equation model for P, T, we set the following assumptions:

1. There is a one dimensional steady gas flow in the tubing, and all featured parameters are homochromatic at any transverse cross section.
2. The one dimensional heat transfer is steady from the tubing to the second interface, but the heat transfer in one dimension is unsteady from the second interface to the surrounding formation.
3. The heat losses from the tubing and the surrounding earth are radial.
4. The vertical formation temperature is linearly distributed based on known geothermal gradients.

7.2.1 **Physical model**

The physical model that underlies the equations describing the well-bore heat transmission consists of a straight, cased well that is cemented to the formation and equipped with tubing for the transfer to the surface. Both the casing and tubing have a constant diameter. The diameter of the tubing is small with respect to its length. Initially, the tubing is filled with a fluid in thermal equilibrium with the formation. At zero, fluid starts flowing from the bottom of the tubing to the top at a constant flow rate. Heat conduction in the flow direction and frictional heating in the tubing are negligible. The fluid that is initially present in the tubing is the same as the fluid that enters from the bottom of the tubing. The temperature of the fluid is equal to the formation temperature at the bottom. The flow in the tubing is 1D (i.e., temperature and fluid velocity depend only on the distance along the tubing). As the fluid moves up the tubing, it loses heat to the colder formation. Heat losses to the formation take place through heat conduction in a radial direction only. The effect of the tubing wall, the annular space between the casing and the tubing, the casing wall and the cement zone, on the heat transmission is included using a single, steady-state heat-transfer coefficient. The initial temperature of the formation increases linearly with depth, reflecting a constant geothermal gradient.

7.2.2 **Mathematical model**

The tubing is surrounded by a homogenous rock formation that extends to infinity. The distant temperatures in the formation increase linearly with depth reflecting geothermal temperature. The gas enters the flow tubing at a given pressure and temperature. Within the tubing, the gas flow takes place under turbulent flow conditions. Consider the flow system depicted in Figure 7.1.

A straight cylindrical flow tube with an inclination angle θ, a constant cross-sectional flow area A, a hydraulic diameter, and a total length Z. Through this tubing gas flows from the bottom to the top with a mass flow rate W. The distance co-ordinate in the flow direction along the tubing is denoted z.

Mass, momentum, and energy balances, along with the dryness fraction of gas, pressure and the temperature in relation to the gas, are used to generate the constitutive equations.

1. Pressure gradient model

   According to momentum balances, the pressure gradient is a combination of the gas lift gradient, friction gradient, acceleration gradient:

$$\frac{dP}{dz} = \left(\frac{dP}{dz}\right)_{lift} + \left(\frac{dP}{dz}\right)_{fri} + \left(\frac{dP}{dz}\right)_{acc},$$

[1]Predicting dryness fraction of gas, pressure and temperature in HTHP injection wells

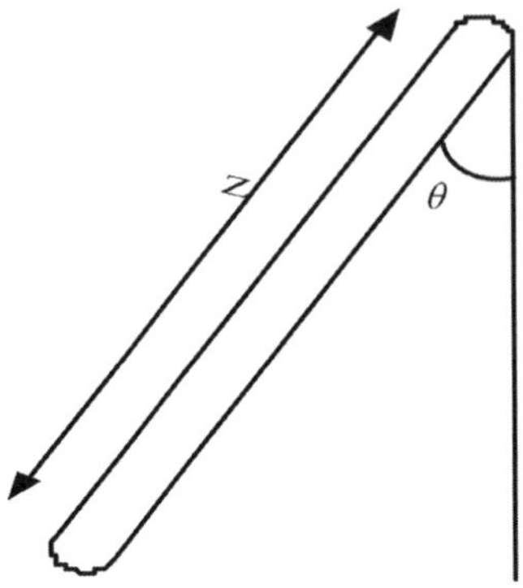

Figure 7.1 The physical figure.

thus,

$$\frac{dP}{dz} = -\rho_m g \cos\theta - \frac{f_m \rho_m v_m^2}{2d} - \rho_m v_m \frac{dv_m}{dz}. \tag{7.1}$$

2. Dryness fraction of gas model

From the energy balances equation, we have:

$$\frac{dq_m}{dz} = -\frac{dh_m}{dz} - v_m \frac{dv_m}{dz} - g\cos\theta, \tag{7.2}$$

where,

$$h_m = h_g x + h_l(1-x), \quad v_m = v_g + v_l = \frac{Wx}{\rho_g A} + \frac{W(1-x)}{\rho_l A}.$$

Since $h_m = h_g x + h_l(1-x)$, we have:

$$\frac{dh_m}{dz} = h_g \frac{dx}{dz} + x\frac{dh_g}{dz} - h_l \frac{dx}{dz} + (1-x)\frac{dh_l}{dz},$$

i.e.,

$$\frac{dh_m}{dz} = (h_g - h_l)\frac{dx}{dz} + x\frac{dh_g}{dz} + (1-x)\frac{dh_l}{dz}. \tag{7.3}$$

From $v_m = v_g + v_l = \frac{Wx}{\rho_g A} + \frac{W(1-x)}{\rho_l A}$, we can get:

$$\frac{dv_m}{dz} = \frac{W}{A}\left(\frac{1}{\rho_g}\frac{dx}{dz} - \frac{x}{\rho_g^2}\frac{d\rho_g}{dz} - \frac{1}{\rho_l}\frac{dx}{dz} - \frac{1-x}{\rho_l^2}\frac{d\rho_l}{dz}\right),$$

thus,

$$\frac{dv_m}{dz} = \frac{W}{A}\left[\left(\frac{1}{\rho_g} - \frac{1}{\rho_l}\right)\frac{dx}{dz} - \left(\frac{x}{\rho_g^2}\frac{d\rho_g}{dz} + \frac{1-x}{\rho_l^2}\frac{d\rho_l}{dz}\right)\right].$$

By means of the relationship before ρ_g, ρ_l and pressure P, we have:

$$\frac{dv_m}{dz} = \frac{W}{A}\left[\left(\frac{1}{\rho_g} - \frac{1}{\rho_l}\right)\frac{dx}{dz} - \left(\frac{x}{\rho_g^2}\frac{d\rho_g}{dP}\frac{dP}{dz} + \frac{1-x}{\rho_l^2}\frac{d\rho_l}{dP}\frac{dP}{dz}\right)\right],$$

thus,

$$\frac{dv_m}{dz} = \frac{W}{A}\left[\left(\frac{1}{\rho_g} - \frac{1}{\rho_l}\right)\frac{dx}{dz} - \left(\frac{x}{\rho_g^2}\frac{d\rho_g}{dP} + \frac{1-x}{\rho_l^2}\frac{d\rho_l}{dP}\right)\frac{dP}{dz}\right].$$

Letting:

$$B = \frac{W}{A}\left(\frac{1}{\rho_g} - \frac{1}{\rho_l}\right), \quad C = \frac{W}{A}\left(\frac{x}{\rho_g^2}\frac{d\rho_g}{dP} + \frac{1-x}{\rho_l^2}\frac{d\rho_l}{dP}\right),$$

thus,

$$\frac{dv_m}{dz} = B\frac{dx}{dz} - C\frac{dP}{dz}. \tag{7.4}$$

Therefore 7.1 can be written as:

$$\frac{dP}{dz} = -\rho_m g\cos\theta - \frac{f_m\rho_m v_m^2}{2d} - \frac{W}{A}\left(B\frac{dx}{dz} - C\frac{dP}{dz}\right) \tag{7.5}$$

i.e.,

$$\left(1 - \frac{CW}{A}\right)\frac{dP}{dz} = -\rho_m g\cos\theta - \frac{f_m\rho_m v_m^2}{2d} - \frac{BW}{A}\frac{dx}{dz},$$

therefore, we can get the coupled differential equation of pressure P and dryness fraction of gas x:

$$\frac{dP}{dz} = -\frac{\rho_m g\cos\theta + \frac{f_m\rho_m v_m^2}{2d} + \frac{BW}{A}\frac{dx}{dz}}{1 - \frac{CW}{A}}. \tag{7.6}$$

Similarly, (7.2) can be written as:

$$\frac{dq_m}{dz} = -(h_g - h_l)\frac{dx}{dz} - x\frac{dh_g}{dz} - (1-x)\frac{dh_l}{dz} - v_m\left(B\frac{dx}{dz} - C\frac{dP}{dz}\right) - g\cos\theta,$$

i.e.,

$$\frac{dq_m}{dz} = -(h_g - h_l + v_m B)\frac{dx}{dz} - x\frac{dh_g}{dz} - (1-x)\frac{dh_l}{dz} + v_m C\frac{dP}{dz} - g\cos\theta,$$

thus,

$$\frac{dx}{dz} = -\frac{\frac{dq_m}{dz} + x\frac{dh_g}{dz} + (1-x)\frac{dh_l}{dz} - v_m C\frac{dP}{dz} + g\cos\theta}{h_g - h_l + v_m B}. \tag{7.7}$$

Because of:

$$\frac{dh_g}{dz} = \left(\frac{\partial h_g}{\partial P}\right)_T\frac{dP}{dz} + \left(\frac{\partial h_g}{\partial T}\right)_P\frac{dT}{dz},$$

where,

$$\left(\frac{\partial h_g}{\partial T}\right)_P = C_{P_g}, \quad \left(\frac{\partial h_g}{\partial P}\right)_T = \frac{\partial h_g}{\partial T}\frac{\partial T}{\partial P} = -C_{J_g}C_{P_g}.$$

Thus,

$$\frac{dh_g}{dz} = C_{P_g}\frac{dT}{dz} - C_{J_g}C_{P_g}\frac{dP}{dz}.$$

Similarly,

$$\frac{dh_l}{dz} = C_{P_l}\frac{dT}{dz} - C_{J_l}C_{P_l}\frac{dP}{dz}.$$

Therefore, (7.7) can be changed into:

$$\frac{dx}{dz} = -\frac{\frac{dq_m}{dz} + x\left[C_{P_g}\frac{dT}{dz} - C_{J_g}C_{P_g}\frac{dP}{dz}\right] + (1-x)\left[C_{P_l}\frac{dT}{dz} - C_{J_l}C_{P_l}\frac{dP}{dz}\right] - v_m C\frac{dP}{dz} + g\cos\theta}{h_g - h_l + v_m B},$$

i.e.,

$$\frac{dx}{dz} = -\frac{\frac{dq_m}{dz} + [xC_{P_g} + (1-x)C_{P_l}]\frac{dT}{dz} - [xC_{J_g}C_{P_g} + (1-x)C_{J_l}C_{P_l} + v_m C]\frac{dP}{dz} + g\cos\theta}{h_g - h_l + v_m B}. \tag{7.8}$$

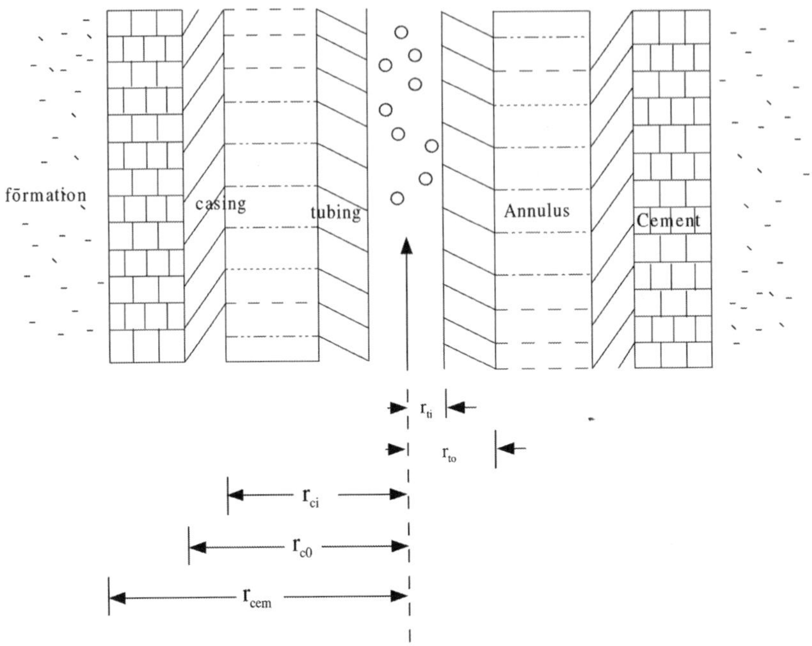

Figure 7.2 The radial transfer of heat.

3. The $\frac{dq_m}{dz}$ model

Ramey and Willhite discussed the radial transfer of heat between the fluid and the earth in detail. Over the differential element dz as shown in Figure 7.2, the radial transfer of heat from the fluid to the cement/earth interface can be described by:

$$dq_m = \frac{2\pi r_{to} U_{to}}{W}(T - T_k)dz \tag{7.9}$$

the radial transfer of heat from the cement/earth interface to the surrounding earth is:

$$dq_m = \frac{2\pi K_e(T_k - T_e)}{Wf(t_D)}dz. \tag{7.10}$$

Combining Equation (7.9) and (7.10) gives the equation for the radial transfer of heat between the fluid and the surrounding earth:

$$\frac{dq_m}{dz} = \frac{2\pi r_{to} U_{to} K_e}{W[k_e + f(t_D) r_{to} U_{to}]}(T - T_e). \tag{7.11}$$

Let:

$$a = \frac{2\pi r_{to} U_{to} K_e}{W[k_e + f(t_D) r_{to} U_{to}]},$$

then:

$$\frac{dq_m}{dz} = a(T - T_e). \tag{7.12}$$

From the relationship between the temperature and pressure of the saturated steam, we have:

$$\frac{dT}{dz} = 44.15P^{-0.79}\frac{dP}{dz}. \tag{7.13}$$

Combining (7.6), (7.7), (7.8), (7.12) and (7.13), we have:

$$\begin{cases} \dfrac{dx}{dz} = -\dfrac{\frac{dq_m}{dz} + [xC_{P_g} + (1-x)C_{P_l}]\frac{dT}{dz} - [xC_{J_g}C_{P_g} + (1-x)C_{J_l}C_{P_l} + v_mC]\frac{dP}{dz} + g\cos\theta}{h_g - h_l + v_mB} \\ \dfrac{dP}{dz} = -\dfrac{\rho_m g\cos\theta + \frac{f_m\rho_m v_m^2}{2d} + \frac{BW}{A}\frac{dx}{dz}}{1 - \frac{CW}{A}} \\ \dfrac{dq_m}{dz} = a(T - T_e) \\ \dfrac{dT}{dz} = 44.15P^{-0.79}\dfrac{dP}{dz} \end{cases} . \quad (7.14)$$

The primal conditions:

$$P(z_0) = P_0, \quad T(z_0) = T_0, \quad x(z_0) = x_0.$$

Therefore,

$$\begin{cases} \dfrac{dx}{dz} = -\dfrac{a(T-T_e) + [xC_{P_g} + (1-x)C_{P_l}]\frac{dT}{dz} - [xC_{J_g}C_{P_g} + (1-x)C_{J_l}C_{P_l} + v_mC]\frac{dP}{dz} + g\cos\theta}{h_g - h_l + v_mB} \\ \dfrac{dP}{dz} = -\dfrac{\rho_m g\cos\theta + \frac{f_m\rho_m v_m^2}{2d} + \frac{BW}{A}\frac{dx}{dz}}{1 - \frac{CW}{A}} \\ \dfrac{dT}{dz} = 44.15P^{-0.79}\dfrac{dP}{dz} \end{cases} . \quad (7.15)$$

The primal conditions:

$$P(z_0) = P_0, \quad T(z_0) = T_0, \quad x(z_0) = x_0.$$

Equation (7.15) can be rewritten as follows,

$$\begin{cases} \dfrac{dx}{dz} = -\left(1 - \dfrac{CW}{A}\right)\dfrac{a(T - T_e)}{h_g - h_l + v_mB} - \left(1 - \dfrac{CW}{A}\right)\dfrac{g\cos\theta}{h_g - h_l + v_mB} \\ \qquad + \dfrac{[44.15xC_{P_g}P^{-0.79} + 44.15(1-x)C_{P_l}P^{-0.79} - xC_{J_g}C_{P_g} - (1-x)C_{J_l}C_{P_l} - v_mC] \times \left(\rho_m g\cos\theta + \frac{f_m\rho_m v_m^2}{2d}\right)}{\left(1 - \frac{cw}{A}\right)(h_g - h_l + v_mB) - [44.15xC_{P_g}P^{-0.79} + 44.15(1-x)C_{P_l}P^{-0.79} - xC_{J_g}C_{P_g} - (1-x)C_{J_l}C_{P_l} - v_mC] \times \frac{Bw}{A}} \\ \dfrac{dP}{dz} = -\dfrac{\rho_m g\cos\theta + \frac{f_m\rho_m v_m^2}{2d} + \frac{BW}{A}\frac{dx}{dz}}{1 - \frac{CW}{A}} \\ \dfrac{dT}{dz} = 44.15P^{-0.79}\dfrac{dP}{dz} \end{cases} . \quad (7.16)$$

The primal conditions:

$$P(z_0) = P_0, \quad T(z_0) = T_0, \quad x(z_0) = x_0.$$

7.2.3 Solution to the model

Let:

$$\frac{dx}{dz} = f_1(z; x, P, T), \quad \frac{dP}{dz} = f_2(z; x, P, T), \quad \frac{dT}{dz} = f_3(z; x, P, T).$$

Denote:

$$F=(f_1,f_2,f_3)^T, \quad y=(x,P,T)^T, \quad y(0)=y(z_0)=(x(z_0),P(z_0),T(z_0))^T,$$

then, the systems of ordinary differential equations can be written as:

$$y'=F(z;y), \quad y(0)=y(z_0).$$

The norm of vector f is given as follows:

$$\|F\|=\max(|f_1|,|f_2|,|f_3|).$$

For f_1,f_2,f_3, there is:

$$|f_1|\le\left|1-\frac{CW}{A}\right|\left|\frac{a(T-T_e)}{h_g-h_l+v_mB}\right|+\left|1-\frac{CW}{A}\right|\left|\frac{g\cos\theta}{h_g-h_l+v_mB}\right|$$
$$+\left|\frac{[44.15xC_{P_g}P^{-0.79}+44.15(1-x)C_{P_l}P^{-0.79}-xC_{J_g}C_{P_g}-(1-x)C_{J_l}C_{P_l}-v_mC]\times\left(\rho_m g\cos\theta+\frac{f_m\rho_m v_m^2}{2d}\right)}{\left(1-\frac{cw}{A}\right)(h_g-h_l+v_mB)-[44.15xC_{P_g}P^{-0.79}+44.15(1-x)C_{P_l}P^{-0.79}-xC_{J_g}C_{P_g}-(1-x)C_{J_l}C_{P_l}-v_mC]\times\frac{Bw}{A}}\right|$$

i.e.,

$$|f_1|\le\left[1+\left|\frac{cw}{A}\right|\right]\left[\frac{|a(T-T_e)|}{|h_g-h_l+v_mB|}+\frac{|g\cos\theta|}{|h_g-h_l+v_mB|}\right]$$
$$+\frac{\left|[44.15xC_{P_g}P^{-0.79}+44.15(1-x)C_{P_l}P^{-0.79}-xC_{J_g}C_{P_g}-(1-x)C_{J_l}C_{P_l}-v_mC]\times\left(\rho_m g\cos\theta+\frac{f_m\rho_m v_m^2}{2d}\right)\right|}{\left|\left(1-\frac{cw}{A}\right)(h_g-h_l+v_mB)-[44.15xC_{P_g}P^{-0.79}+44.15(1-x)C_{P_l}P^{-0.79}-xC_{J_g}C_{P_g}-(1-x)C_{J_l}C_{P_l}-v_mC]\times\frac{Bw}{A}\right|}$$
$$\le\left[1+\left|\frac{cw}{A}\right|\right]\left[\frac{|a(T-T_e)|}{|h_g-h_l+v_mB|}+\frac{g}{|h_g-h_l+v_mB|}\right]$$
$$+\frac{|44.15xC_{P_g}P^{-0.79}+44.15(1-x)C_{P_l}P^{-0.79}-xC_{J_g}C_{P_g}-(1-x)C_{J_l}C_{P_l}-v_mC|\times\left|\rho_m g\cos\theta+\frac{f_m\rho_m v_m^2}{2d}\right|}{\left|\left(1-\frac{cw}{A}\right)(h_g-h_l+v_mB)-[44.15xC_{P_g}P^{-0.79}+44.15(1-x)C_{P_l}P^{-0.79}-xC_{J_g}C_{P_g}-(1-x)C_{J_l}C_{P_l}-v_mC]\times\frac{Bw}{A}\right|}$$

Let:

$$K_1=|h_g-h_l+v_mB|$$
$$K_2=\left|\left(1-\frac{cw}{A}\right)(h_g-h_l+v_mB)-[44.15P^{-0.79}(xC_{P_g}+(1-x)C_{P_l})-xC_{J_g}C_{P_g}-(1-x)C_{J_l}C_{P_l}-v_mC]\times\frac{Bw}{A}\right|.$$
$$K_3=|a(T-T_e)|+g$$
$$K_4=44.15P^{-0.79}(|xC_{P_g}|+|(1-x)C_{P_l}|)+|xC_{J_g}C_{P_g}|+|(1-x)C_{J_l}C_{P_l}|+|v_mC|$$
$$K_5=|\rho_m g|+\frac{|f_m\rho_m v_m^2|}{2d}$$

Thus, we have:

$$|f_1| \leq \frac{K_3}{K_1} + \frac{K_4 \times K_5}{K_2}.$$

Since all parameters are bounded quantities, therefore K_1, K_2, K_3, K_4, K_5 are bounded. Let:

$$N_1 = \sup\left\{\frac{K_3}{K_1} + \frac{K_4 \times K_5}{K_2}\right\},$$

then,

$$|f_1| \leq N_1.$$

Similarly,

$$|f_2| = \left| -\frac{\rho_m g \cos\theta + \frac{f_m \rho_m v_m^2}{2d} + \frac{BW}{A} f_1}{1 - \frac{CW}{A}} \right| \leq \frac{\left|\rho_m g + \frac{f_m \rho_m v_m^2}{2d} + \frac{BW}{A} f_1\right|}{\left|1 - \frac{CW}{A}\right|}.$$

Let:

$$N_2 = \sup\left\{\frac{\left|\rho_m g + \frac{f_m \rho_m v_m^2}{2d} + \frac{BW}{A} N_1\right|}{\left|1 - \frac{CW}{A}\right|}\right\}, \text{ then, there is } |f_2| \leq N_2.$$

Similarly,

$$|f_3| = \left|44.15P^{-0.79}\frac{dP}{dz}\right| \leq 44.15P^{-0.79}N_2$$

Let:

$$N_3 = \sup\{44.15P^{-0.79}N_2\},$$

then,

$$|f_3| \leq N_3.$$

The partial differential of f_1, f_2, f_3 about x, P, T are bounded is discussed as follows:

$$\frac{\partial f_1}{\partial x} = \frac{[44.15C_{P_g}P^{-0.79} - 44.15C_{P_l}P^{-0.79} - C_{J_g}C_{P_g} + C_{J_l}C_{P_l}]f_2}{h_g - h_l + Bv_m} + \frac{[44.15xC_{P_g}P^{-0.79} + 44.15(1-x)C_{P_l}P^{-0.79} - xC_{J_g}C_{P_g} - (1-x)C_{J_l}C_{P_l} - v_m C]\frac{\partial f_2}{\partial x}}{h_g - h_l + Bv_m}$$

$$\frac{\partial f_1}{\partial P} = \frac{[-34.8785C_{P_g}xP^{-1.79} - 34.8785(1-x)C_{P_l}P^{-1.79}]f_2}{h_g - h_l + Bv_m} + \frac{[44.15xC_{P_g}P^{-0.79} + 44.15(1-x)C_{P_l}P^{-0.79} - xC_{J_g}C_{P_g} - (1-x)C_{J_l}C_{P_l} - v_m C]\frac{\partial f_2}{\partial P}}{h_g - h_l + Bv_m}$$

$$\frac{\partial f_1}{\partial T} = \frac{-a}{h_g - h_l + Bv_m} + \frac{[44.15xC_{P_g}P^{-0.79} + 44.15(1-x)C_{P_l}P^{-0.79} - xC_{J_g}C_{P_g} - (1-x)C_{J_l}C_{P_l} - v_m C]\frac{\partial f_2}{\partial T}}{h_g - h_l + Bv_m}.$$

Similarly,

$$\begin{cases} \dfrac{\partial f_2}{\partial x} = \dfrac{Bw}{cW - A}\dfrac{\partial f_1}{\partial x} \\ \dfrac{\partial f_2}{\partial P} = \dfrac{Bw}{cW - A}\dfrac{\partial f_1}{\partial P} \\ \dfrac{\partial f_2}{\partial T} = \dfrac{Bw}{cW - A}\dfrac{\partial f_1}{\partial T} \end{cases} \tag{7.17}$$

and

$$\begin{cases} \dfrac{\partial f_3}{\partial x} = 44.15P^{-0.79}\dfrac{\partial f_2}{\partial x} \\ \dfrac{\partial f_3}{\partial P} = -34.8785P^{-1.79}f_2 + 44.15P^{-0.79}\dfrac{\partial f_2}{\partial P} \\ \dfrac{\partial f_3}{\partial T} = 44.15P^{-0.79}\dfrac{\partial f_2}{\partial T} \end{cases}. \tag{7.18}$$

By (7.17), we have:

$$\begin{cases} \dfrac{\partial f_1}{\partial x} = \dfrac{\left(1 - \frac{cw}{A}\right)[44.15C_{P_g}P^{-0.79} - 44.15C_{P_l}P^{-0.79} - C_{J_g}C_{P_g} + C_{J_l}C_{P_l} - v_mC]f_2}{\left(1 - \frac{cw}{A}\right)(h_g - h_l + v_mB) - [44.15xC_{P_g}P^{-0.79} + 44.15(1-x)C_{P_l}P^{-0.79} \\ \quad - xC_{J_g}C_{P_g} - (1-x)C_{J_l}C_{P_l} - v_mC] \times \frac{Bw}{A}} \\ \dfrac{\partial f_1}{\partial P} = \dfrac{\left(1 - \frac{cw}{A}\right)[-34.8785xC_{P_g}P^{-1.79} - 34.8785(1-x)C_{P_l}P^{-1.79}]f_2}{\left(1 - \frac{cw}{A}\right)(h_g - h_l + v_mB) - [44.15xC_{P_g}P^{-0.79} + 44.15(1-x)C_{P_l}P^{-0.79} \\ \quad - xC_{J_g}C_{P_g} - (1-x)C_{J_l}C_{P_l} - v_mC] \times \frac{Bw}{A}} \\ \dfrac{\partial f_1}{\partial T} = \dfrac{\left(\frac{cw}{A} - 1\right)a}{\left(1 - \frac{cw}{A}\right)(h_g - h_l + v_mB) - [44.15xC_{P_g}P^{-0.79} + 44.15(1-x)C_{P_l}P^{-0.79} \\ \quad - xC_{J_g}C_{P_g} - (1-x)C_{J_l}C_{P_l} - v_mC] \times \frac{Bw}{A}} \end{cases}. \tag{7.19}$$

Then,

$$\begin{aligned} \left|\frac{\partial f_1}{\partial x}\right| &= \left|\frac{\left(1 - \frac{cw}{A}\right)[44.15C_{P_g}P^{-0.79} - 44.15C_{P_l}P^{-0.79} - C_{J_g}C_{P_g} + C_{J_l}C_{P_l} - v_mC]f_2}{\left(1 - \frac{cw}{A}\right)(h_g - h_l + v_mB) - [44.15xC_{P_g}P^{-0.79} + 44.15(1-x)C_{P_l}P^{-0.79} \\ - xC_{J_g}C_{P_g} - (1-x)C_{J_l}C_{P_l} - v_mC] \times \frac{Bw}{A}}\right| \\ &\le \frac{\left(1 + \left|\frac{cw}{A}\right|\right)|[44.15C_{P_g}P^{-0.79} - 44.15C_{P_l}P^{-0.79} - C_{J_g}C_{P_g} + C_{J_l}C_{P_l} - v_mC]f_2|}{K_2} \\ &\le \frac{\left(1 + \left|\frac{cw}{A}\right|\right)[|44.15C_{P_g}P^{-0.79}| + |44.15C_{P_l}P^{-0.79}| + |C_{J_g}C_{P_g}| + |C_{J_l}C_{P_l}| + |v_mC|]N_2}{K_2} \end{aligned}.$$

Let:

$$M_{11} = \sup\left\{\frac{\left(1 + \left|\frac{cw}{A}\right|\right)[|44.15C_{P_g}P^{-0.79}| + |44.15C_{P_l}P^{-0.79}| + |C_{J_g}C_{P_g}| + |C_{J_l}C_{P_l}| + |v_mC|]N_2}{K_2}\right\}$$

then,

$$\left|\frac{\partial f_1}{\partial x}\right| \le M_{11}.$$

Similarly,

$$\left|\frac{\partial f_1}{\partial P}\right| = \left|\frac{\left(1-\frac{cw}{A}\right)[-34.8785xC_{P_g}P^{-1.79} - 34.8785(1-x)C_{P_l}P^{-1.79}]f_2}{\left(1-\frac{cw}{A}\right)(h_g - h_l + v_m B) - [44.15xC_{P_g}P^{-0.79} + 44.15(1-x)C_{P_l}P^{-0.79} - xC_{J_g}C_{P_g} - (1-x)C_{J_l}C_{P_l} - v_m C] \times \frac{Bw}{A}}\right|.$$

$$\leq \frac{\left(1+\left|\frac{cw}{A}\right|\right)[|34.8785xC_{P_g}P^{-1.79}| + |34.8785(1-x)C_{P_l}P^{-1.79}|]N_2}{K_2}$$

Let:

$$M_{12} = \sup\left\{\frac{\left(1+\left|\frac{cw}{A}\right|\right)[|34.8785xC_{P_g}P^{-1.79}| + |34.8785(1-x)C_{P_l}P^{-1.79}|]N_2}{K_2}\right\},$$

thus:

$$\left|\frac{\partial f_1}{\partial P}\right| \leq M_{12}.$$

Similarly,

$$\left|\frac{\partial f_1}{\partial T}\right| \leq \frac{\left(\left|\frac{cw}{A}\right|+1\right)|a|}{K_2}$$

Let:

$$M_{13} = \sup\left\{\frac{\left(\left|\frac{cw}{A}\right|+1\right)|a|}{K_2}\right\},$$

then,

$$\left|\frac{\partial f_1}{\partial T}\right| \leq M_{13}.$$

According to a similar method, there are:

$$\begin{cases} \left|\frac{\partial f_2}{\partial x}\right| = \left|\frac{Bw}{cW-A}\right|\left|\frac{\partial f_1}{\partial x}\right| \\ \left|\frac{\partial f_2}{\partial P}\right| = \left|\frac{Bw}{cW-A}\right|\left|\frac{\partial f_1}{\partial P}\right| \\ \left|\frac{\partial f_2}{\partial T}\right| = \left|\frac{Bw}{cW-A}\right|\left|\frac{\partial f_1}{\partial T}\right| \end{cases}. \tag{7.20}$$

Thus,

$$\left|\frac{\partial f_2}{\partial x}\right| \leq \left|\frac{Bw}{cW-A}\right| M_{11}.$$

Let:

$$M_{21} = \sup\left\{\left|\frac{Bw}{cW-A}\right| M_{11}\right\},$$

thus,

$$\left|\frac{\partial f_2}{\partial \rho}\right| \leq M_{21}.$$

Similarly,

$$\frac{\partial f_2}{\partial P} \leq \left|\frac{Bw}{cW-A}\right| M_{12}.$$

Let:

$$M_{22} = \sup\left\{\left|\frac{Bw}{cW-A}\right| M_{12}\right\},$$

then,

$$\left|\frac{\partial f_2}{\partial v}\right| \leq M_{22}.$$

Similarly,

$$\frac{\partial f_2}{\partial T} \leq \left|\frac{Bw}{cW - A}\right| M_{13}.$$

Let:

$$M_{23} = \sup\left\{\left|\frac{Bw}{cW - A}\right| M_{13}\right\},$$

then,

$$\left|\frac{\partial f_2}{\partial T}\right| \leq M_{23}.$$

Repeating the above method, we can get:

$$\begin{cases} \left|\frac{\partial f_3}{\partial x}\right| = |44.15P^{-0.79}| \left|\frac{\partial f_2}{\partial x}\right| \\ \left|\frac{\partial f_3}{\partial P}\right| = \left|-34.8785P^{-1.79} f_2 + 44.15P^{-0.79}\frac{\partial f_2}{\partial P}\right| \\ \left|\frac{\partial f_3}{\partial T}\right| = \left|44.15P^{-0.79}\frac{\partial f_2}{\partial T}\right| \end{cases}.$$

Thus,

$$\begin{cases} \left|\frac{\partial f_3}{\partial x}\right| \leq |44.15P^{-0.79}| M_{21} \\ \left|\frac{\partial f_3}{\partial P}\right| \leq |34.8785P^{-1.79}| N_2 + |44.15P^{-0.79}| M_{22} \\ \left|\frac{\partial f_3}{\partial T}\right| \leq |44.15P^{-0.79}| M_{23} \end{cases}.$$

Taking $M_{31} = \sup\{|44.15P^{-0.79}|M_{21}\}$, $M_{32} = \sup\{|34.8785P^{-1.79}|N_2 + |44.15P^{-0.79}|M_{22}\}$, $M_{33} = \sup\{\leq |44.15P^{-0.79}|M_{23}\}$, then:

$$\begin{cases} \left|\frac{\partial f_3}{\partial x}\right| \leq M_{31} \\ \left|\frac{\partial f_3}{\partial P}\right| \leq M_{32} \\ \left|\frac{\partial f_3}{\partial T}\right| \leq M_{33} \end{cases}.$$

The Lipschitz condition is very important in discussing the solution of the system of differential equations, thus we firstly consider the Lipschitz condition of $F(z; y)$. We write the primal problem again as follows:

$$\frac{dx}{dz} = f_1(z; x, P, T), \quad \frac{dP}{dz} = f_2(z; x, P, T), \quad \frac{dT}{dz} = f_3(z; x, P, T).$$

Its can be written as:

$$x' = f_1(z; x, P, T), \quad P' = f_2(z; x, P, T), \quad T' = f_3(z; x, P, T).$$

The primal condition is: $x(z_0) = x_0$, $P(z_0) = P_0$, $T(z_0) = T_0$.

Using the Euler method, for $i = 0, 1, 2, \ldots$, we have:

$$x_{i+1} = x_i + (z_{i+1} - z_i) f_1(z_i; x_i, P_i, T_i), \quad P_{i+1} = P_i + (z_{i+1} - z_i) f_2(z_i; x_i, P_i, T_i),$$

$$T_{i+1} = T_i + (z_{i+1} - z_i) f_3(z_i; x_i, P_i, T_i).$$

Here x_i, P_i, T_i are intended approximately to $\rho(z_i), P(z_i), T(z_i)$, where $z_0 < z_1 < z_2 \ldots$ is subdivision of the interval of integration. Let $y_i = (x_i, P_i, T_i)^T$, then:

$$y_{i+1} = y_i + (z_{i+1} - z_i) F(z_i; y_i), \quad i = 0, 1, 2, \ldots, n-1.$$

If set $h_i = z_{i+1} - z_i$, then for the subdivision above can be written as:

$$h = (h_0, h_1, \ldots, h_{n-1}).$$

If we connect y_0 and y_1, y_1 and $y_2, \ldots$ by straight lines we obtain the Euler polygon:

$$y_h(z) = y_i + (z - z_i) f(z_i; y_i) \quad \text{for } z_i \le z \le z_{i+1}.$$

Theorem 7.1. For $\|F(z; y)\| \le N = \max\{N_1, N_2, N_3\}$, then there is for x_i, P_i, T_i defined by above way the estimate:

$$\|y_i - y_0\| \le N|z_i - z_0|,$$

where $y_i = (x_i, P_i, T_i)^T$.

For $\left|\frac{\partial f_k}{\partial x}\right| \le M_{k1}, \quad \left|\frac{\partial f_k}{\partial v}\right| \le M_{k2}, \quad \left|\frac{\partial f_k}{\partial P}\right| \le M_{k3}$, then

$$\|F(z; y) - F(z; \hat{y})\| \le L\|y - \hat{y}\|,$$

where $k = 1, 2, 3$; $L = \max\limits_k \left(\sum\limits_{i=1}^{3} M_{ki}\right)$.

Proof. (1) From $x_{i+1} = x_i + (z_{i+1} - z_i) f_1(z_i; x_i, P_i, T_i)$ and definition of $\|F(z; x, P, T)\|$, we have:

$$|x_{i+1} - x_i| = |z_{i+1} - z_i||f_1(z_i; x_i, P_i, T_i)| \le N(z_{i+1} - z_i).$$

Therefore,

$$|x_i - x_{i-1}| \le N(z_i - z_{i-1}), \ldots, |x_2 - x_1| \le N(z_2 - z_1), \ |x_1 - x_0| \le N(z_1 - z_0).$$

Thus,

$$|x_i - x_{i-1}| + \cdots + |x_2 - x_1| + |x_1 - x_0| \le N(z_i - z_0).$$

Since:

$$|x_i - x_{i-1} + \cdots + x_2 - x_1 + \rho_1 - x_0| \le |x_i - x_{i-1}| + \cdots + |x_2 - x_1| + |x_1 - x_0|,$$

so,

$$|x_i - x_0| \le N(z_i - z_0).$$

Similarly,

$$|P_i - P_0| \le N(z_i - z_0),$$

$$|T_i - T_0| \le N(z_i - z_0).$$

From definition of $\|y_i - y_0\|$, there is:

$$\|y_i - y_0\| \le N(z_i - z_0).$$

(2) For $f_1(z;y), f_2(z;y), f_3(z;y), y=(x,P,T)^T$,

$$f_1(z;\hat{y}) - f_1(z;y) = \frac{\partial f_1}{\partial x}(\hat{x}-x) + \frac{\partial f_1}{\partial P}(\hat{P}-P) + \frac{\partial f_1}{\partial T}(\hat{T}-T).$$

Thus,

$$|f_1(z;\hat{y}) - f_1(z;y)| \le \left|\frac{\partial f_1}{\partial x}\right| |\hat{x}-x| + \left|\frac{\partial f_1}{\partial P}\right| |\hat{P}-P| + \left|\frac{\partial f_1}{\partial T}\right| |\hat{T}-T|.$$

Let $\triangle y = \max\{|\hat{\rho}-\rho|, |\hat{P}-P|, |\hat{T}-T|\}$, then:

$$|f_1(z;\hat{y}) - f_1(z;y)| \le \left(\left|\frac{\partial f_1}{\partial x}\right| + \left|\frac{\partial f_1}{\partial P}\right| + \left|\frac{\partial f_1}{\partial T}\right|\right)\triangle y.$$

Similarly,

$$|f_2(z;\hat{y}) - f_2(z;y)| \le \left(\left|\frac{\partial f_2}{\partial x}\right| + \left|\frac{\partial f_2}{\partial P}\right| + \left|\frac{\partial f_2}{\partial T}\right|\right)\triangle y,$$

$$|f_3(z;\hat{y}) - f_3(z;y)| \le \left(\left|\frac{\partial f_3}{\partial x}\right| + \left|\frac{\partial f_3}{\partial P}\right| + \left|\frac{\partial f_3}{\partial T}\right|\right)\triangle y.$$

From definition of norm, there has:

$$\|F(z;\hat{y}) - F(z;y)\| = \max\{|f_1(z;\hat{y}) - f_1(z;y)|, |f_2(z;\hat{y}) - f_2(z;y)|, |f_3(z;\hat{y}) - f_3(z;y)|\}.$$

Let $L = \max_k\left(\sum_{i=1}^{3} M_{ki}\right)$ and $\|\hat{y}-y\| = \triangle y$, then:

$$\|F(z;\hat{y}) - F(z;y)\| \le L\|\hat{y}-y\|.$$

We consider a subdivision of the interval of integration:

$$z_0, z_1, \ldots, z_{n-1}, z_n = Z,$$

Theorem 7.2. For a fixed subdivision h, let $y_h(x)$ and $\hat{y}_h(x)$ are the Euler polygons corresponding to the initial values y_0 and $\hat{y}_0$, respectively, for $F(z;y)$ satisfies the Lipschitz condition of Theorem 7.1. in a convex region which contains $(z;y_h(z))$ and $(z;\hat{y}_h(x))$ for $z_0 \le z \le Z$, then:

$$\|y_h(z) - \hat{y}_h(z)\| \le e^{L(z-z_0)}\|y_0 - \hat{y}_0\|.$$

Proof. From $y_{i+1} = y_i + (z_{i+1} - z_i)F(z_i;y_i)$, we have:

$$y_1 - y_0 = (z_1 - z_0)F(z_0;y_0), \quad \hat{y}_1 - \hat{y}_0 = (z_1 - z_0)F(z_0;\hat{y}_0).$$

Thus, we get:

$$y_1 - \hat{y}_1 = (y_0 - \hat{y}_0)(z_1 - z_0)[F(z_0;y_0) - F(z_0;\hat{y}_0)].$$

From Theorem 7.1, we have:

$$\|F(z;y) - F(z;\hat{y}\| \le L\|y - \hat{y}\|.$$

Therefore,

$$\|F(z_0;y_0) - F(z_0;\hat{y}_0\| \le L\|y_0 - \hat{y}_0\|.$$

Thus,

$$\|y_1 - \hat{y}_1\| \leq (1 + L(z_1 - z_0))\|y_0 - \hat{y}_0\|.$$

From $(1 + L(z_1 - z_0)) \leq e^{L(z_1 - z_0)}$, we get:

$$\|y_1 - \hat{y}_1\| \leq e^{L(z_1 - z_0)}\|y_0 - \hat{y}_0\|.$$

If we repeat the same argument for $y_2 - \hat{y}_2$, $y_3 - \hat{y}_3$, and so on, we finally obtain:

$$\|y_h(z) - (\hat{y})_h(z)\| \leq e^{L(z - z_0)}\|y_0 - \hat{y}_0\|.$$

Theorem 7.3. Let $F(z;y)$ be continuous, and $\|F(z;y)\| \leq N$ and satisfy the Lipschitz condition of Theorem 7.1 on $D = \{(z;y)|z_0 \leq z \leq Z,\ \|y - y_0\| \leq b\}$. If $Z - z_0 \leq \frac{b}{N}$, then we have:

(1) For $|h| = \max_{i=0,1,2,\ldots,n-1}(|z_{i+1} - z_i|) \to 0$, the Euler polygons $y_{|h|}(z) = (x_{|h|}(z), P_{|h|}(z), T_{|h|}(z))^T$ converge uniformly to a continuous vector function $\phi(z)$.
(2) $\phi(z)$ is continuously differential and is a solution of primal problem on $z_0 \leq z \leq Z$.
(3) There exists no other solution of primal problem on $z_0 \leq z \leq Z$.

Proof. (1) Take an $\varepsilon > 0$. Since F is uniformly continuous on the compact D, there exists a $\delta > 0$, such that

$$|z_2 - z_1| \leq \delta \quad \text{and} \quad \|y_1 - y_2\| \leq N\delta,$$

imply

$$\|F(z_2; y_2) - F(z_1; y_1)\| \leq \varepsilon.$$

Suppose that the subdivision h satisfies

$$|z_{i+1} - z_i| \leq \delta, \quad \text{i.e. } |h| \leq \delta.$$

We study first the effects of adding new mesh-points. In a first step, we consider a subdivision $h(1)$, which is obtained by adding new points only to the first subinterval. We have the estimate $\|y_{h(1)}(z_1) \quad y_h(z_1)\| \leq \varepsilon|z_1 - z_0|$. Since subdivision h and $h(1)$ are identical on $z_1 \leq z \leq Z$, we can obtain by Theorem 7.2:

$$\|y_{h(1)}(z) - y_h(z)\| \leq e^{L(z - z_1)}\|y_{h(1)}(z_1) - y_h(z_1)\|.$$

Thus,

$$\|y_{h(1)}(z) - y_h(z)\| \leq e^{L(z - z_1)}|z_1 - z_0|\varepsilon, \quad \text{for } z_1 \leq z \leq Z.$$

We next add further points to the subinterval (z_1, z_2) and denote the new subdivision by $h(2)$. In the same way as above this lead to:

$$\|y_{h(2)}(z_1) - y_{h(1)}(z_1)\| \leq \varepsilon|z_2 - z_1|,$$

and

$$\|y_{h(2)}(z) - y_{h(1)}(z)\| \leq e^{L(z - z_2)}|z_2 - z_1|\varepsilon, \quad \text{for } z_2 \leq z \leq Z.$$

If we denote by $\hat{h}$ the final refinement, we obtain for $z_2 \leq z \leq z_{i+1}$:

$$\begin{aligned}
\|y_{\hat{h}}(z) - y_h(z)\| &\leq \|y_{\hat{h}}(z) - y_{h(i-1)}(z)\| + \|y_{h(i-1)}(z) - y_{h(i-2)}(z)\| + \cdots \\
&\quad + \|y_{h(2)}(z) - y_{h(1)}(z)\| + \|y_{h(1)}(z) - y_h(z)\| \\
&\leq \varepsilon[e^{L(z - z_1)}|z_1 - z_0| + \cdots + e^{L(z - z_i)}|z_i - z_{i+1}|] \leq \varepsilon \int_{z_0}^{z} e^{L(z-s)} ds = \frac{\varepsilon}{L}(e^{L(z - z_0)} - 1)
\end{aligned} \tag{7.21}$$

If we now have different subdivisions h and $\acute{h}$, which both satisfy $|z_2 - z_1| \leq \delta$ and $\|y_1 - y_2\| \leq N\delta$, we introduce a third subdivision $\hat{h}$ which is a refinement of both subdivisions. Applying (7.19) to $\hat{h}$ and $\acute{h}$, we have:

$$\|y_h(z) - y_{\acute{h}}(z)\| \leq \|y_h(z) - y_{\hat{h}}(z)\| + \|y_{\hat{h}}(z) - y_{\acute{h}}(z)\| \leq 2\frac{\varepsilon}{L}(e^{L(z-z_0)} - 1).$$

For $\varepsilon > 0$ small enough, this becomes arbitrarily small and shows the uniform convergence of the Euler polygons to a continuous vector function $\phi(z)$.

(2) Let:

$$\varepsilon(\delta) := \sup\{\|F(z_1; y_1) - F(z_2; y_2)\|; \quad |z_1 - z_2| \leq \delta; \quad \|y_1 - y_2\| \leq N\delta,\ (z_i; y_i) \in D\}.$$

If z belongs to the subdivision h, then we can obtain:

$$\|y_h(z+\delta) - y_h(z) - \delta F(z; y_h(z))\| \leq \varepsilon(\delta)\delta.$$

By the limit $|h| \to 0$,

$$\|\phi(z+\delta) - \phi(z) - \delta F(z; \phi(z))\| \leq \varepsilon(\delta)\delta.$$

Since $\varepsilon(\delta) \to 0$ for $\delta \to 0$, this proves the differentiability of $\phi(z)$ and $\phi'(z) = F(z; \phi(z))$.

(3) From (2), the solution of the primal problem exists. Let $\phi(z)$ and $\varphi(z)$ be solutions of the primal problem, then:

$$\phi(z) = y_0 + \int_{z_0}^{z} F(x; \phi(x))dx \tag{7.22}$$

and

$$\varphi(z) = y_0 + \int_{z_0}^{z} F(x; \varphi(x))dx. \tag{7.23}$$

From (7.22), (7.23) and Lipschitz condition of Theorem 7.1, we have:

$$\|\phi(z) - \varphi(z)\| \leq L\left|\int_{z_0}^{z} \|\phi(x) - \varphi(x)\|dx\right|. \tag{7.24}$$

Let:

$$g(z) = \int_{z_0}^{z} \|\phi(x) - \varphi(x)\|dx, \quad z \geq z_0.$$

Thus,

$$g'(z) = \|\phi(z) - \varphi(z)\|.$$

Therefore, (7.24) can be revised as:

$$g'(z) \leq Lg(z),$$

so,

$$(e^{-L(z-z_0)}g(z))' \leq 0,$$

thus,

$$e^{-L(z-z_0)}g(z) \leq g(z_0) = 0.$$

When $z \geq z_0,\ g(z) \geq 0$, thus:

$$g(z) \equiv 0, \quad z \geq z_0.$$

Therefore,

$$\phi(z) = \varphi(z).$$

7.2.4 Solving model

To simplify the calculation, we divided the wells into several short segments of the same length h. The segment varies depending on variations in wall thickness, hole diameter, the fluid density inside and outside the pipe and well geometry. The model begins with a calculation at the top of the pipe. Then the dryness fraction of the gas, gas pressure and temperature calculations are performed for each successive "segment" of the pipe down to the bottom. Based on the aforementioned discussion, we use the fourth order Runge-Kutta method to solve the model, with an algorithm designed as follows.

Step 1. Obtain each point's inclination angle:

$$\theta_j = \theta_{j-1} + \frac{\theta_k - \theta_{k-1}}{\Delta s_k} \Delta s_j,$$

where j represents a selected segment point, Δs_k represents the measurement of the depth of the inclination angle θ_k and θ_{k-1}, Δs_j is step length of calculation.

Step 2. Calculating the gas condensation parameter Z_g:
If ($p < 35$ MPa)

$$Z_g = 1 + \left(0.31506 - \frac{1.0467}{T_{pr}} - \frac{0.5783}{T_{pr}^3}\right)\rho_{pr} + \left(0.053 - \frac{0.6123}{T_{pr}}\right)\rho_{pr}^2 + 0.6815\frac{\rho_{pr}^2}{T_{pr}^3},$$

where, $\rho_{pr} = 0.27\frac{P_{pr}}{T_{pr}}$, $T_{pr} = \frac{T}{T_{pc}}$, $P_{pr} = \frac{P}{P_{pc}}$.

Else

$$\begin{aligned} Z_g &= (90.7x - 242.2x^2 + 42.4x^3)y^{(1.18+2.82x)} - (14.76x - 9.76x^2 + 4.58x^3)y \\ &\quad + \frac{1 + y + y^2 + y^3}{(1-y)^3}, \end{aligned}$$

where,

$$\begin{aligned} F(y) &= -0.06125P_{pr}xe^{-1.2(1-x)^2} + (90.7x - 242.2x^2 + 42.4x^3)y^{2.18+2.82x} \\ &\quad + \frac{y + y^2 + y^3 - y^4}{(1-y^3)} - (14.76x - 9.76x^2 + 4.58x^3)y^2 \\ &= 0 \end{aligned}$$

$$x = \frac{1}{T_{pr}}$$

Step 3. Calculating parameter GWR:

$$GWR = 829.88 \cdot \frac{\gamma_l}{\gamma_g} \cdot \frac{x}{1-x},$$

where x is the steam dryness fraction of the gas, γ_g and γ_l are the gas and liquid relative density respectively.

Step 4. Calculating the gas-liquid mixture's density ρ_m:

$$\rho_m = \frac{M_t}{V_t},$$

where,

$$M_t = 1000\gamma_l + 1.205GWR \cdot \gamma_g, \quad V_t = 1 + 0.0003458 \cdot \frac{Z_g T}{P} \cdot GWR.$$

Step 5. Calculating parameter B:

$$B = \frac{W}{A} \cdot \left(\frac{1}{\rho_g} - \frac{1}{\rho_l}\right),$$

where,

$$\rho_g = 3484.48\gamma_g \cdot \frac{P}{Z_g T}, \quad \rho_l = 1000\gamma_l.$$

Step 6. Calculating parameter C:

$$C = \frac{W}{A} \cdot \frac{x}{\rho_g^2} \cdot \frac{3484.48\upsilon_g}{Z_g T}.$$

Step 7. Calculating parameter $h_g - h_l$:

$$h_g - h_l = 8.3143 \cdot T_c \cdot T_{br} \frac{3.978T_{br} - 3.938 + 1.555\ln(10P_c)}{1.07 - T_{br}},$$

where, T_c is the gas-liquid mixture's critical temperature, P_c is the gas-liquid mixture's critical pressure.

Step 8. Obtain the gas-liquid mixture's velocity v_m:

$$v_m = \frac{q_t}{A},$$

where,

$$q_t = \left(1 + 0.0003458 \cdot \frac{Z_g T}{P}\right) \cdot \frac{Q_{gsc}}{GWR},$$

A is the area of the pipe, Q_{gsc} is the quantity of the gas-liquid mixture.

Step 9. Calculating the gas heat ratio C_{P_g}:

$$C_{P_g} = 1697.5107P^{0.0661}T^{0.0776}.$$

Step 10. Calculating the gas-liquid mixture heat ratio C_{P_m}:

$$C_{P_m} = C_{P_g}x + C_{P_l}(1-x),$$

where, C_{pl} is the liquid heat ratio, $C_{pl} = 4.2\,\text{kJ/(kg}\cdot{}^\circ\text{C)}$.

Step 11. Let the right side of the coupled differential equations be function F_i, where ($i = 1, 2, 3, 4$). Then we can obtain a system of coupled functions as follows:

$$\begin{cases} F_1 = -\dfrac{\rho_m g\cos\theta + \frac{f_m\rho_m v_m^2}{2d} + \frac{Qgsc}{A}BF_4}{1 - \frac{Qgsc}{A}C} \\ F_2 = \dfrac{2\pi Ke}{Qgsc}\left[\dfrac{T - T_e}{f(t_D) + \frac{Ke}{r_{to}U_{to}}}\right] \\ F_3 = 44.15P^{-0.79} \cdot F_1 \\ F_4 = -\dfrac{F_2 + [xC_{pg} + (1-x)C_{pl}]F_3 - [xC_{jg}C_{pg} + (1-x)C_{jl}C_{pl}]F_1 - V_mCF_1 + g\cos\theta}{h_g - h_l + V_mB} \end{cases},$$

where,

$$\alpha = \frac{2\pi r_{to} U_{to} Ke}{r_{to} U_{to} f(t_D) + Ke},$$

$$U_{ti}^{-1} = \frac{1}{hc + hr} + \frac{r_{ti} \ln\left(\frac{r_{cem}}{r_{co}}\right)}{k_{cem}} + \frac{r_{ti} \ln\left(\frac{r_{ci}}{r_{to}}\right)}{k_{ang}}$$

and

$$C_J = \frac{R}{C_P} \frac{(2r_A - r_B T - 2r_B BT)Z - (2r_A B + r_B AT)}{3Z^2 - 2Z + A - B - B^2)T}$$

$$A = \frac{r_A P}{T}, \quad B = \frac{r_B P}{T}, \quad r_A = \frac{0.42747^2_{pc}}{P_{pc}}, \quad r_B = \frac{0.08664 C_b T_{pc}}{P_{pc}},$$

$$\alpha_i = [1 + m(1 - T_{pr}^{0.5})]^2, \quad m = 0.48 + 1.574w - 0.176w^2$$

and

$$\begin{cases} f(t_D) = 1.1281\sqrt{t_D}\left(1 - 0.3\sqrt{t_D}\right) & (t_D \le 1.5) \\ f(t_D) = \left(1 + \frac{0.6}{t_D}\right)[0.4063 + 0.5\ln(t_D)] & (t_D > 1.5) \end{cases}$$

$$t_D = \frac{t\alpha}{r_{wb}^2}$$

and

$$\frac{1}{\sqrt{f_m}} = 1.14 - 2\lg\left(\frac{0.00001524}{r_{ti}} + \frac{21.25}{Re^{0.9}}\right).$$

Step 12. Assume that P, Q, T, x to be y_i $(i = 1, 2, 3, 4)$, respectively. Then we can obtain some basic parameters as follows:

$$\begin{cases} a_i = F_i[y_1, y_2, y_3, y_4] \\ b_i = F_i\left[y_1 + \frac{h}{2}a_1, y_2 + \frac{h}{2}a_2, y_3 + \frac{h}{2}a_3, y_4 + \frac{h}{2}a_4\right] \\ c_i = F_i\left[y_1 + \frac{h}{2}b_1, y_2 + \frac{h}{2}b_2, y_3 + \frac{h}{2}b_3, y_4 + \frac{h}{2}b_4\right] \\ d_i = F_i[y_1 + hc_1, y_2 + hc_2, y_3 + hc_3, y_4 + hc_4] \end{cases}$$

Step 13. Calculating the gas-liquid mixture's dry degree, pressure and temperature at point $(j + 1)$:

$$y_i^{(j+1)} = y_i^j + \frac{h}{6}(a_i + 2b_i + 2c_i + d_i) \quad (i = 1, 2, 3, 4) \quad (j = 1, 2, \ldots, n).$$

Step 14. Repeat steps 1 to 13 until y_i^n is calculated.

7.2.5 Numerical simulation

As it was described previously, the algorithm begins withe a calculation for the pipe at the top of the pipe. The calculations are performed for each successive "segment" of the pipe down to the bottom.

Table 7.1 Pipe parameters.

Diameter [mm]	Thickness [mm]	Weight [kg]	Expansion coefficient [m/m · K]	Young's Modulus [GPa]	Poisson's ratio [–]	Using length [m]
88.9	9.53	18.9	0.0000115	215	0.3	1400
88.9	7.34	15.18	0.0000115	215	0.3	750
88.9	6.45	13.69	0.0000115	215	0.3	4200
73	7.82	12.8	0.0000115	215	0.3	600
73	5.51	9.52	0.0000115	215	0.3	150

Table 7.2 Well parameters.

Measured depth [m]	Internal diameter [mm]	External diameter [mm]
4325.69	168.56	193.7
6301.7	168.3	193.7
7100	121.42	146.1

Table 7.3 Azimuth, inclination and vertical depth.

Number	Measured depth [m]	Azimuth [°]	Inclination [°]	Vertical depth [m]
1	0	0	120.33	0
2	303	1.97	121.2	302.87
3	600	1.93	120.28	599.73
4	899	0.75	126.57	898.59
5	1206	1.25	124.9	1205.45
6	1505	1.04	124.62	1504.32
7	1800	0.49	123.75	1799.18
8	2105	2.49	125.27	2104.04
9	2401	1.27	123.13	2399.91
10	2669	2.44	120.12	2667.79
11	3021	0.14	127.39	3019.63
12	3299	1.18	122.6	3297.5
13	3605	2.05	123.25	3603.36
14	3901	0.16	121.45	3899.22
15	4183	2.92	121.24	4181.09
16	4492	2.73	129.22	4489.95
17	4816.07	1.98	121.61	4813.87
18	5099.07	2.74	129.93	5096.74
19	5394.07	0.13	120.46	5391.61
20	5706.07	0.63	129.59	5703.47
21	5983.07	2.09	120.14	5980.34
22	6302.07	2.69	122.91	6299.19
23	6597.07	2.45	129.41	6594.06
24	6911.12	0.15	124.88	6907.96
25	7100	1.15	123.2	7085.88

In this simulation, we study a pipe in X well, which is in Sichuan Province, China. All the needed parameters are given as following:

Internal fluid density = 1000 kg/m^3
External fluid density = 1000 kg/m^3
Depth of the well = 7100 m

Table 7.4 Dryness, temperature and pressure.

Number	Depth [m]	Dryness fraction [–]	Temperature [°C]	Pressure [MPa]
1	1	0.8000	200.0000	16.0000
2	251	0.7070	197.4870	15.9720
3	501	0.6460	195.4760	15.9500
4	751	0.5970	193.6780	15.9300
5	1001	0.5580	191.9960	15.9110
6	1251	0.5240	190.3830	15.8930
7	1501	0.4930	188.8120	15.8760
8	1751	0.4660	187.2680	15.8590
9	2001	0.4420	185.7340	15.8420
10	2251	0.4190	184.2020	15.8250
11	2501	0.3970	182.6660	15.8080
12	2751	0.3770	181.1160	15.7910
13	3001	0.3580	179.5480	15.7740
14	3251	0.3400	177.9550	15.7560
15	3501	0.3230	176.3330	15.7380
16	3751	0.3060	174.6750	15.7200
17	4001	0.2900	172.9770	15.7020
18	4251	0.2750	171.2310	15.6820
19	4501	0.2600	169.4320	15.6630
20	4751	0.2450	167.5720	15.6420
21	5001	0.2310	165.6420	15.6210
22	5251	0.2170	163.6350	15.5990
23	5501	0.2040	161.5410	15.5770
24	5751	0.1900	159.3460	15.5530
25	6001	0.1780	157.0400	15.5280
26	6251	0.1650	154.6050	15.5010
27	6501	0.1560	152.7660	15.4810
28	6751	0.1490	151.3670	15.4660
29	6958	0.1430	150.1710	15.4530

Table 7.5 Calculation results.

Location	Dryness fraction [–]	Temperature [°C]	Pressure [MPa]
Well head	0.8000	200.0000	16.0000
Well bottom	0.1430	150.1710	15.4530

Friction coefficient = 1.2
Ground temperature = 16°C
Ground thermal conductivity parameter = 2.06 W/m · K
Geothermal gradient = 2.18°C/100 m
Gas outputs = 500000 m^3/day
Length of one segment = 1 m

Parameters of pipes, inclined well, inclination, azimuth and vertical depth are given as follows:

Main results

After calculations are performed, we obtain a series of results for this well. The detailed source code can be seen in **Appendix 3**. The gas-liquid mixture's dryness fraction, pressure and temperature are shown in Table 7.4.

At the same time, we obtain a comparative result with a measurement value.

Relative errors are as follows in Table 7.7.

Table 7.6 Measurement result.

Location	Dryness fraction [–]	Temperature [°C]	Pressure [MPa]
Well head	0.8000	200.0000	16.0000
Well bottom	0.1380	147.6230	14.8710

Table 7.7 Relative error.

Dryness [–]	Temperature [°C]	Pressure [MPa]
0.03623	0.01726	0.03914

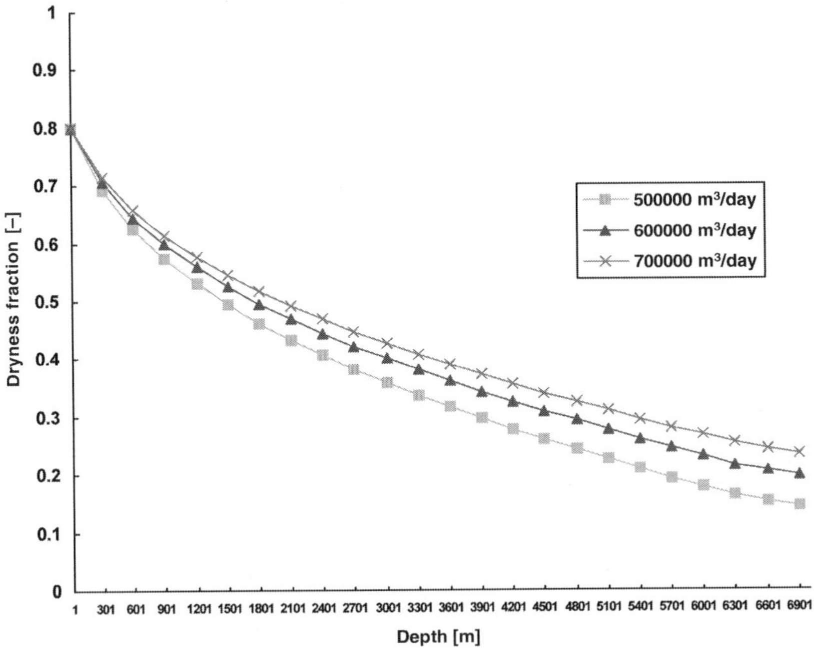

Figure 7.3 Dryness distribution with an injection volume variation.

7.2.6 **Sensitivity analysis**

To study different gas-liquid mixture injection volumes and the geothermal gradients to determine how it could be possible to influence the gas-liquid mixture's dryness fraction of gas, pressure and temperature; we used different gas-liquid mixture injection volumes and geothermal gradients. Using the algorithm and through a simulation, we obtained a series of results.

Firstly, we used three different gas-liquid mixture injection volumes: 500000, 600000 and 700000 m^3/day, with the other parameters remaining the same. Then we obtained the dryness fraction for the gas, pressure and temperature distribution figures as shown in Figures 7.3, 7.4 and 7.5.

From the figures we can see that the fraction of gas in the gas-liquid mixture's dryness decreases from the top to the bottom of the pipe. At the same pipe depth, if the gas-liquid mixture injection volumes increase, then the dryness fraction of gas also increases. Also, the pressure of the gas-liquid mixture decreases from the top to the bottom of the pipe. At the same pipe depth, if the

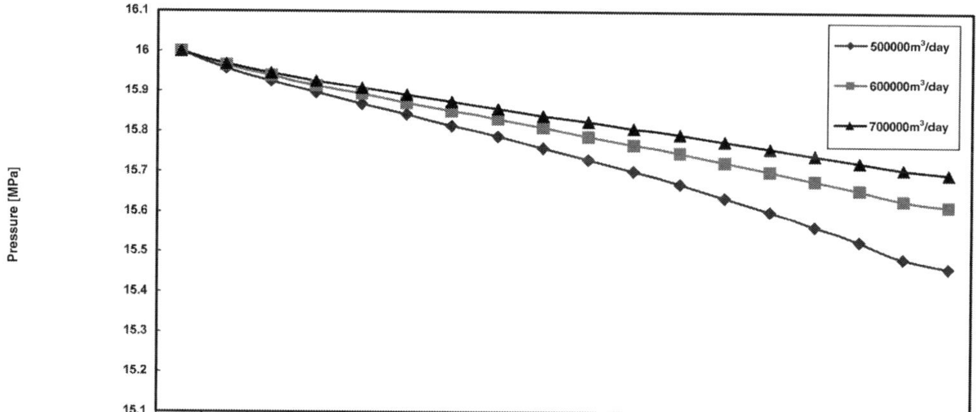

	1	401	801	1201	1601	2001	2401	2801	3201	3601	4001	4401	4801	5201	5601	6001	6401	6801
500000m3/day	16	15.958	15.926	15.897	15.869	15.842	15.815	15.788	15.76	15.731	15.702	15.671	15.638	15.604	15.567	15.528	15.487	15.463
600000m3/day	16	15.965	15.939	15.915	15.893	15.872	15.851	15.831	15.811	15.79	15.769	15.748	15.727	15.704	15.681	15.656	15.632	15.617
700000m3/day	16	15.97	15.947	15.927	15.908	15.891	15.874	15.857	15.841	15.825	15.809	15.793	15.776	15.759	15.742	15.725	15.708	15.696

Depth [m]

Figure 7.4 Pressure distribution with an injection volume variation.

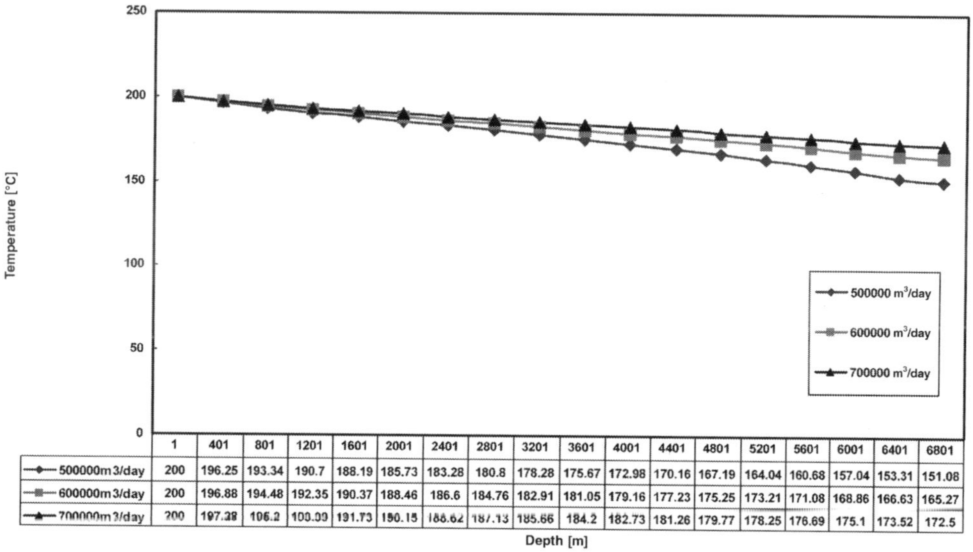

	1	401	801	1201	1601	2001	2401	2801	3201	3601	4001	4401	4801	5201	5601	6001	6401	6801
500000m3/day	200	196.25	193.34	190.7	188.19	185.73	183.28	180.8	178.28	175.67	172.98	170.16	167.19	164.04	160.68	157.04	153.31	151.08
600000m3/day	200	196.88	194.48	192.35	190.37	188.46	186.6	184.76	182.91	181.05	179.16	177.23	175.25	173.21	171.08	168.86	166.63	165.27
700000m3/day	200	197.38	195.2	[illegible]	191.73	190.15	188.62	187.13	185.66	184.2	182.73	181.26	179.77	178.25	176.69	175.1	173.52	172.5

Depth [m]

Figure 7.5 Temperature distribution with an injection volume variation.

gas-liquid mixture injection volume increases, the pressure also increases. From Figure 7.5 we can see that the temperature decreases from the top to the bottom of the pipe. At the same pipe depth, if the gas-liquid mixture inputs increase, then the temperature also increases.

Secondly, we used three different geothermal gradients; 1.6, 2.18 and 2.6°C/100 m, with the other parameters remaining the same. We obtained the dryness fraction of gas, pressure and temperature distribution figures, as shown in Figures 7.6, 7.7 and 7.8.

From the figures we can see that the dryness fraction of gas decreases from the top to the bottom of the pipe. At the same pipe depth, if the geothermal gradient increases, then the dryness fraction of gas decreases. Also, the pressure decreases from the top to the bottom of the pipe. At the same pipe depth, if the geothermal gradient increases, then the pressure decreases. From

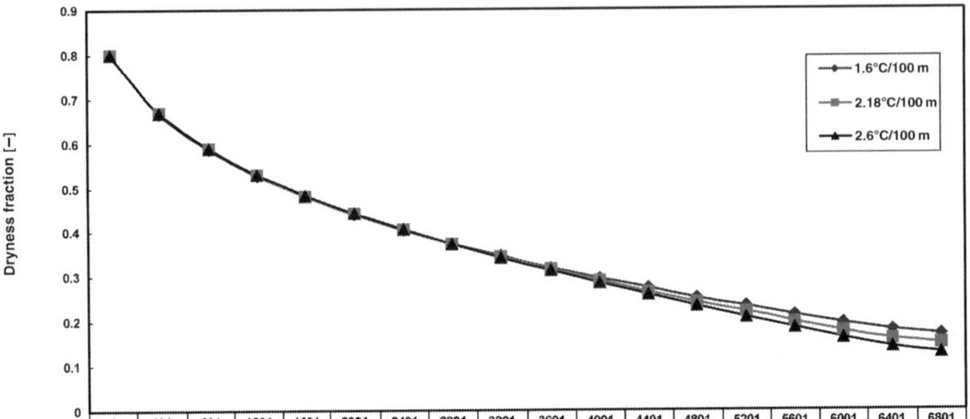

Depth [m]	1	401	801	1201	1601	2001	2401	2801	3201	3601	4001	4401	4801	5201	5601	6001	6401	6801
1.6	0.8	0.667	0.587	0.528	0.48	0.44	0.405	0.374	0.346	0.32	0.296	0.274	0.253	0.233	0.214	0.196	0.18	0.171
2.18	0.8	0.668	0.589	0.53	0.482	0.442	0.406	0.373	0.344	0.316	0.29	0.266	0.242	0.22	0.198	0.178	0.159	0.148
2.6	0.8	0.669	0.59	0.531	0.483	0.442	0.406	0.372	0.342	0.313	0.285	0.259	0.233	0.209	0.185	0.162	0.142	0.129

Depth [m]

Figure 7.6 Dryness distribution with a geothermal gradient (°C/100 m) variation.

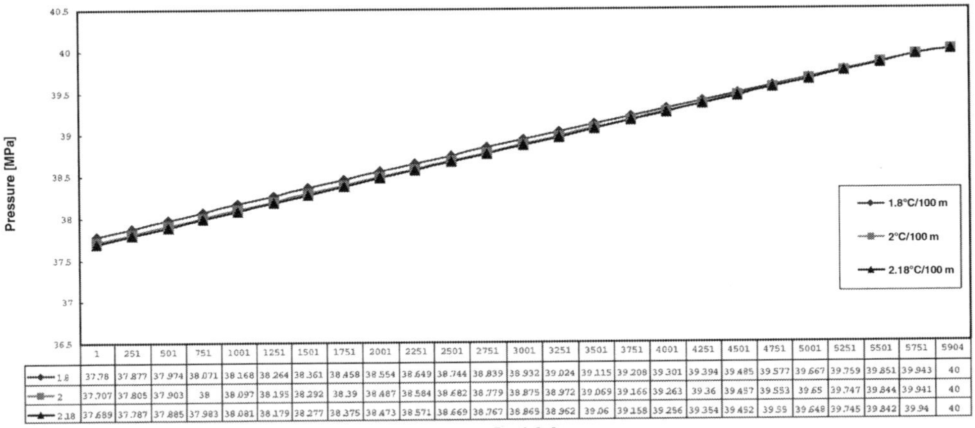

Depth [m]	1	251	501	751	1001	1251	1501	1751	2001	2251	2501	2751	3001	3251	3501	3751	4001	4251	4501	4751	5001	5251	5501	5751	5904
1.8	37.78	37.877	37.974	38.071	38.168	38.264	38.361	38.458	38.554	38.649	38.744	38.839	38.932	39.024	39.115	39.208	39.301	39.394	39.485	39.577	39.667	39.759	39.851	39.943	40
2	37.707	37.805	37.903	38	38.097	38.195	38.292	38.39	38.487	38.584	38.682	38.779	38.875	38.972	39.069	39.166	39.263	39.36	39.457	39.553	39.65	39.747	39.844	39.941	40
2.18	37.689	37.787	37.885	37.983	38.081	38.179	38.277	38.375	38.473	38.571	38.669	38.767	38.865	38.962	39.06	39.158	39.256	39.354	39.452	39.55	39.648	39.745	39.842	39.94	40

Depth [m]

Figure 7.7 Pressure distribution with a geothermal gradient variation.

Figure 7.7 we can see that the temperature is decreasing from the top to the bottom of the pipe. At the same pipe depth, if the geothermal gradient increases, the temperature decreases.

7.3 PDPT-SIBUHT[2]

Assumptions for the steam and the heat transmission are as follows:

1. The physical properties of the fluid and the formation are independent of depth and temperature.
2. All parameters for the steam (the velocity, pressure, temperature and the dryness fraction of gas) are constant.

[2] Predicting dryness fraction of gas, pressure and temperature for steam injection based on unsteady heat transmission

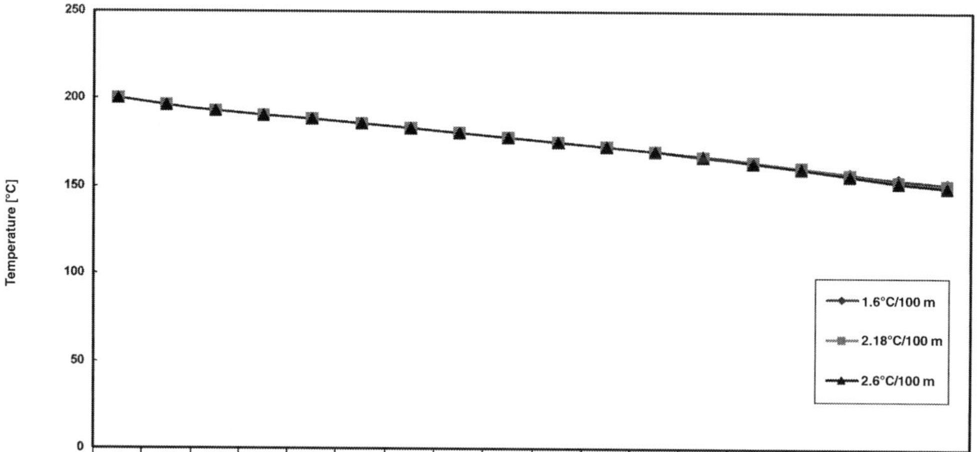

	1	401	801	1201	1601	2001	2401	2801	3201	3601	4001	4401	4801	5201	5601	6001	6401	6801
1.6	200	196.254	193.35	190.723	188.218	185.761	183.309	180.832	178.306	175.712	173.031	170.244	167.331	164.27	161.035	157.599	154.14	152.096
2.18	200	196.246	193.335	190.701	188.192	185.734	183.281	180.804	178.276	175.674	172.977	170.158	167.192	164.044	160.675	157.04	153.312	151.082
2.6	200	196.241	193.324	190.686	188.175	185.715	183.263	180.785	178.254	175.644	172.93	170.082	167.063	163.831	160.331	156.494	152.487	150.052

Depth [m]

Figure 7.8 Temperature distribution with a geothermal gradient variation.

3. The heat transfer from the tube to the second interface is considered steady, but the heat transfer from the second interface to the stratum is unsteady.
4. The steam flow is regarded as a one-dimensional two-phase homogeneous flow.
5. The physical properties of other materials except for the steam and some thermal insulation materials are independent of time and temperature.
6. The thermal conductivity of the thermal insulating materials are is in a linear relationship with the temperature.
7. The geothermal gradient is constant.
8. The frictional losses and the kinetic energy effects are considered.

Assumptions for the temperature field of the stratum are as follows:

1. The temperature around the well-bore in the stratum has an axisymmetric distribution.
2. The stratum away from the axes of the well-bore has the same temperature when injecting the steam in the well-bore.
3. The temperature field in the stratum has no endogenous pyrogen and has transient conduction.

7.3.1 Mathematical model

Pressure gradient. Since the steam injection is a constant-specific mass flow, that is, the mass flowing in the microelement is equal to the one flowing out of the microelement, it follows from the mass conservation equation that:

$$M = \rho_1 v_1 A = \rho_2 v_2 A = \rho_m v_m A. \tag{7.25}$$

It follows that:

$$\rho_m \frac{dv_m}{dz} + v_m \frac{d\rho_m}{dz} = 0.$$

The impulse the infinitesimal receives during the time dt is:

$$Fdt = P_1Adt + \rho_m Ag\cos\theta dz\,dt - P_2Adt - \tau_f dt,$$

and the change of the momentum during the time dt is:

$$\rho_2 v_2 Adt.v_2 - \rho_1 v_1 Adt.v_1 = \Delta(mv).$$

It follows from momentum theorem that:

$$\frac{dP}{dz} = \rho_m g\cos\theta - \rho_m v_m \frac{dv_m}{dz} - f_m \frac{\rho_m v_m^2}{4r_{ti}}. \tag{7.26}$$

The dryness fraction of gas. Given the loss of the energy from the steam flow friction, we have the energy conservation equation:

$$\frac{dQ}{dz} + \frac{dW}{dz} = -M\frac{dH_m}{dz} - M\frac{dv_m}{dz} + mg\cos\theta, \tag{7.27}$$

where dW is the loss of the energy caused by the friction between the steam and the tube wall and expresses the enthalpy of the mixed flow which is defined as:

$$H_m = H_s X + H_w(1 - X), \tag{7.28}$$

where H_s is the enthalpy of the saturated steam and H_w is the saturated water. It follows from that:

$$\frac{dH_m}{dz} = \left(\frac{dH_s}{dz} - \frac{dH_w}{dz}\right)X + (H_s - H_w)\frac{dX}{dz} + \frac{dH_w}{dz}.$$

As:

$$\frac{dH_s}{dz} = C_{Ps}\frac{dT}{dz} - C_{Js}C_{Ps}\frac{dP}{dz}$$

and

$$\frac{dH_w}{dz} = C_{Pw}\frac{dT}{dz} - C_{Jw}C_{Pw}\frac{dP}{dz}$$

it follows that:

$$\frac{dH_m}{dz} = \left[(C_{Ps} - C_{Pw})\frac{dT}{dz} - (C_{Js}C_{Ps} - C_{Jw}C_{Pw})\frac{dP}{dz}\right]X + (H_s - H_w)\frac{dX}{dz} + C_{Pw}\frac{dT}{dz} - C_{Jw}C_{Pw}\frac{dP}{dz}.$$

In addition, for the velocity of the mixed flow, we have:

$$v_m = v_s + v_w = \frac{MX}{\rho_s A} + \frac{M(1-X)}{\rho_w A}. \tag{7.29}$$

Thus,

$$\frac{dv_m}{dz} = B\frac{dX}{dz} - C\frac{dP}{dz}, \tag{7.30}$$

where:

$$B = \frac{M}{A}\left(\frac{1}{\rho_s} - \frac{1}{\rho_w}\right)$$

and

$$C = \frac{M}{A}\left(\frac{X}{\rho_s^2}\frac{d\rho_s}{dP} + \frac{1-X}{\rho_w^2}\frac{d\rho_w}{dP}\right).$$

Since the steam flow and the friction force directions are opposite, the loss of the energy dW caused by the friction between the steam and the tube wall is negative. The power caused by the friction in dz during the unit time dt is:

$$dW = \frac{\tau_f dz}{dt} = \frac{\tau_f dz}{dz/v_m} = \tau_f v_m. \tag{7.31}$$

The definition of mass is introduced in (7.25) and so we have the dryness fraction of gas model as follows:

$$A_1\frac{dX}{dz} + A_2X + A_3 = 0, \tag{7.32}$$

where,

$$A_1 = H_s - H_w + Bv_M, \quad A_2 = (C_{Ps} - C_{Pw})\frac{dT}{dz} - (C_{Js}C_{Ps} - C_{Jw}C_{Pw})\frac{dP}{dz}$$

$$A_3 = C_{Pw}\frac{dT}{dz} - (C_{Jw}C_{Pw} + v_mC)\frac{dP}{dz} - g\cos\theta + \frac{1}{M}\left(\frac{dQ}{dz} + \frac{\tau_f v_m}{dz}\right)$$

The heat transmission in the well-bore. Note that the heat transfer from the tube to the second interface is considered steady. It follows that:

$$\frac{dQ}{dz} = \pi D_{to}U_{to}(T - T_{ref}). \tag{7.33}$$

The radial heat transfer from the second interface to the surrounding earth is:

$$\frac{dQ}{dz} = \frac{\pi K_e}{f(t_D)}(T_{ref} - T_e). \tag{7.34}$$

Combine Equations (7.33) and (7.34), so then the heat transmission model is built between the steam flow and the surrounding earth as follows:

$$\frac{dQ}{dz} = \frac{\pi D_{to}U_{to}K_e}{\frac{1}{2}D_{to}U_{to}f(t_D) + K_e}(T - T_e). \tag{7.35}$$

Let $a = \frac{\pi D_{to}U_{to}K_e}{\frac{1}{2}D_{to}U_{to}f(t_D)+K_e}$, then:

$$\frac{dQ}{dz} = a(T - T_e). \tag{7.36}$$

The heat transmission in the stratum. From the assumption that the heat transfer from the second interface to the stratum is unsteady, the heat transmission model in the stratum is as follows:

$$\frac{\partial T_e}{\partial t} = \frac{\lambda_e}{C_{Pe}\rho_e}\left(\frac{\partial^2 T_e}{\partial r^2} + \frac{1}{r}\frac{\partial T_e}{\partial r}\right). \tag{7.37}$$

Initial condition:

$$T_e = T_a + \gamma z, \qquad t = 0.$$

Boundary conditions:

$$\frac{\partial T_e}{\partial r} = 0, \qquad r \to \infty; \quad dQ = -2\pi r_{cem}\frac{\partial T_e}{\partial r}dz\bigg|_{r=r_{cem}}.$$

It follows from the dimensionless variables $r_D = \frac{r}{r_{cem}}$ and $t_D = \frac{\lambda_e t}{r_{cem}^2}$ that (7.37) is changed into:

$$\frac{\partial T_e}{\partial t_D} = \frac{\lambda_e}{C_{Pe}\rho_e}\left(\frac{\partial^2 T_e}{\partial r_D^2} + \frac{1}{r_D}\frac{\partial T_e}{\partial r_D}\right). \tag{7.38}$$

The boundary conditions are converted into:

$$\left.\frac{\partial T_e}{\partial r_D}\right|_{r_D=1} = -\frac{dQ}{dz}(2\pi\lambda_e)^{-1},$$

$$\left.\frac{\partial T_e}{\partial r_D}\right|_{r_D\to\infty} = 0.$$

In addition, Liao and Feng (2005) proposed a relationship between the temperature and pressure of the saturated steam:

$$\frac{dT}{dz} = 44.15P^{-0.79}\frac{dP}{dz}. \tag{7.39}$$

Combining (7.26), (7.32), (7.35), (7.38) and (7.39), we have a coupled system model of differential equations on pressure, dryness and temperature as follows:

$$\begin{cases} \dfrac{dP}{dz} = \rho_m g\cos\theta - \rho_m v_m\left(B\dfrac{dX}{dz} - C\dfrac{dP}{dz}\right) - f_m\dfrac{\rho_m v_m^2}{4r_{t_i}} \\ \dfrac{dX}{dz} = -\dfrac{A_2}{A_1}X - \dfrac{A_3}{A_1} \\ \dfrac{dT}{dz} = 44.15P^{-0.79}\dfrac{dP}{dz} \\ \dfrac{dQ}{dz} = a(T - T_e) \\ \dfrac{\partial T_e}{\partial t_D} = \dfrac{\lambda_e}{C_{Pe}\rho_e}\left(\dfrac{\partial^2 T_e}{\partial r_D^2} + \dfrac{1}{r_D}\dfrac{\partial T_e}{\partial r_D}\right) \end{cases}. \tag{7.40}$$

The primal conditions:

$$P(z_0) = P_0, \quad T(z_0) = T_0, \quad x(z_0) = x_0$$

the boundary conditions:

$$\left.\frac{\partial T_e}{\partial r_D}\right|_{r_D=1} = -\frac{dQ}{dz}(2\pi\lambda_e)^{-1},$$

$$\left.\frac{\partial T_e}{\partial r_D}\right|_{r_D\to\infty} = 0.$$

Equation (7.40) can be rewritten as follows:

$$\begin{cases} \dfrac{dP}{dz} = \dfrac{\rho_m g\cos\theta - \rho_m v_m B\frac{dX}{dz} - \frac{f_m\rho_m v_m^2}{4r_{t_i}}}{1 - \rho_m v_m C} \\ \dfrac{dX}{dz} = \dfrac{\begin{array}{c}[C_{Pw}(X-1) - C_{Ps}X]\frac{dT}{dz} + [C_{Jw}C_{Pw}(1-X) + C_{Js}C_{Ps}X + v_m C]\frac{dP}{dz} \\ + g\cos\theta - \frac{a(T-T_e)}{M} - \frac{\tau_f v_m}{dz}\end{array}}{H_s - H_w + Bv_m} \\ \dfrac{dT}{dz} = 44.15P^{-0.79}\dfrac{dP}{dz} \\ \dfrac{\partial T_e}{\partial t_D} = \dfrac{\lambda_e}{C_{Pe}\rho_e}\left(\dfrac{\partial^2 T_e}{\partial r_D^2} + \dfrac{1}{r_D}\dfrac{\partial T_e}{\partial r_D}\right) \end{cases}. \tag{7.41}$$

The primal conditions:

$$P(z_0)=P_0, \quad T(z_0)=T_0, \quad x(z_0)=x_0$$

the boundary conditions:

$$\left.\frac{\partial T_e}{\partial r_D}\right|_{r_D=1} = -a(T-T_e)(2\pi\lambda_e)^{-1},$$

$$\left.\frac{\partial T_e}{\partial r_D}\right|_{r_D\to\infty} = 0.$$

7.3.2 **Solution of the model**

Firstly, we consider the solution of the following second-order parabolic equation:

$$\frac{\partial T_e}{\partial t_D} = \frac{\lambda_e}{C_{Pe}\rho_e}\left(\frac{\partial^2 T_e}{\partial r_D^2} + \frac{1}{r_D}\frac{\partial T_e}{\partial r_D}\right). \tag{7.42}$$

Equation (7.42) can be rewritten as:

$$\frac{\lambda_e}{C_{Pe}\rho_e}\frac{\partial^2 T_e}{\partial r_D^2} + \frac{\lambda_e}{C_{Pe}\rho_e r_D}\frac{\partial T_e}{\partial r_D} - \frac{\partial T_e}{\partial t_D} = 0. \tag{7.43}$$

Let $a(r_D,t_D)=\frac{\lambda_e}{C_{Pe}\rho_e}$, $b(r_D,t_D)=\frac{\lambda_e}{C_{Pe}r_D\rho_e}$, $c(r_D,t_D)=0$, then:

$$a(r_D,t_D)\frac{\partial^2 T_e}{\partial r_D^2} + b(r_D,t_D)\frac{\partial T_e}{\partial r_D} - \frac{\partial T_e}{\partial t_D} = 0. \tag{7.44}$$

A function $T_e(r_D,t_D;\xi,\tau)$ is called a fundamental solution of Equation (7.44) in the E_1 if it has the following properties:

(1) The function $T_e(r_D,t_D;\xi,\tau)$ in the domain $\{0\le\gamma<t_D\le T, r_D\in E_1, \xi\in E_1\}$ is continuous jointly in r_D, t_D, ξ, and τ together with its derivatives $\frac{\partial^2 T_e}{\partial r_D^2}$ and $\frac{\partial T_e}{t_D}$ and satisfies Equation (7.44) with respect to variable r_D and t_D. The function $T_e(r_D,t_D;\xi,\tau)$ is bounded in any domain $t_D-\tau+|r_D-\xi|\ge\delta$, where $\delta>0$.
(2) For any continuous and bounded function $\varphi(\xi)$ in E_1 and for all $r_D\in E_1$, $\tau\in[0,T]$ the relation

$$\lim_{t_D\to\tau+0}\int_{E_1} T_e(r_D,t_D;\xi,\tau)\varphi(\xi)d\xi=\varphi(r_D) \tag{7.45}$$

holds, and the integral tends to the limit uniformly with respect to r_D in any bounded domain of the space E_1.

Now, we discuss the solution existence and uniqueness of Equation (7.44).

Theorem 7.4. Suppose all the coefficients of Equation (7.44) are bounded and continuous in E_1 jointly in r_D and t_D and are Hölder continuous with respect to r_D:

$$\begin{aligned}|a(r'_D,t_D)-a(r_D,t_D)| &\le M_1|r'_D-r_D|^\lambda \\ |b(r'_D,t_D)-b(r_D,t_D)| &\le M_1|r'_D-r_D|^\lambda\end{aligned}, \tag{7.46}$$

where $\lambda>0$, $M_1>0$. Suppose additionally that the coefficients $a(r_D,t_D)$ in E_1 are Hölder continuous with respect to t_D:

$$|a(r_D,t'_D)-a(r_D,t_D)|\le M_2|t'_D-t_D|^\lambda. \tag{7.47}$$

Let us also suppose that the following inequality holds:

$$a(r_D, t_D))| \geq \mu, \quad \mu > 0. \tag{7.48}$$

Then there exists a unique fundamental solution $T_e(r_D, t_D; \xi, \tau)$ of Equation (7.44). For $T_e(r_D, t_D; \xi, \tau)$ the estimates:

$$\begin{aligned}
|T_e(r_D, t_D; \xi, \tau)| &< M_2(t_D - \tau)^{-\frac{1}{2}} e^{-\frac{\mu_1 (r_D - \xi)^2}{t_D - \tau}} \\
\left| \frac{\partial T_e(r_D, t_D; \xi, \tau)}{\partial r_D} \right| &< M_2(t_D - \tau)^{-1} e^{-\frac{\mu_1 (r_D - \xi)^2}{t_D - \tau}} \\
\left| \frac{\partial^2 T_e(r_D, t_D; \xi, \tau)}{\partial r_D^2} \right| &< M_2(t_D - \tau)^{-\frac{3}{2}} e^{-\frac{\mu_1 (r_D - \xi)^2}{t_D - \tau}} \\
\left| \frac{\partial T_e(r_D, t_D; \xi, \tau)}{\partial t_D} \right| &< M_2(t_D - \tau)^{-\frac{3}{2}} e^{-\frac{\mu_1 (r_D - \xi)^2}{t_D - \tau}}
\end{aligned} \tag{7.49}$$

hold; here M_2 and μ_1 are positive constants.

Proof. We shall try to find the fundamental solution in the form:

$$T_e(r_D, t_D; \xi, \tau) = W_{\xi,\tau}(r_D, t_D; \xi, \tau) + \int_{\tau}^{t_D} d\theta \int_{E_1} W_{\zeta,\theta}(r_D, t_D; \zeta, \theta) \phi(\zeta, \theta; \xi, \tau) d\zeta, \tag{7.50}$$

where the function $W_{\xi,\tau}(r_D, t_D; \xi, \tau)$ is given in the following form and $\phi(r_D, t_D; \xi, \tau)$ is found to be:

$$W_{\xi,\tau}(r_D, t_D; \xi, \tau) = [4\pi(t_D - \tau)]^{-\frac{1}{2}} [a(r_D, t_D)]^{-\frac{1}{2}} e^{-\frac{\frac{1}{a(r_D, t_D)}(r_D - \xi)^2}{4(t_D - \tau)}}. \tag{7.51}$$

First we shall study the properties of the integral:

$$V(r_D, t_D; \xi, \tau) = \int_{\tau}^{t_D} d\theta \int_{E_1} W_{\zeta,\theta}(r_D, t_D; \zeta, \theta) \phi(\zeta, \theta; \xi, \tau) d\zeta. \tag{7.52}$$

According to Equation (7.51), for the function $W_{\xi,\tau}(r_D, t_D; \xi, \tau)$ we have the following estimate:

$$|W_{\xi,\tau}(r_D, t_D; \xi, \tau)| < M_3(t_D - \tau)^{-\frac{1}{2}} e^{-\frac{\mu_1 (r_D - \xi)^2}{(t_D - \tau)}}, \quad \mu_1 > 0. \tag{7.53}$$

Using the following elementary inequality,

$$q^{v} e^{-\mu' q} \leq K_v e^{-\tilde{\mu} q}, \quad (0 \leq q < +\infty, 0 < \tilde{\mu} < \mu') \tag{7.54}$$

which holds for all $v > 0$, we can obtain estimate for the derivatives of $W_{\xi,\tau}(r_D, t_D; \xi, \tau)$:

$$\frac{\partial W_{\xi,\tau}(r_D, t_D; \xi, \tau)}{\partial r_D} = -\frac{1}{4\sqrt{\pi}} a^{-\frac{3}{2}} (t_D - \tau)^{-1} \left[\frac{(r_D - \xi)^2}{t_D - \tau} \right]^{\frac{1}{2}} e^{-\frac{\frac{1}{a}(r_D - \xi)^2}{4(t_D - \tau)}}. \tag{7.55}$$

Let $q = \frac{(r_D - \xi)^2}{t_D - \tau}$, $v = \frac{1}{2}$, $\mu' = -\frac{1}{4a}$, then Equation (7.55) can be written as:

$$\frac{\partial W_{\xi,\tau}(r_D, t_D; \xi, \tau)}{\partial r_D} = -\frac{1}{4\sqrt{\pi}} a^{-\frac{3}{2}} (t_D - \tau)^{-1} q^{v} e^{-\frac{\mu' (r_D - \xi)^2}{t_D - \tau}}. \tag{7.56}$$

Take $K_v = a^{\frac{3}{2}}$, we have:

$$\left|\frac{\partial W_{\xi,\tau}(r_D,t_D;\xi,\tau)}{\partial r_D}\right| < M_3(t_D-\tau)^{-1}e^{-\frac{\mu_2(r_D-\xi)^2}{t_D-\tau}}. \tag{7.57}$$

Similarly,

$$\begin{aligned}\frac{\partial^2 W_{\xi,\tau}(r_D,t_D;\xi,\tau)}{\partial r_D^2} &= -\frac{1}{4\sqrt{\pi}}a^{-\frac{3}{2}}(t_D-\tau)^{-\frac{3}{2}}e^{-\frac{\frac{1}{a}(r_D-\xi)^2}{4(t_D-\tau)}}\\ &\quad+\frac{1}{8\sqrt{\pi}}a^{-\frac{5}{2}}(t_D-\tau)^{-\frac{3}{2}}\frac{(r_D-\xi)^2}{t_D-\tau}e^{-\frac{\frac{1}{a}(r_D-\xi)^2}{4(t_D-\tau)}}\end{aligned} \tag{7.58}$$

thus,

$$\begin{aligned}\left|\frac{\partial^2 W_{\xi,\tau}(r_D,t_D;\xi,\tau)}{\partial r_D^2}\right| &\le \left|\frac{1}{4\sqrt{\pi}}a^{-\frac{3}{2}}(t_D-\tau)^{-\frac{3}{2}}e^{-\frac{\frac{1}{a}(r_D-\xi)^2}{4(t_D-\tau)}}\right|\\ &\quad+\left|\frac{1}{8\sqrt{\pi}}a^{-\frac{5}{2}}(t_D-\tau)^{-\frac{3}{2}}\frac{(r_D-\xi)^2}{t_D-\tau}e^{-\frac{\frac{1}{a}(r_D-\xi)^2}{4(t_D-\tau)}}\right|\end{aligned}. \tag{7.59}$$

Take $q=\frac{(r_D-\xi)^2}{t_D-\tau}$, $v=1$, $K_v=a$, by means the elementary inequality, we can obtain:

$$\left|\frac{1}{8\sqrt{\pi}}a^{-\frac{5}{2}}(t_D-\tau)^{-\frac{3}{2}}\frac{(r_D-\xi)^2}{t_D-\tau}e^{-\frac{\frac{1}{a}(r_D-\xi)^2}{4(t_D-\tau)}}\right| \le \frac{1}{8\sqrt{\pi}}a^{-\frac{3}{2}}(t_D-\tau)^{-\frac{3}{2}}e^{-\frac{\mu_2(r_D-\xi)^2}{t_D-\tau}}. \tag{7.60}$$

Therefore,

$$\left|\frac{\partial^2 W_{\xi,\tau}(r_D,t_D;\xi,\tau)}{\partial r_D^2}\right| \le M_3(t_D-\tau)^{-\frac{3}{2}}e^{-\frac{\mu_2(r_D-\xi)^2}{t_D-\tau}}. \tag{7.61}$$

Similarly,

$$\frac{\partial W_{\xi,\tau}(r_D,t_D;\xi,\tau)}{\partial r_D} = -\frac{1}{4\sqrt{\pi}}a^{-\frac{1}{2}}(t_D-\tau)^{-\frac{3}{2}}e^{-\frac{\frac{1}{a}(r_D-\xi)^2}{4(t_D-\tau)}}+\frac{1}{8\sqrt{\pi}}a^{-\frac{3}{2}}(t_D-\tau)^{-\frac{3}{2}}\frac{(r_D-\xi)^2}{t_D-\tau}e^{-\frac{\frac{1}{a}(r_D-\xi)^2}{4(t_D-\tau)}}. \tag{7.62}$$

Take $q=\frac{(r_D-\xi)^2}{t_D-\tau}$, $v=1$, $K_v=a$, by means the elementary inequality, we can obtain:

$$\left|\frac{1}{8\sqrt{\pi}}a^{-\frac{3}{2}}(t_D-\tau)^{-\frac{3}{2}}\frac{(r_D-\xi)^2}{t_D-\tau}e^{-\frac{\frac{1}{a}(r_D-\xi)^2}{4(t_D-\tau)}}\right| \le \frac{1}{8\sqrt{\pi}}a^{-\frac{1}{2}}(t_D-\tau)^{-\frac{3}{2}}e^{-\frac{\mu_2(r_D-\xi)^2}{t_D-\tau}}. \tag{7.63}$$

Therefore,

$$\left|\frac{\partial W_{\xi,\tau}(r_D,t_D;\xi,\tau)}{\partial t_D}\right| < M_3(t_D-\tau)^{-\frac{3}{2}}e^{-\frac{\mu_2(r_D-\xi)^2}{t_D-\tau}}. \tag{7.64}$$

Suppose that the function $\phi(r_D,t_D;\xi,\tau)$ is continuous jointly r_D, t_D, ξ, τ and for any r_D, ξ and can be estimated in the following way:

$$|\phi(r_D,t_D;\xi,\tau)| < M_4(t_D-\tau)^{-\frac{3}{2}+\lambda_1}e^{-\frac{\mu_3(r_D-\xi)^2}{t_D-\tau}}, \tag{7.65}$$

where $\lambda_1 > 0$ and $\mu_3 > 0$. Moreover, we shall suppose that for $(t'_D - t_D)^2 < \alpha(t_D - \tau)$, where $\alpha > 0$ is a constant, we have:

$$|\phi(r'_D, t_D; \xi, \tau) - \phi(r_D, t_D; \xi, \tau)| \le M_4(r'_D - r_D)^{\lambda_2}(t_D - \tau)^{-\frac{3}{2}+\lambda_3} e^{-\frac{\mu_3(r_D-\xi)^2}{t_D-\tau}}, \tag{7.66}$$

where $\lambda_2 > 0$ and $\lambda_3 > 0$. Let us show that under such suppositions the function (7.52) is continuous and has continuous derivatives $\frac{\partial V}{\partial r_D}, \frac{\partial^2 V}{\partial r_D^2}, \frac{\partial V}{\partial t_D}$ for $t_D > \tau$ and any r_D, ξ.

According to our suppositions, the function:

$$J(r_D, t_D, \theta; \xi, \tau) = \int_{E_1} W_{\zeta,\theta}(r_D, t_D; \zeta, \theta)\Phi(\zeta, \theta; \xi, \tau)d\zeta \tag{7.67}$$

is continuous with respect to all its arguments and has derivatives of any order with respect to r_D and t_D for $\tau < \theta < t_D$ and any r_D, ξ. Taking into account (7.53) and (7.65), we obtain:

$$\begin{aligned}
&|J(r_D, t_D, \theta; \xi, \tau)| \\
&\quad < M_4 \int_{E_1} (t_D - \theta)^{-\frac{1}{2}} e^{-\frac{\mu_1(r_D-\zeta)^2}{t_D-\theta}} (\theta - \tau)^{-\frac{3}{2}+\lambda_1} e^{-\frac{\mu_3(r_D-\xi)^2}{t_D-\tau}} \\
&\quad = M_4(\theta - \tau)^{\lambda_1 - 1} \int_{-\infty}^{+\infty} [(t_D - \theta)(\theta - \tau)]^{-\frac{1}{2}} e^{-\mu_4\left[\frac{(r_D-\xi)^2}{t_D-\tau} + \frac{(\zeta-\xi)^2}{\theta-\tau}\right]} d\zeta \\
&\quad = M_4(\theta - \tau)^{\lambda_1 - 1} \int_{-\infty}^{+\infty} \frac{1}{\sqrt{2}}(\theta - \tau)^{-\frac{1}{2}} e^{-\frac{(\sqrt{2}(\zeta-\xi))^2}{2\sqrt{\frac{\theta-\tau}{\mu_4}}}} d\sqrt{2}(\zeta - \xi) \cdot (t_D - \theta)^{-\frac{1}{2}} e^{-\frac{\mu_4(r_D-\xi)^2}{t_D-\theta}} \\
&\quad = M_4(\theta - \tau)^{\lambda_1 - 1} \frac{1}{\sqrt{2}}(\theta - \tau)^{-\frac{1}{2}} \sqrt{2\pi} \sqrt{\frac{\theta - \tau}{\mu_4}} \cdot (t_D - \theta)^{-\frac{1}{2}} e^{-\frac{\mu_4(r_D-\xi)^2}{t_D-\theta}} \\
&\quad = M_4(\theta - \tau)^{\lambda_1 - 1} \pi^{\frac{1}{2}} [(\theta - \tau)^{-\frac{1}{2}}][\mu_4(t_D - \theta)]^{-\frac{1}{2}} e^{-\frac{\mu_4(r_D-\xi)^2}{t_D-\theta}} \\
&\quad < M_5(\theta - \tau)^{\lambda_1 - 1}(t_D - \tau)^{-\frac{1}{2}} e^{-\frac{\mu_4(r_D-\xi)^2}{t_D-\theta}}.
\end{aligned} \tag{7.68}$$

Therefore, for $t_D - \tau \ge \delta > 0$, where $\delta > 0$ is arbitrary, and for any r_D, ξ, the inequality:

$$|J(r_D, t_D, \theta; \xi, \tau)| < M_\delta^1 (\theta - \tau)^{\lambda - 1} \tag{7.69}$$

holds. This allows us to establish the uniform convergence of the integral $\int_\tau^{t_D} J(r_D, t_D, \theta; \xi, \tau)d\theta$ with respect to r_D, t_D, ξ, τ. Thus, we have established that the function $V(r_D, t_D; \xi, \tau)$ is continuous jointly in all its arguments for $t_D > \tau$ and any r_D, ξ. According to (7.68), we have:

$$\begin{aligned}
|V(r_D, t_D; \xi, \tau)| &\le \int_\tau^{t_D} |J(r_D, t_D, \theta; \xi, \tau)| d\theta \\
&< M_5 \int_\tau^{t_D} (\theta - \tau)^{\lambda_1 - 1}(t_D - \theta)^{-\frac{1}{2}} e^{-\mu_4 \frac{(r_D-\xi)^2}{t_D-\tau}} d\theta \\
&= \frac{M_5}{\lambda_1}(\theta - \tau)^{\lambda_1}(t_D - \tau)^{-\frac{1}{2}} e^{-\mu_4 \frac{(r_D-\xi)^2}{t_D-\tau}} \Bigg|_\tau^{t_D} \\
&= \frac{M_5}{\lambda_1}(t_D - \tau)^{\lambda_1}(t_D - \tau)^{-\frac{1}{2}} e^{-\mu_4 \frac{(r_D-\xi)^2}{t_D-\tau}} \\
&< M_6(t_D - \tau)^{-\frac{1}{2}+\lambda_1} e^{-\mu_4 \frac{(r_D-\xi)^2}{t_D-\tau}}.
\end{aligned} \tag{7.70}$$

Using (7.53) and (7.65), with the help of the similarly arguments we obtain the estimate:

$$\begin{aligned}
\left|\frac{\partial J(r_D,t_D,\theta;\xi,\tau)}{\partial r_D}\right| &\leq \int_{E_1}\left|\frac{\partial W_{\zeta,\theta}(r_D,t_D;\zeta,\theta)}{\partial r_D}\right|\cdot|\Phi(\zeta,\theta;\xi,\tau)|d\zeta\\
&< \int_{E_1} M_3(t_D-\theta)^{-1}e^{-\mu_2\frac{(r_D-\zeta)^2}{t_D-\theta}}\cdot M_4(\theta-\tau)^{-\frac{2}{3}+\lambda_1}e^{-\mu_3\frac{(\zeta-\xi)^2}{\theta-\tau}}d\zeta\\
&= M_7(\theta-\tau)^{\lambda_1-\frac{1}{2}}\int_{-\infty}^{+\infty}[(t_D-\theta)(\theta-\tau)]^{-1}e^{\left[-\mu_2\frac{(r_D-\zeta)^2}{t_D-\theta}-\mu_3\frac{(\zeta-\xi)^2}{\theta-\tau}\right]}d\zeta\\
&< M_7(\theta-\tau)^{\lambda_1-\frac{1}{2}}\int_{-\infty}^{+\infty}[(t_D-\theta)(\theta-\tau)]^{-1}e^{-\mu_4\left[\frac{(r_D-\xi)^2}{t_D-\theta}+\frac{(\zeta-\xi)^2}{\theta-\tau}\right]}d\zeta\\
&= M_7(\theta-\tau)^{\lambda_1-\frac{1}{2}}\int_{-\infty}^{+\infty}\frac{1}{\sqrt{2}}[(t_D-\theta)(\theta-\tau)]^{-1}e^{-\mu_4\frac{(r_D-\xi)^2}{t_D-\theta}}\\
&\quad\cdot e^{-\frac{[\sqrt{2}(\zeta-\xi)]^2}{2\sqrt{\frac{\theta-\tau}{\mu_4}}}}d\sqrt{2}(\zeta-\xi)\\
&= M_7(\theta-\tau)^{\lambda_1-\frac{1}{2}}\frac{1}{\sqrt{2}}[(t_D-\theta)(\theta-\tau)]^{-1}\cdot\sqrt{2\pi}\sqrt{\frac{\theta-\tau}{\mu_4}}e^{-\mu_4\frac{(r_D-\xi)^2}{t_D-\theta}}\\
&= M_7(\theta-\tau)^{\lambda_1-\frac{1}{2}}\sqrt{\pi}\mu_4^{-\frac{1}{2}}(t_D-\theta)^{-1}e^{-\mu_4\frac{(r_D-\xi)^2}{t_D-\tau}}\\
&< M_8(\theta-\tau)^{\lambda_1-1}(t_D-\theta)^{-1}e^{-\mu_4\frac{(r_D-\xi)^2}{t_D-\tau}}.
\end{aligned}\tag{7.71}$$

Hence

$$\left|\frac{\partial J(r_D,t_D,\theta;\xi,\tau)}{\partial r_D}\right| < M_\delta^2(\theta-\tau)^{\lambda_1-1}\tag{7.72}$$

for $t_D-\iota\geq\delta>0$ and any r_D,ξ. Therefore the integral $\int_\tau^{t_D}\frac{\partial J(r_D,t_D,\theta;\xi,\tau)}{\partial r_D}d\theta$ converge uniformly with respect to r_D,t_D,ξ,τ in the above-mentioned domain. It follows that for $t_D>\tau$, and for any r_D,ξ, the derivatives:

$$\frac{\partial V(r_D,t_D;\xi,\tau)}{\partial r_D}=\int_\tau^{t_D}d\theta\int\frac{\partial W_{\zeta,\theta}(r_D,t_D;\zeta,\theta)}{\partial r_D}\Phi(\zeta,\theta;\xi,\tau)d\zeta\tag{7.73}$$

exist and are continuous. According to the inequality (7.71) for the derivatives $\frac{\partial V}{\partial r_D}$ we have the estimate:

$$\begin{aligned}
\left|\frac{\partial V(r_D,t_D;\xi,\tau)}{\partial r_D}\right| &< \int_\tau^{t_D}M_8(\theta-\tau)^{\lambda_1-1}(t_D-\tau)^{-1}e^{-\mu_4\frac{(r_D-\xi)^2}{t_D-\tau}}d\theta\\
&= \frac{M_8}{\lambda_1}(\theta-\tau)^{\lambda_1}(t_D-\tau)^{-1}e^{-\mu_4\frac{(r_D-\xi)^2}{t_D-\tau}}\Bigg|_\tau^{t_D}\\
&= \frac{M_8}{\lambda_1}(t_D-\tau)^{\lambda_1-1}e^{-\mu_4\frac{(r_D-\xi)^2}{t_D-\tau}}\\
&< M_9(t_D-\tau)^{\lambda_1-1}e^{-\mu_4\frac{(r_D-\xi)^2}{t_D-\tau}}.
\end{aligned}\tag{7.74}$$

To prove the existence of the derivatives $\frac{\partial^2 V}{\partial r_D^2}$, consider for $t_D - \tau \geq \delta > 0$, the integral:

$$F(r_D, t_D; \xi, \tau) = \int_\tau^{t_D} d\theta \int \frac{\partial^2 W_{\zeta,\theta}(r_D, t_D; \zeta, \theta)}{\partial r_D^2} \Phi(\zeta, \theta; \xi, \tau) d\zeta$$
$$= \int_\tau^{t_D} \frac{\partial^2 J(r_D, t_D, \theta; \xi, \tau)}{\partial r_D^2} d\theta \tag{7.75}$$

can be represented as a sum:

$$F(r_D, t_D; \xi, \tau) = \int_\tau^{\tau+\frac{\delta}{2}} \frac{\partial^2 J(r_D, t_D, \theta; \xi, \tau)}{\partial r_D^2} d\theta + \int_{\tau+\frac{\delta}{2}}^{t_D} \frac{\partial^2 J(r_D, t_D, \theta; \xi, \tau)}{\partial r_D^2} d\theta$$
$$= F_1(r_D, t_D; \xi, \tau) + F_2(r_D, t_D; \xi, \tau). \tag{7.76}$$

Using the estimates (7.61) and (7.65), we get:

$$\left| \frac{\partial^2 J(r_D, t_D, \theta; \xi, \tau)}{\partial r_D^2} \right| < M_9 \int_{E_1} (t_D - \theta)^{-\frac{3}{2}} (\theta - \tau)^{\lambda_1 - \frac{3}{2}} e^{\left[-\mu_2 \frac{(r_D - \zeta)^2}{t_D - \theta} - \mu_3 \frac{(\zeta - \xi)^2}{\theta - \tau} \right]} d\zeta$$
$$< M_{10}(\theta - \tau)^{\lambda_1} \int_{-\infty}^{+\infty} [(t_D - \theta)(\theta - \tau)]^{-\frac{3}{2}} e^{-\mu_4 \left[\frac{(r_D - \xi)^2}{t_D - \theta} + \frac{(\zeta - \xi)^2}{\theta - \tau} \right]} d\zeta$$
$$= M_{10}(\theta - \tau)^{\lambda_1} \int_{-\infty}^{+\infty} \frac{1}{\sqrt{2}} [(t_D - \theta)(\theta - \tau)]^{-\frac{3}{2}} e^{-\mu_4 \frac{(r_D - \xi)^2}{t_D - \theta}}$$
$$\cdot e^{-\frac{[\sqrt{2}(\zeta - \xi)]^2}{2\sqrt{\frac{\theta - \tau}{\mu_4}}}} d\sqrt{2}(\zeta - \xi)$$
$$= M_{10}(\theta - \tau)^{\lambda_1} [(t_D - \theta)(\theta - \tau)] - \frac{3}{2} \cdot \sqrt{\pi} \sqrt{\frac{\theta - \tau}{\mu_4}} e^{-\mu_4 \frac{(r_D - \xi)^2}{t_D - \theta}}$$
$$= M_{10}(\theta - \tau)^{\lambda_1 - 1} \sqrt{\pi} \mu_4^{-\frac{1}{2}} (t_D - \theta)^{-\frac{3}{2}} e^{-\mu_4 \frac{(r_D - \xi)^2}{t_D - \tau}}$$
$$< M_{11}(\theta - \tau)^{\lambda_1 - 1} (t_D - \tau)^{-\frac{3}{2}} e^{-\mu_4 \frac{(r_D - \xi)^2}{t_D - \tau}}. \tag{7.77}$$

Therefore, for $\tau < \theta < \tau + \frac{\delta}{2}$ the inequality:

$$\left| \frac{\partial^2 J(r_D, t_D, \theta; \xi, \tau)}{\partial r_D^2} \right| < M_\delta^3 (\theta - \tau)^{\lambda_1} \tag{7.78}$$

holds, and it follows that the integral F_1 converges uniformly in the domain $\{t_D - \tau \geq \delta > 0, r_D \in E_1, \xi \in E_1\}$.

To estimate the function under the integral sign in F_2, we shall first transform $\frac{\partial J(r_D,t_D,\theta;\xi,\tau)}{\partial r_D}$ in the following way:

$$
\begin{aligned}
\frac{\partial J(r_D,t_D,\theta;\xi,\tau)}{\partial r_D} &= \int_{E_1} \frac{\partial W_{\zeta,\theta}(r_D,t_D;\zeta,\theta)}{\partial r_D}\Phi(\zeta,\theta;\xi,\tau)d\zeta \\
&= \Phi(y,\theta;\xi,\tau)\int_{E_1} \frac{\partial W_{y,\theta}(r_D,t_D;y,\theta)}{\partial r_D}d\zeta \\
&+ \Phi(y,\theta;\xi,\tau)\int_{E_1} \left[\frac{\partial W_{\zeta,\theta}(r_D,t_D;\zeta,\theta)}{\partial r_D} - \frac{\partial W_{y,\theta}(r_D,t_D;y,\theta)}{\partial r_D}\right]d\zeta \\
&+ \int_{E_1} [\Phi(\zeta,\theta;\xi,\tau) - \Phi(y,\theta;\xi,\tau)]\frac{\partial W_{\zeta,\theta}(r_D,t_D;\zeta,\theta)}{\partial r_D}d\zeta
\end{aligned} \tag{7.79}
$$

here y is an arbitrary fixed point. Let r_D be located inside the ball K of radius $\frac{1}{2}$ with center at an arbitrary point r_D^0. In the first term of the right-hand side of (7.79), we can single out the integral over a ball K_1 of radius 1, concentric to the ball K, and apply to this integral the Gauss-Ostrogradski formula, taking into account that:

$$
\frac{\partial W_{y,\theta}(r_D,t_D;y,\theta)}{\partial r_D} = -\frac{\partial W_{y,\theta}(r_D,t_D;y,\theta)}{\partial \zeta}
$$

we obtain:

$$
\begin{aligned}
\frac{\partial J(r_D,t_D,\theta;\xi,\tau)}{\partial r_D} &= -\Phi(y,\theta;\xi,\tau)\int_{\sum_1} W_{y,\theta}(r_D,t_D;\zeta,\theta)\cos(v,\zeta)d\zeta \\
&+ \int_{E_1\backslash K_1} \frac{\partial W_{y,\theta}(r_D,t_D;\zeta,\theta)}{\partial r_D}d\zeta \\
&+ \Phi(y,\theta;\zeta,\tau)\int_{E_1} \left[\frac{\partial W_{\zeta,\theta}(r_D,t_D;\zeta,\theta)}{\partial r_D} - \frac{\partial W_{y,\theta}(r_D,t_D;\zeta,\theta)}{\partial r_D}\right]d\zeta \\
&+ \int_{E_1} [\Phi(\zeta,\theta;\xi,\tau) - \Phi(y,\theta;\xi,\tau)]\frac{\partial W_{\zeta,\theta}(r_D,t_D;\zeta,\theta)}{\partial r_D}d\zeta.
\end{aligned} \tag{7.80}
$$

Here $\sum_1$ is the boundary of the ball K_1 and v is the outer normal to $\sum_1$. Differentiating (7.80) with respect to r_D and then setting $y = r_D$, we get:

$$
\begin{aligned}
\frac{\partial^2 J(r_D,t_D,\theta;\xi,\tau)}{\partial r_D^2} &= \overbrace{-\Phi(r_D,\theta;\zeta,\tau)\int_{\sum_1} \left.\frac{\partial W_{y,\theta}(r_D,t_D;\zeta,\theta)}{\partial r_D}\right|_{y=r_D} \cos(v,\zeta)d\zeta} \\
&+ \overbrace{\Phi(r_D,\theta;\xi,\tau)\int_{E_1\backslash K_1} \left.\frac{\partial^2 W_{y,\theta}(r_D,t_D;\zeta,\theta)}{\partial r_D^2}\right|_{y=r_D} d\zeta} \\
&+ \overbrace{\Phi(r_D,\theta;\zeta,\tau)\int_{E_1} \left.\left[\frac{\partial^2 W_{\zeta,\theta}(r_D,t_D;\zeta,\theta)}{\partial r_D^2} - \frac{\partial^2 W_{y,\theta}(r_D,t_D;\zeta,\theta)}{\partial r_D^2}\right]\right|_{y=r_D} d\zeta} \\
&+ \overbrace{\int_{E_1} [\Phi(\zeta,\theta;\xi,\tau) - \Phi(r_D,\theta;\xi,\tau)]\frac{\partial^2 W_{\zeta,\theta}(r_D,t_D;\zeta,\theta)}{\partial r_D^2}d\zeta} \\
&= I_1 + I_2 + I_3 + I_4.
\end{aligned} \tag{7.81}
$$

Since the inequality (7.57) holds, we have for the function under the integral sign in I_1 for $r_D \in K$, $\zeta \in \sum_1$, the estimate:

$$\left|\left[\frac{\partial W_{y,\theta}(r_D,t_D;\zeta,\theta)}{\partial r_D}\right]\Bigg|_{y=r_D}\cos(\nu,\zeta)\right| < M_3(t_D-\theta)^{-1}e^{-\frac{\mu_2(r_D-\zeta)^2}{t_D-\theta}} < M_{12}(t_D-\theta)^{-1}e^{-\frac{\mu_2}{4(t_D-\theta)}} < M_{13}. \tag{7.82}$$

Taking into account additionally (7.65), we get:

$$|I_1| < M_{13}\omega_1 \cdot M_4(\theta-\tau)^{-\frac{3}{2}+\lambda_1}e^{-\frac{\mu_3(r_D-\xi)^2}{\theta-\tau}} < M_{13}(\theta-\tau)^{-\frac{3}{2}+\lambda_1}e^{-\frac{\mu_3(r_D-\xi)^2}{\theta-\tau}}. \tag{7.83}$$

(Here ω_1 is the area of the unit sphere in the space) Hence:

$$|I_1| < M_\delta^4, \qquad \text{for } \tau+\frac{\delta}{2}<\theta<t_D, \;\; r_D\in K, \;\; \xi\in E_1. \tag{7.84}$$

Similarly, we can estimate $|I_2|$. Using (7.61), (7.65) and the elementary inequality (7.54), we have:

$$\begin{aligned}
|I_2| &< |\Phi(r_D,\theta;\xi,\tau)|\left|\int_{E_1\backslash K_1}\left[\frac{\partial^2 W_{y,\theta}(r_D,t_D;\zeta,\theta)}{\partial r_D^2}\right]_{y=r_D} d\zeta\right| \\
&< M_4(\theta-\tau)^{-\frac{3}{2}+\lambda_1}e^{-\frac{\mu_3(r_D-\xi)^2}{\theta-\tau}}\cdot\int_{E_1\backslash K_1} M_3(t_D-\tau)^{-\frac{3}{2}}e^{-\frac{\mu_2(r_D-\xi)^2}{t_D-\tau}}d\zeta \\
&< M_{14}(\theta-\tau)^{-\frac{3}{2}+\lambda_1}e^{-\frac{\mu_3(r_D-\xi)^2}{\theta-\tau}}\cdot\int_{r_D-\zeta\geq\frac{1}{2}}(t_D-\theta)^{-\frac{3}{2}}e^{-\frac{\mu_2(r_D-\xi)^2}{t_D-\theta}}d\zeta \\
&= M_{14}(\theta-\tau)^{-\frac{3}{2}+\lambda_1}e^{-\frac{\mu_3(r_D-\xi)^2}{\theta-\tau}}\cdot\int_{\frac{1}{2}}^{+\infty}(t_D-\theta)^{-\frac{3}{2}}e^{-\frac{\mu_2 r^2}{t_D-\theta}}dr \\
&< 4M_{14}(\theta-\tau)^{-\frac{3}{2}+\lambda_1}e^{-\frac{\mu_3(r_D-\xi)^2}{\theta-\tau}}\cdot\int_{\frac{1}{2}}^{+\infty}(t_D-\theta)^{-\frac{3}{2}}r^2e^{-\frac{\mu_2 r^2}{t_D-\theta}}dr \\
&= 4M_{14}(\theta-\tau)^{-\frac{3}{2}+\lambda_1}e^{-\frac{\mu_3(r_D-\xi)^2}{\theta-\tau}}\cdot(t_D-\theta)^{-\frac{1}{2}}\int_{\frac{1}{2}}^{+\infty}\frac{r^2}{t_D-\theta}e^{-\mu_2\frac{r^2}{t_D-\theta}}dr \\
&< 4M_{14}(\theta-\tau)^{-\frac{3}{2}+\lambda_1}e^{-\frac{\mu_3(r_D-\xi)^2}{\theta-\tau}}\cdot(t_D-\theta)^{-\frac{1}{2}}\int_{\frac{1}{2}}^{+\infty}e^{-\mu_2'\frac{r^2}{t_D-\theta}}dr \\
&< 4M_{14}(\theta-\tau)^{-\frac{3}{2}+\lambda_1}e^{-\frac{\mu_3(r_D-\xi)^2}{\theta-\tau}}\cdot(t_D-\theta)^{-\frac{1}{2}}\int_{0}^{+\infty}e^{-\mu_2'\frac{r^2}{t_D-\theta}}dr \\
&= 4M_{14}(\theta-\tau)^{-\frac{3}{2}+\lambda_1}e^{-\frac{\mu_3(r_D-\xi)^2}{\theta-\tau}}\cdot(t_D-\theta)^{-\frac{1}{2}}\cdot\frac{\sqrt{\pi}}{2}\sqrt{\frac{t_D-\theta}{\mu_2'}} \\
&< M_{15}(\theta-\tau)^{-\frac{3}{2}+\lambda_1}e^{-\frac{\mu_3(r_D-\xi)^2}{\theta-\tau}}.
\end{aligned} \tag{7.85}$$

It follows that:

$$|I_2| < M_\delta^5, \qquad \text{for } \tau+\frac{\delta}{2}<\theta<t_D, \;\; r_D\in K, \;\; \xi\in E_1. \tag{7.86}$$

Applying the mean-value theorem and using the elementary inequality (7.54), (7.61), we come to the following estimate for the function under the integral sign in I_3:

$$\begin{aligned}
\int_{E_1}\left|\frac{\partial^2 W_{\zeta,\theta}(r_D,t_D;\zeta,\theta)}{\partial r_D^2}-\frac{\partial^2 W_{y,\theta}(r_D,t_D;\zeta,\theta)}{\partial r_D^2}\right|d\zeta &= \int_{E_1}\left|\frac{\partial\frac{\partial^2 W_{\zeta,\theta}(r_D,t_D;\zeta,\theta)}{\partial r_D^2}}{\partial\zeta}(y-\xi)\right|d\zeta\\
&= \left|\frac{\partial^2 W_{\zeta,\theta}(r_D,t_D;\zeta,\theta)}{\partial r_D^2}(r_D-\xi)\right|d\zeta\\
&< M_3(t_D-\theta)^{-\frac{3}{2}}e^{-\frac{\mu_2(r_D-\xi)^2}{t_D-\theta}}(r_D-\xi)\\
&= M_3(t_D-\theta)^{-\frac{1}{2}}\frac{r_D-\xi}{t_D-\theta}e^{-\frac{\mu_2(r_D-\xi)^2}{t_D-\theta}}\\
&= M_3(t_D-\theta)^{-1}\left[\frac{(r_D-\xi)^2}{t_D-\theta}\right]\frac{1}{2}e^{-\frac{\mu_2(r_D-\xi)^2}{t_D-\theta}}\\
&< M_{16}(t_D-\theta)^{-1}(t_D-\theta)^{\frac{\lambda}{2}}e^{-\frac{\mu_2''(r_D-\xi)^2}{t_D-\theta}}.
\end{aligned}\tag{7.87}$$

From (7.87), (7.65), we obtain:

$$\begin{aligned}
|I_3| &< M_4(\theta-\tau)^{-\frac{3}{2}+\lambda_1}e^{-\frac{\mu_3(r_D-\xi)^2}{\theta-\tau}}\cdot M_{16}(t_D-\theta)^{-1}(t_D-\theta)^{\frac{\lambda}{2}}e^{-\frac{\mu_2''(r_D-\xi)^2}{t_D-\theta}}\\
&< M_{17}(\theta-\tau)^{-\frac{3}{2}+\lambda_1}(t_D-\theta)^{\frac{\lambda}{2}-1}e^{-\frac{\mu_3(r_D-\xi)^2}{\theta-\tau}}.
\end{aligned}\tag{7.88}$$

It follows that:

$$|I_3|<M_\delta^6(t_D-\theta)^{\frac{\lambda}{2}-1},\qquad \text{for } \tau+\frac{\delta}{2}<\theta<t_D,\quad r_D\in K,\quad \xi\in E_1.\tag{7.89}$$

The integral I_4 can be represented as the sum $I_4=I_4^{(1)}+I_4^{(2)}$, where the integral $I_4^{(1)}$ is taken over the domain defined by the inequality $(\zeta-r_D)^2<l(\theta-\tau)\le l(t_D-\theta)$. Taking into account (7.61), (7.66), we obtain:

$$\begin{aligned}
|I_4^{(1)}| &< \int_{(\zeta-r_D)^2<l(\theta-\tau)\le l(t_D-\theta)} M_3(t_D-\theta)^{-\frac{3}{2}}e^{-\frac{\mu_2(r_D-\zeta)^2}{t_D-\theta}}\cdot M_4|\zeta-r_D|^{\lambda_2}(\theta-\tau)^{-\frac{3}{2}+\lambda_3}e^{-\frac{\mu_3(\zeta-\xi)^2}{\theta-\tau}}d\zeta\\
&< M_{18}\int_{E_1}(t_D-\theta)^{-\frac{3}{2}+\frac{\lambda_2}{2}}(\theta-\tau)^{-\frac{3}{2}+\lambda_3}e^{-\mu_4\left[\frac{(r_D-\zeta)^2}{t_D-\theta}+\frac{(\zeta-\xi)^2}{\theta-\tau}\right]}d\zeta.
\end{aligned}\tag{7.90}$$

Using arguments similar to those used when we deduced the inequality (7.68), we get:

$$|I_4^{(1)}|<M_{19}(t_D-\theta)^{\frac{\lambda_2}{2}-1}(\theta-\tau)^{\lambda_3-1}(t_D-\tau)^{-\frac{1}{2}}e^{-\frac{\mu_4(r_D-\xi)^2}{t_D-\tau}}.\tag{7.91}$$

Therefore, for $\tau+\frac{\delta}{2}<\theta<t_D,\ r_D\in K,\xi\in E_1$, the estimate:

$$|I_4^{(1)}|<M_\delta^7(t_D-\theta)^{\frac{\lambda_2}{2}-1}\tag{7.92}$$

is valid. The integral $I_4^{(2)}$ can represented with the help of inequalities (7.61), (7.65):

$$\begin{aligned}
|I_4^{(2)}| &< M_{19}\int_{l(t_D-\theta)\geq|\zeta-r_D|^2\geq l(\theta-\tau)} (t_D-\theta)^{-\frac{3}{2}}e^{-\frac{\mu_2(r_D-\zeta)^2}{t_D-\theta}}\cdot(\theta-\tau)^{-\frac{3}{2}+\lambda_1}e^{-\frac{\mu_3(\zeta-\xi)^2}{\theta-\tau}}d\zeta \\
&< M_{20}\int_{E_1}(\theta-\tau)^{-\frac{3}{2}+\lambda_1}e^{-\frac{\mu_3(\zeta-\xi)^2}{\theta-\tau}}\cdot l^{-\frac{\lambda_1}{2}}|\zeta-r_D|^{\lambda_1}\cdot(t_D-\theta)^{-\frac{3}{2}}e^{-\frac{\mu_2(r_D-\zeta)^2}{t_D-\theta}}d\zeta \\
&< M_{21}\int_{E_1}(\theta-\tau)^{-\frac{3}{2}+\lambda_1}(t_D-\theta)^{-\frac{3}{2}+\frac{\lambda_1}{2}}e^{-\mu_4\left[\frac{(r_D-\zeta)^2}{t_D-\theta}+\frac{(\zeta-\xi)^2}{\theta-\tau}\right]}d\zeta \\
&< M_{22}(t_D-\theta)^{\frac{\lambda_2}{2}-1}(\theta-\tau)^{\frac{\lambda_1}{2}-1}(t_D-\tau)^{-\frac{1}{2}}e^{-\frac{\mu_4(r_D-\xi)^2}{t_D-\tau}}.
\end{aligned} \tag{7.93}$$

If $\tau+\frac{\delta}{2}<\theta<t_D, r_D\in K, \xi\in E_1$, this lead to:

$$|I_4^{(2)}|<M_\delta^8(t_D-\theta)^{\frac{\lambda_2}{2}-1}. \tag{7.94}$$

It follows from (7.94), (7.92), (7.89), (7.86), (7.84) that the integral:

$$F_2(r_D,t_D;\zeta,\theta)=\int_{\tau+\frac{\delta}{2}}^{t_D}\frac{\partial^2 J(r_D,t_D,\theta;\zeta,\theta)}{\partial r_D^2}d\theta$$

converges uniformly in the domain $\{t_D-\tau\geq\delta, r_D\in K, \xi\in E_1\}$. We have shown before that the integral:

$$F_1(r_D,t_D;\zeta,\theta)=\int_{\tau}^{\tau+\frac{\delta}{2}}\frac{\partial^2 J(r_D,t_D,\theta;\zeta,\theta)}{\partial r_D^2}d\theta$$

also converges uniformly in this domain. Therefore in this domain there exist continuous derivatives:

$$\frac{\partial^2 V(r_D,t_D;\xi,\tau)}{\partial r_D^2}=\int_\tau^{t_D}d\theta\int\frac{\partial^2 W_{\zeta,\theta}(r_D,t_D;\zeta,\theta)}{\partial r_D^2}\Phi(\zeta,\theta;\xi,\tau)d\zeta. \tag{7.95}$$

Since the number $\delta>0$ and the point x^0 (the center of the ball K) are arbitrary, equality (7.95) holds everywhere in the domain $\{0\leq\tau<t\leq t_D, r_D\in E_1, \xi\in E_1\}$.

Because of inequalities (7.93), (7.91), (7.88), (7.85) and (7.83), we have:

$$\begin{aligned}
\left|\frac{\partial^2 V(r_D,t_D;\xi,\tau)}{\partial r_D^2}\right| &\leq \int_\tau^{\tau+\frac{\delta}{2}}\left|\frac{\partial^2 J(r_D,t_D,\theta;\xi,\tau)}{\partial r_D^2}\right|d\theta+\int_{\tau+\frac{\delta}{2}}^{t_D}\left|\frac{\partial^2 J(r_D,t_D,\theta;\xi,\tau)}{\partial r_D^2}\right|d\theta \\
&< M_{23}(t_D-\tau)^{-\frac{3}{2}+\lambda_4}e^{-\frac{\mu_4(r_D-\xi)^2}{t_D-\tau}},
\end{aligned} \tag{7.96}$$

where $\lambda_4>0$.

Now we shall show that for $t_D>\tau$ and any r_D and ξ the function $V(r_D,t_D;\xi,\tau)$ has a continuous derivative $\frac{\partial V}{\partial t_D}$, which can be calculated by the following formula:

$$\begin{aligned}
\frac{\partial V(r_D,t_D;\xi,\tau)}{\partial t_D} &= \Phi(\zeta,\theta;\xi,\tau)+\int_\tau^{t_D}d\theta\int\frac{\partial W_{\zeta,\theta}(r_D,t_D;\zeta,\theta)}{\partial t_D}\Phi(\zeta,\theta;\xi,\tau)d\zeta \\
&= \Phi(\zeta,\theta;\xi,\tau)+\int_\tau^{t_D}\int\frac{\partial J(r_D,t_D,\theta;\zeta,\theta)}{\partial t_D}d\theta.
\end{aligned} \tag{7.97}$$

According to Equation (7.44), for $W_{\zeta,\theta}(r_D, t_D; \zeta, \theta)$ we have:

$$\frac{\partial J(r_D, t_D; \xi, \tau)}{\partial t_D} = \int_{E_1} a(\zeta, \theta) \frac{\partial^2 W_{\zeta,\theta}(r_D, t_D; \zeta, \theta)}{\partial r_D^2} \Phi(\zeta, \theta; \xi, \tau) d\zeta \tag{7.98}$$

if $\tau < \theta < t_D$. Therefore the integral:

$$\int_\tau^{t_D} \frac{\partial J(r_D, t_D; \xi, \tau)}{\partial t_D} d\theta \tag{7.99}$$

is similar to the integral (7.95) we have studied before, and as before, we can see that the integral (7.98) converges absolutely and uniformly in the domain $\{t_D - \tau \geq \delta >), r_D \in E_1, \xi \in E_1\}$.

Let us consider the difference:

$$\begin{aligned}
&\frac{V(r_D, t_D + \Delta t_D; \xi, \tau) - V(r_D, t_D; \xi, \tau)}{\Delta t_D} - \Phi(r_D, t_D; \xi, \tau) - \int_\tau^{t_D} \frac{\partial J(r_D, t_D, \theta; \xi, \tau)}{\partial t_D} d\theta \\
&= \int_\tau^{t_D} \left[\frac{J(r_D, t_D + \Delta t_D; \xi, \tau) - J(r_D, t_D; \xi, \tau)}{\Delta t_D} - \frac{\partial J(r_D, t_D, \theta; \xi, \tau)}{\partial t_D} \right] d\theta \\
&+ \frac{1}{\Delta t_D} \int_{t_D}^{t_D + \Delta t_D} J(r_D, t_D + \Delta t_D, \theta; \xi, \tau) d\theta - \Phi(r_D, t_D; \xi, \tau) \\
&= \{J(r_D, t_D + \Delta t_D, t_D'; \xi, \tau) - \Phi(r_D, t_D; \xi, \tau)\} - \left[\int_\tau^{\tau+\eta_1} + \int_{t_D - \eta_2}^{t_D} \right] \frac{\partial J(r_D, t_D, \theta; \xi, \tau)}{\partial t_D} d\theta \\
&+ \left[\int_\tau^{\tau+\eta_1} + \int_{t_D - \eta_2}^{t_D} \right] \frac{\partial J(r_D, t_D'', \theta; \xi, \tau)}{\partial t_D} d\theta + \int_{\tau+\eta_1}^{t_D - \eta_2} \left[\frac{\partial J(r_D, t_D'', \theta; \xi, \tau)}{\partial t_D} - \frac{\partial J(r_D, t_D, \theta; \xi, \tau)}{\partial t_D} \right] d\theta,
\end{aligned} \tag{7.100}$$

where t_D' and t_D'' are some value between t_D and $t_D + \Delta t_D$. We shall suppose that $\Delta t_D > 0$. According this relation we get:

$$\lim_{\Delta t_D \to 0} \{J(r_D, t_D + \Delta t_D, t_D'; \xi, \tau) - \Phi(r_D, t_D; \xi, \tau)\} = 0.$$

Since the integral (7.99) converges uniformly with respect to t_D, for sufficiently small $\eta_1 > 0$ and $\eta_2 > 0$ the first-three integrals in the right-hand side of (7.100) can be made arbitrarily small in absolute value independently of Δt_D. The fourth integral can be estimated in the following way:

$$\left| \int_{t_D - \eta_2}^{t_D} \frac{\partial J(r_D, t_D'', \theta; \xi, \tau)}{\partial t_D} d\theta \right| \leq \int_{t_D - \eta_2}^{t_D} \left| \frac{\partial J(r_D, t_D'', \theta; \xi, \tau)}{\partial t_D} \right| d\theta.$$

It follows that this integral can be made arbitrarily small for sufficiently small $\eta_2 > 0$ and Δt_D. Finally for fixed $\eta_1 > 0$ and $\eta_2 > 0$ the last integral in the right-hand side of (7.100) tends to zero as $\Delta t_D \to 0$, because the function under the integral sign tends to zero and is continuous in θ and t_D'' in the closed domain $\{\tau + \eta_1 \leq \theta \leq t_D - \eta_2, t_D \leq t_D'' \leq T\}$.

Thus, the left-hand side of the equality (7.100) tends to zero as $\Delta t_D \to +0$. In a similar way we can consider the case where $\Delta t_D \to -0$.

The equality (7.97) is proved.

Now we pass to the immediate construction of the fundamental solution $T_e(r_D, t_D; \xi, \tau)$ in the form (7.50).

Suppose that the function $\Phi(r_D, t_D; \xi, \tau)$ satisfies the conditions (7.65) and (7.66). Using formulae (7.50), (7.73), (7.95), (7.97), we get:

$$\begin{aligned} L_{r_D,t_D}(T_e) &= L_{r_D,t_D}(W_{\xi,\tau}(r_D, t_D; \xi, \tau)) \\ &\quad + \int_{\tau}^{t_D} d\theta \int_{E_1} L_{r_D,t_D}(W_{\zeta,\theta}(r_D, t_D; \zeta, \theta))\Phi(\zeta, \theta; \xi, \tau)d\zeta - \Phi(r_D, t_D; \xi, \tau). \end{aligned} \tag{7.101}$$

The requirement that $L_{r_D,t_D}(T_e)$ be zero leads to the following equation for Φ:

$$\Phi(r_D, t_D; \xi, \tau) = L_{r_D,t_D}(W_{\xi,\tau}(r_D, t_D; \xi, \tau)) + \int_{\tau}^{t_D} d\theta \int_{E_1} L_{r_D,t_D}(W_{\zeta,\theta}(r_D, t_D; \zeta, \theta)\Phi(\zeta, \theta; \xi, \tau)d\zeta. \tag{7.102}$$

We seek a solution of the integral equation (7.102) as a series:

$$\Phi(r_D, t_D; \xi, \tau) = \sum_{m=1}^{\infty} \Phi(r_D, t_D; \xi, \tau), \tag{7.103}$$

where

$$\begin{aligned} \Phi_1(r_D, t_D; \xi, \tau) &= L_{r_D,t_D}(W_{\xi,\tau}(r_D, t_D; \xi, \tau)) \\ \Phi_{m+1}(r_D, t_D; \xi, \tau) &= \int_{\tau}^{t_D} d\theta \int_{E_1} L_{r_D,t_D}(W_{\zeta,\theta}(r_D, t_D; \zeta, \theta)\Phi_m(\zeta, \theta; \xi, \tau)d\zeta \quad (m = 1, 2, \dots). \end{aligned} \tag{7.104}$$

Let us show that the series (7.103) converges uniformly for $t_D > \tau$. Taking into account (7.103), we have for $W_{\xi,\tau}(r_D, t_D; \xi, \tau)$:

$$\Phi_1(r_D, t_D; \xi, \tau) = [a(r_D, t_D) - a(\xi, t_D)]\frac{\partial^2 W_{\xi,\tau}(r_D, t_D; \xi, \tau)}{\partial r_D^2} + b(r_D, t_D)\frac{\partial W_{\xi,\tau}(r_D, t_D; \xi, \tau)}{\partial r_D}$$

By inequalities (7.46), (7.47) and (7.57), (7.61) we get estimate:

$$|\Phi_1(r_D, t_D; \xi, \tau)| = |L_{r_D,t_D}(W_{\xi,\tau}(r_D, t_D; \xi, \tau))| < M_{24}(t_D - \tau)^{-\frac{3}{2}+\lambda} e^{-\frac{\mu_5(r_D-\xi)^2}{t_D-\tau}}, \tag{7.105}$$

where $\mu_5 > 0$. Then we have:

$$\begin{aligned} &|\Phi_1(r_D, t_D; \xi, \tau)| \\ &< M_{24}^2 \int_{\tau}^{t_D} d\theta \int_{E_1} (t_D - \theta)^{-\frac{3}{2}+\lambda} e^{-\frac{\mu_5(r_D-\zeta)^2}{t_D-\theta}} (\theta - \tau)^{-\frac{3}{2}+\lambda} e^{-\frac{\mu_5(\zeta-\xi)^2}{\theta-\tau}} d\zeta \\ &= M_{24}^2 \int_{\tau}^{t_D} (t_D - \theta)^{\lambda-1}(\theta - \tau)^{\lambda-1} d\theta \int_{-\infty}^{+\infty} [(t_D - \theta)(\theta - \tau)]^{-\frac{1}{2}} e^{-\mu_5\left[\frac{(r_D-\zeta)^2}{t_D-\theta} + \frac{(\zeta-\xi)^2}{\theta-\tau}\right]} d\zeta \\ &= M_{24}^2 \left[\frac{\pi}{\mu_5(t_D - \tau)}\right]^{\frac{1}{2}} e^{-\mu_5\frac{(r_D-\xi)^2}{t_D-\tau}} \int_{\tau}^{t_D} (t_D - \theta)^{\lambda-1}(\theta - \tau)^{\lambda-1} d\theta \\ &= M_{24}^2 \left[\frac{\pi}{\mu_5(t_D - \tau)}\right]^{\frac{1}{2}} e^{-\mu_5\frac{(r_D-\xi)^2}{t_D-\tau}} (t_D - \tau)^{2\lambda-1} \int_0^1 s^{\lambda-1}(1 - s)^{\lambda-1} ds \\ &= \frac{\Gamma^2(\lambda)}{\Gamma(2\lambda)} M_{24}^2 \left[\frac{\pi}{\mu_5}\right]^{\frac{1}{2}} (t_D - \tau)^{2\lambda-\frac{3}{2}} e^{-\mu_5\frac{(r_D-\xi)^2}{t_D-\tau}}, \end{aligned} \tag{7.106}$$

where $\Gamma(h)$ is the gamma-function.

It is easy to show by induction with respect to m that:

$$|\Phi_m(r_D, t_D; \xi, \tau)| < \frac{\Gamma^m(\lambda)}{\Gamma(m\lambda)} M_{24}^m \left[\frac{\pi}{\mu_5}\right]^{\frac{m-1}{2}} (t_D - \tau)^{m\lambda - \frac{3}{2}} e^{-\mu_5 \frac{(r_D-\xi)^2}{t_D-\tau}} \quad (m = 1, 2, \ldots) \tag{7.107}$$

Since we have $\Gamma(m\lambda) \geq [m\lambda - 1]!$ for $m\lambda > 2$, it follows from (7.107) that series (7.103) converges absolutely and uniformly for $t_D > \tau$ and the inequality (7.65) holds for the function $\Phi(r_D, t_D; \xi, \tau)$. From the uniform convergence of the series (7.103) it follows that the function $\Phi(r_D, t_D; \xi, \tau)$ is continuous jointly in all its arguments in the domain $\{0 \leq \tau < t_D \leq T, r_D \in E_1, \xi \in E_1\}$.

Let us show that the function $\Phi(r_D, t_D; \xi, \tau)$ we have constructed satisfies the inequality (7.66) for $|r'_D - r_D|^2 < \alpha(t_D - \tau)$. It is sufficient to prove that both terms of the right-hand of Equation (7.103) satisfy the inequality (7.66).

Consider first the difference $L(W_{\xi,\tau}(r'_D, t_D; \xi, \tau)) - L(W_{\xi,\tau}(r_D, t_D; \xi, \tau))$. We shall concentrate on finding bounds only for the terms most difficult to estimate, that is, terms containing second derivatives of the function $W_{\xi,\tau}$. Using the inequalities (7.46), (7.47) and (7.57), (7.61) we obtain:

$$\begin{aligned}
&\left|[a(r'_D, t_D) - a(\xi, \tau)]\frac{\partial^2 W_{\xi,\tau}(r'_D, t_D; \xi, \tau)}{\partial r_D^2} - [a(r_D, t_D) - a(\xi, \tau)]\frac{\partial^2 W_{\xi,\tau}(r_D, t_D; \xi, \tau)}{\partial r_D^2}\right| \\
&\leq \left|[a(r'_D, t_D) - a(r_D, t_D)]\frac{\partial^2 W_{\xi,\tau}(r'_D, t_D; \xi, \tau)}{\partial r_D^2}\right| \\
&\quad + \left|[a(r_D, t_D) - a(\xi, \tau)]\left[\frac{\partial^2 W_{\xi,\tau}(r'_D, t_D; \xi, \tau)}{\partial r_D^2} - \frac{\partial^2 W_{\xi,\tau}(r_D, t_D; \xi, \tau)}{\partial r_D^2}\right]\right| \\
&< M_{25}|r'_D - r_D|^\lambda (t_D - \tau)^{-\frac{3}{2}} e^{-\mu_2 \frac{(r'_D-\xi)^2}{t_D-\tau}} + M_{26}|r'_D - r_D|[|r_D - \xi|^\lambda + |t_D - \tau|^\lambda] \\
&\quad \cdot (|r_D - \xi| + |r'_D - \xi|)(t_D - \tau)^{-\frac{3}{2}} \left\{e^{-\mu_2 \frac{(r_D-\xi)^2}{t_D-\tau}} + \left[1 + \frac{(r_D - \xi)^2}{t_D - \tau}\right] e^{-\mu_4 \frac{(\bar{r}_D-\xi)^2}{t_D-\tau}}\right\},
\end{aligned} \tag{7.108}$$

where $\bar{r}_D$ belongs to the interval jointing r_D and r'_D. We consider first such ξ that satisfy the condition $|r_D - \xi| \leq 2|r'_D - r_D|$, and then the remaining values of xi. Using the inequality (7.54) we can see that for $|r'_D - r_D|^2 < 2\alpha(t_D - \tau)$ the right-hand of (7.108) is not greater than:

$$M_{27}|r'_D - r_D|^{\lambda_2}(t_D - \tau)^{-\frac{3}{2}+\lambda_3} e^{-\mu_5 \frac{(r_D-\xi)^2}{t_D-\tau}},$$

where $\lambda_2 > 0$ and $\lambda_3 > 0$. Similarly we can estimate the remaining terms of the difference $L(W_{\xi,\tau}(r'_D, t_D; \xi, \tau)) - L(W_{\xi,\tau}(r_D, t_D; \xi, \tau))$. Thus we have:

$$|L(W_{\xi,\tau}(r'_D, t_D; \xi, \tau)) - L(W_{\xi,\tau}(r_D, t_D; \xi, \tau))| < M_{27}|r'_D - r_D|^{\lambda_2}(t_D - \tau)^{-\frac{3}{2}+\lambda_3} e^{-\mu_5 \frac{(r_D-\xi)^2}{t_D-\tau}}. \tag{7.109}$$

Now we pass to the estimate of the term $\psi(r_D, t_D; \xi, \tau)$ of the right-hand side of (7.102). We have:

$$\begin{aligned}
&|\psi(r'_D, t_D; \xi, \tau) - \psi(r_D, t_D; \xi, \tau)| \\
&= \int_{t_D - \frac{1}{2\alpha}|r'_D - r_D|^2}^{t_D} d\theta \int_{E_1} L(W_{\zeta,\tau}(r'_D, t_D; \zeta, \tau))\Phi(\zeta, \tau; \xi, \tau) d\zeta \\
&\quad - \int_{t_D - \frac{1}{2\alpha}|r'_D - r_D|^2}^{t_D} d\theta \int_{E_1} L(W_{\zeta,\tau}(r_D, t_D; \zeta, \tau))\Phi(\zeta, \tau; \xi, \tau) d\zeta \\
&\quad + \int_{\tau}^{t_D - \frac{1}{2\alpha}|r'_D - r_D|^2} d\theta \int_{E_1} \{L(W_{\zeta,\tau}(r'_D, t_D; \zeta, \tau)) - L(W_{\zeta,\tau}(r_D, t_D; \zeta, \tau))\}\Phi(\zeta, \tau; \xi, \tau) d\zeta \\
&= J_1 + J_2 + J_3
\end{aligned} \tag{7.110}$$

By (7.65) and (7.105), for $|r'_D - r_D|^2 < \alpha(t_D - \tau)$ we get:

$$
\begin{aligned}
|J_2| &< M_{28} \int_{t_D - \frac{1}{2\alpha}|r'_D - r_D|^2}^{t_D} d\theta \int_{E_1} (t_D - \theta)^{-\frac{3}{2}+\lambda} (\theta - \tau)^{-\frac{3}{2}+\lambda} e^{-\mu_5\left[\frac{(r_D-\zeta)^2}{t_D-\theta} + \frac{(\zeta-\xi)^2}{\theta-\tau}\right]} d\zeta \\
&= M_{29}(t_D - \tau)^{-\frac{1}{2}} e^{-\mu_5 \frac{(r_D-\xi)^2}{t_D-\tau}} \int_{t_D - \frac{1}{2\alpha}|r'_D - r_D|^2}^{t_D} (t_D - \theta)^{\lambda-1} (\theta - \tau)^{\lambda-1} d\theta \\
&< M_{29}\left(t_D - \tau - \frac{1}{2\alpha}|r'_D - r_D|^2\right)^{\lambda-1} (t_D - \tau)^{-\frac{1}{2}} e^{-\mu_5 \frac{(r_D-\xi)^2}{t_D-\tau}} \int_{t_D - \frac{1}{2\alpha}|r'_D - r_D|^2}^{t_D} (t_D - \theta)^{\lambda-1} d\theta \\
&< M_{29}|r'_D - r_D|^{2\lambda} (t_D - \tau)^{-\frac{3}{2}+\lambda} e^{-\mu_5 \frac{(r_D-\xi)^2}{t_D-\tau}}
\end{aligned} \tag{7.111}
$$

In a similar way we can estimate $|J_1|$. Then, taking into account (7.65) and (7.109), we find:

$$
\begin{aligned}
|J_3| &< M_{30}|r'_D - r_D|^{\lambda_2} \int_{\tau}^{t_D - \frac{1}{2\alpha}|r'_D - r_D|^2} d\theta \int_{E_1} (t_D - \theta)^{-\frac{3}{2}+\lambda} (\theta - \tau)^{-\frac{3}{2}+\lambda} e^{-\mu_5\left[\frac{(r_D-\zeta)^2}{t_D-\theta} + \frac{(\zeta-\xi)^2}{\theta-\tau}\right]} d\zeta \\
&= M_{31}|r'_D - r_D|^{\lambda_2} (t_D - \tau)^{-\frac{1}{2}} e^{-\mu_5 \frac{(r_D-\xi)^2}{t_D-\tau}} \int_{\tau}^{t_D - \frac{1}{2\alpha}|r'_D - r_D|^2} (t_D - \theta)^{\lambda_3-1} (\theta - \tau)^{\lambda-1} d\theta \\
&< M_{32}|r'_D - r_D|^{\lambda_2} (t_D - \tau)^{\lambda+\lambda_3-\frac{3}{2}} e^{-\mu_5 \frac{(r_D-\xi)^2}{t_D-\tau}} \int_{\tau}^{t_D - \frac{1}{2\alpha}|r'_D - r_D|^2} s^{\lambda-1}(1-s)^{\lambda_3-1} ds \\
&< M_{32}|r'_D - r_D|^{\lambda_2} (t_D - \tau)^{-\frac{3}{2}+\lambda+\lambda_3} e^{-\mu_5 \frac{(r_D-\xi)^2}{t_D-\tau}}
\end{aligned} \tag{7.112}
$$

Comparing the inequalities (7.110)–(7.112), we obtain the required estimate (7.66) for the right-hand side of Equation (7.102) and, therefore, for the function $\Phi(r_D, t_D; \xi, \tau)$. Thus we have proved that the function $T_e(r_D, t_D; \xi, \tau)$, given by formula (7.50), for $t_D > \tau$ is continuous jointly the variables r_D, t_D, ξ and τ, together with its derivatives $\frac{\partial T_e}{\partial r_D}$, $\frac{\partial^2 T_e}{\partial r_D^2}$, and $\frac{\partial T_e}{\partial t_D}$. The function $T_e(r_D, t_D; \xi, \tau)$ satisfies Equation (7.44) with respect to the variables r_D, t_D.

It follows from estimate (7.53) for the function $W_{\xi,\tau}(r_D, t_D; \xi, \tau)$ and the estimate (7.70) for the function (7.52) that $T_e(r_D, t_D; \xi, \tau)$ is bounded for $t_D - \tau + |r_D - \xi| \geq \delta > 0$.

Now we show that for $T_e(r_D, t_D; \xi, \tau)$ the relation (7.45) holds.

Let $\varphi(\xi)$ be a continuous and bounded function for E_1. Suppose that r_D is an arbitrary point of E_1 and $0 \leq \tau < T$. Then we have:

$$
\int_{E_1} T_e(r_D, t_D; \xi, \tau)\varphi(\xi) d\xi = \int_{E_1} W_{\xi,\tau}(r_D, t_D; \xi, \tau)\varphi(\xi) d\xi + \int_{E_1} V(r_D, t_D; \xi, \tau)\varphi(\xi) d\xi,
$$

where the function $V(r_D, t_D; \xi, \tau)$ is given by (7.52). Taking into account (7.70), we can see that:

$$
\left| \int_{E_1} V(r_D, t_D; \xi, \tau)\varphi(\xi) d\xi \right| < M_{33} \sup|\varphi| (t_D - \tau)^{\lambda_1}
$$

therefore,

$$
\lim_{t_D \to \tau+0} \int_{E_1} V(r_D, t_D; \xi, \tau)\varphi(\xi) d\xi = 0 \tag{7.113}
$$

uniformly with respect to $r_D \in E_1$. Thus the relation (7.45) holds.

Thus we have constructed the function $T_e(r_D, t_D; \xi, \tau)$ that is a fundamental solution of Equation (7.44).

Let us prove the uniqueness of the fundamental solution of Equation (7.44).

From the properties of the fundamental solution it follows that the function:

$$\mu(r_D,t_D)=\int_{E_1} T_e(r_D,t_D;\xi,\tau)\varphi(\xi)d\xi \tag{7.114}$$

is a bounded solution of the Cauchy problem for Equation (7.44) in the domain $\{0\le\tau<t_D\le T, r_D\in E_1\}$ with primal condition:

$$\mu(r_D,t_D)=\varphi(r_D), \tag{7.115}$$

where $\varphi(r_D)$ is an arbitrary continuous function vanishing outside of some bounded domain. In fact, let $\varphi(r_D)=0$ for $r_D\ge R$; Then:

$$\mu(r_D,t_D)=\int_{|\xi|\le R} T_e(r_D,t_D;\xi,\tau)\varphi(\xi)d\xi. \tag{7.116}$$

For $t_D>\tau$, we have:

$$L(\mu)=\int_{|\xi|\le R} L_{r_D,t_D}(T_e(r_D,t_D;\xi,\tau))\varphi(\xi)d\xi=0.$$

According to (7.45) the function (7.114) for $t_D=\tau$ satisfies the initial condition (7.115). From 7.45 we can also deduce that the function (7.114) is bounded in the domain $\{\tau<t_D\le T, r_D\ge 2R\}$, and the function also is bounded in the domain, we have:

$$|\mu(r_D,t_D)|\le \sup_{|r_D-\xi|\ge R}|T_e(r_D,t_D;\xi,\tau)|\int_{|\xi|\le R}|\varphi(\xi)|d\xi\le M_R.$$

Thus a bounded solution of the Cauchy problem (7.44), (7.115) is unique in the layer $\{0\le\tau<t_D\le T, r_D\in E_1\}$.

Suppose now that besides the fundamental solution $T_e(r_D,t_D;\xi,\tau)$ of Equation (7.44) that we have constructed above, there exists another fundamental solution $\tilde{T}_e(r_D,t_D;\xi,\tau)$ of the same equation. Then, taking into account the uniqueness of the solution for the Cauchy problem (7.44), (7.115), we obtain:

$$\int_{E_1}[T_e(r_D,t_D;\xi,\tau)-\tilde{T}_e(r_D,t_D;\xi,\tau)]\varphi(\xi)d\xi=0$$

for any $r_D\in E_1$ and $0\le\tau<t_D\le T$ and for any continuous function $\varphi(\xi)$ vanishing outside some bounded domain. Since $T_e(r_D,t_D;\xi,\tau)$ and $\tilde{T}_e(r_D,t_D;\xi,\tau)$ are continuous functions for $t_D>\tau$, it follows that $T_e(r_D,t_D;\xi,\tau)\equiv\tilde{T}_e(r_D,t_D;\xi,\tau)$.

Since the solution of the equation (7.44) exists and is unique, thus using a similar method as before, we can prove the existence and uniqueness of the Equation 8.30.

7.3.3 Solution model

Before solving the model, we introduce the treatment of some parameters. The details are as follows:

1. Velocity of the steam v_m:

$$v_m=v_s+v_w=\frac{MX}{\rho_s A}+\frac{M(1-X)}{\rho_w A}. \tag{7.117}$$

2. Density of the steam ρ_m. Since the flow of the water vapour in is a gas-liquid two-phase flow, we apply the Beggs-Brill method (Beggs and Brill, 1973) to calculate the average density of the mixture.

3. Friction force τ_f. This can be calculated using the following:

$$\tau_f = \frac{1}{4}\pi f_m r_{ti} \rho_m v_m^2 dz. \tag{7.118}$$

4. Friction factor of gas-liquid mixture. f_m is a function as regards the Reynold number and the absolute roughness ε:

$$\begin{cases} f_m = \dfrac{R_e}{64}, & \text{if } R_e \leq 2000 (t_D \leq 1.5) \\ f_m = \left[1.14 - 2\ln\left(\dfrac{\varepsilon}{2r_{ti}} + 21.25 Re^{-o.9}\right)\right]^{-2}, & \text{if } R_e > 2000 \end{cases}$$

5. Heat transfer coefficient. U_{to} is measured from different positions of the well-bore axis to the second surface:

$$\frac{1}{U_{to}} = r_{ti}\frac{1}{\lambda_{ins}}\ln\left(\frac{r_{ci}}{r_{ti}}\right) + \frac{1}{h_c + h_r} + r_{ti}\frac{1}{\lambda_{cem}}\ln\left(\frac{r_{cem}}{r_{co}}\right). \tag{7.119}$$

   On the right of (7.119), they are the thermal resistances for the insulated tubing, the hohlraum and the cement sheath, respectively. λ_{ins} and λ_{cem} are the heat conductivities for the heat insulating material and the cement sheath, respectively and h_c and h_r are the coefficients for the convection heat transfer and the radiating heat transfer.
6. Dimensionless time function. This can be calculated by:

$$\begin{cases} f(t_D) = 1.1281\sqrt{t_D}\left(1 - 0.3\sqrt{t_D}\right) & (t_D \leq 1.5) \\ f(t_D) = \left(1 + \dfrac{0.6}{t_D}\right)[0.4063 + 0.5\ln(t_D)] & (t_D > 1.5) \end{cases}. \tag{7.120}$$

Since the coupled system model of differential equations contains not only ordinary differential equations but also partial differential equations, it is recommended that the four-order Runge-Kutta method and the finite difference method be interactively applied to solve the problem. The detailed algorithm can be summarized as follows.

Step 1. Give the initial values θ_0, T_0, P_0, X_0, and T_{e0}, respectively.
Step 2. Compute all the coefficients in the coupled system.
Step 3. Let the differentiations be functions f_i $(i = 1, 2, 3, 4)$. Then we obtain a system of coupled function as follows:

$$\begin{cases} f_1 = \rho_m g\cos\theta - \rho_m v_m (Rf_2 - Sf_1) - f_m \dfrac{\rho_m v_m^2}{4r_{ti}} \\ f_2 = -\dfrac{A_2}{A_1}X - \dfrac{A_3}{A_1} \\ f_3 = a(T - T_e) \\ f_4 = 44.15P^{-0.79}f_1 \end{cases}. \tag{7.121}$$

Step 4. Solve the above system of equations at P_k, X_k, Q_k, T_k and T_{ek}, then we get the coefficients $a_j = f_j(P_k, X_k, Q_k, T_k), j = 1, 2, 3, 4$.

Step 5. Let $T = T_k + ha_k$, then we get the T_k by solving the following equations:

$$\begin{cases} \dfrac{\partial T_e}{t_D} - \dfrac{\lambda_e}{C_{Pe\rho e}} \dfrac{\partial^2 T_e}{\partial r_D^2} - \dfrac{\lambda_e}{r_D C_{Pe\rho e}} \dfrac{\partial T_e}{\partial r_D} = 0 \\ T_e|_{t_D=0} = T_0 + rz \\ \left.\dfrac{T_e}{r_D}\right|_{r_D=1} = -a(T - T_e)(2\pi\lambda_e)^{-1} \\ \left.\dfrac{T_e}{r_D}\right|_{r_D\to\infty} = 0 \end{cases} . \tag{7.122}$$

Let T_i^j be the temperature at the time j and radial i at the depth z, $i = 1, 2, \ldots, M; j = 1, 2, \ldots, N$, where M and N express the last node of time and radial, respectively. By applying the finite different method, (7.122) is discretized as follows:

$$\frac{T_j^{i+1} - T_j^i}{\tau} - \frac{\lambda_e}{C_{Pe\rho e}} \frac{T_{j+1}^{i+1} - 2T_j^{i+1} + T_{j-1}^{i+1}}{\varsigma^2} - \frac{\lambda_e}{r_D C_{Pe\rho e}} \frac{T_{j+1}^{i+1} - T_j^{i+1}}{\varsigma} = 0, \tag{7.123}$$

where τ is the interval of time and ς is the interval of radial, respectively. It can be transformed into the standard form as follows:

$$-\left(\frac{\tau\lambda_e}{C_{Pe\rho_e}} + \frac{\tau\zeta\lambda_e}{r_D C_{Pe\rho_e}}\right) T_{j+1}^{i+1} + \left(\frac{2\tau\lambda_e}{C_{Pe\rho_e}} + \frac{\tau\zeta\lambda_e}{r_D C_{Pe\rho_e}}\right) T_j^{i+1} - \frac{\tau\lambda_e}{C_{Pe\rho_e}} T_{j-1}^{i+1} = \zeta^2 T_j^i. \tag{7.124}$$

Then the difference method is used to discretize the boundary condition. For $r_D = 1$, we have:

$$\left.\frac{\partial T_e}{\partial r_{D_e}}\right|_{r_D=1} - \left.\frac{aT_e}{2\pi\lambda_e}\right|_{r_D=1} = -\frac{aT}{2\pi\lambda_e}. \tag{7.125}$$

It follows that:

$$T_2^{i+1} - \left(1 + \frac{a\zeta}{2\pi\lambda_e}\right) T_1^{i+1} = -\frac{aT}{2\pi\lambda_e}. \tag{7.126}$$

For $r_D = N$, we have:

$$T_N^{i+1} - T_{N-1}^{i+1} = 0. \tag{7.127}$$

Combining (7.124), (7.126) and (7.127), we can compute the numerical solution of the temperature T_e of the stratum.

Step 6. Take the T_e at $r_D = 1$ into the system (7.121) and we get:

$$b_j = f_j(P_k + ha_1, X_k + ha_2, Q_k + ha_3, T_k + ha_4), \quad j = 1, 2, 3, 4.$$

Step 7. Calculate the gas-liquid mixture dryness fraction of gas, pressure and temperature at the point $k + 1$:

$$P_{k+1} = P_k + \frac{h(a_1 + 2b_1 + 2c_1 + d_1)}{6}, \quad X_{k+1} = X_k + \frac{h(a_2 + 2b_2 + 2c_2 + d_2)}{6}$$

$$Q_{k+1} = Q_k + \frac{h(a_3 + 2b_3 + 2c_3 + d_3)}{6}, \quad T_{k+1} = T_k + \frac{h(a_4 + 2b_4 + 2c_4 + d_4)}{6}.$$

Step 8. Take $T = T_{k+1}$ into the boundary condition of (6.36) and we get $T_{e,k+1}$ by finite difference method.

Step 9. Repeat Step 2 to Step 8 until P, X, Q, T and $T_{e,n}$ are calculated.

Table 7.8 Dryness, pressure and temperature.

Depth [m]	Pressure [MPa]	Dryness fraction [–]	T (in wellbore) [°C]	T (in stratum) [°C]
1	20	0.8	200	22.4986
251	19.9823	0.7791	199.2425	26.678
501	19.9603	0.7028	197.476	31.4519
751	19.94	0.6815	196.8	37.4286
1001	19.9225	0.6032	195.3245	41.1397
1251	19.901	0.5713	194.409	47.0054
1501	19.8723	0.5509	192.2	53.8953
1751	19.8401	0.5401	190.45	56.9012
2001	19.82	0.5285	189.6238	62.0089
2251	19.7965	0.4996	187.79	67.1357
2501	19.7734	0.4765	186.6753	71.8903
2751	19.7421	0.4501	184.3901	76.7761
3001	19.7217	0.4289	182.567	82.0014
3251	19.7023	0.4067	180.983	86.0096
3501	19.6776	0.3865	178.134	91.8989
3751	19.6495	0.3621	176.889	97.5623
4001	19.6109	0.3501	175.351	102.2509
4251	19.5605	0.3378	173.867	106.9761
4501	19.519	0.3019	172.2579	112.2489
4751	19.489	0.2864	169.3665	116.9865
5001	19.438	0.2698	168.0116	123.0004
5251	19.4073	0.2501	167.775	126.7963
5501	19.3795	0.2386	166.6689	132.1474
5751	19.3493	0.2072	166.2017	137.0982
6001	19.3107	0.1964	166.1178	141.987
6251	19.2702	0.1861	166.0367	148.2009
6501	19.2478	0.1632	165.94	152.7891
6751	19.2206	0.1498	165.8743	158.0036
7001	19.1989	0.1467	165.6412	163.4286
7100	19.1687	0.1402	165.2675	165.1795

7.3.4 **Numerical simulation**

As described previously, the algorithm begins with a calculation for the pipe at the top of the pipe. The calculations are performed for each successive "segment" of the pipe from the top to the bottom.

7.3.4.1 *Parameters*

Some parameters of X well in Sichuan province, China are presented as follows:

Internal fluid density $= 1000\,\text{kg/m}^3$
External fluid density $= 1000\,\text{kg/m}^3$
Depth of the well $= 7100\,\text{m}$
Ground temperature $= 20°\text{C}$
Ground thermal conductivity parameter $= 2.06\,\text{W/m}\cdot\text{K}$
Geothermal gradient $= 2.18°\text{C}/100\,\text{m}$
Length of one segment $= 1\,\text{m}$

Furthermore, parameters of pipes, inclined well, inclination, azimuth and vertical depth are given as in Table 7.1, Table 7.2 and Table 7.3, respectively.

7.3.4.2 *Main results*

Based on the Runge-Kutta method and the finite difference method, we obtained a series of results for this well. The dryness fraction of steam, pressure and temperature in the well-bore, and the temperature in the stratum are shown in Table 7.8.

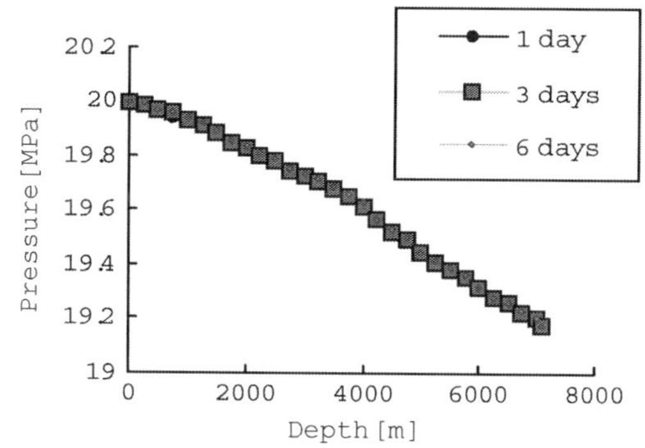

Figure 7.9 Pressure under different injection times.

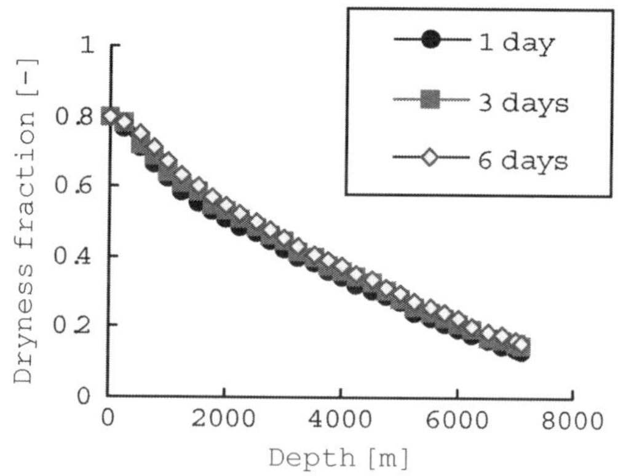

Figure 7.10 Dryness under different injection times.

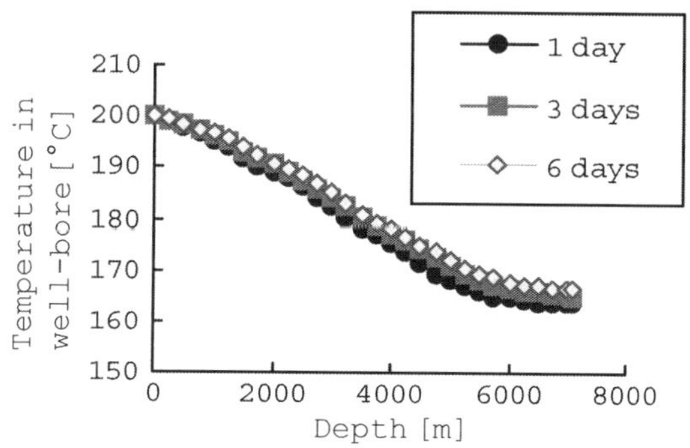

Figure 7.11 Temperature in the well-bore.

7.3.4.3 *Sensitivity analysis*

The main results are shown in Figures 7.9, 7.10, 7.11 and 7.12.

The pressure, temperature and the dryness fraction of steam increase as the injection time increases. In fact, this is very close to a real-world situation. As the amount of the injected steam increases, the water in the well-bore gradually decreases, so the dryness of the steam increases. At the same time, as the heat transfer increases, the temperature of the stratum

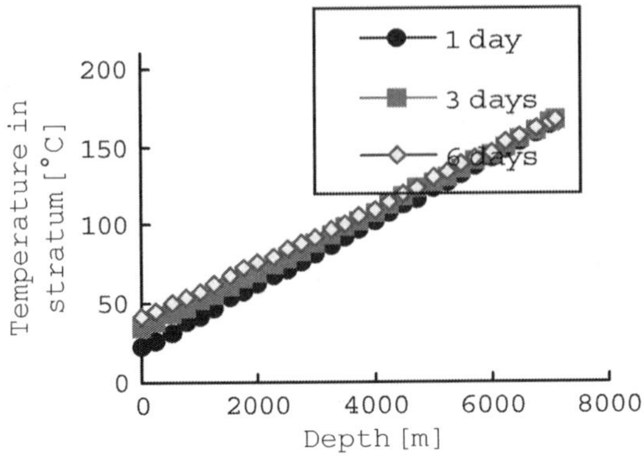

Figure 7.12 Temperature in the stratum.

also increases and the speed of the increment becomes slower. To study the impact of the gas-liquid mixture injection volumes and the geothermal gradients on the dryness fraction of gas, pressure and temperature, we used different gas-liquid mixture injection volumes and geothermal gradients to redo the above calculations from which we obtained the following series of results.

7.4 PTPTF-IWLFM[3]

Considering the partial differential equation model for the density, velocity, pressure, and temperature (ρ, v, P, T), we assume the following:

1. (ρ, v, P, T) is transient along the depth of the well at different times. The parameters are homochromatic at any transverse cross-section.
2. The heat transfer is steady from the tubing to the second interface and unsteady in the dimension from the second interface to the surrounding formation.
3. The heat loss from the tubing and surrounding formation is radial and the heat-transmission within the well-bore and from the well-bore to formation is transient.
4. Pressure and temperature are considered to be linear distributions with respect to the known geothermal gradients.

Consider the flow system depicted in Figure 7.1, a straight cylindrical flow tube with an inclination angle θ, a constant cross-sectional flow area A, a hydraulic diameter d, and a total length Z. Through this tubing gas flows from the bottom to the top with a mass flow rate w. The distance co-ordinate in the flow direction along the tubing is denoted z. Mass, momentum, and energy balances, along with the density, velocity, pressure, and temperature (ρ, v, P, T), are used to generate the constitutive equations.

7.4.1 **Model building**

Mass balance. Consider the flow model shown at Figure 7.13; according to how the fluid moves through the fixed control volume, the fundamental physical principle mass is conserved which means that the net mass flow out of the control volume through surface S equals the time rate of decrease of the mass inside the control volume (Anderson, 1995). Thus we have:

$$\iint_s \rho v ds = -\frac{\partial}{\partial t} \iiint_v \rho dv. \tag{7.128}$$

[3]Pressure and temperature prediction of transient flow in HTHP injection wells by Lax-Friedrichs method

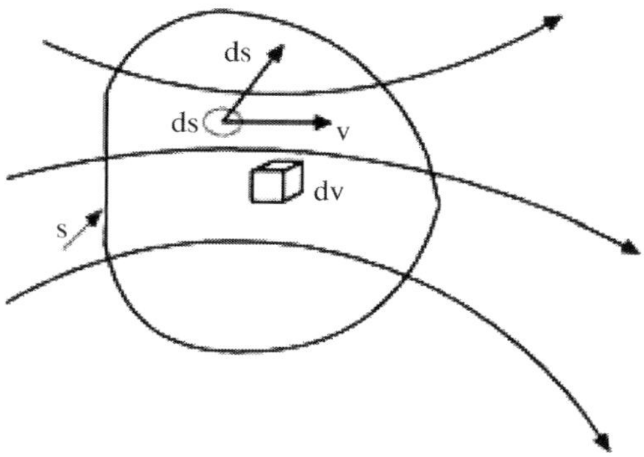

Figure 7.13 Control volume 1.

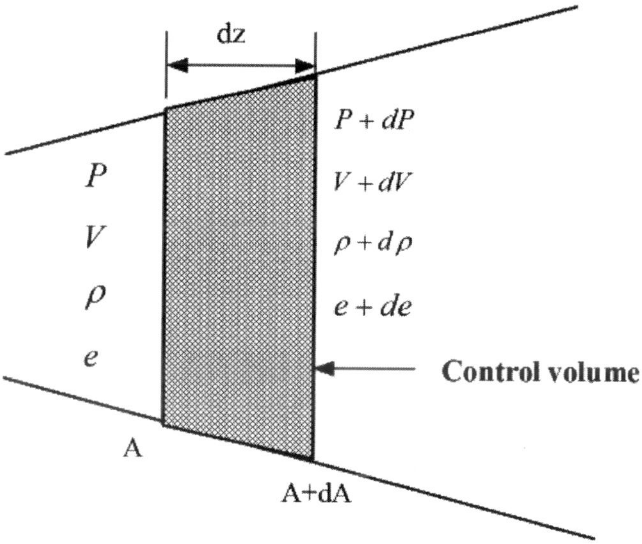

Figure 7.14 Control volume 2.

This is shown under transient conditions, for a partial differential depth dz, time dt of the well in terms of gas density ρ, at cross-section area A, and applied to the control volume in Figure 7.14, the volume integral in (7.128) becomes, in the limit as dz becomes very small,

$$\iint_s \rho ds = -\frac{\partial}{\partial t}\iiint_v \rho dv. \tag{7.129}$$

The surface integral becomes:

$$\iint_s \rho V ds = -\rho VA + (\rho + d\rho)(V + dV)(A + dA). \tag{7.130}$$

Expanding the triple product term in (7.119),

$$\iint_s \rho V ds = -\rho VA + \rho VA + \rho V dA + \rho A dV + \rho dV dA + AV d\rho + V dA d\rho + A dV d\rho + d\rho dA dV. \tag{7.131}$$

In the limit as dz tends toward zero, the terms involving the differentials products in (7.131), such as $\rho dVdA, AVd\rho, Vd\rho dA, d\rho dAdV$, move toward zero much faster than those terms involving only one differential. Hence,

$$\iint_s \rho V ds = \rho V dA + \rho A dV + AV d\rho = d(\rho VA). \tag{7.132}$$

Substituting (7.129) and (7.132) into (7.128),

$$\frac{\partial}{\partial t}(\rho A dz) + d(\rho AV) = 0. \tag{7.133}$$

Dividing (7.133) by dx and noting that this is, as dz vanishes, recalling the definition of the partial derivative with respect to z,

$$\frac{\partial \rho A}{\partial t} + \frac{\partial \rho AV}{\partial z} = 0. \tag{7.134}$$

By defining the mass flow $G = \rho V$, in order to simplify the following calculation, the above equation can be simplified as:

$$\frac{\partial \rho}{\partial t} + \frac{\partial G}{\partial z} = 0. \tag{7.135}$$

The momentum balance. For a transient flow system, the integral form of the z component of the momentum equation, can be written as below with the external forces:

$$\frac{\partial}{\partial t}\iiint_V \rho u dV + \iint_s \rho u V ds = -\iint_s (\rho ds)_z - \rho g \cos\theta A dz - \frac{\lambda \rho v^2}{2d} A dz, \tag{7.136}$$

where $\rho g \cos\theta A dz$ is the force of gravity, $\frac{\lambda \rho v^2}{2} A dz$ is the shear stress, and $(\rho ds)_z$ denotes the z component of the vector Pds. As shown in Figure 7.15:

$$\frac{\partial}{\partial t}\iiint_V \rho u dv = \frac{\partial}{\partial t}(\rho VA dz) \tag{7.137}$$

$$\iint_s (\rho uv) ds = -\rho V^2 A + (\rho + d\rho)(V + dV)^2 (A + dA) \tag{7.138}$$

$$-\iint_s (Pds)_z = -PA + (P + dP)(A + dA) - 2P\left(\frac{dA}{2}\right). \tag{7.139}$$

Substituting (7.137), (7.138), (7.139) into (7.136), we obtain:

$$\begin{aligned} &\frac{\partial}{\partial t}(\rho VA dz) - \rho V^2 A + (\rho + d\rho)(V + dV)^2 (A + dA) \\ &= PA - (P + dP)(A + dA) + PdA - \rho g \cos\theta A dz - \frac{\lambda \rho v^2}{2d} A dz. \end{aligned} \tag{7.140}$$

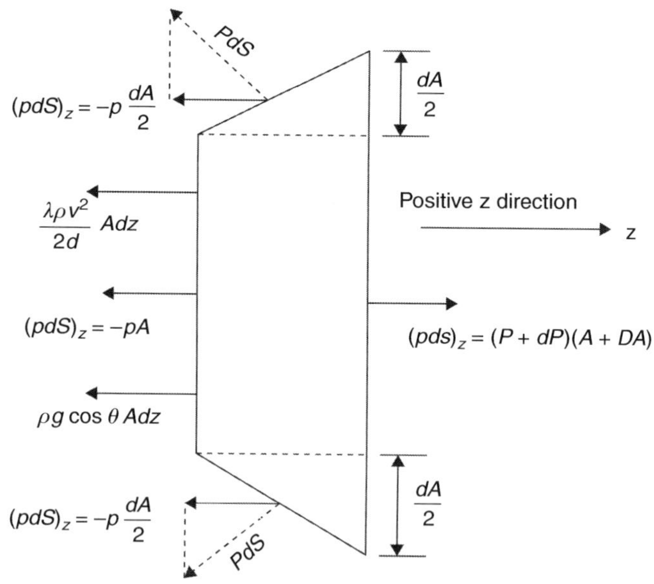

Figure 7.15 The forces in the z direction on the control volume.

Cancelling like terms and ignoring the differentials products, (7.140) becomes the limit of dz becoming simple as:

$$\frac{\partial}{\partial t}(\rho VAdz) - \rho V^2 A + d(\rho V^2 A) = -AdP - \rho g \cos\theta Adz - \frac{\lambda \rho v^2}{2d}Adz. \tag{7.141}$$

Dividing (7.141) by dz and A, and taking the limit as dz moves toward zero, we obtain the partial differential equation:

$$\frac{\partial}{\partial t}(\rho V) + \frac{\partial}{\partial z}(\rho V^2) = -\frac{\partial P}{\partial z} - \rho g \cos\theta - \frac{\lambda \rho v^2}{2d}. \tag{7.142}$$

The momentum balance equation can be obtained,

$$\frac{\partial}{\partial t}(\rho V) + \frac{\partial}{\partial z}(P + \rho V^2) = -\rho g \cos\theta - \frac{\lambda \rho v^2}{2d} \tag{7.143}$$

which can be simplified as:

$$\frac{\partial G}{\partial t} + \frac{P + G^2/\rho}{\partial z} = -\rho g \cos\theta - \frac{\lambda G|G|}{2\rho d}. \tag{7.144}$$

The energy balance. For a transient flow, this is a type of energy equation in terms of temperature. As shown in Figure 7.2, we consider the heat-transmission within the well-bore and from the well-bore to the formation as transient. In particular, for the flowing gas control volume, we derive the following energy transfer functions: heat of inflow control volume:

$$Q_m = (WC_P T_e)z \tag{7.145}$$

heat of outflow control volume:

$$Q_{out} = (WC_P T_e)(z + dz) \tag{7.146}$$

heat transferring to the second dimensions:

$$Q_r = 2\pi r_{to} U_{to}(T_e - T_r)dz \tag{7.147}$$

and heat variation flow control volume:

$$\Delta Q = \frac{\partial(\rho C_P T_e)}{\partial t} Adz. \tag{7.148}$$

According to energy balance law, by combining the functions (7.145), (7.146), (7.148), (7.147), we determine the energy balance equation for the transient flow:

$$(WC_P T_e)z - (WC_P T_e)(z+dz) - 2\pi r_{to} U_{to}(T_e - T_r)dz = \frac{\partial(\rho C_P T_e)}{\partial t} Adz, \tag{7.149}$$

where the parameter

$$T_r = \frac{K_e T_{ei} + r_{to} U_{to} T_e T_{WbD}}{K_e + r_{to} U_{to} T_{WbD}},$$

and the mass flow of gas

$$W = \rho VA.$$

Thus, the Equation (7.149) equals the following equation:

$$\frac{\partial(\rho V T_e)}{\partial z} + \frac{\partial(\rho T_e)}{\partial t} = (WC_P T_e)z - (WC_P T_e)(z+dz) - 2\pi r_{to} U_{to}(T_e - T_r)dz,$$

which can be simplified as:

$$\frac{\partial(GT_e)}{\partial z} + \frac{\partial(\rho T_e)}{\partial t} = \frac{2\pi K_e r_{to} U_{to}(T_{ei} - T_e)}{C_P(k_e + r_{to} U_{to} T_{WbD})}. \tag{7.150}$$

In addition, the stated equation for the gas is:

$$\rho = \frac{MPr_g}{ZRT_e}. \tag{7.151}$$

Finally, we obtain the partial differential equation coupled system model about the gas density, velocity, temperature and pressure with initial condition:

$$\begin{cases} \dfrac{\partial \rho}{\partial t} + \dfrac{\partial G}{\partial z} = 0 \\ \dfrac{\partial G}{\partial t} + \dfrac{\partial(\rho + \frac{G^2}{\rho})}{\partial z} = \dfrac{-\rho g \cos\theta - \lambda G|G|}{2\rho d} \\ \dfrac{\partial(GT_e)}{\partial z} + \dfrac{\partial(\rho T_e)}{\partial t} = \dfrac{2\pi K_e r_{to} U_{to}(T_{ei} - T_e)}{C_P(k_e + r_{to} U_{to} T_{WbD})} \\ \rho = \dfrac{MPr_g}{ZRT_e} \\ G = \rho V \\ (P, T, G, \rho) = (P_0, T_0, G_0, \rho_0), \quad T = 0, \quad z = 0 \end{cases} . \tag{7.152}$$

7.4.2 **Model solution**

7.4.2.1 *Format construction*

The unified conservation Equations (7.135), (7.136), (7.137) are put into the following formation:

$$\frac{\partial A_m}{\partial t} + \frac{\partial B_m}{\partial z} = C_m, \quad m = 1, 2, 3. \tag{7.153}$$

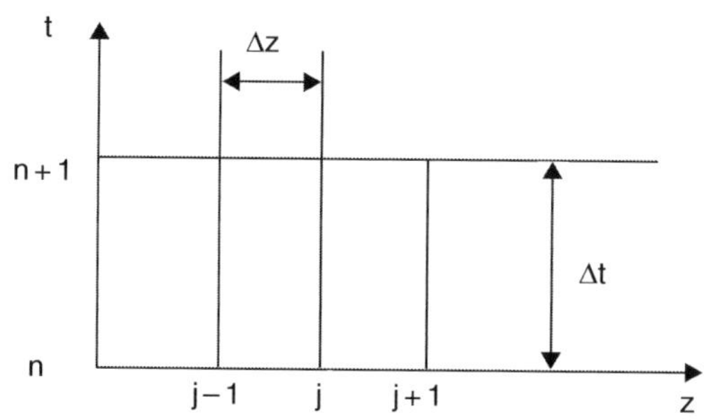

Figure 7.16 Grid for Lax scheme.

When $m = 1, 2, 3$, we get:

$$\begin{cases} A_1 = \rho \\ B_1 = G \\ C_1 = 0 \end{cases}$$

$$\begin{cases} A_2 = G \\ B_2 = \rho + \dfrac{G^2}{\rho} \\ C_2 = \dfrac{-\rho g \cos\theta - \lambda G|G|}{2\rho d} \end{cases}$$

$$\begin{cases} A_3 = \rho T_e \\ B_3 = G T_e \\ C_3 = \dfrac{2\pi K_e r_{to} U_{to}(T_{ei} - T_e)}{C_P(k_e + r_{to} U_{to} T_{WbD})} \end{cases}.$$

Defining z, t as two-dimensional coordinates in Euclidean space E2. τ, h is the step size of E2. According to the divergence discrete theorem, we discretize the definite region using a finite discrete grid as shown in Figure 7.16.

Then we use a difference quotient instead of derivative in the partial differential equations, and finally solve the system of difference equations with a coupled iterative solution method.

Using Lax-Friedrichs scheme to discretize the conservation Equation (7.152), we in particular discretize the part on the right side of the equation with a central difference format. Then, we obtain the following deference forms for (7.153):

$$\frac{(A_m)_j^{n+1} - \frac{1}{2}[(A_m)_{j-1}^n + (A_m)_{j+1}^n]}{\tau} + \frac{(B_m)_{j+1}^n - (B_m)_{j-1}^n}{2h} = \frac{(C_m)_{j+1}^{n+1}}{2} + \frac{(C_m)_j^{n+1}}{2}. \tag{7.154}$$

Transform (7.153) as below:

$$\frac{\Delta A_m}{\tau} + \frac{\Delta B_m}{2h} = \frac{\Delta C_m}{2}, \quad m = 1, 2, 3. \tag{7.155}$$

Then we determine the difference forms of (7.135), (7.146), (7.150) as follows. The form of the Equation (7.136) becomes:

$$\begin{cases} \Delta A_1 = \rho_j^{n+1} - \dfrac{1}{2}(\rho_{j-1}^n + \rho_{j+1}^n) \\ \Delta B_1 = G_{j+1}^n - G_{j-1}^n \\ \Delta C_1 = 0 \end{cases}. \tag{7.156}$$

The form of Equation (7.146) becomes:

$$\begin{cases} \Delta A_2 = G_j^{n+1} - \dfrac{1}{2}(G_{j-1}^n + G_{j+1}^n) \\ \Delta B_2 = \rho_{j+1}^n - \rho_{j-1}^n + \dfrac{(G_{j+1}^n)^2}{\rho_{j+1}^n} - \dfrac{(G_{j-1}^n)^2}{\rho_{j-1}^n} \\ \Delta C_2 = -\rho_{j+1}^n \cos\theta_{j+1}^n + \rho_j^n \cos\theta_{j+1}^n - \dfrac{\lambda G_{j+1}^n |G_{j+1}^n|}{2d\rho_{j+1}^n} + \dfrac{\lambda G_j^n |G_j^n|}{2d\rho_j^n} \end{cases}. \tag{7.157}$$

The form of Equation (7.150) is:

$$\begin{cases} \Delta A_3 = G_j^{n+1} T_{ej}^{n+1} - \dfrac{1}{2}(G_{j-1}^n T_{ej-1}^n + G_{j+1}^n T_{ej+1}^n) \\ \Delta B_3 = \rho_{j+1}^n T_{ej+1}^n - \rho_{j-1}^n T_{ej-1}^n + \dfrac{(G_{j+1}^n)^2}{\rho_{j+1}^n} - \dfrac{(G_{j-1}^n)^2}{\rho_{j-1}^n}, \\ \Delta C_3 = \dfrac{T_{eij+1}^{n+1} - T_{ej+1}^{n+1}}{a} + \dfrac{T_{eij}^{n+1} - T_{ej}^{n+1}}{a} \end{cases} \tag{7.158}$$

where

$$a = \frac{C_P(K_e + r_{to}U_{to}T_{WbD})}{2\pi K_e r_{to} U_{to}}.$$

In addition, to the stated equation for gas is: $\rho = \frac{MPr_g}{ZRT_e}$, its finite difference form is:

$$\rho_{j+1}^{n+1} = \frac{MP_{j+1}^{n+1} r_g}{Z_{j+1}^{n+1} RT_{ej+1}^{n+1}}. \tag{7.159}$$

7.4.2.2 *Solution process*

Combining the three conservation equations and the stated equation above, their corresponding finite difference form equations construct a nonlinear equation system which can be solved in the following steps:

Step 1. Set the step length of the depth and the time. $h = 1$ (m), $t = 60$ (s), $\varepsilon = 3\%$.

Step 2. Obtain each point's inclination $\theta_j = \theta_{j-1} + (\theta_{j'} - \theta_{j'-1})h/\Delta s_{j'-1}$, where $\Delta s_{j'}$ represents the measurement depth of inclination angle $\theta_{j'}, \theta_{j'-1}$. Calculate the parameter λ from function $\frac{1}{\sqrt{\lambda}} = -2\log\left[\frac{\varepsilon/d}{3.7} - \frac{5.02}{R_e}\log\left(\frac{\varepsilon/d}{3.7} + \frac{13}{R_e}\right)\right]$.

Step 3. Calculate the initial conditions. The initial conditions comprise the distribution of the pressure, temperature and density along the well at the initial time. From the following functions we can calculate $(\rho_j^n)^0$, $(V_j^n)^0$, $(G_j^n)^0$:

$$\rho_j^n = \frac{0.000001 \times 3484.48\gamma_g P_j^n}{ZT_j^n}, \quad V_j^n = 101000 \times 300000T_j^n(293 \times 8643000P_j^n A).$$

Step 4. Calculate the boundary conditions, which comprise the bottom-hole temperature at any every time segment: $T_r(z,\infty,t) = T_{sur0} + T_{grad}\int_0^z \cos\theta dz$.

Step 5. Discretize the finite solution region and set $j = 0, n = 0$.

Step 6. Calculate ρ_{j+1}^{n+1} from formula of (6.68) and (6.69).

Step 8. Repeat from step 6 to step 7, if $\frac{|\rho_{j+1}^{n+1}-(\rho_{j+1}^{n+1})^0|}{(\rho_{j+1}^{n+1})^0} \leq \varepsilon$, let $j=j+1$, $n=n+1$ and return to step 9. Otherwise, return to step 6.

Step 9. Calculate the following parameters: $C_P = 1697.5107P^{0.0661}T_e^{0.776}$,

$$U_{to} = \left(\frac{r_{to}}{r_{ti}h_r} + \frac{r_{to}\ln(r_{to}/r_{ti})}{r_{ti}h_r} + \frac{1}{h_c+h_r} + \frac{r_{to}\ln(r_{co}/r_{ci})}{r_{cas}} + \frac{r_{to}\ln(r_h/r_{co})}{r_{cem}}\right)^{-1}.$$

Calculate T_{ej+1}^{n+1} from formula of (6.68) and (6.71), which is calculated by considering the heat-transmission both within well-bore and from well-bore to formation as transient. In the engineering calculation, T_{WbD} can be calculated using the following function:

$$\begin{cases} T(WbD) = 1.1281\sqrt{t_D}\left(1-0.3\sqrt{t_D}\right) & (t_D \leq 1.5) \\ T(WbD) = \left(1+\frac{0.6}{WbD}\right)[0.4063+0.5\ln(t_D)] & (t_D > 1.5) \end{cases}$$

However, this calculation is only suitable for a steady heat transfer. In actual conditions, the heat flow from the well-bore to formation decreases with an increase in the time, especially in the initial production process, when the temperature differences between the well-bore and formation are large. This heat loss process can be broken up into several variation procession time sections, in which the heat flow temperature is considered to be fixed. According to the superposition principle that the heat discretized on the variation procession time sections can be superposed and averaged (Liao and Feng, 2005), the temperature at the well-bore can be calculated using the following function:

$$T_{Wb} = T_{ei} + \frac{1}{2\pi K_e}\sum_{j=1}^{m}(Q_j - Q_{j-1})T_{WbD_j}(t_{D_j}),$$

where $t_{D_j} = \frac{3.6K_e(t_j - t_{j-1})}{C_e P_e}$. If the time section is equidistance, it is simplified as:

$$T_{WbD} = \frac{2\pi K_e(T_{ei} - T_{Wb})}{\sum\limits_{j=1}^{m}(Q_j - Q_{j-1})}.$$

7.4.3 **Examples calculation**

As was described previously, the algorithm begins with a calculation for the pipe at the bottom end of the pipe. The calculations are performed for each successive segment of the pipe up to the surface.

7.4.3.1 *Parameters*

In this simulation, we study a pipe at *X* well in the Sichuan basin, South-West China. The needed parameters are as follows:

Internal fluid density = 1000 kg/m^3
External fluid density = 1000 kg/m^3
Depth of the well = 7100 m
Friction coefficient = 1.2
Ground temperature = 16°C
Ground thermal conductivity parameter = 2.06 W/m · K
Geothermal gradient = 2.18°C/100 m
Length of one segment = 1 m

The parameters for the pipes, inclined well, inclination, azimuth and vertical depth are as in Tables 7.1, 7.2 and 7.3.

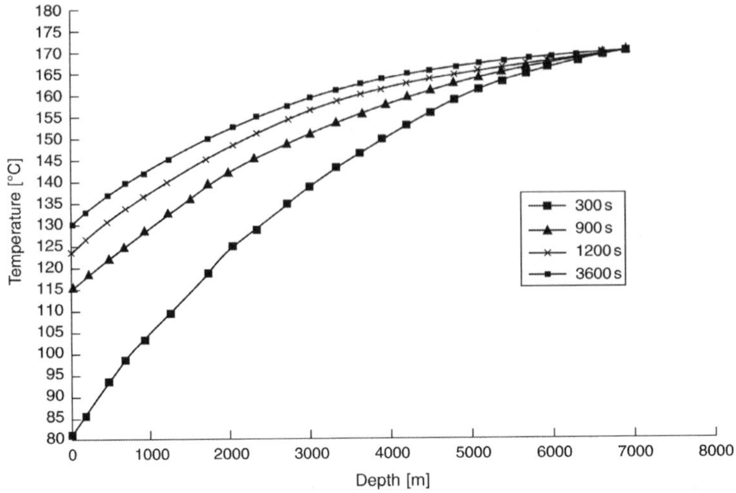

Figure 7.17 Temperature at 300 s, 900 s, 1200 s, and 3600 s.

Table 7.9 Temperatures [°C] at different times.

Well depth [m]	300 s	900 s	1200 s	3600 s
0	81.27	117.24	123.14	130.1
300	85.63	121.32	127.55	133.41
600	91.72	125.43	131.36	136.66
900	96.44	129.34	134.88	138.84
1200	101.08	133.38	137.39	141.97
1500	106.42	136.33	140.57	143.05
1800	111.6	140.2	143.03	145.1
2100	116.34	143.95	146.28	147.09
2400	121.23	145.6	148.47	149.03
2700	126.78	148.13	150.58	151.91
3000	131.09	151.54	152.61	153.82
3300	136.15	153.83	154.56	155.36
3600	140.98	155.01	156.44	157.19
3900	144.65	157.08	158.78	159.83
4200	148.15	159.04	160.67	161.42
4500	152.72	161.89	161.62	162.95
4800	155.65	162.65	162.62	163.43
5100	159.51	163.31	163.45	164.86
5400	162.35	164.88	164.9	165.23
5700	164.1	165.46	165.72	166.55
6000	166.34	166.95	167.58	167.82
6300	167.93	168.06	168.91	169.05
6600	169.22	169.24	169.86	169.99
6900	170.12	170.33	170.48	170.69

7.4.3.2 *Result analysis*

To study whether the different gas outputs and thermal conductivity parameters influence the pressure and the temperature, we used different times; 300 s, 900 s, 1200 s, and 3600 s; and thermal conductivity parameters; 0.52, 1.35 and 2.57 W/m · K. In particular, we considered the varying temperatures along with the variations in the bottom pressure. Using the algorithm and

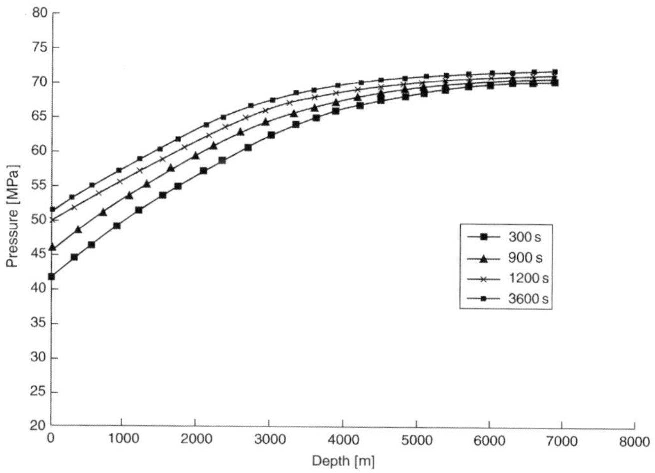

Figure 7.18 Pressure at 300 s, 900 s, 1200 s, and 3600 s.

the simulation, we obtained a series of results in tables and figures for temperature, pressure with a variation in products, and thermal conductivity parameters. We analyse these results below.

Temperature

When the bottom pressure is 70 MPa, temperatures for the different gas outputs are shown in Figure 7.17 and Table 7.9. From Figure 7.17 and Table 7.9, it can be seen that when the depth is constant, the temperature increases along with increasing time. When the output is constant, the temperature decreases as the well depth decreases. This is because as the gas output increases, the frictional heat leads to an increase in the well head temperature, while the temperature difference between the well head and the bottom decreases. The maximal relative error between the results is in Table 7.9 and the results calculated using the steady-state calculation in Xu *et al.* (2011) was 2.39%, indicating that a temperature calculated model considering heat-transmission is more accurate.

Pressure

From Figure 7.18 and Table 7.10, it can be established that when the depth is constant, the pressure increases with an increase in the time. When the output is constant, the pressure increases as the well depth decreases. This is because the whole heat transfer coefficient becomes larger with an increase as the heat transmission capacity increases and the gas temperature decreases. The maximal relative error between the results is shown in Table 7.5 and the results calculated using the steady-state calculation in Xu *et al.* (2011) was 3.22%, indicating that this new model is of higher accuracy.

7.5 PTPD-IGWTE[4]

Before establishing the differential equation model for pressure and temperature in the formation and the well-bore, we set the following assumptions:

1. The porous medium is horizontal, isotropic, homogeneous and isothermal with a constant thickness, porosity and permeability.

[4]Prediction of temperature and pressure distribution in HTHP injection gas wells considering thermal effect of wellbore

Table 7.10 Pressure [MPa] at different times.

Well depth [m]	300 s	900 s	1200 s	3600 s
0	42.23	46.33	50.12	51.21
300	43.45	47.43	50.42	52.23
600	44.16	48.31	50.43	53.44
900	45.02	49.19	50.43	54.65
1200	46.88	50.07	51.44	54.86
1500	48.63	51.94	53.44	55.06
1800	49.47	52.8	54.43	56.22
2100	50.32	53.66	55.41	57.38
2400	51.23	56.52	54.39	57.53
2700	53.5	57.37	55.36	58.67
3000	54.57	58.23	56.34	59.82
3300	55.34	59.08	57.31	60.97
3600	56.11	59.43	58.37	61.21
3900	57.87	60.78	59.27	62.21
4200	58.63	61.63	60.25	63.3
4500	59.39	62.48	61.23	64.45
4800	60.15	63.33	62.21	64.71
5100	61.91	64.18	63.19	64.84
5400	63.66	65.04	64.17	65
5700	64.42	65.89	65.15	66.17
6000	65.17	66.74	66.14	67.39
6300	66.92	67.59	67.12	67.44
6600	67.71	68.52	68.23	68.73
6900	69.21	69.45	69.66	69.72

2. The flow is radial, so there is no fluid flow across the horizontal boundaries and negligible gravitational effect.
3. The thermal conductivity of the thermal insulating materials is in a linear relationship with the temperature.
4. The heat transfer is one dimensional and is steady from the tubing to the second interface, but the heat transfer in the one dimension is unsteady from the second interface to the surrounding formation.
5. There is no heat transfer by conduction in the vertical directions either in the flowing steam or in the formation.
6. The pressure gradients are small, so the square of the pressure gradient with respect to the radial distance is negligible.

7.5.1 Mathematical model of heat transmission in the well-bore

When fluid flows to the well-bore or the pipe, depending on the different physical properties of the rock, there is a difference in the temperature between the fluid and the formation, thus a heat transfer exists. The heat transfer process is complex, which includes a steady heat transmission in the well-bore and an unsteady heat conduction in the formation. As in Figure 7.19, the tubing is surrounded by a homogenous rock formation that extends to infinity. The distant temperatures in the formation increase linearly with the depth as reflected in geothermal temperatures. The gas enters the flow tubing at a given pressure and temperature. Within the tubing, the gas flow takes place under turbulent flow conditions. Consider the flow system depicted in Figure 7.2. Ramey (1962) and Willhite (1967) discussed the radial transfer of heat between the fluid and the formation in detail. The heat transferred from the fluid to the formation leads to the control equation for thermal exchange between the well-bore and the formation. In injected fluid, over

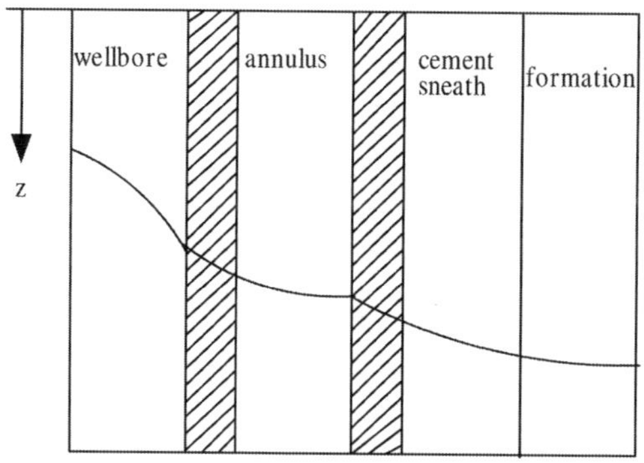

Figure 7.19 Temperature distribution of well-bore and formation.

the different depths, the radial heat transfer from the fluid to the cement/earth interface can be described by:

$$dQ = 2\pi r_{to} U_{to}(T_w - T_{ref})dz. \tag{7.160}$$

The rate of radial heat transfer from the cement/earth interface to the surrounding formation can be described by:

$$dQ = \frac{2\pi\lambda_f(T_{ref} - T_r)}{f(t_D)}dz. \tag{7.161}$$

Combining Equation (7.160) and (7.161) gives the equation for the rate of radial heat transfer between the fluid and the surrounding earth:

$$\frac{dQ}{dz} = \frac{2\pi r_{to} U_{to}\lambda_f}{r_{to} U_{to} f(t_D) + \lambda_f}(T_w - T_r). \tag{7.162}$$

Let $\eta = \frac{2\pi r_{to} U_{to}\lambda_f}{r_{to} U_{to} f(t_D) + \lambda_f}$, then:

$$\frac{dQ}{dz} = \eta(T_w - T_r).$$

Taking into account the effect of friction heating and the change in fluid, the heat conservation equation from the fluid can be written:

$$\pi r_w^2 \rho_w C_w \frac{\partial T_w}{\partial t} + GC_w\left(\frac{\partial T_w}{\partial z} - \frac{g}{C_w}\right) = q, \tag{7.163}$$

where mass flow rate $G = \pi r_w^2 \rho_w v$, and is the rate of heat per unit depth transferred to the fluid by conduction from the formation.

According to the over-all heat-transfer coefficient U_{to} given by Ramey (1962), we have:

$$q = -U_{to}(T_w - T_r|_{r=r_w}), \tag{7.164}$$

where $T_r(z, r_w, t) = T_w(z, t)$.

Unsteady heat conduction model in the formation. According to the theory of heat conduction through the medium and the above assumption, over the same differential element of depth dz, the radial heat conduction equation for the formation can be described by:

$$\rho_f C_f \frac{\partial T_r}{\partial t} = \lambda_f\left(\frac{\partial^2 T_r}{\partial r^2} + \frac{1}{r}\frac{\partial T_r}{\partial r}\right). \tag{7.165}$$

The well-bore temperature T_w and the formation temperature T_r must satisfy the following initial condition and the boundary conditions.

Initial condition. When the gas is injected, the formation temperature field is the initial formation temperature:

$$T_r(z,r,t=0)=T_e+T_{grad}z\cos\theta. \tag{7.166}$$

Boundary conditions. According to Fourier's law of heat conduction, on the interface between the well-bore and the formation, the heat flux exchange through the unit depth of the well-bore and the surrounding formation is:

$$dQ=2\pi\lambda_f dzr\left.\frac{\partial T_r}{\partial r}\right|_{r=r_w} \tag{7.167}$$

and

$$T_w(z=0,t)=T_c. \tag{7.168}$$

At infinity distance, the temperature distributions for the formation stay at the initial temperature distributions for the formation and are independent of time:

$$\left.\frac{\partial T_r}{\partial r}\right|_{r\to\infty}=0. \tag{7.169}$$

The boundary condition that couples Equation (7.163) and Equation (7.165) is:

$$q=2\pi\lambda_f\left.\left(r\frac{\partial T_r}{r}\right)\right|_{r=r_w}. \tag{7.170}$$

The temperature of the fluid and the earth can be explained by the above two differential equations and the correspondent initial and boundary conditions. To facilitate a solution, the following dimensionless variables are defined:

$$r_D=\frac{r}{r_w},\quad t_D=\frac{\lambda_f}{(\rho C)_f}\frac{t}{r_w^2},\quad Z_D=\frac{z}{L},\quad \beta=\frac{(\rho C)_w}{2(\rho C)_f},\quad r=\frac{GC_w}{2\pi L\lambda_f}.$$

then exchange variables r,z,t into the dimensionless variables r_D,z_D,t_D, and Equations (7.163) and (7.165) can be rewritten as:

$$\beta\frac{\partial T_w}{\partial t_D}+r\left(\frac{\partial T_w}{\partial z_D}-\frac{gL}{C_w}\right)=q_D \tag{7.171}$$

$$\frac{\partial^2 T_r}{\partial r_D^2}+\frac{1}{r_D}\frac{\partial T_r}{\partial r_D}=\frac{\partial T_r}{\partial t_D}. \tag{7.172}$$

The corresponding boundary conditions (7.167) and (7.168) are exchanged:

$$q_D=\left.\frac{\partial T_r}{\partial r_D}\right|_{r_D=1}=-\frac{1}{2\pi\lambda_f}\frac{dQ}{dz} \tag{7.173}$$

$$\left.\frac{T_r}{\partial r_D}\right|_{r_D\to\infty}=0. \tag{7.174}$$

From the above, the heat exchange in the well-bore includes a steady heat transmission in the well-bore and an unsteady heat conduction in the formation, with both coupled at the interface between the earth/cement and the formation, and this coupled system exists under the condition (7.170). The coupled differential model for the temperatures of the well-bore and the formation are as follows:

$$\begin{cases} \dfrac{\partial^2 T_r}{\partial r_D^2} + \dfrac{1}{r_D}\frac{\partial T_r}{\partial r_D} = \dfrac{\partial T_r}{\partial t_D} \\ \beta\dfrac{\partial T_w}{\partial t_D} + r\left(\dfrac{\partial T_w}{\partial z_D} - \dfrac{gL}{C_w}\right) = -\dfrac{1}{2\pi\lambda_f}\eta(T_w - T_r) \\ \left.\dfrac{T_r}{\partial r_D}\right|_{r_D\to\infty} = 0 \\ T_r(z_D, r_D, t_D=0) = T_0 + T_{grad}Lz_D\cos\theta, \quad T_w(z_D=0, t_D) = T_c \end{cases} \tag{7.175}$$

7.5.2 Pressure in the well-bore mathematical model

For the formation, based on the above assumptions, the well-known diffusivity equation derived from the continuity equation and Darcy's law (Darcy, 1856) can be used. For a transient fluid flow in porous media, the radial flow model for the gas reservoir can be established:

$$\frac{1}{r}\frac{\partial}{\partial r}\left(\frac{k\rho}{\mu}\frac{r\partial P}{\partial r}\right) = \phi C_g \rho \frac{\partial P}{\partial t}. \tag{7.176}$$

Substituting the gas equation of state

$$P = \frac{\rho ZRT}{M}$$

into Equation (7.176) leads to:

$$\frac{1}{r}\frac{\partial}{\partial r}\left(\frac{k}{P}\mu Z\frac{M}{RT}\frac{r\partial P}{\partial r}\right) = \frac{\phi C_g P}{Z}\frac{M}{RT}\frac{\partial P}{\partial t}. \tag{7.177}$$

For the isothermal flow, the $\frac{M}{RT}$ is constant, then (7.177) can be written:

$$\frac{1}{r}\frac{\partial}{\partial r}\left(\frac{k}{P}\mu Z\frac{r\partial P}{\partial r}\right) = \frac{\phi C_g P}{Z}\frac{\partial P}{\partial t}. \tag{7.178}$$

Initial condition. For a reservoir initially at a constant pressure P_i, the pressure of the formation is the initial formation pressure, and the initial condition can be given by:

$$P(r, t=0) = P_i, \quad 0 \le r < +\infty. \tag{7.179}$$

Boundary conditions. At infinity distance, the pressure distributions of the formation stay at the initial pressure distributions for the formation and are independent of time:

$$P(r,t) = P_i, \quad r \to \infty, \ t > 0. \tag{7.180}$$

With regard to the well-bore skin effect, the flowing bottom-hole pressure $P_w(t)$, the pressure of the well-bore $P(r_w, t)$ and the pressure affected by the temperature $P_T(t)$ can be expressed as:

$$P_w(t) = P(r_w, t) - S\left(r\frac{\partial P}{\partial r}\right)_{r=r_w} - P_r(t). \tag{7.181}$$

$P_T(t)$ can be obtained by combining the Equation (7.175) and the gas state equation.

For single phase flow, rate of flow q concludes two parts: well-bore storage part q_c and inflow from formation part q_f, so we have $q = q_c + q_f$. If the cross-sectional flow area of well-bore is A_w, then the volume of fluid $V_w = A_w z$. According to the definition of the isothermal compressibility:

$$C_g = -\frac{1}{V_w}\frac{dV_w}{dP_w} = -\frac{1}{V_w}\frac{dV_w/dt}{dP_w/dt} \tag{7.182}$$

then:

$$-q_c = \frac{\partial V_w}{\partial t} = C_g V_w \frac{\partial P_w}{\partial t} \tag{7.183}$$

and

$$q_f = q - q_c = q + C_g V_w \frac{dP_w}{dt}. \tag{7.184}$$

Define the well-bore storage constant:

$$C = \frac{dV_w}{dP_w}. \tag{7.185}$$

Substituting (7.182) into (7.185) leading to. And according to Darcy law (1856), the flow into the well-bore can be written:

$$q_f = \frac{2\pi kh}{\mu}\left(r\frac{\partial P}{\partial r}\right)_{r=r_w}. \tag{7.186}$$

Substituting (7.186) into (7.184) leading to the form:

$$\frac{2\pi kh}{\mu}\left(r\frac{\partial P}{\partial r}\right)_{r=r_w} = q + C\frac{dP_w}{dt}. \tag{7.187}$$

Substituting (7.181) into (7.187) leading to the couple inner boundary condition considered skin factor, well storage constant and temperature effect:

$$\left[\frac{2\pi kh}{\mu}\left(r\frac{\partial P}{\partial r}\right) - C\frac{dP}{dt} + CS\frac{d}{dt}\left(r\frac{\partial P}{\partial r}\right)\right]_{r=r_w} = q - C\frac{dP_t}{dt}. \tag{7.188}$$

Combining the (7.179), (7.178), (7.180) and (7.186), the coupled differential model on pressure of well-bore and formation as follows:

$$\begin{cases} \dfrac{1}{r}\dfrac{\partial}{\partial r}\left(\dfrac{k}{P}\mu Z\dfrac{r\partial P}{\partial r}\right) = \dfrac{\phi C_g P}{Z}\dfrac{\partial P}{\partial t}, \quad 0 \le r < +\infty, \ t > 0 \\ \left[\dfrac{2\pi kh}{\mu}\left(r\dfrac{\partial P}{\partial r}\right) - C\dfrac{dP}{dt} + CS\dfrac{d}{dt}\left(r\dfrac{\partial P}{\partial r}\right)\right]_{r=r_w} = q - C\dfrac{dP_t}{dt}. \\ P(r,t) = P_i, \quad r \to \infty, \ t > 0 \\ P(r,t=0) = P_i, \quad 0 \le r < +\infty \end{cases} \tag{7.189}$$

7.5.3 **Model solution**

To simplify the calculation, we divided the well into several short segments of the same length h. The segments vary depending on variations in well thickness, hole diameter, the fluid density inside and outside the pipe and well geometry. The model begins with a calculation at the top of the pipe. Then the gas pressure and temperature calculations are performed for each successive "segment" of the pipe down to the bottom. We use the finite difference method to solve the models. Based on the aforementioned discussion, an algorithm is designed as follows: for the well temperature and pressure models coupled partial differential equations and ordinary differential equations and a finite difference method are used interactively to solve the coupled system. The detailed algorithm is as follows:

Step 1. Obtain each point's inclination:

$$\theta_j = \theta_{j-1} + \frac{\theta_k - \theta_{k-1}}{\Delta s_k}\Delta s_j,$$

where j represents the segment point of calculation, Δs_k represents the measurement depth of inclination angle θ_k and θ_{k-1}, Δs_j the step length of calculation.

Step 2. Given the initial values T_r^0, T_w^0, respectively;

Step 3. Calculating the temperature of formation T_r by simulating the coupled model system:

$$\begin{cases} \dfrac{\partial^2 T_r}{\partial r_D^2} + \dfrac{1}{r_D}\dfrac{\partial T_r}{\partial r_D} = \dfrac{\partial T_r}{\partial t_D} \\ \left.\dfrac{\partial T_r}{\partial r_D}\right|_{r_D=1} = -\dfrac{1}{2\pi\lambda_f}\eta(T_w - T_r) \\ \left.\dfrac{T_r}{\partial r_D}\right|_{r_D\to\infty} = 0 \\ T_r(z_D, r_D, t_D = 0) = T_0 + T_{grad}Lz_D\cos\theta \end{cases}, \tag{7.190}$$

where

$$\begin{cases} f(t_D) = 1.1281\sqrt{t_D}\left(1 - 0.3\sqrt{t_D}\right) \quad (t_D \le 1.5) \\ f(t_D) = \left(1 + \dfrac{0.6}{t_D}\right)[0.4063 + 0.5\ln(t_D)] \quad (t_D > 1.5) \end{cases}$$

$t_D = \frac{t\alpha}{r_{wb}^2}$ and $\frac{1}{U_{to}} = r_{ti}\frac{1}{\lambda_{ins}}\ln\left(\frac{r_{ci}}{r_{to}}\right) + \frac{1}{h_c+h_r} + r_{ti}\frac{1}{\lambda_{cem}}\ln\left(\frac{r_{cem}}{r_{co}}\right)$.

To solve this 1D radial model, we use logarithmic transformation $x = \ln(r/r_w)$ to exchange the circle region for the linear region, and then the finite difference method is used to discretize the problem:

$$\frac{T_{r_{i-1}}^{n+1} - 2T_{r_i}^{n+1} + T_{r_{i+1}}^{n+1}}{(\Delta x)^2} = r_D^2\frac{T_{r_i}^{n+1} - T_{r_i}^n}{\Delta t}, \tag{7.191}$$

where Δx is the interval of length along radial and Δt is the interval of time. It can be simplified as follows:

$$T_{r_{i-1}}^{n+1} - [2 + r_D^2(\Delta x)^2/\Delta t]T_{r_i}^{n+1} + T_{r_{i+1}}^{n+1} = -[r_D^2(\Delta x)^2/\Delta t]T_{r_i}^n. \tag{7.192}$$

For boundary condition, we also change the polar coordinate into rectangular coordinate. For outer boundary condition, as $i = N$, use a pseudo-node on space $i = N + 1$, so we have:

$$\frac{T_{r_{N+1}}^{n+1} - T_{r_{N-1}}^{n+1}}{2\Delta x} = 0. \tag{7.193}$$

For inner boundary condition,

$$\left.\frac{\partial T_r}{\partial x}\right|_{x=0} = -\frac{\eta(T_w - T_r)}{2\pi\lambda_f} \tag{7.194}$$

so it follows that:

$$T_{r_1}^{n+1} - T_{wk}^{n+1} = -\frac{\eta(T_{wk}^n - T_{r1}^n)}{2\pi\lambda_f}, \tag{7.195}$$

where T_{wk}^n is the temperature of well-bore at the space point $k\Delta z$, which can be calculated from following steps, Δz is the step size of well-bore.

Step 4. Calculating the temperature of well-bore T_w. T_w can be obtained by solving following equation:

$$\begin{cases} \beta\dfrac{\partial T_w}{\partial t_D} + r\left(\dfrac{\partial T_w}{\partial z_D} - \dfrac{gL}{C_w}\right) = -\dfrac{1}{2\pi\lambda_f}\eta(T_w - T_r) \\ T_w(z_D = 0, t_D) = T_c \end{cases}. \tag{7.196}$$

The finite difference approximation can be written:

$$\beta\frac{T_{wk+1}^{n+1} - T_{wk+1}^{n}}{\Delta t} + \gamma\left[\frac{T_{wk+1}^{n+1} - T_{wk+1}^{n}}{\Delta z} - \frac{gL}{C_w}\right] = -\frac{\eta}{2\pi\lambda_f}(T_{wk}^{n+1} - T_{r1}^{n+1}) \tag{7.197}$$

it can be simplified as follows:

$$\left(\frac{\beta}{\Delta t}\right)T_{wk+1}^{n+1} + \left[\frac{\eta}{2\pi\lambda_f} - \frac{r}{\Delta z}\right]T_{wk}^{n+1} = \frac{\beta}{\Delta t}T_{wk+1}^{n} + \frac{\eta}{2\pi\lambda_f}T_{r1}^{n+1} + \frac{gL\gamma}{C_w}. \tag{7.198}$$

Step 5. Calculating the pressure which effected by the temperature of well-bore. According to the gas state equation and the temperature of well-bore calculated from step 4, pressure P_T can be obtained:

$$P_T = \frac{\rho ZRT_w}{M}. \tag{7.199}$$

Step 6. Calculating all the coefficients in coupled differential model.
Step 7. Calculating the pressure of well-bore P. It can be obtained by solving equation:

$$\begin{cases} \dfrac{1}{r}\dfrac{\partial}{\partial r}\left(\dfrac{k}{P}\mu Z\dfrac{r\partial P}{\partial r}\right) = \dfrac{\phi C_g P}{Z}\dfrac{\partial P}{\partial t}, \quad 0 \le r < +\infty,\ t > 0 \\ \left[\dfrac{2\pi kh}{\mu}\left(r\dfrac{\partial P}{\partial r}\right) - C\dfrac{dP}{dt} + CS\dfrac{d}{dt}\left(r\dfrac{\partial P}{\partial r}\right)\right]_{r=r_w} = q - C\dfrac{dP_t}{dt} \\ P(r,t) = P_i, \quad r \to \infty,\ t > 0 \\ P(r,t=0) = P_i, \quad 0 \le r < +\infty \end{cases}. \tag{7.200}$$

We use logarithmic transformation $x = \ln(r/r_w)$ to exchange the circle region for the linear region, and the finite difference approximation can be obtained:

$$\left[\frac{kP}{\mu Z}\right]_i^n P_{i-1}^{n+1} - \left[\left(\frac{kP}{\mu Z}\right)_i^n + \left(\frac{kP}{\mu Z}\right)_i^n\right]P_i^{n+1} + \left(\frac{kP}{\mu Z}\right)_{i+1}^n = \frac{\Phi C_g V^2(\Delta x)^2 P_i^n}{z}\frac{P_i^{n+1} - P_i^n}{\Delta t}. \tag{7.201}$$

It can be simplified as follows:

$$\left[\frac{kP}{\mu Z}\right]_i^n P_{i-1}^{n+1} - \left[\left(\frac{kP}{\mu Z}\right)_i^n + \left(\frac{kP}{\mu Z}\right)_i^n - \frac{\Phi C_g V^2(\Delta x)^2 P_i^n}{Z\Delta t}\right]P_i^{n+1} + \left(\frac{kP}{\mu Z}\right)_{i+1}^n = -\frac{\Phi C_g V^2(\Delta x)^2 P_i^n}{Z\Delta t}P_i^n. \tag{7.202}$$

For boundary condition, we also change the polar coordinate into rectangular coordinate:

$$\left[\frac{2\pi kh}{\mu}\frac{\partial P}{\partial x} - C\frac{dP}{dt} + CS\frac{d}{dt}\frac{\partial P}{\partial x}\right]_{x=0} = q - C\frac{dP_t}{dt} \tag{7.203}$$

the difference approximation can be written:

$$\frac{2\pi kh}{\mu}\frac{P_1^{n+1} - P_w^{n+1}}{\Delta x} - C\frac{P_1^{n+1} - P_1^n}{\Delta t} + CS\frac{P_1^{n+1} - P_1^n - P_w^{n+1} + P_w^n}{\Delta x\Delta t} = q - C\frac{P_t^{n+1} - P_t^n}{\Delta t} \tag{7.204}$$

It can be simplified as follows:

$$\left[\frac{2\pi kh\Delta t}{\mu} - C\Delta x + CS\right]P_1^{n+1} - \left[\frac{2\pi kh\Delta t}{\mu} + CS\right]P_w^{n+1}$$
$$= q\Delta x\Delta t + C(s - \Delta x)P_1^n - CSP_w^n - C\Delta x(P_T^{n+1} - P_T^n). \quad (7.205)$$

Step 8. Repeat the second to seventh steps before calculating T_r, T_w and P.

7.5.4 Numerical simulation

As was described previously, the algorithm begins with a calculation at the bottom of the pipe. The calculations are performed for each successive "segment" of the pipe from the bottom to the top.

7.5.4.1 *Parameters*

In this simulation, we study a pipe in X well located in Sichuan province, China. The needed parameters are as follows:

Internal fluid density = 1000 kg/m^3
External fluid density = 1000 kg/m^3
Depth of the well = 7100 m
Ground temperature = 16°C
Ground thermal conductivity parameter = 2.06 W/m · K
Geothermal gradient = 2.18°C/100 m
Gas injection volumes = 700000 m^3/day
Length of one segment = 1 m

More detailed data, such as the parameters of pipes, the inclined well, the inclination, the azimuth and the vertical depth are given in Tables 7.1 to 7.3.

7.5.4.2 *Main results*

From the numerical simulation, we obtained a series of results for this well. The pressure and temperature in the well-bore and the temperature in the formation are shown in Table 7.11.

7.5.4.3 *Results analysis*

To study how the different geothermal gradients and gas injection volumes influence gas pressure, we used different geothermal gradients and gas injection volumes. Through the numerical simulation, we obtained a series of results.

Firstly, we used three different geothermal gradients, 0.0218, 0.024 and 0.026°C/m, with the other parameters unchanged. We obtained the gas pressure and the temperature distribution figures as shown in Figures 7.20, 7.21 and 7.22.

From Figures 7.20, 7.21 and 7.22, we can see that as the temperature of the formation increases from the top to the bottom of the pipe, the temperature of the well-bore and the gas pressure decreases from the top to the bottom. At the same pipe depth, if the geothermal gradient increases, then the temperature of the formation also increases, while the temperature of the well-bore and the gas pressure decrease.

Also, we considered four different injection volumes; 500000, 700000, 900000 and 1100000 m^3/day, with the other parameters remaining the same. Then we obtained the gas pressure and temperature distributions for the formation and the well-bore as shown in Figure 7.23.

From Figure 7.23, we can see that the gas pressure decreases from the top to the bottom of the pipe. At the same pipe depth, if the injection volume increases, the pressure also decreases.

Table 7.11 Results of pressure and temperature.

Well depth [m]	T (formation) [°C]	T (well-bore) [°C]	Pressure [MPa]
1	16.0138	199.9946	49.7415
251	21.4611	198.6327	49.3948
501	26.9083	197.2709	49.0480
751	32.3556	195.9091	48.7013
1001	37.8029	194.5473	48.3545
1251	43.2502	193.1855	48.0077
1501	48.6974	191.8236	47.6610
1751	54.1447	190.4618	47.3142
2001	59.5920	189.1000	46.9674
2251	65.0393	187.7382	46.6207
2501	70.4865	186.3764	46.2739
2751	75.9338	185.0145	45.9272
3001	81.3811	183.6527	45.5804
3251	86.8284	182.2909	45.2336
3501	92.2756	180.9291	44.8869
3751	97.7229	179.5673	44.5401
4001	103.1702	178.2055	44.1933
4251	108.6175	176.8436	43.8466
4501	114.0647	175.4818	43.4998
4751	119.5120	174.1200	43.1531
5001	124.9593	172.7582	42.8063
5251	130.4066	171.3964	42.4595
5501	135.8538	170.0345	42.1128
5751	141.3011	168.6727	41.7660
6001	146.7484	167.3109	41.4192
6251	152.1957	165.9491	41.0725
6501	157.6429	164.5873	40.7257
6751	163.0902	163.2254	40.3790
7001	168.5375	161.8636	40.0322
7100	170.6946	161.3243	39.8949

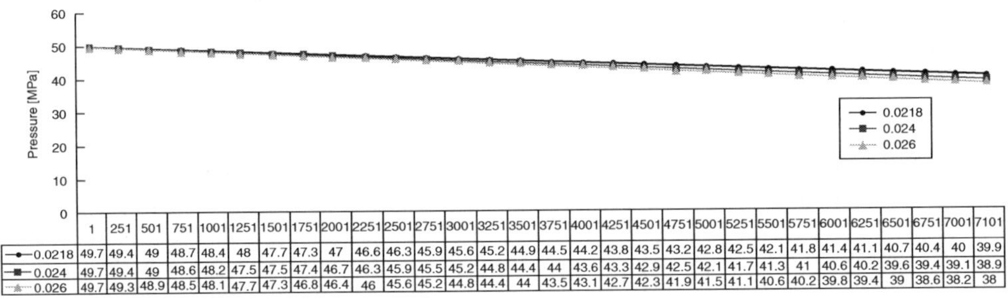

	1	251	501	751	1001	1251	1501	1751	2001	2251	2501	2751	3001	3251	3501
0.0218	49.7	49.4	49	48.7	48.4	48	47.7	47.3	47	46.6	46.3	45.9	45.6	45.2	44.9
0.024	49.7	49.4	49	48.6	48.2	47.5	47.5	47.4	46.7	46.3	45.9	45.5	45.2	44.8	44.4
0.026	49.7	49.3	48.9	48.5	48.1	47.7	47.3	46.8	46.4	46	45.6	45.2	44.8	44.4	44

	3751	4001	4251	4501	4751	5001	5251	5501	5751	6001	6251	6501	6751	7001	7101
0.0218	44.5	44.2	43.8	43.5	43.2	42.8	42.5	42.1	41.8	41.4	41.1	40.7	40.4	40	39.9
0.024	44	43.6	43.3	42.9	42.5	42.1	41.7	41.3	41	40.6	40.2	39.6	39.4	39.1	38.9
0.026	43.5	43.1	42.7	42.3	41.9	41.5	41.1	40.6	40.2	39.8	39.4	39	38.6	38.2	38

Depth [m]

Figure 7.20 Pressure distribution of formation.

7.6 DFA-SIPVF[5]

To discuss the dryness fraction for the steam injection process in HTHP wells with varied (T, P) fields, we set the following assumptions.

[5]Dryness fraction analysis for steam injection process of HTHP wells in the varied (T, P) fields

	1	251	501	751	1001	1251	1501	1751	2001	2251	2501	2751	3001	3251	3501	3751	4001	4251	4501	4751	5001	5251	5501	5751	6001	6251	6501	6751	7001	7101
0.0218	16	21.5	26.9	32.4	37.8	43.3	48.7	54.1	59.6	65	70.5	75.9	81.4	86.8	92.3	97.7	103	109	114	120	125	130	138	141	147	152	158	163	169	171
0.024	16	22	28	34	40	46	52	58	64	70	76	82	88	94	100	106	112	116	124	130	136	142	148	154	160	166	172	178	184	186
0.026	16	22.5	29	35.5	42	48.5	55	61.5	60	74.5	81	87.5	94	100	107	113	120	126	133	139	140	152	159	165	172	170	185	191	190	200

Depth [m]

Figure 7.21 Temperature distribution of formation.

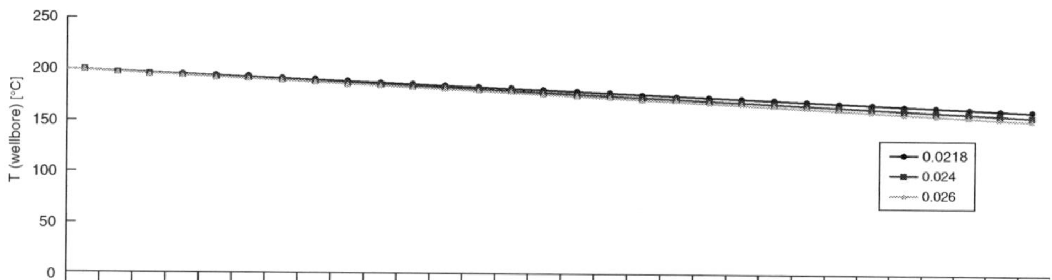

	1	251	501	751	1001	1251	1501	1751	2001	2251	2501	2751	3001	3251	3501	3751	4001	4251	4501	4751	5001	5251	5501	5751	6001	6251	6501	6751	7001	7101
0.0218	200	199	197	196	195	193	192	190	189	188	186	185	184	182	181	180	178	177	175	174	173	171	170	169	167	166	165	163	162	161
0.024	200	198	197	195	194	192	191	189	188	187	185	184	182	181	179	178	176	175	173	172	170	169	167	166	164	163	161	160	158	157
0.026	200	198	197	195	193	192	190	189	187	185	184	182	181	179	177	176	174	172	171	169	168	166	164	163	161	159	158	156	155	154

Depth [m]

Figure 7.22 Temperature distribution of well-bore.

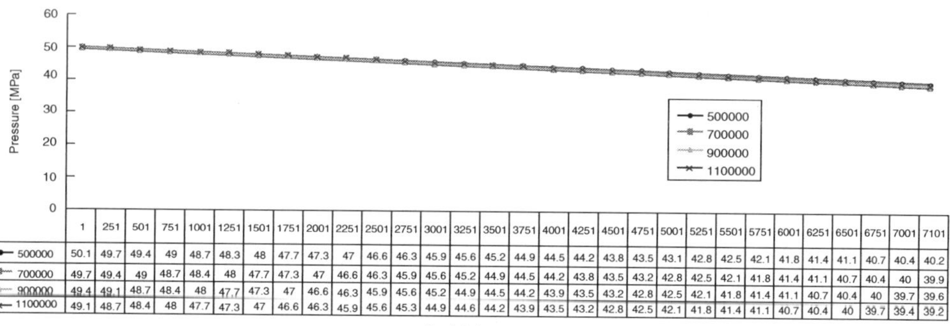

	1	251	501	751	1001	1251	1501	1751	2001	2251	2501	2751	3001	3251	3501	3751	4001	4251	4501	4751	5001	5251	5501	5751	6001	6251	6501	6751	7001	7101
500000	50.1	49.7	49.4	49	48.7	48.3	48	47.7	47.3	47	46.6	46.3	45.9	45.6	45.2	44.9	44.5	44.2	43.8	43.5	43.1	42.8	42.5	42.1	41.8	41.4	41.1	40.7	40.4	40.2
700000	49.7	49.4	49	48.7	48.4	48	47.7	47.3	47	46.6	46.3	45.9	45.6	45.2	44.9	44.5	44.2	43.8	43.5	43.2	42.8	42.5	42.1	41.8	41.4	41.1	40.7	40.4	40	39.9
900000	49.4	49.1	48.7	48.4	48	47.7	47.3	47	46.6	46.3	45.9	45.6	45.2	44.9	44.5	44.2	43.9	43.5	43.2	42.8	42.5	42.1	41.8	41.4	41.1	40.7	40.4	40	39.7	39.6
1100000	49.1	48.7	48.4	48	47.7	47.3	47	46.6	46.3	45.9	45.6	45.3	44.9	44.6	44.2	43.9	43.5	43.2	42.8	42.5	42.1	41.8	41.4	41.1	40.7	40.4	40	39.7	39.4	39.2

Depth [m]

Figure 7.23 Gas pressure distribution.

1. The physical properties of the fluid and the formation are independent of depth and temperature.
2. The bottom of the insulated tubing is a packer setting to ensure steam cannot flow into the space between the tube and casing, which is full of air.
3. The steam flow is one dimensional and in a steady state along the well-bore; the pressure and temperature in the same horizontal cross section is constant.
4. The physical properties of the other materials, except for the steam and some thermal insulating materials, are independent of time and temperature.
5. The temperature around the well-bore in the stratum has an axisymmetric distribution.
6. The heat transfer from the inner well-bore to the cement mantle external boundary is stable, while an unstable transfer occurs from the cement mantle external boundary to the geological formation. The heat transfer that occurs in the longitudinal direction is ignored.

7.6.1 The model dryness fraction in the varied (*T, P*) fields

Within the tubing, the flow is under turbulent conditions. Consider the flow system depicted in Figure 7.1, a straight cylindrical flow tube with an inclination angle θ, a constant cross-sectional flow area A, a hydraulic diameter d, and a total length Z. Through this tubing gas flows from the bottom to the top with a mass flow rate W. The distance co-ordinate in the flow direction along the tubing is denoted z.

An energy equation is introduced to construct the dryness model. From the principles of fluid mechanics and thermodynamics, any flow of fluid at some state parameter (P, T) at some location including energy are as follows: inner energy, pressure energy, kinetic energy and potential energy:

$$E = U + \frac{1}{2}mv_m^2 + mgz + Pv_m. \tag{7.206}$$

Here, $U, \frac{1}{2}mv_m^2, mgz, Pv_m$ mean the thermodynamic energy, the kinetic energy, the potential energy, the pressure energy, respectively. The inner energy and the pressure energy are collectively referred to as enthalpy:

$$h = U + \frac{1}{2}mv_m^2. \tag{7.207}$$

From the energy balance rule, we know that the energy of the fluid flowing into the distance element equals the sum of energy losses from the distance element and from the steam flow friction. The infinitesimal energy equation can be written as follows:

$$\frac{dQ}{dz} + \frac{dW}{dz} + m_k g\cos\theta = -m_k\frac{dh_m}{dz} - m_k v_m\frac{dv_m}{dz}. \tag{7.208}$$

Divide by m_k,

$$\frac{dq}{dz} + \frac{dw}{dz} + g\cos\theta = -\frac{dh_m}{dz} - v_m\frac{dv_m}{dz} \tag{7.209}$$

h_m expresses the enthalpy of the gas liquid two phase flow which is defined as:

$$h_m = h_g x + h_w(1-x), \tag{7.210}$$

where h_g is the enthalpy of the saturated steam and h_w is the saturated water. It follows from (7.209) that:

$$\frac{dh_m}{dz} = \left(\frac{dh_g}{dz} - \frac{dh_w}{dz}\right)x + (h_g - h_w)\frac{dx}{dz} + \frac{dh_w}{dz}. \tag{7.211}$$

The enthalpy of the saturated steam is a function of pressure. That is $h = f(P)$. Hence:

$$\frac{dh_k}{dz} = \frac{dh_k}{dP}\frac{dP}{dz}. \tag{7.212}$$

The subscript k means anything such as water, steam or mixture. Substituting the (7.212) into the (7.211), the following equation can be obtained:

$$\frac{dh_m}{dz} = (h_g - h_w)\frac{dx}{dz} + \frac{dh_w}{dz}\frac{dP}{dz} + \left(\frac{dh_g}{dP} - \frac{dh_w}{dP}\right)\frac{dP}{dz}x. \tag{7.213}$$

We can also apply the mass equation $m_k = \rho_k v_k A$. Derivative at both ends,

$$\frac{dv_k}{dz} = \frac{d}{dz}\left(\frac{m_k}{\rho_k A}\right) = \frac{1}{\rho_k A}\frac{dm_k}{dz} + \frac{m_k}{A}\frac{d}{dz}\left(\frac{1}{\rho_k}\right). \tag{7.214}$$

Taking in account the mist state of the steam at the injection process and the large gas flow, the gas state equation can be applied:

$$\rho_g = \frac{MP}{RZ_g T}.$$

Derivative at both ends,

$$d\left(\frac{1}{\rho_g}\right)=\frac{RZ_g}{PM}dT-\frac{RTZ_g}{P^2M}dP+\frac{R}{PM}dZ_g=\frac{1}{\rho_g}\left(\frac{1}{T}\frac{dT}{dP}+\frac{1}{Z_g}\frac{dZ_g}{dP}-\frac{1}{P}\right)dP. \tag{7.215}$$

The gas deviation factor gradient is normally small and often neglected. Then:

$$d\left(\frac{1}{\rho_g}\right)=\frac{1}{\rho_g}\left(\frac{1}{T}\frac{dT}{dP}-\frac{1}{P}\right)dP. \tag{7.216}$$

Hence, substituting the (7.216) into (7.209), the following equation is determined:

$$\frac{dq}{dz}+\frac{dw}{dz}+gz\cos\theta=-\left[(h_g-h_w)\frac{dx}{dz}+\frac{dh_w}{dz}\frac{dP}{dz}+\left(\frac{dh_g}{dP}-\frac{dh_w}{dP}\right)\frac{dP}{dz}x\right]-v_m^2\left(\frac{1}{T}\frac{dT}{dP}-\frac{1}{P}\right)\frac{dP}{dz}. \tag{7.217}$$

Assume,

$$A=(h_g-h_w),\quad B=\left(\frac{dh_g}{dP}-\frac{dh_w}{dP}\right)\frac{dP}{dz},$$
$$C=\frac{dq}{dz}+\frac{dw}{dz}+gz\cos\theta+\frac{dh_w}{dz}\frac{dP}{dz}+v_m^2\left(\frac{1}{T}\frac{dT}{dP}-\frac{1}{P}\right)\frac{dP}{dz}$$

the (7.217) can be changed as:

$$A\frac{dx}{dz}+Bx+C=0\quad\Longleftrightarrow\quad\frac{dx}{dz}+\frac{B}{A}x=-\frac{C}{A}. \tag{7.218}$$

A, B, C are constant at fixed depth. Thus, the Equation (7.218) is a first order linear ordinary differential equation. From this, the dryness computing model can be obtained:

$$\begin{cases} x=e^{-\frac{Bz}{A}}\left(-\frac{C}{B}e^{-\frac{Bz}{A}}+x_0-\frac{C}{B}\right)\\ A=(h_g-h_w)\\ B=\left(\frac{dh_g}{dP}-\frac{dh_w}{dP}\right)\frac{dP}{dz}\\ C=\frac{dq}{dz}+\frac{dw}{dz}+gz\cos\theta+\frac{dh_w}{dz}\frac{dP}{dz}+\frac{v_m}{\rho_m}\left(\frac{1}{T}\frac{dT}{dP}-\frac{1}{P}\right)\frac{dP}{dz}\\ x|_{z=0}=x_0 \end{cases}. \tag{7.219}$$

7.6.2 **Varied (*T*, *P*) fields analysis**

While dryness modelling, we find the dryness numerical values are affected by temperature and pressure. In fact, the two parameters vary according to depth and time. As a result, the varied (T, P) fields need to be researched.

Material balance

Since steam injection is a constant-specific mass flow, that is, the mass flowing in the infinitesimal equals the one flowing out of the infinitesimal, so it follows from the mass conservation equation that:

$$\frac{d(\rho_m v_m A)}{dz}=0. \tag{7.220}$$

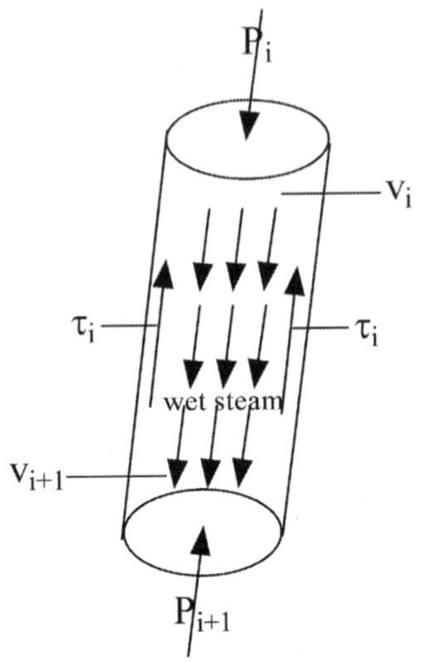

Figure 7.24 Flow analysis of infinitesimal section in deviated well-bore.

Momentum balance
As shown in Figure 7.24, there are surface and mass forces. The mass force is $\rho_m g\cos\theta A dz$. The surface forces are: the upper pressure P_i, the lower pressure P_{i+1}, and the friction force τ_i. Because of the flow changes between the upper and lower element, the momentum changes. When considering these momentum balances, the equation is built as follows:

$$P_{i+1}A - P_iA - \tau_i + \rho_m g\cos\theta A dz = \Delta(m_i v_i) = (\rho_{i+1}A v_{m(i+1)})v_{m(i+1)} - (\rho_i A v_{m(i)})v_{m(i)}. \tag{7.221}$$

Divide by A,

$$dP = -\frac{\tau_i}{A} + \rho_m g\cos\theta dz - d(\rho_m v_m^2). \tag{7.222}$$

Substituting (7.220) into (7.222) and divide by dz,

$$\frac{dP}{dz} = -\frac{1}{A}\frac{\tau_i}{dz} + \rho_m g\cos\theta - \rho_m v_m\frac{dv_m}{dz}. \tag{7.223}$$

7.6.2.1 *The temperature field component processing*
There are some unknown parameters in (7.209), which must be must researched to analyse the changing temperature and pressure fields.

Heat transmission in the well-bore In Equation (7.209), dq denotes the radial heat transfer between the gas and the surrounding earth tube. Ramey (1962) and Willhite and Dietrich (1967) discussed the radial transfer of heat between the fluid and the earth in detail. The heat transfer from the tube to the second interface is considered steady, but the heat transfer from the second interface to the stratum is unsteady. Therefore, determining the interface between the cement mantle and the stratum (wall of a well) is a key problem.

The overall heat transmission coefficient in the well case is proportional to the temperature difference and the cross section perpendicular in the direction of heat transfer:

$$q = UA\Delta T. \tag{7.224}$$

As Figure 7.25 shows, using the (7.224), the radial heat transfer from the fluid to the wall interface can be described by:

$$\frac{dq_1}{dz} = 2\pi r_{to}U_{to}(T - T_{ref}). \tag{7.225}$$

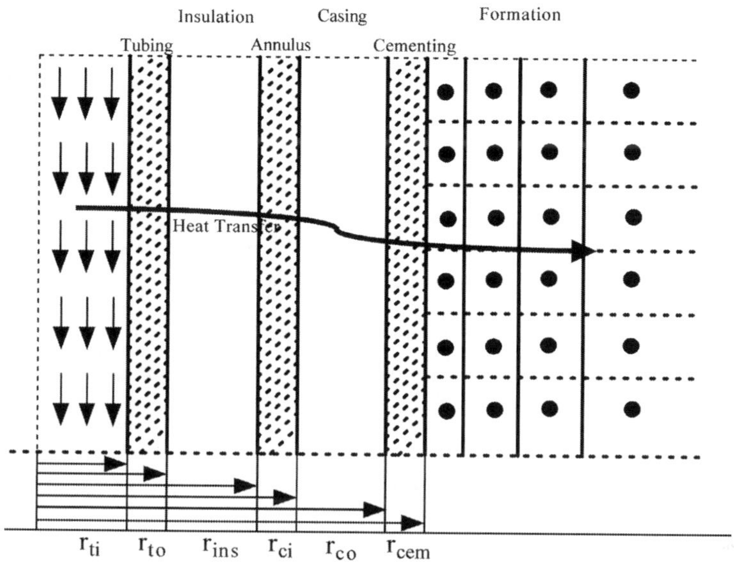

Figure 7.25 Steam injection string structure scheme of wells.

The heat loss from the cement mantle external boundary to the geological formation is an unsteady-state heat transfer, Ramey (1962) gave a model to describe this phenomenon:

$$\frac{dq_2}{dz} = \frac{2\pi K_e(T_{ref} - T_e)}{f(t_D)}. \tag{7.226}$$

Combine Equations (7.225) and (7.226) and the heat transmission model between the steam flow and the surrounding earth is as follows:

$$\frac{dq}{dz} = \frac{2\pi r_{to} U_{to} K_e}{r_{to} U_{to} f(t_D) + K_e}(T - T_e). \tag{7.227}$$

Let $a = \frac{2\pi r_{to} U_{to} K_e}{r_{to} U_{to} f(t_D) + K_e}$, then:

$$dq = a(T - T_e)dz. \tag{7.228}$$

This heat loss model has been widely used in steam and other injection wells, and provides the distribution of the fluid parameters and temperature in the tube. The temperature T_e is commonly considered a geothermal gradient. However, in actual engineering practice, the steam injection parameters are not constant and the temperature at the wall changes with time and depth.

Considering the heat transmission in the well-bore, according to the theory of heat conduction through the medium and the above assumption, over the same infinitesimal of depth, the radial heat conduction equation for the formation can be described as:

$$\rho_f C_f \frac{\partial T_e}{\partial t} = \lambda_f \left(\frac{\partial^2 T_e}{\partial r^2} + \frac{1}{r}\frac{\partial T_e}{\partial r} \right). \tag{7.229}$$

Initial condition. When steam is injected, the formation temperature field is at the initial formation:

$$T_{z,r,0} = T_0 + \gamma z \cos\theta. \tag{7.230}$$

Boundary conditions. The inner boundary condition: according to Fourier's law of heat conduction, on the interface between the well-bore and the formation, the heat flux exchange through a unit well-bore depth and the surrounding formation is:

$$dq = 2\pi\lambda_f r dz \left.\frac{\partial T_r}{\partial r}\right|_{r=r_w} \tag{7.231}$$

and

$$T_w(z=0,t) = T_0 \tag{7.232}$$

for the injection well is the surface temperature.

Outer boundary condition. At infinity distance, the formation temperature distributions stay at the initial formation temperature distributions and are independent of time:

$$\frac{\partial T_r}{\partial r} = 0, \qquad \text{if } r \to \infty. \tag{7.233}$$

The temperature of the fluid and the earth can be explained using the above two differential equations and the correspondent initial and boundary conditions. To facilitate a solution, the following dimensionless variables are defined: $r_D = r/r_{cem}$, $t_D = \lambda_e t/(\rho_f C_f r_{cem}^2)$: then, the variables r, t are exchanged into dimensionless variables r_D, t_D, (7.229) is exchanged into:

$$\frac{\partial T_e}{\partial t_D} = \left(\frac{\partial^2 T_e}{\partial r_D^2} + \frac{1}{r_D}\frac{\partial T_e}{\partial r_D}\right). \tag{7.234}$$

The boundary conditions are converted into:

$$\left.\frac{\partial T_e}{\partial r_D}\right|_{r_D=1} = -\frac{1}{2\pi\lambda_f}\frac{dq}{dz} \left.\frac{\partial T_e}{\partial r_D}\right|_{r_D\to\infty} = 0.$$

Enthalpy drop. h_m, which satisfies the following relation, denoting specific enthalpy:

$$\frac{dh_m}{dz} = C_{pm}\frac{dT}{dz} - \eta_m C_{pm}\frac{dP}{dz} \tag{7.235}$$

η_m is the Joule-Thomson coefficient defined as:

$$\begin{cases} \eta_m = 0, & m = g \\ \eta_m = -\dfrac{1}{C_{pm}\rho_m}, & m = o, \omega \end{cases}.$$

Because of the mist state of the steam at the injection process and the large gas flow, we denote the mixture as a gas state.

Friction force doing work. As the direction of the friction force is contrary to the steam flow, and has a negative work as steam flows, thus work done by friction force on infinite length in unit time can be described by:

$$dw = \frac{\tau_i dz}{dt} = \frac{\tau_i dz}{2dz/(v_{mi} + v_{mi+1})} = \frac{\tau_i(v_{mi} + v_{mi+1})}{2} \approx \tau_i v_m. \tag{7.236}$$

The solution of friction uses a calculation method introduced in fluid dynamics:

$$\tau_i = \frac{\pi f r_{ti}\rho_m v_m^2}{4} dz. \tag{7.237}$$

The friction factor f of gas-liquid mixture. f is the function regarding the Reynolds number Re and the absolute roughness ε:

$$f = \begin{cases} Re/64, & \text{if } Re \le 2000 \\ \left[1.14 - 2\ln\left(\dfrac{\varepsilon}{2r_{ti}} + 21.25Re^{-0.9}\right)\right]^{-2}, & \text{if } Re > 2000 \end{cases} \qquad Re = \frac{2rv\rho}{\mu}.$$

7.6.2.2 *The coupled model of the varied (T, P) fields*

By means of the relationship of (7.209), (7.235), (7.236) and (7.237), the gas energy equation is obtained:

$$C_{Pg}\frac{dT}{dz} + v_m\frac{dv_m}{dz} + g\cos\theta + \frac{\pi f r_{ti}\rho_m v_m^3}{4} - a(T - T_e) = 0. \tag{7.238}$$

In addition, for the velocity of the mixed flow, we have:

$$v_m = v_g + v_w = \frac{mx}{\rho_g A} + \frac{m(1-x)}{\rho_w A}. \tag{7.239}$$

Thus,

$$\frac{dv_m}{dz} = R\frac{dx}{dz} - S\frac{dP}{dz}, \tag{7.240}$$

where $R = \frac{m}{A}\left(\frac{1}{\rho_g} - \frac{1}{\rho_w}\right)$ and $S = \frac{m}{A}\left(\frac{x}{\rho_g^2}\frac{d\rho_g}{dP} + \frac{1-x}{\rho_w^2}\frac{d\rho_w}{dP}\right) = \frac{m^2 x}{ARZ_g T\rho_g^2}$.

Substituting the (7.240) into (7.223) and (7.238), respectively, we have:

$$\frac{dP}{dz} = \frac{-\frac{\tau_i}{A} + \rho_m g\cos\theta + \frac{m}{A}R\frac{dx}{dz}}{1 - \frac{m}{A}S} \tag{7.241}$$

$$\frac{dT}{dz} = -\frac{v_m}{C_{Pg}}\left(R\frac{dx}{dz} - S\frac{dP}{dz}\right) - \frac{g\cos\theta}{C_{Pg}} - \frac{\pi f r_{ti}\rho_m v_m^3}{4C_{Pg}} + \frac{a(T - T_e)}{C_{Pg}}. \tag{7.242}$$

Substituting the (7.238) and (7.241), the coupled equations can be obtained:

$$\begin{cases} \dfrac{dP}{dz} = \dfrac{-\frac{\tau_i}{A} + \rho_m g\cos\theta + \frac{m}{A}R\frac{dx}{dz}}{1 - \frac{m}{A}S} \\ \dfrac{dT}{dz} = -\dfrac{v_m}{C_{Pg}}\left(R\dfrac{dx}{dz} - S\dfrac{dP}{dz}\right) - \dfrac{g\cos\theta}{C_{Pg}} - \dfrac{\pi f r_{ti}\rho_m v_m^3}{4C_{Pg}} + \dfrac{a(T - T_e)}{C_{Pg}} \cdot \\ P(z_0) = P_0, \quad T(z_0) = T_0, \quad dx(z_0) = dx_0, \quad x(z_0) = z_0 \end{cases} \tag{7.243}$$

7.6.2.3 *Calculation of some parameters*

In this section, we demonstrate the calculation method for some parameters.

1. Each point's inclination:

$$\theta_j = \theta_{j-1} + \frac{\theta_k - \theta_{k-1}}{\Delta s_k}\Delta s_j,$$

where j represents segment point of calculation, Δs_k represents measurement depth of inclination angle θ_k, θ_{k-1}, Δs_j step length of calculation.

2. Transient heat transfer function:

$$f(t_D) = \begin{cases} 1.128\sqrt{t_D}(1 - 0.3\sqrt{t_D}), & t_D \le 1.5 \\ (0.4063 + 0.5\ln t_D)(1 + 0.6/t_D), & t_D > 1.5 \end{cases}.$$

3. Gas condensing parameter:

$$Z_g = \begin{cases} 1 + (0.31506 - 1.0467/T_{pr} - 0.5783/T_{pr}^3)\rho_{pr} + (0.053 - 0.6123/T_{pr})\rho_{pr}^2 \\ \quad + 0.6815\rho_{pr}^2/T_{pr}^3, \qquad \text{if } P < 35\,\text{MPa}; \\ (90.7x - 242x^2 + 42.4x^3)y^{1.18+2.82x} - (14.76x - 9.76x^2 + 4.58x^3)y \\ \quad + (1 + y + y^2 + y^3/(1-y)^3, \qquad \text{otherwise}. \end{cases}$$

$$F(y) = -0.06125P_{pr}x\exp[-1.2(1-x)^2] + (90.7x - 242.2x^2 + 42.4x3)y^{1.18+2.82x}$$
$$+\frac{1+y+y^2+y^3}{(1-y)^3} - (14.76x - 9.76x^2 + 4.58x3)y^2 = 0, \quad x = 1/T_{pr}.$$

4. Wet steam density

   Since the water vapour flow is a gas-liquid two-phase flow, there has been significant research into this problem (see Beggs and Brill (1973), Mukherjee and Brill (1985)). Here, we adopt the M-B model to calculate the average density.
5. The heat transfer coefficient U_{to} from different positions of the axis of the well-bore to the second surface.

   These resistances include the tubing wall, the possible insulation around the tubing, the annular space (possibly filled with a gas or liquid but sometimes a vacuum), the casing wall, and the cement behind the casing as illustrated in Figure 7.3:

$$\frac{1}{U_{to}} = r_{ti}\frac{1}{\lambda_{ins}}\ln\left(\frac{r_{ci}}{r_{to}}\right) + \frac{1}{h_c + h_r} + r_{ti}\frac{1}{\lambda_{cem}}\ln\left(\frac{r_{cem}}{r_{co}}\right)$$

λ_{ins} and λ_{cem} are the heat conductivity of the heat insulating material and the cement sheath, respectively. h_c and h_r are the coefficients of the convection heat transfer and the radiation heat transfer.

7.6.3 Algorithm steps

To simplify the calculation, we divided the well into several short segments of the same length. The length of a segment varies depending on variations in wall thickness, hole diameter, the fluid density inside and outside the pipe and well geometry. The model begins with a calculation at the top of the pipe. Then pressure and temperature calculations are performed for each successive "segment" of the pipe up to the surface.

Step 1. Set depth step length. In addition, we denote the relatively tolerant error as ε. The smaller h, ε is, the more accurate the results are. However, this leads to a rapid increase in the calculation time. In our paper, we set $h = 1$ (m), and $\varepsilon = 5\%$.

Step 2. Given the initial conditions.

Step 3. Compute each point's inclination.

Step 4. Compute the parameters under the initial conditions or the last depth variables.

Step 5. Let $T = T_k$, then we can determine the T_e by solving the following equation:

$$\begin{cases} \dfrac{\partial T_e}{\partial t_D} = \left(\dfrac{\partial^2 T_e}{\partial r_D^2} + \dfrac{1}{r_D}\dfrac{\partial T_e}{\partial r_D}\right) \\ T_e|_{t_D=0} = T_0 + \gamma z\cos\theta \\ \left.\dfrac{\partial T_e}{\partial r_D}\right|_{r_D=1} = -\dfrac{1}{2\pi\lambda_f}\dfrac{dq}{dz} \\ \left.\dfrac{\partial T_e}{\partial r_D}\right|_{r_D\to\infty} = 0 \end{cases}$$

Let $T_{e,i}^j$ be the temperature at the injection time j and radial i at the depth z. We apply the finite difference method to discretize the equations as follows:

$$\frac{T_{e,j}^{i+1} - T_{e,j}^i}{\varphi} = \frac{T_{e,j+1}^{i+1} - 2T_{e,j+1}^j + T_{e,j+1}^{i-1}}{\xi^2} - \frac{T_{e,j+1}^{i+1} - T_{e,j}^{i+1}}{r_D\varphi},$$

where φ is the interval of time and ξ is the interval of radial, respectively. It can be transformed into the standard form as follows:

$$-\left(\varphi+\frac{\varphi\xi}{r_D}\right)T_{e,j+1}^{i+1}+\left(2\varphi+\frac{\varphi\xi}{r_D}\right)T_{e,j}^{i+1}-\varphi T_{e,j-1}^{i+1}=\xi^2 T_{e,j}^{i}$$

Then a different method is used to discretize the boundary condition.

For $r_D=1$, we have:

$$T_{e,2}^{e,i+1}-\left(1+\frac{a\xi}{2\pi\lambda_f}\right)T_{e,1}^{i+1}=\frac{aT_k}{2\pi\lambda_f}$$

For $r_D=N$, we have:

$$T_{e,n}^{i+1}-T_{e,n-1}^{i+1}=0$$

we can compute the symbolic solution to the temperature T_e of the stratum. In this step, we determine the discrete distribution of T_e using the following matrix:

$$A=\begin{bmatrix} T_{e,1}^1 & T_{e,1}^2 & \cdots & T_{e,1}^i & \cdots \\ T_{e,2}^1 & T_{e,2}^2 & \cdots & T_{e,2}^i & \cdots \\ \cdots & \cdots & \cdots & \cdots & \cdots \\ T_{e,j}^1 & T_{e,j}^2 & \cdots & T_{e,j}^i & \cdots \\ \cdots & \cdots & \cdots & \cdots & \cdots \\ T_{e,N}^1 & T_{e,N}^2 & \cdots & T_{e,N}^i & \cdots \end{bmatrix},$$

where i represents the injection time and j represents the radial.

Step 6. Let the right side of the coupled differential equations to be functions F_i, where $(i=1,2)$. Then we can obtain a system of coupled functions as follows:

$$\begin{cases} F_1=\dfrac{-\frac{\tau_i}{A}+\rho_m g\cos\theta+\frac{m}{A}R\frac{dx}{dz}}{1-\frac{m}{A}S} \\ F_2=-\dfrac{v_m}{C_{Pg}}\left(R\dfrac{dx}{dz}-S\dfrac{dP}{dz}\right)-\dfrac{g\cos\theta}{C_{Pg}}-\dfrac{\pi f r_{ti}\rho_m v_m^3}{4C_{Pg}}+\dfrac{a(T-T_e)}{C_{Pg}} \end{cases},$$

where, T_e at $r_D=1$

Step 7. Assume P, T to be y_i $(i=1;\ 2)$, respectively. Then we obtain some basic parameters as follows:

$$\begin{cases} a_i=f_i(y_1,y_2) \\ b_i=f_i(y_1+ha_1/2, y_2+ha_2/2) \\ c_i=f_i(y_1+hb_1/2, y_2+hb_2/2) \\ d_i=f_i(y_1+hc_1, y_2+hc_2) \end{cases}$$

Step 8. Calculate the pressure and temperature at point: $(j+1)$:

$$y_i^{j+1}=y_i^j+h(a_i+2b_i+2c_i+d_i)/6, \qquad i=1,2 \quad j=1,2,\ldots,n$$

Step 9. Calculate the dryness at point $(j+1)$:

$$x_{j+1}=e^{-\frac{Bz}{A}}\left(-\frac{C}{B}e^{-\frac{Bz}{A}}+x_0-\frac{C}{B}\right)$$

(The value of A, B, C can be determined from the value of the pressure and temperature at j point).

Step 10. Calculate the dryness drop: $\frac{dx}{dz}=\frac{x_{j+1}-x_j}{h}$.

Step 11. Repeat the step 3 to step 10 until y_n is calculated.

Table 7.12 Parameters of pipes.

Diameter [mm]	Thickness [mm]	Weight [kg]	Expansion coefficient [m/m · K]	Young's modulus [GPa]	Poisson's ratio [–]	Used length [m]
88.9	12.95	23.791	0.0000115	215	0.3	700
88.9	9.53	18.28	0.0000115	215	0.3	2850
88.9	7.34	15.034	0.0000115	215	0.3	1430
88.9	6.45	13.582	0.0000115	215	0.3	950
73	5.51	9.493	0.0000115	215	0.3	185

Table 7.13 Well parameters.

Measured depth [m]	Internal diameter [mm]	External diameter [mm]
3301.7	154.78	193.7
5936.83	152.5	177.8
6115	108.62	127

7.6.4 **Simulation and discussion**

As described previously, the algorithm begins with a calculation for the pipe at the bottom of the pipe. The calculations are then performed for each successive segment of the pipe up to the surface.

7.6.4.1 *Parameters*

In this simulation, we study a pipe in *X* well, which is located in China. The needed parameters are as follows:

Depth of the well = 6115 m
Ground thermal conductivity parameter = 2.06 W/m · K
Ground temperature = 16°C
Geothermal gradient = 2.18°C/100 m
Roughness of the inner surface of the well = 0.000015

The parameters of the pipes, the inclined well, the inclination, the azimuth and the vertical depth are given in

7.6.4.2 *Main results*

From this calculation, we obtained the results for this well. The dryness, casing temperatures, steam temperatures and pressures are shown in Table 7.15. At the same time, we obtained a comparative result with a measurement value as shown in Table 7.16.

7.6.4.3 *Trend analysis*

A trend analysis was conducted to ensure that the developed model was physically correct. To test the developed model, the pressure distribution, temperature and dryness in the well were determined.

Dryness fraction trend under changing temperature and pressure fields

From Figure 7.26 it can be seen that the steam dryness fraction decreases at an increasing rate with an increase in depth. The reason for this is that the mass flow of steam becomes smaller in the steam injection oil reservoir, and as this flow rate decreases, the heat loss in the well-bore increases. As a consequence, the steam dryness fraction decreases significantly.

Table 7.14 Azimuth, inclination and vertical depth.

Number	Measured depth [m]	Inclination [°]	Azimuth [°]	Vertical depth [m]
1	1000	2.82	240.84	999.88
2	1200	2.28	237.69	1199.53
3	1300	1.13	213.69	1299.49
4	2800	1.19	26.21	2799.41
5	3000	1.74	44.39	2999.25
6	3400	1.92	190.95	3399.21
7	3900	1.98	268.9	3899.14
8	4000	2.00	297.38	3999.11
9	4100	4.68	324.34	4098.96
10	4200	1.97	302.88	4198.74
11	4300	1.03	204.57	4298.72
12	4400	1.54	164.16	4398.68
14	4500	2.37	195.11	4498.61
15	4600	2.12	214.67	4598.54
16	4700	1.96	216.31	4698.47
17	4800	3.04	229.14	4798.38
18	4900	3.59	243.86	4898.23
19	5000	5.79	366.45	4997.87
20	5100	8.14	258.61	5097.01
21	5200	7.01	236.71	5196.12
22	5300	5.78	239.1	5295.51
23	5400	5.05	244.42	5395.04
24	5500	3.92	228.03	5494.72
25	5600	4.44	233.71	5594.49
26	5700	5.03	234.87	5694.17
27	5800	5.13	233.21	5793.77
28	5900	4.53	234.82	5893.44
29	6000	3.67	232.4	5993.21
30	6115	4.94	233.11	6107.88

Varied (*T, P*) fields trend

From Figure 7.27, it can be seen that steam pressure is not a uniform distribution but decreases at a decreasing rate with increasing depth. The reason for this is that as the velocity and the steam mass decrease, the pressure drop caused by the frictional force decreases. Hence, the change in the pressure of the steam slows.

The steam temperature is a function of pressure, so the changes in the steam show a similar pattern to those of pressure as shown in Figure 7.28.

At the same time, as the heat transfer increases, the temperature of the casing also increases at a decreasing rate.

7.6.5 Sensitivity analysis

For this analysis, we considered the two parameters that could have influenced our results; the steam injection pressure at the top of the well and the injection rate. We used different pressures, 20 MPa, 22 MPa or 24 MPa; to study the difference between the models, with the three different injection rates; 90000 kg/day, 80000 kg/day or 70000 kg/day. From the algorithm and through simulation, we obtained the results. To better visualize the trend sensitivity, a two-dimensional projection chart was developed.

Table 7.15 The value of dryness fraction, temperature, and pressure.

Measured depth [m]	Dryness fraction [–]	Casing temperature [°C]	Steam temperature [°C]	Pressure [MPa]
0	0.800	130.2	350.0	22.0000
1000	0.763	130.3	349.8	21.9845
1200	0.743	130.7	349.4	21.9605
1300	0.732	131.2	347.2	21.9432
2800	0.721	131.4	346.2	21.9315
3000	0.713	132.3	345.7	21.9156
3400	0.701	133.1	344.9	21.8845
3900	0.682	134.3	343.2	21.8451
4000	0.671	135.8	341.9	21.8252
4100	0.657	137.4	340.5	21.8023
4200	0.645	139.2	339.5	21.7774
4300	0.624	140.6	337.6	21.7469
4400	0.615	142.5	335.8	21.7256
4500	0.583	143.6	334.3	21.7074
4600	0.565	145.3	332.5	21.6806
4700	0.545	146.8	330.3	21.6711
4800	0.516	147.5	328.6	21.6532
4900	0.482	149.7	327.1	21.6148
5000	0.431	150.8	325.6	21.5662
5100	0.381	152.6	324.5	21.5231
5200	0.334	154.3	322.4	21.4934
5300	0.306	155.5	320.8	21.4442
5400	0.276	156.9	319.3	21.4123
5500	0.247	158.2	318.3	21.3833
5600	0.220	159.9	317.3	21.3537
5700	0.191	160.7	316.4	21.3149
5800	0.172	161.8	316.2	21.2762
5900	0.146	162.2	315.7	21.2583
6000	0.106	164.4	315.3	21.2261
6115	0.097	165.8	315.2	21.2034

Table 7.16 Comparative results.

Well-at 100 m depth	Pressure [MPa]	Casing temperature [°C]	Steam temperature [°C]	Dryness fraction [–]
Calculation result	21.9845	130.3	349.8	0.7635
Measurement results	21.8987	127.8	355.9	0.7595
Relative error	0.003	0.02	0.02	0.005

7.6.5.1 *Dryness fraction sensitivity at different injection pressures*

The steam dryness fraction at the bottom of the well-bore, which determines the amount of steam going into the oil reservoir, is an important indicator, on which pressure has a significant impact. From Figure 7.30, it can be seen that the dryness fraction decreases along with an increase in injection pressure. The temperature of the steam increases with increased steam pressure, so then the heat transfer becomes larger causing a rapid decline in the steam dryness fraction.

7.6.5.2 *Temperature field sensitivity at different injection pressures*

The distribution of temperature in the casing along the well-bore is determined using the heat transfer and heat transfer coefficient. At a fixed heat transfer coefficient, the higher the heat transfer quantity, the higher the temperature in the stratum. At the same time, the heat transfer

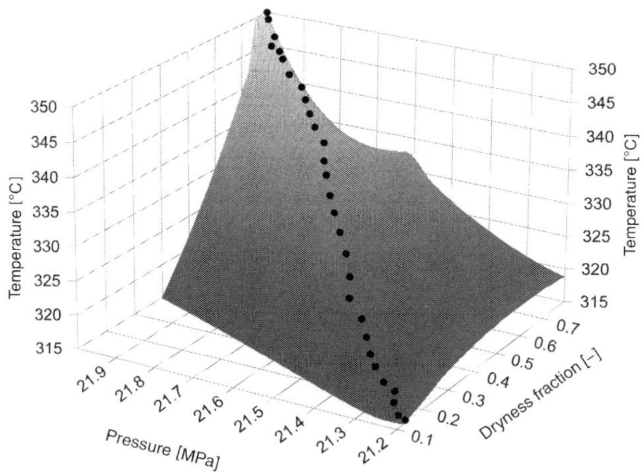

Figure 7.26 Dryness curve under change temperature and change pressure fields.

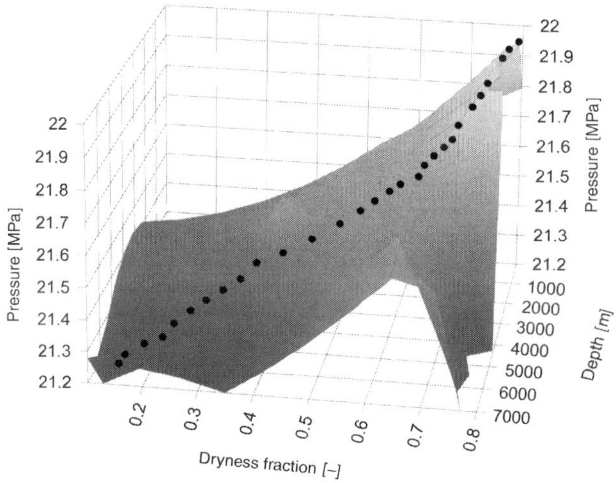

Figure 7.27 Distribution of varied pressure.

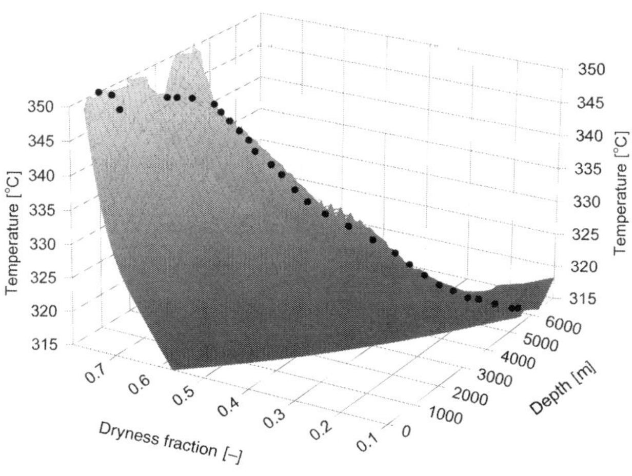

Figure 7.28 Distribution of varied temperature.

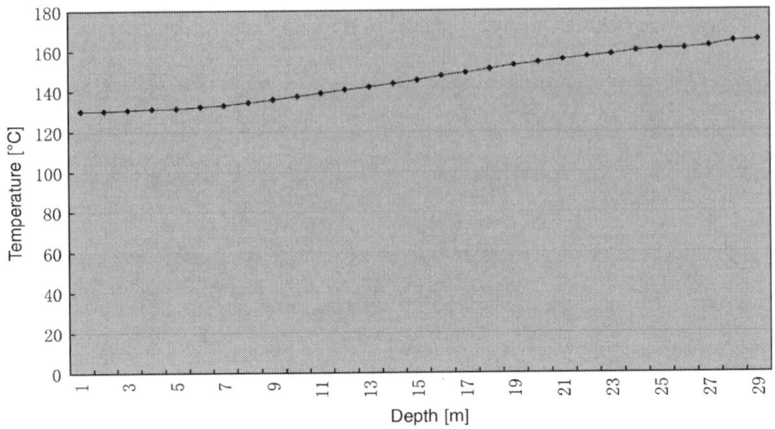

Figure 7.29 Distribution of casing temperature.

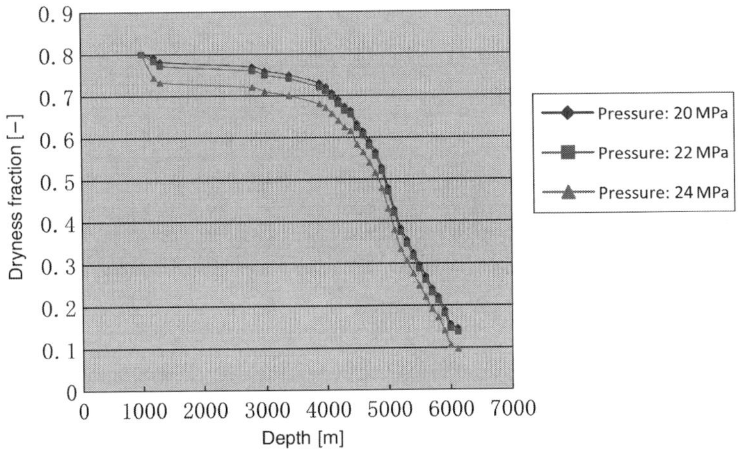

Figure 7.30 Dryness fraction for different injection pressures.

quantity depends on the temperature difference between the steam and the stratum. The stratum temperature is assumed to be the geothermal gradient, so the higher the steam temperature, the more heat is transferred. At saturation, the pressure decides the temperature. The temperature in the casing increases along with the increasing injection pressure but the changes are not linear. This is mainly because of the quality of the saturated steam. The higher the steam pressure, the larger the temperature, which in turn increases the heat transfer. This relationship is described in Figure 7.31.

7.6.5.3 *Dryness fraction sensitivity at different steam injection rates*

The steam injection speed is a very important parameter in steam injection design. It not only involves oil extraction efficiency, but also impacts steam injection costs.

At a fixed steam injection pressure, the steam dryness fraction decreases as the rate of steam injection increases. This is mainly because a drop in the steam dryness fraction is supplemented

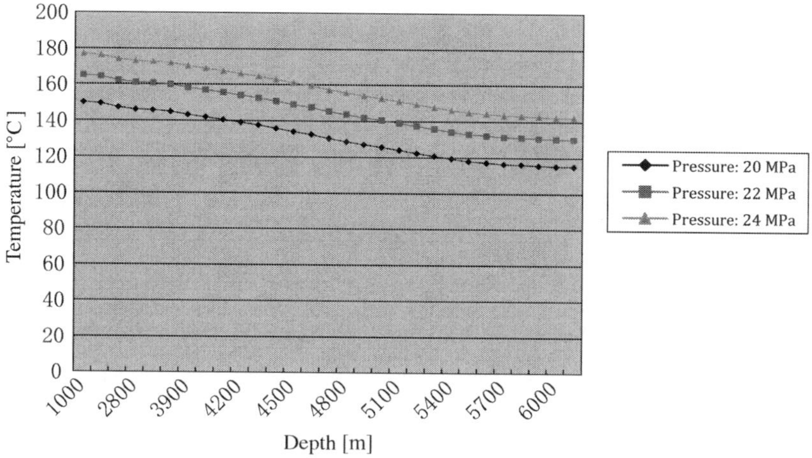

Figure 7.31 Casing temperature for different injection pressures.

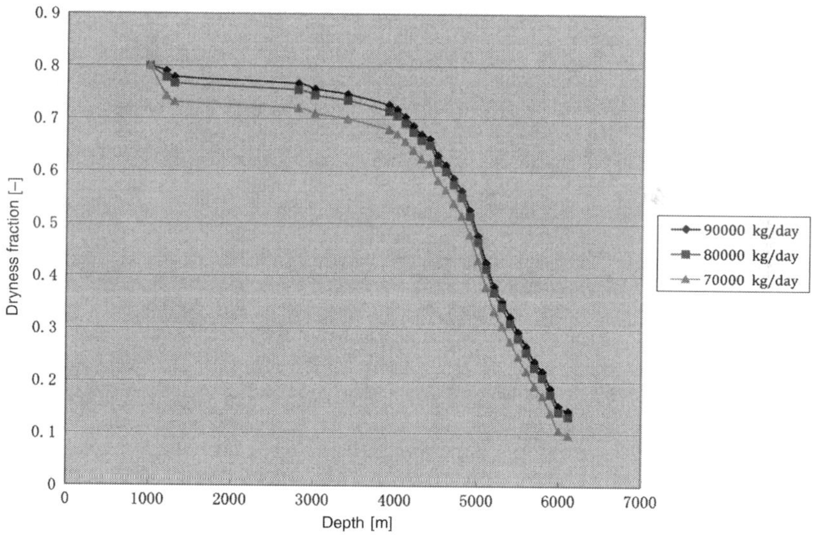

Figure 7.32 Dryness for different steam injection rates.

by new steam because of the rapid steam injection. At the same time, the steam heat transfer is small and the enthalpy drop is also small, which causes the decreases in the steam dryness fraction to slow.

The steam dryness fraction at the bottom of the well-bore increases as the steam injection velocity increases. The steam dryness fraction at the bottom has an increasing tendency and lessens when the steam velocity increases. From Figure 7.32, we can see that when the steam injection velocity is 80000 kg/day, the steam dryness fraction at the bottom does not change significantly, so steam injection is recommended.

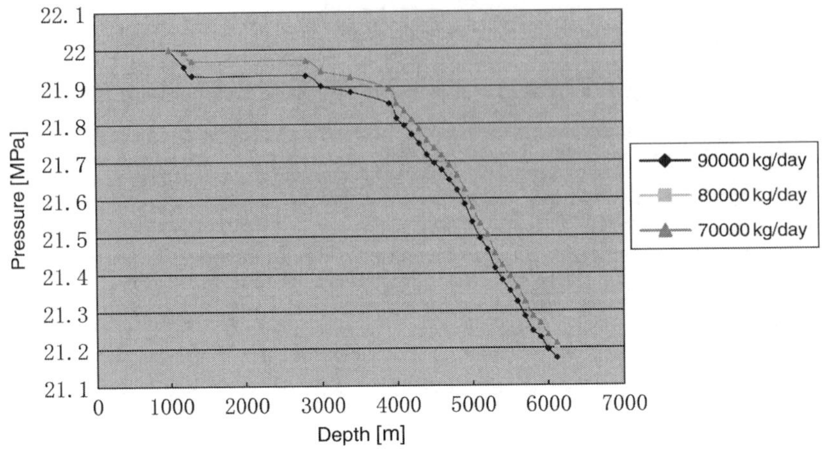

Figure 7.33 Pressure for different steam injection rates.

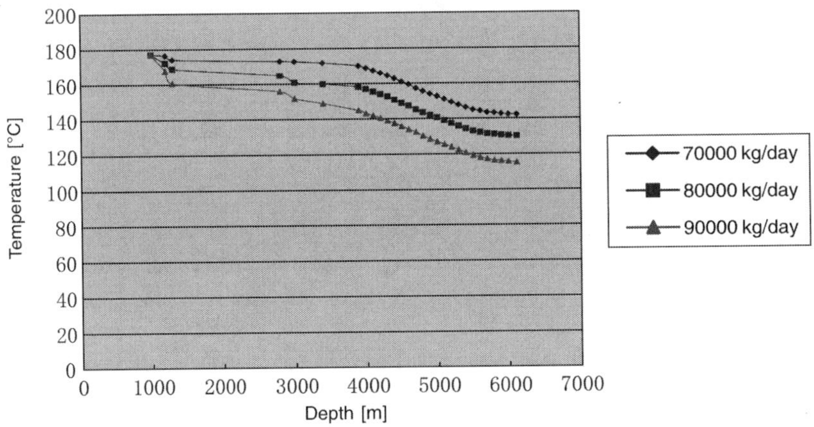

Figure 7.34 Temperature for different steam injection rates.

7.6.5.4 *Pressure field sensitivity at different steam injection rates*

From the changing pressure curve for the different steam injection rates shown in Figure 7.33, it can be seen that the larger the steam velocity, the smaller the pressure drop. This is because high velocity steam injection decreases the heat loss in the well-bore and increases friction drag.

7.6.5.5 *Casing temperature field sensitivity at different steam injection rates*

The temperature effect in the casing is significantly influenced by the injection steam rate. As shown in Figure 7.34, with the same steam injection pressure and different steam injection speeds, the casing temperatures are distinctly dissimilar. The temperature in the casing at the top of the well case is the same but is different at the bottom for any steam injection speed. The faster the steam rate, the greater the temperature drop in the casing. This is chiefly because with a faster steam rate, there is a larger friction force and therefore a quicker steam pressure drop along the well case. The steam temperature drop along the well case increases and causes the heat transfer between the casing and the formation and the heat absorbed by the casing along the depth of the well case to decrease. Therefore, temperature falls with an increase in the depth of the casing.

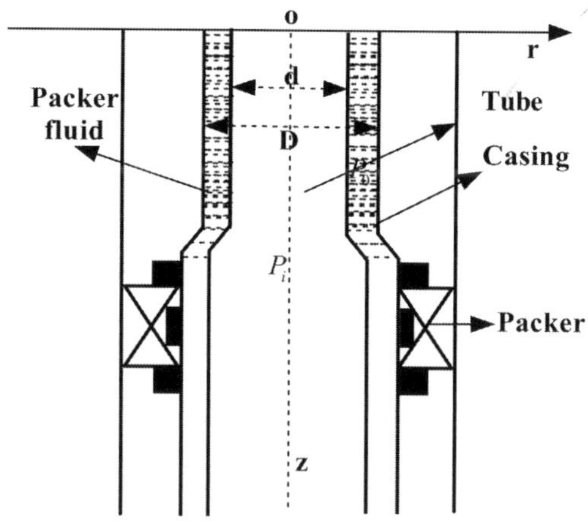

Figure 7.35 The physical figure of forces analysis on tube.

7.7 AASDT-SITP[6]

Before analysing the force on the infinitesimal, some assumptions are introduced;

1. The hole curvature of the considered modular section is constant.
2. On the upperside or underside of the section which is the point of contact of the pipe and tube wall, the curvature is the same as the hole curvature.
3. The radius of the steam injection string, in contrast to the curvature of the borehole, is insignificant.
4. The string has a linear elastic relationship state.

7.7.1 **Force analysis on the tubular string**

The tubular string forces are shown in Figure 7.35. There is a constant cross-sectional flow area A, an inner diameter d, an outer diameter D, a material density ρ_1, a packer fluid density ρ_2 and a total length Z. Through this tubing gas flows from the bottom to the top with a mass flow rate W. The distance co-ordinate in the flow direction along the tubing is denoted z. The cylindrical coordinate system $r\theta z$, the origin of which is in wellhead, and the Z axis as the borehole axis, are used.

As shown in Figure 7.35, the tubular string is mainly affected by the steam injection process forces:

1. Initial axial force. The initial axial force of the tubular includes the deadweight, the buoyant weight and the initial pull force.
2. Thermal stress. With steam injection, the temperature stress acts differently on the tubular at various temperatures.
3. The axial force of the internal and external pressure varies. Because of the varied internal and external pressures, the tubular is affected by a bending force, a piston force and other axial forces.
4. Steam injection friction drag. With steam injection, the flow in the tubular produces a viscous flow, which causes friction drag.

[6]Analyzing axial stress and deformation of tubular for steam injection process in HTHP wells based on the varied (T, P) fields

7.7.2 The tubular axial load and axial stress

7.7.2.1 *Initial axial load and initial axial stress of the steam injection tubular.*
The section $z(m)$, which is the distance from the wellhead was considered. The axial static load considered was the deadweight of the tubular:

$$N_{qz} = \int_z^L q\cos\alpha dz = \frac{\pi}{4}\rho_1 g(D^2 - d^2)\int_z^L \cos\alpha dz, \tag{7.244}$$

where, N_{qz} is the tubular deadweight, q is the average unit length weight of the tubing, L is the tubular length, ρ_1 is the tubular density, α is the inclination angle.

When the axial static load considered is the buoyant weight:

$$N_{bz} = -\rho_2 g A_2 \int_z^L \cos\alpha dz = -\rho_2 g z\pi\left(\frac{D}{2}\right)^2 \int_z^L \cos\alpha dz, \tag{7.245}$$

where, N_{bz} is the tubular buoyant weight, ρ_2 is the packer fluid density.

When we calculate the axial load using the steam injection pressure:

$$N_{pz} = P_{z1}\pi d^2 z/4. \tag{7.246}$$

P_{z1} represents the inner pressure at this section.

Therefore, summing the Equations (7.244), (7.245), (7.246), the axial forces in the section are obtained as follows:

$$F_z = N_{qz} + N_{bz} + N_{pz}. \tag{7.247}$$

Thus the initial axial stress is:

$$\sigma_{zi} = \frac{4F_z}{\pi(D^2 - d^2)} \tag{7.248}$$

7.7.2.2 *Steam injection tubular axial thermal stress*
In the steam injection process, the tubular temperature changes with time and depth, which causes the tubular to deform:

$$\sigma_{zt} = E\beta(T_{z1} - T_{z0}) = E\beta\Delta T. \tag{7.249}$$

E represents the steel elastic modulus of the tubular, β is the warm balloon coefficient for the tubular string, and ΔT is the temperature change before and after the steam injection.

7.7.2.3 *Axial stress of steam injection tubular with a change in pressure*
The effect of the pressure change on the tubular is called the ballooning effect. The ballooning effect is produced because of pressure on the inner and outer sides of the tube. Generally, there are two kinds of tubulars in oil wells. One is a tubular with an outer diameter of 88.9 mm, an inner diameter of 76 mm, and a tube of 6.5 mm ($\frac{\delta}{d/2} = 17.1\% > 5\%$); the other is the tubular with an outer diameter of 114.3 mm, an inner diameter of 100.5 mm, and a tube thickness of 6.9 mm ($\frac{\delta}{d/2} = 13.7\% > 5\%$). Neither is a thin-wall problem. Therefore, this needs to be solved using the Lame formula (Li, 2008). The radial and tangential stresses in the thick-wall cylinder are shown in Figure 7.36. The two can be calculated:

$$\begin{cases} \sigma_{rz} = \dfrac{d^2P_{z1} - D^2P_{z0}}{D^2 - d^2} - \dfrac{(P_{z1} - P_{z0})D^2d^2}{(D^2 - d^2)4r^2} \\ \sigma_{\theta z} = \dfrac{d^2P_{z1} - D^2P_{z0}}{D^2 - d^2} + \dfrac{(P_{z1} - P_{z0})D^2d^2}{(D^2 - d^2)4r^2} \end{cases}, \tag{7.250}$$

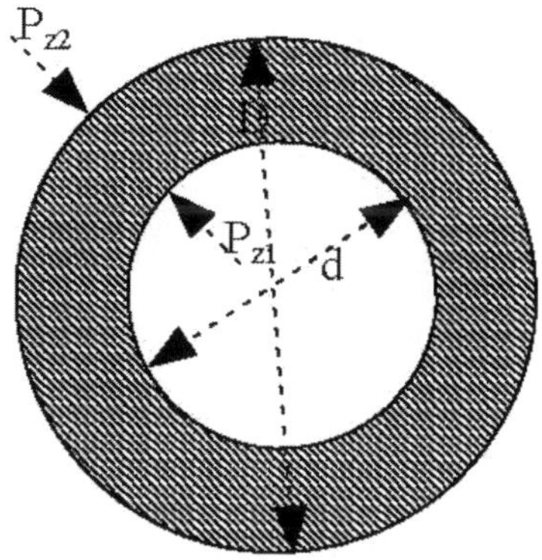

Figure 7.36 The radial and tangential stresses figure of tube.

where, r is the radial stress, θ is the tangential stress, $r(d \leq r \leq D)$ is the radial coordinate, P_{z1} is the tube's internal pressure at z point, and P_{z0} is the tube's external pressure at z point.

7.7.2.4 *Axial stress on the steam injection tubular through the friction loss*

The flow in the tubular is either an oil-gas flow, an oil-gas-water flow, or a multi-flow. For steam injection, the flow results in a friction effect which causes axial stress. Here, we consider the flow to be a gas-liquid mix flow and the liquid head loss is determined using the Darcy-Weisbach formula (Dawson, 1984):

$$h_f = \frac{\lambda(Z - z)v_m^2}{2gd}, \tag{7.251}$$

where, h_f is the liquid flow heat loss, λ is the frictional heat loss coefficient, and v_m is the liquid flow velocity.

The friction drag in the tubular is $N_{fz} = h_f \rho_m g \pi d^2$ (ρ_m is the liquid flow density). The axial stress of the fiction drag can be obtained:

$$\sigma_{zf} = \frac{4N_{fz}}{\pi(D^2 - d^2)}. \tag{7.252}$$

7.7.3 **Analysis of axial deformation**

Based on the studies and analyses mentioned above, the axial deformation of the tubular is made up as follows.

7.7.3.1 *The axial deformation of the axial static stress*

For the tubular infinitesimal dz, the unit deformation caused by the static stress is computed using the generalized Hooke's law:

$$\varepsilon_1 = \frac{1}{E}[\sigma_{zi} - \mu(\sigma_{rz} + \sigma_{\theta z})], \tag{7.253}$$

where, μ represents the Poisson ratios.

The axial deformation at an element can be obtained through an integration of the length of the element:

$$\Delta L_{1i} = \int_{Z_{i-1}}^{Z_i} \frac{1}{E}[\sigma_{zi} - \mu(\sigma_{rz} + \sigma_{\theta z})]dz. \tag{7.254}$$

Therefore, the total axial deformation caused by the static stress can be determined by accumulating each element:

$$\Delta L_1 = \sum_{i=1}^{N} \Delta L_{1i}. \tag{7.255}$$

7.7.3.2 *Axial deformation with temperature change*

For the tubular infinitesimal dz, the unit deformation caused by the temperature change is as follows:

$$\Delta L_{2i} = \int_{Z_{i-1}}^{Z_i} \frac{\sigma_{zt}}{E} dz = \beta \Delta T_i \Delta L_i. \tag{7.256}$$

The same principle is that by accumulating each element the total axial deformation of the varied temperature fields can be determined:

$$\Delta L_2 = \sum_{i=1}^{N} \Delta L_{2i}. \tag{7.257}$$

7.7.3.3 *Axial deformation with the friction drag*

For the tubular infinitesimal dz, the unit deformation caused by the friction force is as follows:

$$\Delta L_3 = \int_0^Z \frac{\sigma_{zf}}{E} dz = \frac{\lambda \rho_m v_m^2 dZ^2}{E(D^2 - d^2)}. \tag{7.258}$$

7.7.3.4 *Axial deformation with the tubular string buckling*

Researchers in general call the buckling, a bending effect. The tubular is freely suspended with an absence of fluid inside as shown in Figure 7.37. A force F is applied at the lower end of the tubular and if the force is large enough, the tubular buckles as shown in Figure 7.37. Lubinski and Althouse (1962) have done much research on the phenomenon. From their work,we can determine the buckling effect. The virtual axial force of the tubing is defined as follows:

$$F_f = A_p(P_1 - P_0), \tag{7.259}$$

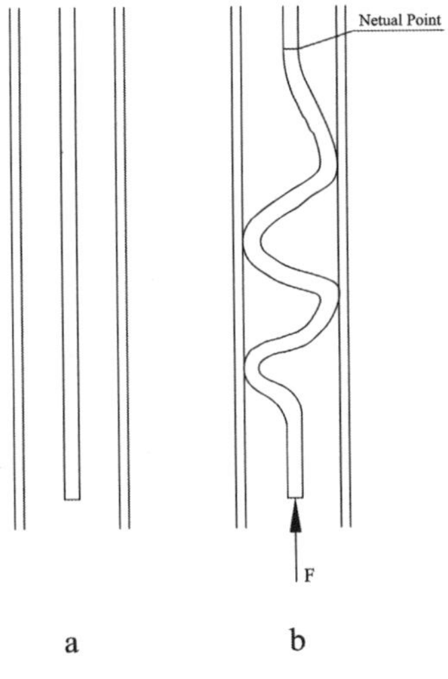

Figure 7.37 Buckling of tubular.

where P_1 is the pressure inside the tubular at the packer length, P_0 is the pressure outside the tubular at the packer length, and A_p is the area corresponding to the packer bore.

From Equation (7.259), it is possible to judge whether or not the tubular will buckle. The string buckles if F_f is positive, and remains straight if F_f is negative or zero. The tubular string buckling axial deformation is:

$$\Delta L_{4i} = -r^2 A_p^2 (\Delta P_{1i} - \Delta P_{0i})^2 / (8EIW), \tag{7.260}$$

where, r is the tubing-to-casing radial clearance, I is the moment of inertia in the tubular cross-section with respect to its diameter ($I = \pi(D^4 - d^4)/64$), Δ denotes the change before and after the injection, and W is the unit weight of the tubing:

$$\Delta L_4 = \sum_{i=1}^{N} \Delta L_{4i}. \tag{7.261}$$

In addition, the neutral point position is needed. The length (n) from the packer to the point can be computed as the following equation:

$$n = F_f / W. \tag{7.262}$$

Generally, the neutral point should be in the tubular ($n \leq Z$). However, with the multi-packers, this point occurs outside the tubing between the dual packers. Here, we assume the latter phenomenon.

After considering all the above, the whole deformation length can be represented as:

$$\Delta L = \Delta L_1 + \Delta L_2 + \Delta L_3 + \Delta L_4. \tag{7.263}$$

7.7.4 Varied (*T*, *P*) fields analysis

When dryness modelling, we find the deformation numerical values are affected by temperature and pressure. In fact, the two parameters vary according to depth and time. As a result, the varied (T, P) fields need to be researched. Under this context, Sichuan University had claimed earlier in the China Sinopec Group Hi-Tech Project "Stress analysis and the optimum design of well completion" in Xu and Wu (2012), that the varied (T, P) fields were deduced strictly based on the mass, momentum and energy balance. The varied (T, P) fields are:

$$\begin{cases} \dfrac{dP}{dz} = \dfrac{-\frac{\tau_i}{A} + \rho_m g \cos\theta + \frac{m}{A} R \frac{dx}{dz}}{1 - \frac{m}{A} S} \\ \dfrac{dT}{dz} = -\dfrac{v_m}{C_{Pg}} \left(R \dfrac{dx}{dz} - S \dfrac{dP}{dz} \right) - \dfrac{g \cos\theta}{C_{Pg}} - \dfrac{\pi f r_{ti} \rho_m v_m^3}{4 C_{Pg}} + \dfrac{a(T - T_e)}{C_{Pg}}. \\ P(z_0) = P_0, \quad T(z_0) = T_0, \quad dx(z_0) = dx_0, \quad x(z_0) = x_0 \end{cases} \tag{7.264}$$

7.7.5 Numerical implementation

In this section, the calculations for some of the parameters are presented.

1. Inclination at each point:

$$\alpha_j = \alpha_{j-1} + (\alpha_k - \alpha_{k-1}) \Delta s_j / \Delta s_k,$$

   where j represents the calculation segment point, and Δs_k represents the measurement depth of the inclination angle α_k, α_{k-1}, Δs_j the step length calculation.
2. Transient heat transfer function:

$$f(t_D) = \begin{cases} 1.128\sqrt{t_D}(1 - 0.3\sqrt{t_D}), & t_D \leq 1.5 \\ (0.4063 + 0.5 \ln t_D)(1 + 0.6/t_D), & t_D > 1.5 \end{cases}.$$

3. Wet steam density
   The flow of water vapour is in a gas-liquid two-phase flow, and there is significant research on this problem (Beggs and Brill, 1973; Hasan and Kabir, 1991). The M-B model was used to calculate the average density of the mixture.
4. The heat transfer coefficient U_{to} is calculated from different positions on the well-bore axis to the second surface.
   These resistances include the tubing wall, the possible insulation around the tubing, the annular space (possibly filled with a gas or liquid but is sometimes a vacuum), the casing wall, and the cement behind the casing:

$$\frac{1}{U_{to}} = r_{ti}\frac{1}{\lambda_{ins}}\ln\left(\frac{r_{ci}}{r_{to}}\right) + \frac{1}{h_c + h_r} + r_{ti}\frac{1}{\lambda_{cem}}\ln\left(\frac{r_{cem}}{r_{co}}\right),$$

   where λ_{ins} and λ_{cem} are the heat conductivity of the heat insulating material and the cement sheath, respectively. h_c and h_r are the coefficients for the convection heat transfer and the radiation heat transfer.

7.7.5.1 *Initial condition*

In order to solve the model, some definite conditions and initial conditions are added. The initial conditions are the distribution of the pressure and temperature at the well top. Here, we adopt the value at the initial time using actual measurement. Before the steam is injected, the temperature of the tubular is the initial temperature of the formation ($T_z = T_0 + \gamma z \cos\alpha$, γ is geothermal gradient). At the same time, the inner tubular pressure is assumed to be equal to the outer tubular pressure before the steam is injected.

7.7.5.2 *Steps in the algorithm*

To simplify the calculation, we divide the well into several short segments of the same length. The length of a segment varies depending on the variations in wall thickness, hole diameter, the fluid density inside and outside the pipe and well geometry. The model begins with a calculation at the top of the pipe.

Step 1. Set the depth step length. In addition, we denote the relative tolerance error by ε. The smaller h, ε is, the more accurate the results are. However, an extended calculation would lead to a rapid increase in the calculation time. Setting $h = 1$ (m), and $\varepsilon = 5\%$.
Step 2. Given the initial conditions.
Step 3. Compute each point's inclination.
Step 4. Compute the parameters under the initial conditions or the last depth variables.
Step 5. Let $T = T_k$, then we can get the T_e by solving the following equation:

$$\begin{cases} \dfrac{\partial T_e}{\partial t_D} = \left(\dfrac{\partial^2 T_e}{\partial r_D^2} + \dfrac{1}{r_D}\dfrac{\partial T_e}{\partial r_D}\right) \\ T_e|_{t_D=0} = T_0 + \gamma z \cos\theta \\ \left.\dfrac{\partial T_e}{\partial r_D}\right|_{r_D=1} = -\dfrac{1}{2\pi\lambda_f}\dfrac{dq}{dz} \\ \left.\dfrac{\partial T_e}{\partial r_D}\right|_{r_D\to\infty} = 0 \end{cases}.$$

Let $T^j_{e,i}$ be the temperature at the injection time j and radial i at the depth z. We apply the finite different method to discretize the equations as follows:

$$\frac{T^{i+1}_{e,j} - T^{i}_{e,j}}{\varphi} = \frac{T^{i+1}_{e,j+1} - 2T^{j}_{e,j+1} + T^{i-1}_{e,j+1}}{\xi^2} - \frac{T^{i+1}_{e,j+1} - T^{i+1}_{e,j}}{r_D\varphi},$$

where φ is the interval of time and ξ is the interval of radial, respectively. It can be transformed into the standard form as follows:

$$-\left(\varphi+\frac{\varphi\xi}{r_D}\right)T_{e,j+1}^{i+1}+\left(2\varphi+\frac{\varphi\xi}{r_D}\right)T_{e,j}^{i+1}-\varphi T_{e,j-1}^{i+1}=\xi^2 T_{e,j}^{i}.$$

Then the different method is used to discretize the boundary condition. For $r_D = 1$, we have:

$$T_{e,2}^{e,i+1}-\left(1+\frac{a\xi}{2\pi\lambda_f}\right)T_{e,1}^{i+1}=\frac{aT_k}{2\pi\lambda_f}.$$

For $r_D = N$, we have:

$$T_{e,n}^{i+1}-T_{e,n-1}^{i+1}=0.$$

Compute the symbolic solution of the temperature T_e of the stratum. In this step, the discrete distribution of T_e is represented as a matrix $[T_{e,j}^{i}]_{n\times\infty}$, where i represents the injection time and j represents the radial.

Step 6. Let the right parts of the coupled differential equations be functions F_k, where $(k = 1, 2)$. Then we obtain a system of coupled functions as follows:

$$\begin{cases} F_1=\dfrac{-\frac{\tau_i}{A}+\rho_m g\cos\theta+\frac{m}{A}R\frac{dx}{dz}}{1-\frac{m}{A}S} \\ F_2=-\dfrac{v_m}{C_{Pg}}\left(R\dfrac{dx}{dz}-S\dfrac{dP}{dz}\right)-\dfrac{g\cos\theta}{C_{Pg}}-\dfrac{\pi f r_{ti}\rho_m v_m^3}{4C_{Pg}}+\dfrac{a(T-T_e)}{C_{Pg}} \end{cases},$$

where, T_e at $r_D = 1$

Step 7. Assume P, T to be y_k $(k = 1;\ 2)$, respectively. Then we obtain some basic parameters as follows:

$$\begin{cases} a_k=F_i(y_1,y_2) \\ b_k=F_i(y_1+ha_1/2,y_2+ha_2/2) \\ c_k=F_i(y_1+hb_1/2,y_2+hb_2/2) \\ d_k=F_i(y_1+hc_1,y_2+hc_2) \end{cases}.$$

Step 8. Calculate the pressure and temperature at point $(j + 1)$:

$$y_k^{j+1}=y_k^{j}+h(a_k+2b_k+2c_k+d_k)/6,\qquad k=1,2\quad j=1,2,\ldots,n.$$

Step 9. Calculate the deformation ΔL_{1j}, ΔL_{2j} and ΔL_{4j} using the above equations.

Step 10. Repeat the third step to the ninth step until the calculated tubular length Z.

Step 11. Calculate the deformation ΔL_3 and total deformation length:

$$\Delta L=\sum_{j=1}^{N}\Delta L_{1j}+\sum_{j=1}^{N}\Delta L_{2j}+\Delta L_3+\sum_{j=1}^{N}\Delta L_{4j}$$

7.7.6 Numerical simulation

7.7.6.1 *Parameters calculation*

In this section, the calculation method for some of the parameters is given.

1. Each point's inclination:

$$\theta_j=\theta_{j-1}+(\theta_c-\theta_{c-1})\Delta s_j/\Delta s_c,$$

where j represents the segment point of calculation, Δs_c represents the measurement depth between inclination angle θ_c and θ_{c-1}, Δs_j is the step length of calculation.

2. Transient heat transfer function (Hasan and Kabir, 1991):

$$f(t_D)=\begin{cases} 1.128\sqrt{t_D}(1-0.3\sqrt{t_D}), & t_D \le 1.5 \\ (0.4063+0.5\ln t_D)(1+0.6/t_D), & t_D > 1.5 \end{cases}.$$

3. Wet steam density

   The problem of the flow of water vapor in the gas-liquid two-phase flow has been previously researched (Beggs and Brill, 1973; Mukherjee and Brill, 1985). The M-B model is adopted to calculate the average density of the mixture.
4. Heat transfer coefficient U_{to} is calculated from different positions of the axis of the well-bore to the second surface.

   These resistances include the tubing wall, possible insulation around the tubing, the annular space (possibly filled with a gas or liquid, but is sometimes a vacuum), the casing wall, and the cementing behind the casing as illustrated in Figure 7.3:

$$\frac{1}{U_{to}}=r_{ti}\frac{1}{\lambda_{ins}}\ln\left(\frac{r_{ci}}{r_{to}}\right)+\frac{1}{h_c+h_r}+r_{ti}\frac{1}{\lambda_{cem}}\ln\left(\frac{r_{cem}}{r_{co}}\right)$$

   λ_{ins} and λ_{cem} represent the heat conductivity of the heat insulating material and the cement sheath, respectively. h_c and h_r are the convection heat transfer and the radiation heat transfer coefficients.
5. Specific enthalpy of saturated steam and saturated water (Tortike, 1989):

$$\begin{aligned} h_L &= 23665.2-366.232T+2.26952T^2-0.00730365T^3+1.3024\times10^{-5}T^4 \\ &\quad -1.22103\times10^{-8}T^5+4.70878\times10^{-12}T^6 \\ h_G &= -22026.9+365.317T-2.25837T^2+0.0073742T^3-1.33437\times10^{-5}T^4 \\ &\quad +1.26913\times10^{-8}T^5-4.9688\times10^{-12}T^6 \end{aligned}.$$

7.7.6.2 *Parameters*

To demonstrate the application of our theory, we study a pipe in X well, which is in Sichuan Province, China. The basic parameters are as follows:

Depth of the well = 6115 m
Ground thermal conductivity parameter = 2.06 W/m · K
Ground temperature = 16°C
Geothermal gradient = 2.18°C/100 m
Roughness of the inner surface of the well is 0.000015

The parameters for the pipes are given in Tables 7.17 and 7.18.

7.7.7 **Main results and analysis**

After calculation, we obtained a series of results for this well as shown in Tables 7.19 and 7.20.

Table 7.17 Parameters of pipes.

Diameter [m]	Thickness [m]	Weight [kg]	Expansion coefficient [m/m · K]
0.0889	0.01295	23.79	0.0000115
0.0889	0.00953	18.28	0.0000115
0.0889	0.00734	15.04	0.0000115
0.0889	0.00645	13.58	0.0000115
0.0889	0.00645	13.58	0.0000115
0.0730	0.00551	9.49	0.0000115

The influence of the outputs on the tubing axial deformation was investigated as shown in Figure 7.38.

From the results shown in Tables 7.19 and 7.20, we can draw some useful conclusions:

1. The amount of steam injected and injection pressure severely affect the stretching force.
2. The tubular deformation length rises with an increase in injection pressure or injection velocity.
3. The length of the tubular deformation increases with an increase in the output, but more slowly.
4. Thermal stress is the main factor influencing tubular deformation. Therefore, the steam injection temperature should not be set too high.
5. Lifting pre-stressed cementing technology could be crucial in a reduction in tubular deformation.
6. The creeping displacement of the downhole strings produces an upward contact which causes the packer to depress or lapse. Therefore, effective measures need to be adopted to control tubular compounding.

Table 7.18 Well parameters.

Measured depth [m]	Internal diameter [m]	External diameter [m]
336.7	0.15478	0.1778
422.6	0.1525	0.1778
1468.8	0.10862	0.127

Table 7.19 The results of the axial force and various deformation by (T, P) fields.

Number	Depth [m]	Axial force	T change [°C]	P change [MPa]
1	135	895244.8	0	0
2	364	706877.3	0.5606	0.046028
3	486	551107.2	1.0212	0.083883
4	628	395301.2	1.3817	0.113541
5	830	251625.2	1.6419	0.134999
6	1089	123051.7	1.8019	0.148253
7	1235	10144.99	1.8577	0.152956
8	1300	9879.191	1.8621	0.153369

Table 7.20 The results of various kinds of deformation length.

Number	Depth [m]	Axial [m]	Buckling [m]	Total [m]
1	135	0	0	0
2	364	0.152	−0.006	0.7523
3	486	0.391	−0.007	1.4896
4	628	0.706	−0.018	2.1828
5	830	1.082	−0.058	2.8014
6	1089	1.562	−0.088	3.4245
7	1235	1.942	−0.149	3.8036
8	1300	2.075	−0.246	3.8445

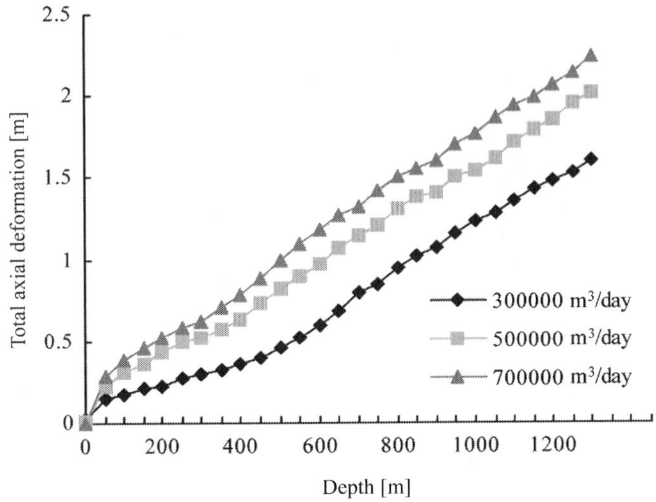

Figure 7.38 The total axial deformation under varied outputs.

Table 7.21 The value of (T, P) fields.

Number	Measured depth [m]	Casing T [°C]	Steam T [°C]	P [MPa]
1	135	130.239	326.3	13.998
2	364	130.62	322.4	13.9845
3	486	130.80	318.2	13.9156
4	628	131.06	314.8	13.8252
5	830	131.42	306.4	13.5662
6	1089	132.11	299.6	13.2762
7	1235	132.68	296.4	13.2583
8	1300	132.81	295.7	13.2261

7.8 NMSQ-DWV[7]

7.8.1 Basic assumptions

(1) The base of the insulated tubing is a packer setting to ensure steam cannot flow into the space between the tube and casing, which is full of air.
(2) The steam flow is one dimensional and is in a steady state along the well-bore; the pressure and temperature in the same horizontal cross section is constant.
(3) The physical properties of other materials and some thermal insulating materials are independent of time and temperature.
(4) The temperature around the well-bore in the stratum has an axisymmetric distribution.
(5) Heat transfer from the inner well-bore to the cement mantle external boundary is stable, while an unstable transfer occurs from the cement mantle external boundary to the geological formation. The heat transfer that occurs longitudinally is ignored.

7.8.2 The steam quality model with variable (*T, P*) fields

Within the tubing, the flow takes place under turbulent flow conditions. Consider the flow system depicted in Figure 7.39; a straight cylindrical flow tube with an inclination angle θ, a constant

[7]Numerical modelling of steam quality in deviated wells with variable (T, P) fields

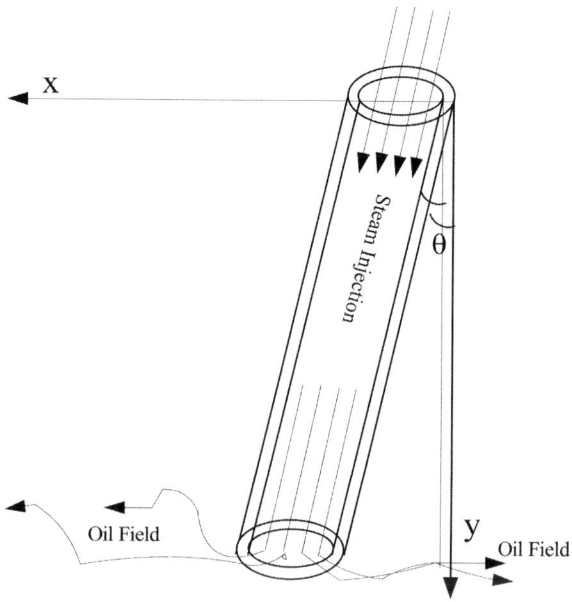

Figure 7.39 The physical figure.

cross-sectional flow area A, and a hydraulic diameter d. The distance co-ordinate in the flow direction along the tubing is denoted z.

Steam quality is the most important parameter in steam injection wells and is constructed using an energy equation. According to the principle of fluid mechanics and thermodynamics, any flow of fluid at some state parameter (P, T) at some location includes the following: internal energy, pressure energy, kinetic energy and potential energy:

$$E = U + \frac{1}{2}mv_{mx}^2 + mgz + Pv_{mx}. \tag{7.265}$$

Here, U, $\frac{1}{2}mv_{mx}^2$, mgz, Pv_{mx} represents thermodynamic energy, kinetic energy, potential energy and pressure energy, respectively. Subscript mx represents the mixture.

The internal energy and pressure energy are collectively referred to as enthalpy:

$$H = U + Pv_{mx}. \tag{7.266}$$

According to the rule of energy balances, the energy of the fluid flowing into the distance element equals the sum of energy losses and the fluid flowing out from the distance element and the loss of energy from the friction of the steam flow.

The energy equation is written as follows:

$$\frac{dQ}{dz} \pm \frac{dW}{dz} + mg\cos\theta = -m\frac{dh_{mx}}{dz} - mv_{mx}\frac{dv_{mx}}{dz}, \tag{7.267}$$

where, the work W is positive if it is done by the system and negative if work is done on the system.

Eliminating mass term from Equation (7.267) gives:

$$\frac{dq}{dz} \pm \frac{dw}{dz} + g\cos\theta = -\frac{dh_{mx}}{dz} - v_{mx}\frac{dv_{mx}}{dz}. \tag{7.268}$$

The equation is very similar to many researchers' model (Gould, 1974; Hasan *et al.*, 2010; Howell *et al.*, 1972).

h_{mx} expresses the enthalpy of the gas-liquid two phase flow which is defined as:

$$h_{mx} = h_G x + h_L(1-x), \tag{7.269}$$

where the enthalpies of dry steam h_G and saturated liquid h_L are expressed as shown in Appendix B. It follows from Equation (7.269) that:

$$\frac{dh_{mx}}{dz} = \left(\frac{dh_G}{dz} - \frac{dh_L}{dz}\right)x + (h_G - h_L)\frac{dx}{dz} + \frac{dh_L}{dz}. \tag{7.270}$$

The enthalpy is the function of pressure. That is $h_k = f(P)$.
Hence,

$$\frac{dh_k}{dz} = \frac{dh_k}{dP}\frac{dP}{dz}. \tag{7.271}$$

The subscript k represents any phase such as water, steam or mixture. Substituting Equation (7.271) in Equation (7.270) gives:

$$\frac{dh_{mx}}{dz} = (h_G - h_L)\frac{dx}{dz} + \frac{dh_L}{dz}\frac{dP}{dz} + \left(\frac{dh_G}{dP} - \frac{dh_L}{dP}\right)\frac{dP}{dz}x. \tag{7.272}$$

The mass equation is applied: $I_{mx} = \rho_{mx} v_{mx} A$.
The following equation is deduced as:

$$\frac{dv_{mx}}{dz} = \frac{d}{dz}\left(\frac{I_{mx}}{\rho_{mx}A}\right) = \frac{1}{\rho_{mx}A}\frac{dI_{mx}}{dz} + \frac{I_{mx}}{A}\frac{d}{dz}\left(\frac{1}{\rho_{mx}}\right). \tag{7.273}$$

The flow mass drop with depth is ignored. Because of the mist state of the steam injection process and the large gas flow, the gas state equation is applied $\rho_{mx} = \frac{MP}{RZ_G T}$ and transformed into:

$$d\left(\frac{1}{\rho_{mx}}\right) = \frac{RZ_G}{PM}dT - \frac{RTZ_G}{P^2M}dP + \frac{R}{PM}dZ_G = \frac{1}{\rho_{mx}}\left(\frac{1}{T}\frac{dT}{dP} + \frac{1}{Z_G}\frac{dZ_G}{dP} - \frac{1}{P}\right)dP. \tag{7.274}$$

With respect to temperature drop or pressure drop, the gas deviation factor gradient is normally small and often ignored. Then, the following equation is obtained:

$$d\left(\frac{1}{\rho_{mx}}\right) = \frac{1}{\rho_{mx}}\left(\frac{1}{T}\frac{dT}{dP} - \frac{1}{P}\right)dP. \tag{7.275}$$

Hence, substituting Equations (7.275), (7.273) and (7.272) in Equation (7.268), the following equation is derived:

$$\begin{aligned}\frac{dq}{dz} \pm \frac{dw}{dz} + g\cos\theta &= (h_L - h_G)\frac{dx}{dz} - \frac{dh_L}{dz}\frac{dP}{dz} - \frac{dh_G}{dP}\frac{dP}{dz}x \\ &\quad + \frac{dh_L}{dP}\frac{dP}{dz}x - \frac{v_{mx}I_{mx}}{\rho_{mx}A}\left(\frac{1}{T}\frac{dT}{dP} - \frac{1}{P}\right)\frac{dP}{dz}.\end{aligned} \tag{7.276}$$

Equation (7.276) can take the following form:

$$C_1\frac{dx}{dz} + C_2 x + C_3 = 0 \iff \frac{dx}{dz} + \frac{C_2}{C_1}x = -\frac{C_3}{C_1}, \tag{7.277}$$

where, C_1, C_2, C_3 are functions of pressure drop, enthalpy drop, heat loss, work and temperature given by:

$$C_1 = (h_G - h_L), \quad C_2 = \left(\frac{dh_G}{dP} - \frac{dh_L}{dP}\right)\frac{dP}{dz} \quad \text{and}$$

$$C_3 = \frac{dq}{dz} \pm \frac{dw}{dz} + g\cos\theta + \frac{dh_L}{dz}\frac{dP}{dz} + \frac{v_{mx}I_{mx}}{\rho_{mx}A}\left(\frac{1}{T}\frac{dT}{dP} - \frac{1}{P}\right)\frac{dP}{dz}$$

C_1, C_2, C_3 are constant at a fixed depth. Thus, Equation (7.277) is a first order linear ordinary differential equation:

From this, the steam quality computing model is obtained:

$$\begin{cases} x = e^{-\frac{C_2 z}{C_1}}\left(-\dfrac{C_3}{C_2}e^{-\frac{C_2 z}{C_1}} + x_0 + \dfrac{C_3}{C_2}\right) \\ C_1 = h_G - h_L \\ C_2 = \left(\dfrac{dh_G}{dP} - \dfrac{dh_L}{dP}\right)\dfrac{dP}{dz} \\ C_3 = \dfrac{dq}{dz} \pm \dfrac{dw}{dz} + g\cos\theta + \dfrac{dh_L}{dz}\dfrac{dP}{dz} + \dfrac{v_{mx} I_{mx}}{\rho_{mx} A}\left(\dfrac{1}{T}\dfrac{dT}{dP} - \dfrac{1}{P}\right)\dfrac{dP}{dz} \\ x|_{z=0} = x_0 \end{cases} \tag{7.278}$$

7.8.3 The analysis of the variable (*T*, *P*) fields

In steam quality modelling, the steam quality numerical values are affected by temperature and pressure. In fact, the two parameters vary according to depth and time.

7.8.3.1 *Material balance*

Since steam injection is a constant-specific mass flow, that is, the mass flowing in equals the one flowing out, it follows from the mass conservation equation that:

$$\frac{d(\rho_{mx} v_{mx} A)}{dz} = 0. \tag{7.279}$$

7.8.3.2 *Momentum balance*

As shown in Figure 7.40, there are both surface and mass forces at the microelement.

The mass force is $\rho_{mx} g \cos\theta A dz$. The surface forces are the upper pressure P_i, the lower pressure P_{i+1} and the friction force τ_i. As the flow between the upper element and lower element changes, the momentum changes. From the momentum balance, the equation is built as follows:

$$P_{i+1}A - P_i A - \tau_i S_i dz + \rho_{mx} g\cos\theta A dz = (\rho_{mx_{i+1}} A v_{mx_{i+1}}) v_{mx_{i+1}} - (\rho_{mx_i} A v_{mx_i}) v_{mx_i}. \tag{7.280}$$

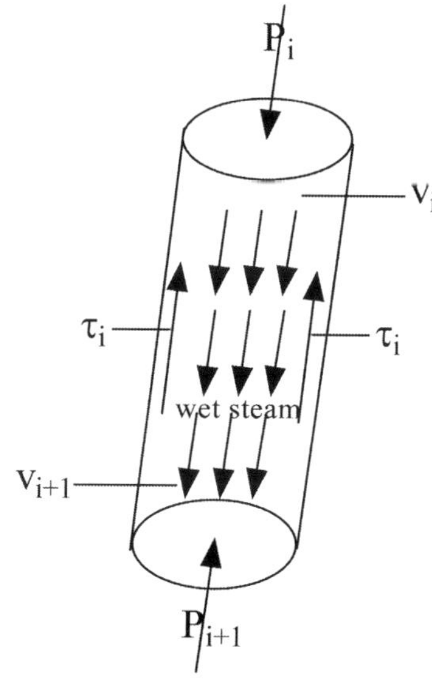

Figure 7.40 Flow analysis of infinitesimal section in deviated well-bore.

Eliminating cross-sectional area term from Equation (7.280) gives:

$$dP = -\frac{\tau_i S_i dz}{A} + \rho_{mx} g \cos\theta dz - d(\rho_{mx} v_{mx}^2). \tag{7.281}$$

Substituting Equation (7.279) in Equation (7.281) gives:

$$\frac{dP}{dz} = -\frac{\tau_i S_i}{A} + \rho_{mx} g \cos\theta - \rho_{mx} v_{mx} \frac{dv_{mx}}{dz}. \tag{7.282}$$

7.8.3.3 *Temperature field component processing*

There are some unknown parameters in Equation (7.268). The effect of changing temperature and changing pressure fields needs to be researched further.

Heat transmission in the well-bore

In Equation (7.268), dq denotes the radial heat transfer between the fluid and the surrounding earthen tube. Ramey (1962) and Willhite (1967) discussed the radial transfer of heat between the fluid and the earth in detail.

The consideration of the heat transferred from the fluid to the formation is the basis of the control equation of the thermal exchange between the well-bore and the formation. The temperature drop was considered in two stages. The first stage is the heat transfer from the fluid to the cement/earth interface (the heat transfer in metal walls was neglected). The second stage is the heat transfer from the cement/earth interface to the formation. Ramey (1962) and Willhite (1967) discussed the radial transfer of heat between the fluid and the earth in detail. The heat transfer from the tube to the casing is considered steady, but the heat transfer from the casing to the formation is unsteady. Therefore, an interface with the cement mantle and the stratum (wall of a well) becomes key in connecting the two.

The overall heat transmission coefficient for steady heat transfer in a well case is proportional to the temperature difference and cross section perpendicular to the direction of the heat transfer:

$$q = UA\Delta T. \tag{7.283}$$

As shown in Figure 7.41, the heat lost transfers through the sequence of heat resistances.

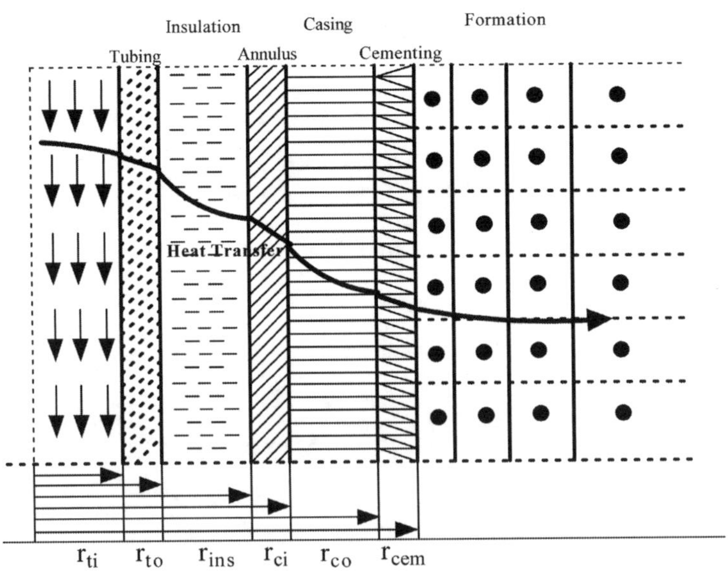

Figure 7.41 Temperature distribution in the well-bore system and formation.

According to Equation (7.283), the radial heat transfer from the fluid to the cement/earth interface is described by:

$$\frac{dq_1}{dz} = 2\pi r_{to} U_{to}(T - T_{ref}). \tag{7.284}$$

Heat loss from the cement/earth interface to the surrounding formation is an unsteady-state heat transfer, but the mixture is treated as a single phase. Ramey (1962) gave a model to describe this phenomenon:

$$\frac{dq_2}{dz} = \frac{2\pi K_e(T_{ref} - T_e)}{f(t_D)}. \tag{7.285}$$

Combining Equations (7.284) and (7.285), the heat transmission model between the steam flow and the surrounding earth gives as follows:

$$\frac{dq}{dz} = \frac{2\pi r_{to} U_{to} K_e}{r_{to} U_{to} f(t_D) + K_e}(T - T_e). \tag{7.286}$$

Let $a = \frac{2\pi r_{to} U_{to} K_e}{r_{to} U_{to} f(t_D) + K_e}$, then:

$$dq = a(T - T_e)dz. \tag{7.287}$$

According to the continuity of heat transfer, the cement/earth interface temperature can be computed as follows:

$$T_{ref} = \frac{K_e T_e + T r_{to} U_{to} f(t_D)}{r_{to} U_{to} f(t_D) + K_e}. \tag{7.288}$$

The casing temperature can also be computed as follows (Aziz *et al.*, 1972):

$$T_r = T_{ref} + \frac{r_{to} U_{to} \ln\frac{r_{cem}}{r_{co}}(T - T_{ref})}{K_{cem}}. \tag{7.289}$$

This heat loss model has been widely used in steam and other injection wells to provide the distribution for the fluid parameters and temperature in the tube. However, from engineering practice, the steam injection parameters are not constant and the temperature at the wall changes with time and depth.

Unsteady heat conduction model in the formation

When considering heat transmission in the well-bore, according to heat conduction theory, over the same differential element of depth, the radial heat conduction equation at the formation can be described as:

$$\rho_f C_f \frac{\partial T_e}{\partial t} = \lambda_f \left(\frac{\partial^2 T_e}{\partial r^2} + \frac{1}{r}\frac{\partial T_e}{\partial r} \right). \tag{7.290}$$

Initial condition: when the steam is injected, the formation temperature field is the initial formation temperature:

$$T_{z,r,0} = T_0 + \gamma z \cos\theta. \tag{7.291}$$

Boundary conditions:

Inner boundary condition: according to Fourier's law of heat conduction, on the interface between the well-bore and the formation, the heat flux exchanges through the unit depth of the well-bore and the surrounding formation is:

$$dq = 2\pi\lambda_f r dz \left.\frac{\partial T_r}{\partial r}\right|_{r=r_{cem}} \tag{7.292}$$

and

$$T_w(z=0, t) = T_0. \tag{7.293}$$

And for the injection well, the temperature is that of the earth's surface.

Outer boundary conditions: at infinity distance the formation temperature stays at the initial formation temperature distributions and is independent of time:

$$\frac{\partial T_r}{\partial r} = 0 \qquad \text{if } r \to \infty. \tag{7.294}$$

The temperature of the fluid and the earth is explained by the above two differential equations and the initial and boundary conditions. To derive the solution, define the following dimensionless variables $r_D = r/r_{cem}$, $t_D = \lambda_e t/(\rho_f C_f r_{cem}^2)$: then, exchange variables r, t into dimensionless variables r_D, t_D, and Equation (7.290) is changed into:

$$\frac{\partial T_e}{\partial t_D} = \left(\frac{\partial^2 T_e}{\partial r_D^2} + \frac{1}{r_D}\frac{\partial T_e}{\partial r_D}\right). \tag{7.295}$$

The boundary conditions are converted into:

$$\left.\frac{\partial T_e}{\partial r_D}\right|_{r_D=1} = -\frac{1}{2\pi\lambda_f}\frac{dq}{dz} \qquad \left.\frac{\partial T_e}{\partial r_D}\right|_{r_D\to\infty} = 0.$$

Enthalpy drop
h_k, which satisfies the following relation, denotes specific enthalpy:

$$\frac{dh_k}{dz} = C_{Pk}\frac{dT}{dz} - \eta_k C_{Pk}\frac{dP}{dz} \tag{7.296}$$

η_k is the Joule-Thomson coefficient defined as:

$$\begin{cases} \eta_k = 0, & k = G \\ \eta_k = -\dfrac{1}{C_{Pk}\rho_k}, & k = L \end{cases}.$$

Friction force doing work
As the direction of the friction is opposite to the steam flow, the friction is negative as the steam flows, thus the work done by the friction on an infinite length in unit time is described by:

$$dw = \frac{\tau_i dz}{dt} = \frac{\tau_i dz}{2dz/(v_{mx_i} + v_{mx_{i+1}})} = \frac{\tau_i(v_{mx_i} + v_{mx_{i+1}})}{2} \approx \tau_i v_m. \tag{7.297}$$

The friction solution uses a fluid dynamics calculation method (White, 1999):

$$\tau_i = \frac{\pi f r_{ti}\rho_{mx} v_{mx}^2}{4} dz. \tag{7.298}$$

Substituting Equation (7.298) in Equation (7.297), the following equation can be determined:

$$dw = \frac{\pi f r_{ti}\rho_{mx} v_{mx}^3}{4}, \tag{7.299}$$

where, f is the friction factor of the gas-liquid mixture and the function regarding the Reynolds number Re and the absolute roughness ε:

$$f = \begin{cases} Re/64, & \text{if } Re \le 2000 \\ \left[1.14 - 2\ln\left(\dfrac{\varepsilon}{2r_{ti}} + 21.25Re^{-0.9}\right)\right]^{-2}, & \text{if } Re > 2000 \end{cases}.$$

7.8.3.4 *The variable* (T, P) *fields coupled model*

From the relationship between Equations (7.268), (7.296), (7.285) and (7.299), the gas energy equation is obtained:

$$C_{PG}\frac{dT}{dz} + v_{mx}\frac{dv_{mx}}{dz} + g\cos\theta + \frac{\pi f r_{ti}\rho_{mx}v_{mx}^3}{4} - a(T - T_e) = 0. \tag{7.300}$$

In addition, for the velocity of the mixed flow, then:

$$v_{mx} = v_G + v_L = \frac{I_{mx}x}{\rho_G A} + \frac{I_{mx}(1-x)}{\rho_L A}. \tag{7.301}$$

Thus,

$$\frac{dv_{mx}}{dz} = R\frac{dx}{dz} - S\frac{dP}{dz}, \tag{7.302}$$

where $R = \frac{I_{mx}}{A}\left(\frac{1}{\rho_G} - \frac{1}{\rho_L}\right)$ and $S = \frac{I_{mx}}{A}\left(\frac{x}{\rho_G^2}\frac{d\rho_G}{dP} + \frac{1-x}{\rho_L^2}\frac{d\rho_L}{dP}\right) = \frac{I_{mx}x}{A\rho_G^2}\frac{d\rho_G}{dP}$ (Generally, the water is treated as an incompressible flow and density change is neglected.)

Substituting Equation (7.302) in Equations (7.282) and (7.300), respectively, the coupled equations are:

$$\begin{cases} \dfrac{dP}{dz} = \dfrac{-\frac{\tau_i S_i}{A} + \rho_{mx}g\cos\theta + \frac{I_{mx}}{A}R\frac{dx}{dz}}{1 - \frac{I_{mx}}{A}S} \\ \dfrac{dT}{dz} = -\dfrac{v_{mx}}{C_{PG}}\left(R\dfrac{dx}{dz} - S\dfrac{dP}{dz}\right) - \dfrac{g\cos\theta}{C_{PG}} - \dfrac{\pi f r_{ti}\rho_{mx}v_{mx}^3}{4C_{PG}} + \dfrac{a(T - T_e)}{C_{PG}} \cdot \\ P(z_0) = P_0, \quad T(z_0) = T_0, \quad dx(z_0) = dx_0, \quad x(z_0) = z_0 \end{cases} \tag{7.303}$$

There are several numerical methods to solve ordinary differential equations, such as the Runge-Kutta methods, linear multi-step methods and predictor-correcting methods. The Runge-Kutta technique is far more widely used than any other technique for the defined initial conditions (Dekker and Verwer, 1984).

7.8.4 **Numerical implementation**

To simplify the calculation, the well is divided into several short segments of the same length. The length of a segment varies depending on variations in wall thickness, hole diameter, the fluid density inside and outside the pipe and well geometry. The model starts with a calculation at the top of the pipe. Then pressure and temperature calculations are performed for each successive "segment" of the pipe up to the surface. Since the coupled system model of differential equations contains not only ordinary differential equations but also partial differential equations, it is recommended that the four-order Runge-Kutta method and the finite difference method be interactively applied to solve the problem.

7.8.4.1 *Algorithm steps*

The proposed algorithmic procedure for solving the problem is designed as follows. The detailed solution algorithm is explained based on the flow diagram in Figure 7.42.

Step 1. Set step length of depth. In addition, the relative tolerance error is denoted by ε. The smaller $\Delta\lambda$, ε is, the more accurate the results. However, this increases calculation time. Here, $\Delta h = 50$ (m), $\Delta\lambda = 1$ and $\varepsilon = 5\%$. ($\Delta\lambda$ denotes RK4 method step. The stability of the numerical algorithm is given by Xu *et al.* (2012b).

Step 2. Give the initial conditions and let $h = 0$.

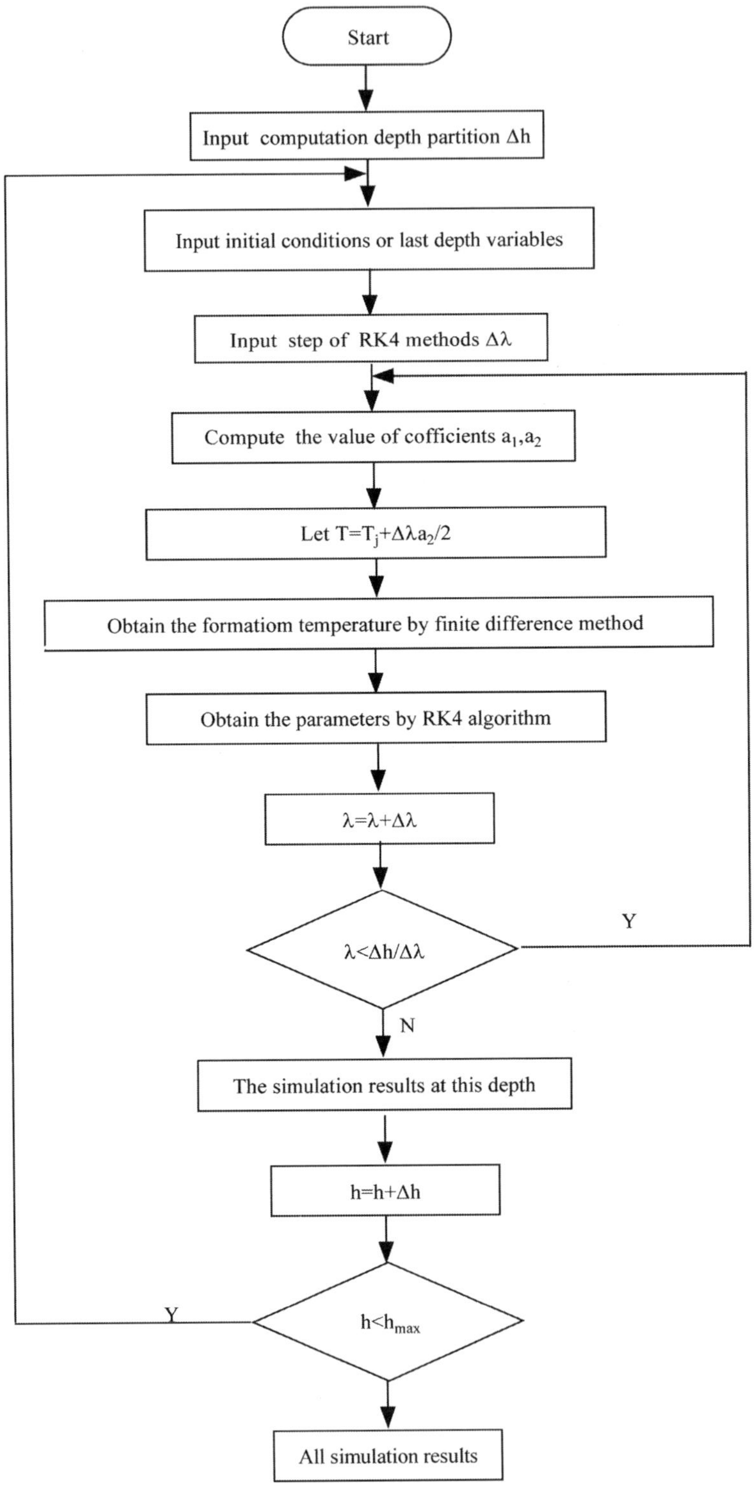

Figure 7.42 The overall procedure of the model.

Step 3. Compute the parameters under the initial conditions or the last depth variables and let $\lambda = 0$.

Step 4. Let the right parts of the coupled differential equations be functions f_n, where $(n = 1, 2)$. A system of coupled functions as follows is then obtained:

$$\begin{cases} f_1 = \dfrac{-\frac{\tau_i}{A} + \rho_{mx} g \cos\theta + \frac{I_{mx}}{A} R \frac{dx}{dz}}{1 - \frac{I_{mx}}{A} S} \\ f_2 = -\dfrac{v_{mx}}{C_{Pg}}\left(R\dfrac{dx}{dz} - S\dfrac{dP}{dz}\right) - \dfrac{g\cos\theta}{C_{Pg}} - \dfrac{\pi f r_{ti} \rho_{mx} v_{mx}^3}{4 C_{Pg}} + \dfrac{a(T - T_e)}{C_{Pg}} \end{cases}.$$

Step 5. Assume P_j, T_j to be y_n^j $(n = 1;\ 2)$, respectively. The basic parameters are obtained as follows: $a_n = f_n(y_1^j, y_2^j)$

Step 6. Let $T = T_j + \frac{\Delta\lambda a_2}{2}$, then solve for the T_e using the following equation:

$$\begin{cases} \dfrac{\partial T_e}{\partial t_D} = \left(\dfrac{\partial^2 T_e}{\partial r_D^2} + \dfrac{1}{r_D}\dfrac{\partial T_e}{\partial r_D}\right) \\ T_e|_{t_D=0} = T_0 + \gamma z \cos\theta \\ \left.\dfrac{\partial T_e}{\partial r_D}\right|_{r_D=1} = -\dfrac{1}{2\pi\lambda_f}\dfrac{dq}{dz} \\ \left.\dfrac{\partial T_e}{\partial r_D}\right|_{r_D\to\infty} = 0 \end{cases}.$$

Let $T_{e,\delta}^{\zeta}$ be the temperature at the injection time ζ and radial δ at depth z. The finite difference method is applied to discretize the equations as follows:

$$\frac{T_{e,\delta}^{\zeta+1} - T_{e,\delta}^{\zeta}}{\varphi} = \frac{T_{e,\delta+1}^{\zeta+1} - 2T_{e,\delta+1}^{\delta} + T_{e,\delta+1}^{\zeta-1}}{\xi^2} - \frac{T_{e,\delta+1}^{\zeta+1} - T_{e,\delta}^{\zeta+1}}{r_D\varphi},$$

where φ is the time interval and ξ is the radial interval, respectively. It can be transformed into the standard form as follows:

$$-\left(\varphi + \frac{\varphi\xi}{r_D}\right)T_{e,\delta+1}^{\zeta+1} + \left(2\varphi + \frac{\varphi\xi}{r_D}\right)T_{e,\delta}^{\zeta+1} - \varphi T_{e,\delta-1}^{\zeta+1} = \xi^2 T_{e,\delta}^{\zeta}.$$

Then a difference method is used to discretize the boundary condition.

For $r_D = 1$:

$$T_{e,2}^{\zeta+1} - \left(1 + \frac{a\xi}{2\pi\lambda_f}\right)T_{e,1}^{\zeta+1} = \frac{aT_K}{2\pi\lambda_f}.$$

For $r_D = N$:

$$T_{e,N}^{\zeta+1} - T_{e,N-1}^{\zeta+1} = 0$$

the symbolic solution of the temperature T_e of the stratum can be computed. In this step, the discrete distribution of T_e is determined as the following matrix:

$$\begin{bmatrix} T_{e,1}^1 & T_{e,1}^2 & \cdots & T_{e,1}^{\zeta} & \cdots \\ T_{e,2}^1 & T_{e,2}^2 & \cdots & T_{e,2}^{\zeta} & \cdots \\ \vdots & \vdots & \cdots & \vdots & \vdots \\ T_{e,\delta}^1 & T_{e,\delta}^2 & \cdots & T_{e,\delta}^{\zeta} & \cdots \\ \vdots & \vdots & \cdots & \vdots & \vdots \\ T_{e,N}^1 & T_{e,N}^2 & \cdots & T_{e,N}^{\zeta} & \cdots \end{bmatrix}.$$

Step 7. Take the T_e at $r_D = 1$, then $(n = 1; 2)$:

$$\begin{cases} b_n = f_n(y_1 + \Delta\lambda a_1/2, y_2 + \Delta\lambda a_2/2) \\ c_n = f_n(y_1 + \Delta\lambda b_1/2, y_2 + \Delta\lambda b_2/2) \\ d_n = f_n(y_1 + \Delta\lambda c_1, y_2 + \Delta\lambda c_2) \end{cases}.$$

Step 8. Calculate the pressure and temperature at point: $(j + 1)$:

$$y_n^{j+1} = y_n^j + \Delta\lambda(a_n + 2b_n + 2c_n + d_n)/6, \qquad n = 1, 2.$$

Step 9. Calculate the steam quality at point $(j + 1)$:

$$x_{j+1} = e^{-\frac{C_2 z}{C_1}}\left(-\frac{C_3}{C_2}e^{-\frac{C_2 z}{C_1}} + x_0 + \frac{C_3}{C_2}\right).$$

(The value of C_1, C_2 and C_3 are obtained from the pressure and temperature values at point j)
Step 10. Calculate the steam quality drop: $\frac{dx}{dz} = \frac{x_{j+1} - x_j}{\Delta h}$.
Step 11. Set $\lambda = \lambda + \Delta\lambda$. Repeat steps 4–9 until $\lambda > h/\Delta\lambda$.
Step 12. Set $h = h + \Delta h$. Repeat steps 3–10 until $h > h_{max}$.

7.8.5 Simulation and discussion

The algorithm begins with a calculation at the top of the pipe. The calculations are performed for each successive segment of the pipe to the surface. The model is used to compare measured steam quality value and some sensitivity studies were also conducted for steam injection. However, as there was a lack of experimental tables about the wells, to validate the model, the results were compared with the measurements.

7.8.5.1 *Parameters*

In this simulation, a pipe in X well in China is studied. The parameters are as follows:

Depth of the well = 1300 m
Ground thermal conductivity parameter = 2.06 W/(m · K)
Ground temperature = 16°C
Geothermal gradient = 2.18°C/100 m
Roughness of the inner surface of the well = 0.000015

The other pipe parameters are given in Tables 7.22–7.24.

7.8.6 Trend analysis

From a practical viewpoint, the model was run with some known parameters (steam pressure was 14 MPa, steam rate was 7 t/h and the injection time was 11 days). Steam quality, steam pressure, steam temperature and casing temperature results are shown in Figures 7.43–7.46, respectively.

Table 7.22 Parameters of pipes.

Diameter [m]	Thickness [m]	Weight [kg]	Expansion coefficient [m/m · K]	Young's modulus [GPa]	Poisson's ratio	Using length [m]
0.0889	0.01295	23.79	0.0000115	215	0.3	270
0.0889	0.00953	18.28	0.0000115	215	0.3	120
0.0889	0.00734	15.04	0.0000115	215	0.3	620
0.0889	0.00645	13.58	0.0000115	215	0.3	290

A trend analysis was conducted to check whether the developed model was physically correct. To test the developed model, the pressure, temperatures in the wells were determined.

From Figure 7.43, the steam quality decreased along with well depth, because the mass flow of steam becomes smaller as steam is injected into the oil reservoir. A 0.78 steam quality in the field reaches 0.52 at a depth of 1300 m.

The steam flow is a gas-liquid two phase and has energy loss during the flow process. Figure 7.44 shows the relationship of pressure along the well depth to the steam quality. From Figure 7.44, it can be seen that the steam pressure does not have a uniform distribution but decreases along with increasing depth. The minimum pressure value reached was approximately 12.6 MPa. As the velocity and the steam mass decrease and the pressure drops, the friction decreases. Hence, the change in the steam pressure slows. Temperature profiles are both a function of pressure, so

Table 7.23 Well parameters.

Measured depth [m]	Internal diameter [m]	External diameter [m]
336.7	0.15478	0.1778
422.6	0.1525	0.1778
1300.0	0.10862	0.127

Table 7.24 Azimuth, inclination and vertical depth.

Number	Measured depth [m]	Inclination [°]	Azimuth [°]	Vertical depth [m]
1	135	2.63	241.01	134.72
2	278	1.23	237.86	277.91
3	364	1.43	213.86	363.82
4	393	2.17	26.38	392.53
5	422	1.85	44.56	421.28
6	450	0.82	191.12	449.62
7	486	2.93	269.07	485.47
8	514	1.03	297.55	513.83
9	543	3.58	324.51	541.74
10	571	2.98	303.05	570.43
11	600	2.03	204.74	599.42
12	628	2.34	164.33	627.28
13	660	1.85	195.28	659.56
14	723	3.14	214.84	721.70
15	782	0.98	216.48	781.30
16	830	2.15	229.31	829.12
17	860	2.67	244.03	859.71
18	908	4.85	266.62	904.08
19	928	6.72	258.78	921.42
20	972	2.03	236.88	971.71
21	1025	4.78	239.27	1021.25
22	1058	4.01	244.59	1055.58
23	1089	4.98	228.2	1084.17
24	1132	3.75	233.88	1129.28
25	1174	5.63	235.14	1168.87
26	1204	4.23	234.38	1200.99
27	1235	3.87	234.99	1232.08
28	1268	4.97	232.57	1263.45
29	1300	8.84	233.28	1284.96

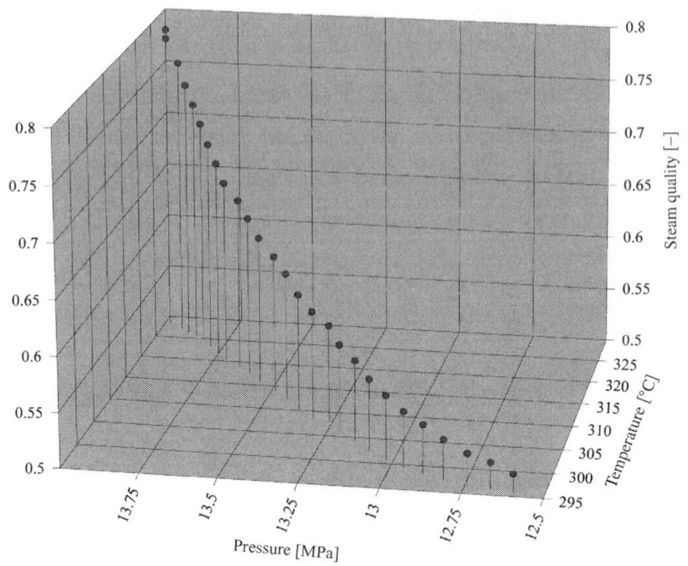

Figure 7.43 Steam quality curve under variable (T, P) fields.

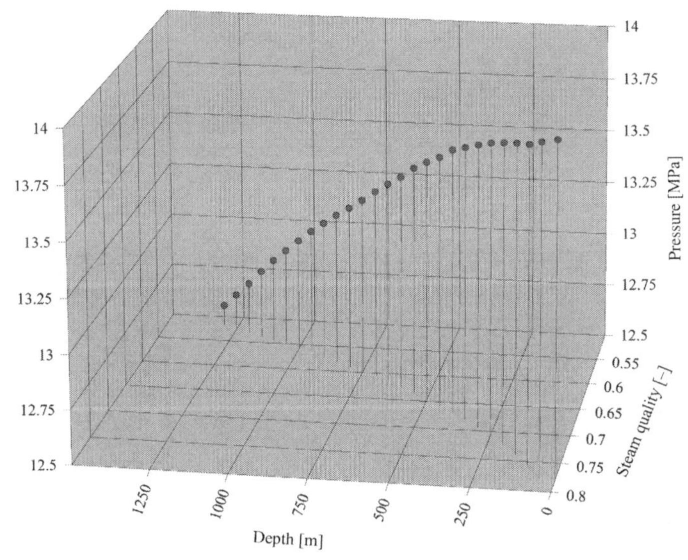

Figure 7.44 Distribution of variable pressure field.

the change in steam effects were as shown in Figure 7.45 and Figure 7.46, and the downhole steam, casing, cement and formation temperature values were 295.7°C, 132.8°C, 102.7°C and 49.5°C, respectively.

7.8.7 **Sensitivity analysis**

The steam injection pressure and the injection rate are two important parameters. Here two parameters that influence the result are examined. One is the steam injection pressure at the top of the well, which can influence inner temperature, steam quality and casing temperature. Proper

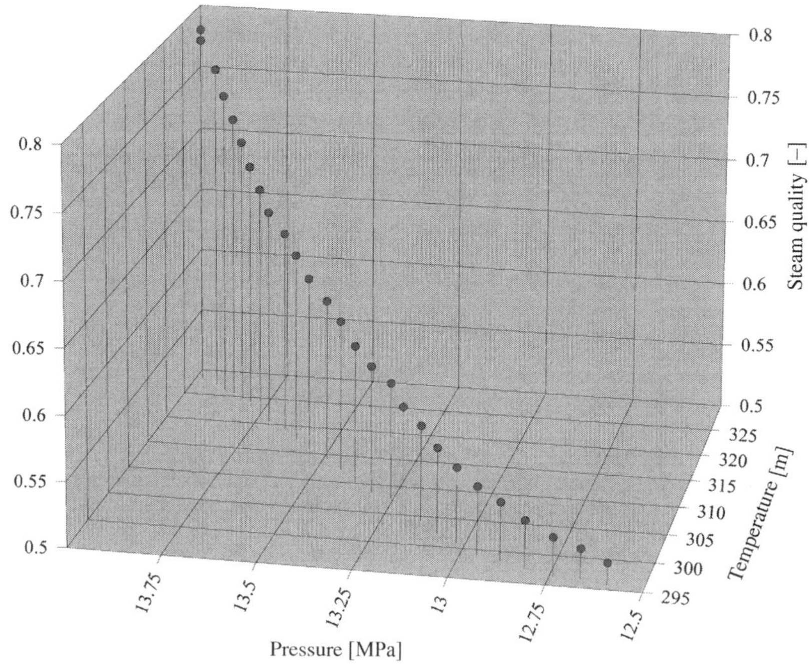

Figure 7.45 Distribution of variable temperature field.

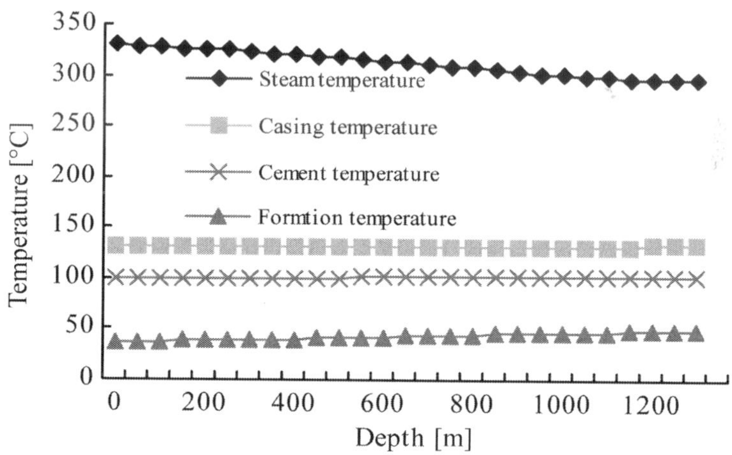

Figure 7.46 Distribution of temperature profiles.

steam pressure selection can have a good effect on crude oil viscosity and prevent casing damage. Different pressures; 12 MPa, 14 MPa or 16 MPa; were used to study the difference between the models with different injection rates the results for which are presented allowing for three rates: 6 t/h, 7 t/h or 8 t/h. The steam flow decreased with an increase in pressure at the top of the casing. Deberne *et al.* (1999) presented a model for calculating the flows in a coordinated fashion and the maximum steam injection flow rate can be also obtained. Therefore, in the steam injection design, if the steam flow is too great, it does not necessarily mean higher recovery from the heavy crude oil. In addition, the tubing size, deviation angle and heat insulation thickness effect was investigated. To better visualize the sensitivity trends, a one-dimensional projection chart was produced.

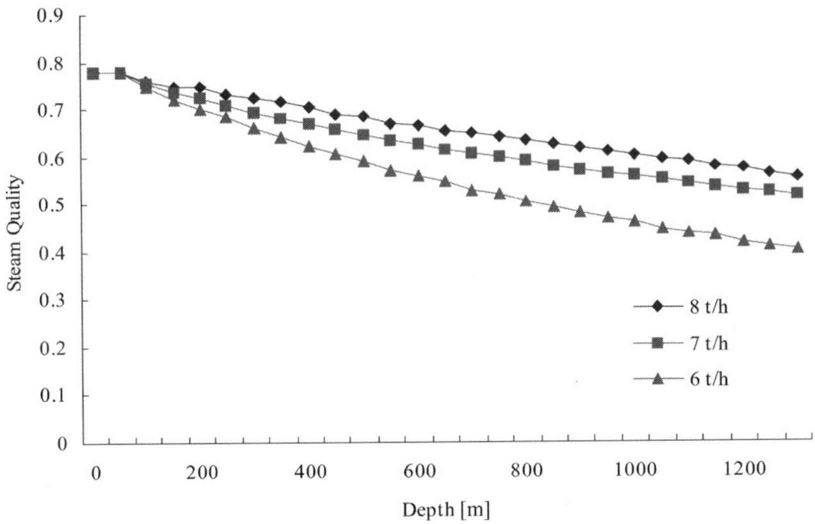

Figure 7.47 Steam quality for different steam injection rates.

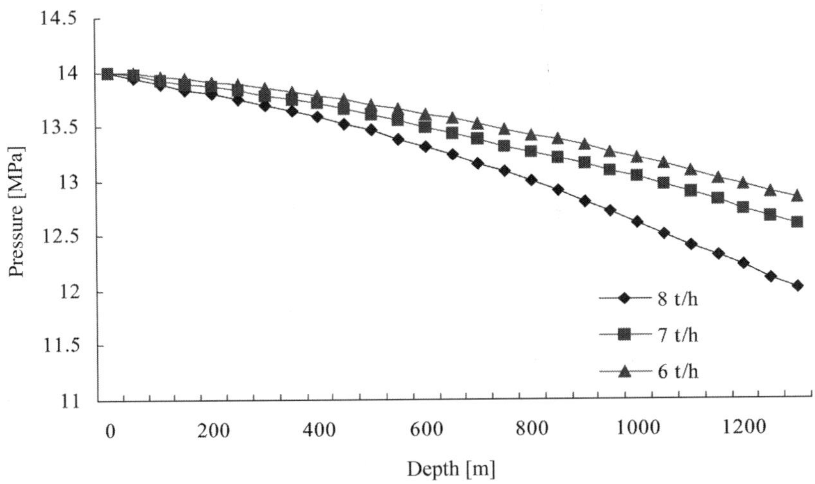

Figure 7.48 Pressure for different steam injection rates.

7.8.7.1 *Effect of injection rate*

In steam injection design, injection speed is a very important parameter as it not only affects oil extraction efficiency, but also impacts cost. This is because, for a fixed steam injection quantity, the heat loss to non-oil producing zones decreases with an increase in steam velocity and the heating radius increases. It is used for an increased yield-increasing effect for oil recovery. However, if the steam injection rate is too high, the oil breaks down and the injection steam channels away from the wells, which causes the formation near the well-bore to be ineffectively heated.

Figures 7.47–7.48 show the variation in steam quality, pressure and casing temperature, respectively, with the injection rate as a parameter. (The wellhead steam quality is 0.78; the injection temperature is 330°C and the injection pressure is 14 MPa.)

At the same depth, steam quality increases as the rate of steam injection increases.

From Figure 7.47, when the injection rate ranged from 6 t/h to 8 t/h, the downhole steam quality ranged from 0.404 to 0.555. This is because the drop in steam quality is supplemented by the

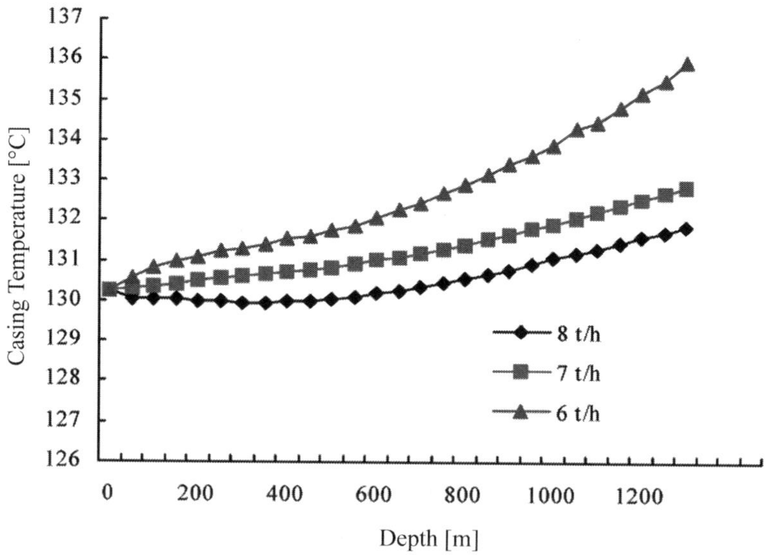

Figure 7.49 Casing temperature for different steam injection rates.

more rapid steam injection. At the same time, the steam heat transfer and the enthalpy drop are both small, causing the drop in the steam quality to slow. As the flow rate decreases, the heat loss in the well-bore increases resulting in a significant decrease in the steam quality. However, the steam quality in the bottom has a tendency to increase with this increase becoming more gentle as the injection rate increases. From Figure 7.47, at 7 t/h of steam flow rate, the steam quality in the bottom did not change dramatically compared to 8 t/h. Therefore, this steam injection rate is recommended.

The pressure at different steam injection rates is shown in Figure 7.48. It can be seen that the larger the steam flow rate, the larger the pressure drop. When the injection rate ranged from 6 t/h to 8 t/h, the downhole pressure ranged from 12.832 MPa to 12.01 MPa, because, in accordance with momentum conservation theory, the total pressure drop includes three major factors: gravity, friction and acceleration. The gravitational effect causes the steam pressure to increase while the other two decrease. An increasing steam flow rate, friction and an acceleration increase causes the pressure drop to increase. In fact, the pressure increases along with depth under a smaller flow. Steam temperature has a similar trend to steam pressure.

The effect of the temperature on the casing is influenced by the steam injection rate. As in Figure 7.49, with the same steam injection pressure but with a difference in steam injection speed, the casing temperature is distinctly dissimilar. The temperature in the casing at the top of the well case is the same but different at the bottom for any steam injection speed. The faster the steam rate the larger the temperature drop in the casing. This is chiefly because as the steam rate increases, the friction increases resulting in a decrease in heat loss. The steam temperature drop along the well case decreases and the heat is absorbed by the casing as the well case depth increases. Therefore, the temperature falls as the depth of the casing increases. When the injection rate ranged from 6 t/h to 8 t/h, the downhole casing temperature ranged from 135.97 to 131.87°C.

7.8.7.2 *Effect of injection pressure*

Steam injection pressure has a great impact on the steam quality and casing temperature. A reasonable choice in steam injection pressure can improve oil flooding performance and prevent

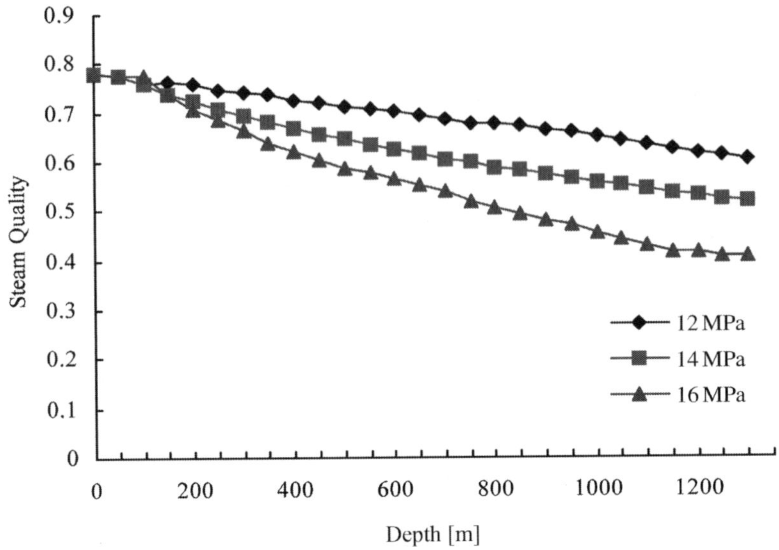

Figure 7.50 Steam quality for different injection pressures.

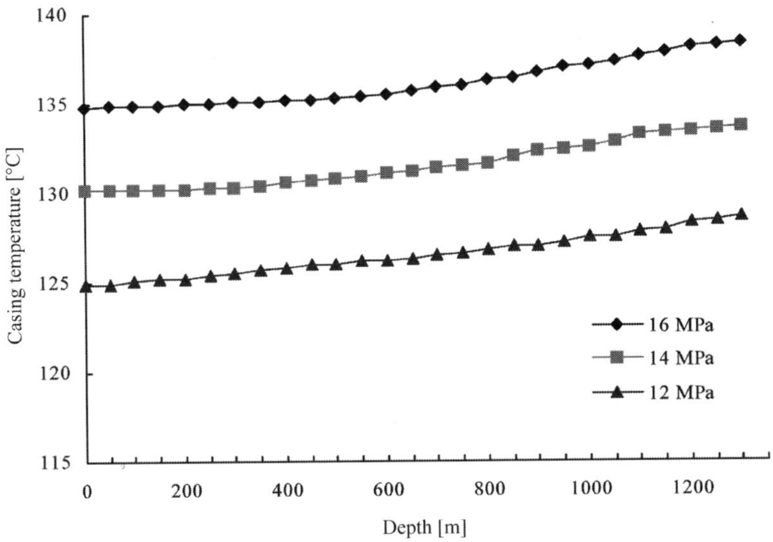

Figure 7.51 Casing temperature for different injection pressures.

casing damage. Figures 7.50 and 7.51 show the variations in steam quality and casing temperature, respectively, with injection pressure as a parameter. (The wellhead steam quality is 0.78 and the injection rate is 7 t/h.)

The quality of the steam at the bottom of the well-bore, which affects the actual amount of steam going into the oil reservoir, is an important indicator and pressure has a significant impact on steam quality. From Figure 7.50, the steam quality decreases along with an increasing injection pressure. The steam quality at the bottom of the well decreased from 0.61 at 12 MPa to 0.41 at 16 MPa, which was demonstrated in Farouq-Ali (1981), because the steam temperature increases with increased steam pressure, and the heat transfer becomes larger causing a rapid decline in steam quality.

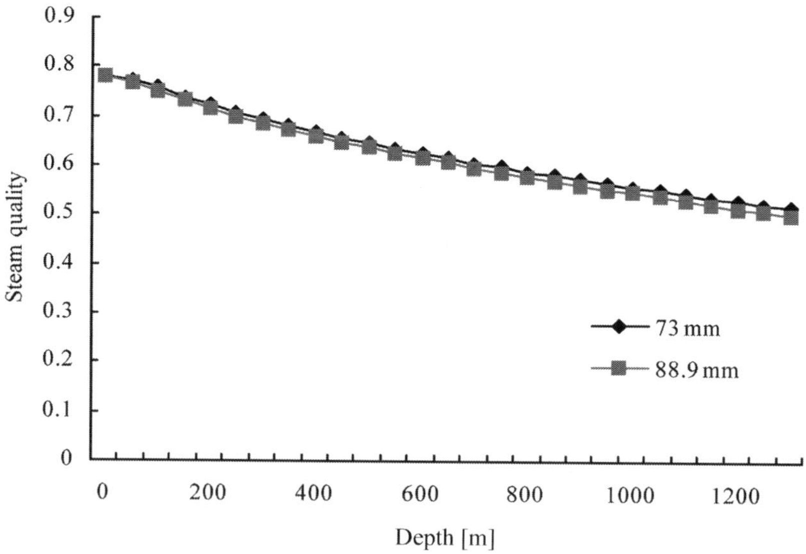

Figure 7.52 Steam quality for different tubing sizes.

The distribution of temperature in the casing along with the well-bore depends on the quantity of heat transfer and the heat transfer coefficient. As with the heat transfer mentioned in Section 2.3, the heat loss from the fluids go through a series of heat resistances and are finally absorbed by the cold formation that surrounds the well-bore system. Some assumptions are made as follows. (Tubing insulation is absent. Both the tubing and casing are made of metals like steel which have high conductivity, so the temperature distribution is neglected. The radiation of heat term in the annulus is negligible.) At a fixed heat transfer coefficient, the more heat transfers, the higher the temperature in the stratum. At the same time, the heat transfer quantity depends on the temperature difference between the steam and the formation. The higher the steam temperature, the more heat is transferred. At saturation, pressure decides the temperature. The temperature in the casing increases along with increasing injection pressure but the changes are not linear mainly because of the quality of the saturated steam. The higher the steam pressure, the higher the temperature and the more heat is transferred. This relationship is illustrated in Figure 7.51, which shows that the higher the steam pressure, the higher the casing temperature. At the same time, the casing temperature increases with increased depth, but at a modest rate. When the steam pressure is 16 MPa, the casing temperature reaches 134.8°C; when the steam pressure is 12 MPa, the casing temperature reaches 124.9°C.

7.8.7.3 *Effect of tubing size*

The effect of tubing size has been studied. Two sizes (88.9 mm and 73 mm) were chosen keeping the value of the other parameters constant. (The wellhead steam quality is 0.78; the injection temperature is 330°C; the injection pressure is 14 MPa, and the injection velocity is 7 t/h.)

Figure 7.52 shows the steam quality profiles with varying tubing sizes. At the same depth, steam quality decreases with an increase in tubing size, a trend which was also observed by Wang *et al.* (2010). This is because the steam flowing velocity decreases and the heat radiation area increases with an increase in tubing size resulting in an increased heat loss. However, from Figure 7.52, the degree of influence reduces. When tubing size is from 88.9 to 73 mm, the downhole steam quality ranges from 0.502 to 0.517. In general, the tubing size has an effect on steam quality, but it is weak.

Figure 7.53 shows the pressure profiles with varying tubing sizes. At the same depth, pressure increases with an increase in tubing size. When tubing size is from 88.9 to 73 mm, the downhole

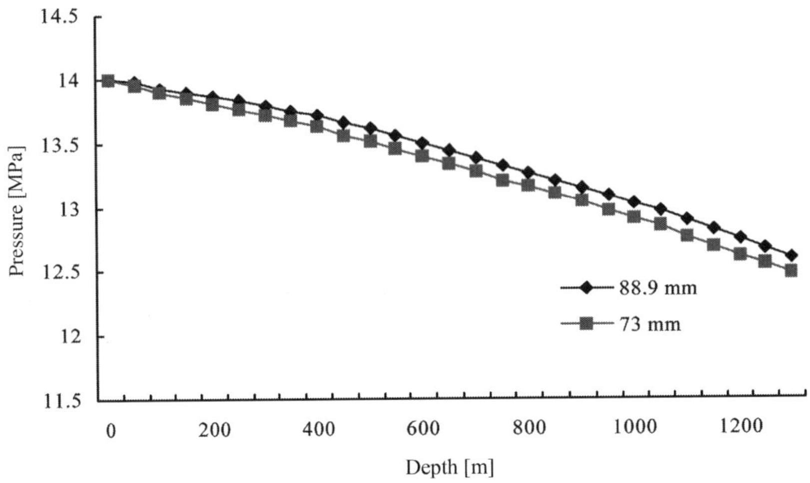

Figure 7.53 Steam pressure for different tubing sizes.

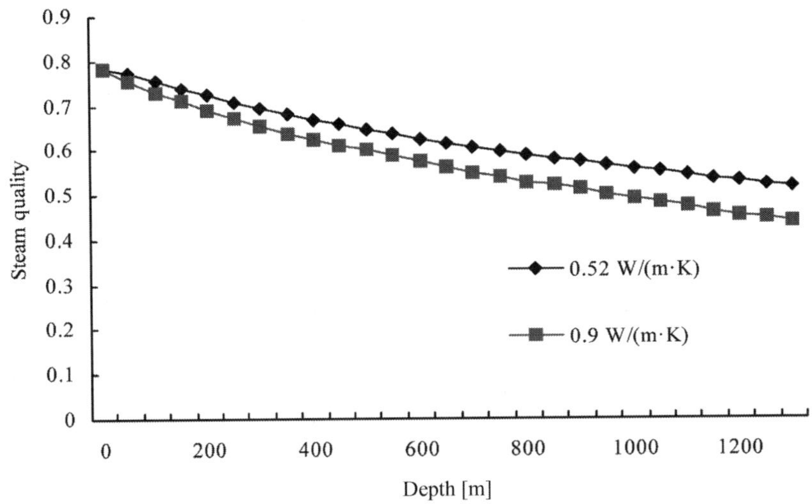

Figure 7.54 Steam quality for different insulating thickness.

pressure ranges from 12.59 to 12.48 MPa because the friction of the pipe decreases with an increase in tubing size resulting in a decreased pressure drop.

7.8.7.4 *Effect of insulating thickness*

The thickness of the heat insulating layer is another factor in steam quality and is significant for steam injection design. (The wellhead steam quality is 0.78; the injection temperature is 330°C; the injection pressure is 14 MPa and the injection velocity is 7 t/h.)

Because thickness is not included in the model, different thermal conductivities are chosen to construct an approximate representation.

As shown in Figure 7.54, when the thermal conductivities of the tubing was from 0.9 W/m · K to 0.52 W/m · K, the downhole steam quality ranges from 0.44 to 0.52, which demonstrates that the insulating thickness has a significant effect on steam quality. In theory, the thicker the insulation, the better the down hole steam quality However, from an engineering view point, when the

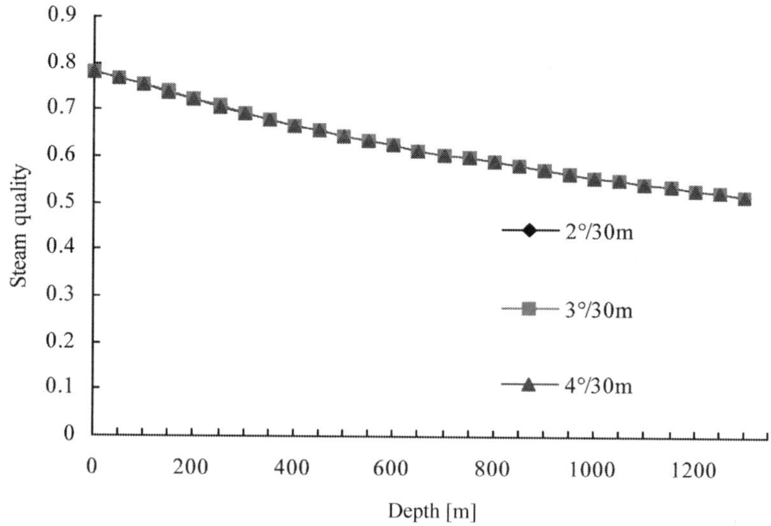

Figure 7.55 Steam quality for different deviation angles.

insulation thickness reaches a certain thickness, the down hole steam quality cannot be improved with increasing thickness and the cost of insulation materials increases.

7.8.7.5 *Effect of deviation angle*

The effect of the deviation angle was investigated. The build-up rate was varied from 2°/30 m to 4°/30 m keeping the other parameters constant. (The wellhead steam quality is 0.78; the injection temperature is 330°C; the injection pressure is 14 MPa and the injection rate is 7 t/h.)

From Figure 7.55, it can be seen that it has a weak effect on steam quality along with the well depth. This is because in the directional well the hole deviation angle is less than 45°, the low build-up rate is low and the dogleg angle is small (from 4°/30 m to 6°/30 m).

7.8.7.6 *Comparison analysis*

In the comparison step, the model of Bahonar *et al.* (2011), which ignored friction and used empirical equations for enthalpy, and our model, which also neglects friction, were compared with the present model. The steam quality value measurement at every depth was obtained and the comparative results are shown in Figure 7.56. (The wellhead steam quality is 0.78; the injection temperature is 330°C; the injection pressure is 14 MPa; the injection rate is 7 t/h.)

From Figure 7.56, the four sets of data show good agreement. From the figures, it can be seen that friction affects steam quality. The interesting feature is that at a fixed depth, the down hole steam quality value was greater than those obtained from the Bahonar *et al.* (2011) model and our model without even considering friction. The reason for this could be negative friction in the opposite direction to the steam flow which causes a decrease in heat loss and increases down hole steam quality. In accordance with Equation 7.297, the effect would become greater as the injection rate increases.

In addition, it is likely that steam quality in the Bahonar *et al.* (2011) model changes linear with the depth.

7.8.8 **Conclusion**

The following conclusions can be derived from the results of this work.

(1) A coupled system model with differential equations for pressure and temperature in deviated steam injection wells according to mass, momentum and energy balances was presented.

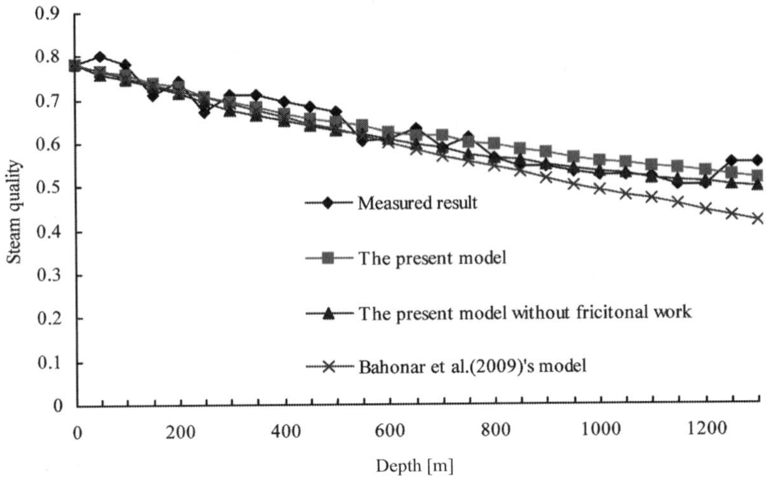

Figure 7.56 Comparison of curves for steam quality.

An algorithm solving the model with a finite difference and a four order Runge-Kutta methods was adopted. Basic data from a well in China was used for case history calculations.

(2) Although the model is based on inclined pipelines, the relevant model is also capable of handling horizontal or vertical wells.
(3) There was good agreement between the measured field data and the model predictions which support the validity of the presented model.
(4) For the trend analysis, steam quality decreases along with the steam temperature and pressure. It also decreases with well depth.
(5) A sensitivity analysis was done on the model. An increase in steam pressure or a decrease in steam rate leads to a decrease in steam quality.
(6) During the steam injection process, the larger the injection rate, the lower the injection pressure, so a higher wellhead steam quality should be chosen, which not only decreases heat loss, but also shortens the steam injection process and even improves thermal efficiency. However, the steam injection pipe and seal component bearing capability should be considered to avoid pipeline leaks or oil reservoir damage.
(7) Further investigations, for example, a tubular dynamics analysis during the steam injection process, and a transit state problem, are desirable.

CHAPTER 8

Predicting of pressure and temperature in HTHP production wells

8.1 INTRODUCTION

One of the earliest works on predicting temperature profiles in a flowing well was presented by Kirkpatrick (1959). He presented a simple flowing temperature gradient chart that could be used to predict gas lift valve temperatures at the injection depth. Much of the classic work in this area was developed by Ramey (1962), who presented approximate methods for predicting the temperature of either a single-phase incompressible liquid or a single-phase ideal gas flow in injection and production wells. Satter (1965) later improved Ramey's method by considering phase changes that occur within steam injection projects. Shiu and Beggs (1980) simplified Ramey's method by correlating for a specific coefficient in Ramey's equation. Hagoort (2004) assessed Ramey's classic method for the calculation of temperatures in injection and production wells. He showed that Ramey's method was an excellent approximation, except for an early transient period in which the calculated temperatures are significantly overestimated. He presented a simple graphical correlation to estimate the length of this early transient period. Willhite (1967) presented a detailed analysis of the overall heat transfer mechanism in an injection well, and Coulter and Bardon (1979) developed a method for predicting temperatures in gas transmission lines. Sagar *et al.* (1991) presented a simple model suitable for calculation by hand to predict temperature profiles in two-phase flowing wells. Hagoort (2007) presented a simple and physically transparent analytical solution for the prediction of wellbore temperatures in gas production wells. In this research, models were built only for the prediction of temperature profiles that assume steady-state conditions, but not for pressure profiles.

Due to the complex nature of the gas-liquid two-phase flows in wells and pipes, many attempts have been made to develop predictive techniques through empirical or semi-empirical methods (such methods have dominated practical design procedures). However, these methods may not be suitable during transient periods for the prediction of pressure and liquid holdup variations.

Since the late 1980s, the trend has shifted toward a more fundamental modelling approach, referred to as the mechanistic approach. The mechanistic approach is based on the fundamentals of multiphase flow and fluid dynamics. The approach begins with local instantaneous conservation equations and systematically develops averaged relations for the variables of interest (pressure, temperature, velocity, and liquid holdup). Many studies (Cazarez-Candia and Vásquez-Cruz, 2005; Grolman and Fortuin, 1997; Hemeida, 1987; Hurlburt and Hanratty, 2002; Ouyang and Aziz, 1999; Taitel *et al.*, 1989) have proposed mechanistic models, assuming the flow to be under steady-state conditions, whereas others (Ouyang and Aziz, 2001; Taitel *et al.*, 1989) have proposed unsteady-state gas-liquid two phase flow models. However, such models predict the pressure profiles but not the temperature profiles.

For pressure research into single-phase flows, Rzasa and Katz (1945) presented an average temperature and average compressibility method, but the method can result in large calculation errors. Su, Z. & Gudmundsson, J.S. only assumed that gas temperature was a constant and could be replaced with an average gas temperature, but its implementation is difficult to compute. In Cullender and Smith (1956), pressure, temperature, and compressibility were all regarded as variables with variations in well depth, a method which has been widely adopted by engineering personnel. For pressure research in two-phase flows, refer to Beggs and Brill (1973), Cazarez-Candia and Vásquez-Cruz (2005), Grolman and Fortuin (1997), Hemeida (1987), Hurlburt and Hanratty (2002), Mukherjee and Brill (1985), Orkiszewski (1967), Ouyang and Aziz (1999; 2000), Taitel *et al.* (1989), and others. However, these models predict pressure profiles but not temperature profiles.

From the above analyses, most research has focused on the calculation of the temperature and pressure distribution predictions separately, and the interdependence is ignored. However, it is well known that there is interdependency in the pressure, temperature, density, and velocity in injection and production wells. In particular, for HTHP deep (superdeep) wells, the pressure, temperature, density, and velocity interdependence must be considered. Thus, it is important to build a coupled differential equations system model for pressure, temperature, density, and velocity to predict the pressure and temperature. However, there is a lack of research in which this method is considered. Wu *et al.* (2011) presented a coupled differential equations system model concerning pressure and temperature in HTHP wells according to mass, momentum, and energy balances and presented an algorithmic solution model, along with the fourth-order Runge-Kutta method, but this model did not consider gas density and velocity.

In the early twentieth century, the gas liquid two-phase flow problem began to be studied. Due to the limited level of experimental methods and basic theoretical research, the factors affecting the gas liquid two-phase flow were unclear. Research in this area has made significant progress since Poettmann and Carpenter published their paper. Since determining that the gas liquid two-phase flow property is an extremely complex problem, the problem used to be solved using experiments and empirical formulas. These empirical formulas were simplified mathematical models reflecting the gas liquid two-phase flow process. At present, there are several empirical formulas which are often used in the oil and gas industry: the Poettmann-Carpenter method (Poettman and Carpenter, 1952), the Orkiszewski method (Orkiszewski, 1967), the Aziz-Govier-Fogarasi method (Aziz *et al.*, 1972), and the Mukherjee and Brill method (Mukherjee and Brill, 1985). These empirical formulas have simple and practical features. However, they may lead to unreliable results if the complex physical mechanisms are neglected. Since the late 1980s, the trend has shifted towards a more fundamental modelling approach, also referred to as the mechanistic approach. This approach begins with local instantaneous conservation equations and systematically develops averaged relations for the variables of interest (pressure, temperature, velocity and liquid hold up). Different researchers have proposed mechanistic models, assuming that the flow is under steady-state conditions, whereas other researchers have proposed unsteady state gas-liquid two phase flow models. However, such models predict the pressure profiles but not the temperature profiles.

In fact, it is well known that pressure and temperature are interdependent, whether it is in injection wells or production wells and therefore should not be predicted separately. Instead, a pressure and temperature prediction model should be built. Wu *et al.* (2011) and Xu *et al.* (2011) presented a coupled differential equations system model concerning pressure and temperature in high temperature-high pressure wells according to mass, momentum and energy balances and presented an algorithmic solution model, along with the four order Runge-Kutta method. This was a steady-state model. However, pressure, temperature, velocity and density were found to not only depend on depth but also on time. In a practical production process, gas and liquid often appear simultaneously. In addition, the gas liquid two-phase is often not steady because of complicated geology or other reasons. If a steady-state model is used to predict the pressure, temperature, velocity and density, it may produce errors, which could lead to economic losses and safety problems. In order to predict the pressure, temperature, velocity and density more accurately, it is necessary to be able to analyse such a transient gas liquid two-phase flow problem.

Oil-water-gas three-phase flows or oil-gas two-phase flows often occur in the petroleum industry during the production and transportation of produced fluids. The prediction of three-phase gas/liquid/liquid flows is therefore of importance to the industry. Bearing in mind that even two-phase gas-liquid flows are highly complex, it is immediately apparent that the addition of a third phase substantially adds to this complexity.

Generally, two-phase research methods are extended to three-phase problems, so here the research covering both two-phase and three-phase problems is reviewed. Flow patterns are key when deciding phase strength. About 14 flow patterns have been observed and several researchers have described three or four of these (Oglesby, 1979). In the past 30 years, there has been significant progress in the analysis of two-phase flow patterns and many new and comprehensive flow patterns have been published (Angeli and Hewitt, 1999; Nädler and Mewes, 1997; Shi, 2001; Shi *et al.*, 1999; 2004; Trallero, 1995; 1997). Because two-phase gas-liquid flows are highly complex, it is apparent that the addition of a third phase will increase this complexity (Cazarez *et al.*, 2010). Experimental observations have shown that the flow structures of a three-phase-pipe flow are much more complicated than that of a two-phase-pipe flow. For example; 10 flow patterns were observed by the experimental study in Açikgöz *et al.* (1992) and Beirute (1991); seven flow patterns were identified for a horizontal gas-oil-water flows (Pan *et al.*, 1995); and 8 flow patterns were identified for vertical air-water-oil flows (Bahonar *et al.*, 2011).

A two-phase flow is the most common flow for fluid in nature, and in developed oil wells in the middle and later stages. The understanding, description, and prediction of the flow characteristics have become an important research focus in the field of two-phase flows. Numerical simulation methods have been found to be an effective way to study two-phase flows. In (Poettman and Carpenter (1952)), a two-phase flow in the vertical tube research for a gas-lift operation design research was studied. The mixture was treated as homogeneous and single phase, and the density was constant, but the viscosity was ignored. This method was found to be suitable for a high flow and low gas-liquid ratio, but large errors were made for other types. A pressure drop computation method was proposed and the oil-liquid two-phase flow was described using dimensional analysis and a mobility pattern distribution map was presented with the experiment, which proved to have very accurate engineering but could not be used for deep or great pressure drop wells. Empirical slip models suitable for a long tube flow-based gas-liquid two-phase slip were built by Hagedorn and Brown (1965). In this research it was established that most energy loss was caused by friction, which could be correlated with a Reynolds number. This friction factor needed to be fixed by calculating the liquid holdup, and although the equation involved empirical correlations, it had a high application accuracy and has been widely used in horizontal multi-phase flows. A two-phase flow pressure drop computation model covering all patterns in a vertical tube was established and a flow pattern identification method was developed in Orkiszewski (1967). This technique has proved to be one of most reliable application methods in the oil industry. A pressure drop model was developed for bubbly flows and slug flows focusing on a gas-liquid two-phase flow, in which a gas volume factor was introduced into the density and friction losses using a gas-liquid two-phase separation effect (Aziz *et al.*, 1972). The relationship between liquid hold-up and the drag coefficient based on a homogeneous flow pressure gradient equation was experimentally studied (Beggs and Brill, 1973). Beggs-Brill's correlations were built and could be used for almost any pipeline inclination. The changing flow pattern conditions from the mechanism were explained and a physical model was proposed (Taitel *et al.*, 1995). This model was important to two-phase flow research, as the pressure drop calculation method was developed from empirical research and mechanistic studies. Early correlations were empirical and were based on experimental studies, and the results were generally satisfactory for the conditions under which each model was developed. Marktos *et al.* has made outstanding contributions to two-phase research. They presented a one-dimensional, isothermal, flow analysis of steam and water mixtures in vertical flow passages in a general form. They extended the model to applications such as fire sprinkler systems, batch sedimentation, granular propellants, pressurized-water-reactor design and airlift pumps simulations, etc. (Latsa *et al.*, 1999; Markatos and Kirkcaldy, 1983; Markatos and Pericleous, 1984; Markatos and Singhal, 1978; Nenes *et al.* 1996). Because of the complex nature of the gas-liquid two-phase flows in wells and pipes, many attempts have been made to develop predictable techniques through empirical or semi-empirical methods (such methods have dominated practical design procedures). However, most research calculates the distribution prediction of temperature and pressure separately (Alves *et al.*, 1992; Hagoort, 2004; Kirkpatrick, 1959; Sagar *et al.*, 1991; Xiao, 1987), and the interdependence is ignored. As stated earlier, it is well-known that there is interdependence between the pressure, temperature, density, and velocity in injection wells and production wells. The characteristics of the oil-water two-phase pipe flow were experimentally studied and it was found that a droplet size distribution could directly affect the accuracy of the model predictions (Vielma *et al.*, 2007). Hasan *et al.* developed a coupled fluid flow model where wellbore/reservoir simulators for modelling the single-phase gas, oil and the two-phase gas-oil flow problems were presented. The transient fluid and heat flow models were solved numerically using finite difference methods to obtain parameters such as pressure and velocity. The numerical algorithms applied a double-iterative procedure on both temperature and pressure to solve the three conservation equations simultaneously (Hasan and Kabir, 1991; Hasan *et al.*, 1998; 2002; 2010; Kabi *et al.*, 1996; Tao *et al.*, 2013; Xu *et al.*, 2012a; 2013c).

Hydrodynamic three-phase modelling is based on flow pattern definitions and some forces are different in each pattern. More flow patterns mean more discontinuities and a greater complexity in the models. Since there are many papers identifying the flow patterns, a model using one pattern can be used to analyse the forces and can then be applied to the other patterns. In the research on limited coupled models, there were two main ideas. One treatment for a three-phase flow is to consider the system of oil, water, and gas as one liquid phase with mixture properties (Cazarez-Candia and Vásquez-Cruz, 2005; Liu *et al.*, 2013; Xu *et al.*, 2011; 2012b; 2013; 2014; Zhao and Xu, 2008). However, the model can predict neither the parameter mentioned nor the volumetric fraction for each phase. The other treatment for a three-phase flow is to combine oil and water into a single liquid phase. Gas-liquid-liquid three-phase flows can be regarded as a special type of gas-liquid two-phase flow if the two liquids are fully mixed, and are probably true of vertical and steeply inclined flows. The physical properties of the liquid mixture can be calculated based on the fractions and the individual physical properties of the two liquids. In this treatment, the slip between the oil and water is ignored and a homogeneous mixture is assumed for the liquid phase (Bonizzi and Issa, 2003; Zhang and Sarica, 2005; Zhang *et al.*, 2003). The other extreme is to a treat three-phase flow as a three-layer stratified flow with gas on the top, oil in the middle, and water at the bottom. This can be done for immiscible liquids flowing in horizontal or slightly inclined pipes with low gas, oil, and water flow rates. The solution is further complicated for two phases and becomes much more complex for three phases. A model was developed to predict the value of the hold-up and pressure gradient for a three-phase stratified flow in a horizontal pipeline (Ghorai *et al.*, 2005). The concept of extended velocity was applied to compute the wall shear stresses. However, the temperature profile was not considered. A three-phase (heavy-oil-water-gas) bubbly flow in upward pipes was simulated using a one dimensional transient two-fluid model in which the continuity and momentum equations for the two liquids (heavy oil and water) were combined to obtain a new equation for the liquid mixture quantities. However, the parameters those models predicted were the pressure profiles rather than the temperature profiles and they failed to consider heat transfer.

8.2 PTP-GW

Considering the differential equation model for P (pressure), T (temperature), we have the following assumptions:

1. There is a steady one dimensional gas flow in the tubing, and all featured parameters are homochromatic at any transverse cross section.
2. The heat transfer in one dimension is steady from the tubing to the second interface, but unsteady from the second interface to the surrounding formation.
3. The heat losses from the tubing and the surrounding earth are radial.
4. The vertical formation temperature is linearly distributed based on known geothermal gradients.

8.2.1 **Physical model**

The physical model that underlies the equations describing well-bore heat transmission consists of a straight, cased well that is cemented to the formation and equipped with tubing for transfer

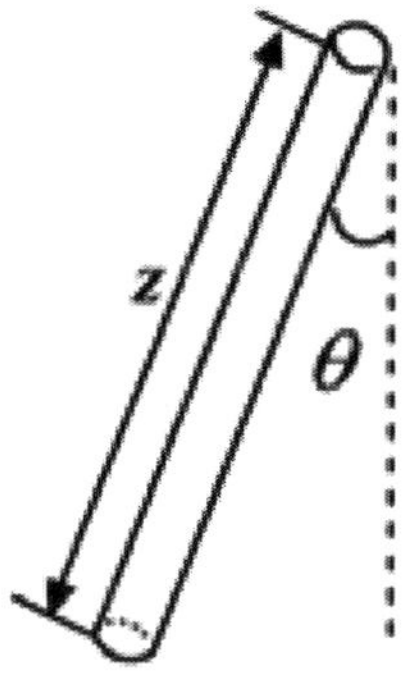

Figure 8.1 The physical figure.

to the surface. Both the casing and tubing have a constant diameter. The tubing diameter is small in respect to its length. Initially, the tubing is filled with a fluid in thermal equilibrium with the formation. At origin, the fluid starts flowing from the bottom of the tubing to the top at a constant flow rate, and the heat conduction in the flow direction and the frictional heating in the tubing are negligible. The fluid that is initially present in the tubing is the same as the fluid that enters from the bottom of the tubing. The temperature of the fluid is equal to the formation temperature at the bottom. The flow in the tubing is 1D (i.e., temperature and fluid velocity depend only on the distance along the tubing). As the fluid moves up the tubing, it loses heat to the colder formation. Heat losses to the formation are through heat conduction in a radial direction only. The effect of the tubing wall, the annular space between the casing and the tubing, the casing wall, and the cement zone on the heat transmission is included in a single, steady-state heat-transfer coefficient. The initial temperature of the formation increases linearly with depth, reflecting a constant geothermal gradient.

The tubing is surrounded by a homogenous rock formation that extends to infinity. The distance temperature in the formation increases linearly with depth reflecting geothermal temperature. The gas enters the flow tubing at a given pressure and temperature and, within the tubing, the gas flow takes place under turbulent flow conditions.

8.2.2 Coupled differential equations system model

Consider the flow system depicted in Figure 8.1:

> a straight cylindrical flow tube with an inclination angle θ, a constant cross-sectional flow area A, a hydraulic diameter, and a total length Z. Through this tubing gas flows from the bottom to the top with a mass flow rate w. The distance co-ordinate in the flow direction along the tubing is denoted z.

Mass, momentum, and energy balances, along with pressure and temperature (P, T) in relation to the gas, are used to generate the constitutive equations.

Material balance: The amount of gas in a given control volume is the product of volume and density. The volume is given by the product of the area (or average area, when the area varies with the length) and the velocity of gas. Thus, under steady-state conditions, the mass balance equation for a differential depth, dz of the well in terms of gas density ρ, and velocity v is written as:

$$\frac{d(\rho_m v_m A)}{dz} = 0$$

i.e.,

$$\rho_m \frac{dv_m}{dz} + v_m \frac{d\rho_m}{dz} = 0. \tag{8.1}$$

Momentum balance: For a steady flow system, the pressure gradient $\frac{dP}{dz}$, is balanced by the static head, $\rho_m g\cos\theta$, the friction head, $\frac{\rho_m f v_m^2}{2d}$, and the kinetic head, $(\rho_m v_m)\left(\frac{dv_m}{dz}\right)$. Thus,

$$\frac{dP}{dz} = -\rho_m v_m \frac{dv_m}{dz} + \rho_m g\cos\theta - \frac{\rho_m f v_m^2}{2d}$$

i.e.,

$$\frac{1}{\rho_m}\frac{dP}{dz} = -v_m \frac{dv_m}{dz} + g\cos\theta - \frac{f v_m^2}{2d}. \tag{8.2}$$

Energy balance: We take the bottom of the well as the origin of the coordinate axis and the vertical as the positive direction. We take the infinitesimal length dz of the tubing, then the fluid containing energy flowing into the infinitesimal tubing: the inner energy $E(z)$, the kinetic energy $\frac{1}{2}M_m v_m^2(z)$, potential energy $M_m gz\cos\theta$ (θ is inclination angle), and pressure energy $P(z)V_m(z)$. Where the sum of inner energy and pressure energy is the flux of fluid $H(z)=E(z)+P(z)V_m(z)$. When fluid flows out from distance element, then energy contains: inner energy $E(z+dz)$, the pressure gradient, $\frac{dP}{dz}$, kinetic energy $\frac{1}{2}M_m v_m^2(z+dz)$, potential energy $M_m g(z+dz)\cos\theta$, and pressure energy $P(z+dz)V_m(z+dz)$. Let dQ be the radial transfer of the heat of the tubing.

According to the rule of energy balance: the energy of fluid flowing into the infinitesimal equals the energy sum of losses and fluid flowing out from the infinitesimal.

In particular, for flowing gas, we have:

$$H(z) + \frac{1}{2}M_m v_m^2(z) - M_m gz\cos\theta = H(z+dz) + \frac{1}{2}M_m v_m^2(z+dz) - M_m g(z+dz)\cos\theta + dQ. \tag{8.3}$$

From 8.3, there is:

$$-\frac{dQ}{dz} = \frac{dH}{dz} + M_m v_m \frac{dv_m}{dz} - M_m g\cos\theta \tag{8.4}$$

(8.4) divided by M_m, then:

$$-\frac{1}{M_m}\frac{dQ}{dz} = \frac{1}{M_m}\frac{dH}{dz} + v_m\frac{dv_m}{dz} - g\cos\theta. \tag{8.5}$$

Let $q_m = \frac{Q}{M_m}$, $h_m = \frac{H_m}{M_m}$, then (8.5) can be written as:

$$\frac{dq_m}{dz} = -\frac{dh_m}{dz} - v_m\frac{dv_m}{dz} + g\cos\theta. \tag{8.6}$$

From:

$$h_m = \frac{W_g}{W_m}h_g + \frac{W_l}{W_m}h_l,$$

there has:

$$\frac{dh_m}{dz} = \frac{W_g}{W_m}\frac{dh_g}{dz} + \frac{W_l}{W_m}\frac{dh_l}{dz}. \tag{8.7}$$

From:

$$\frac{dh}{dz} = C_P\frac{dT}{dz} - C_J C_P\frac{dP}{dz},$$

we have:

$$\frac{dh_g}{dz} = C_{P_g}\frac{dT}{dz} - C_{J_g}C_{P_g}\frac{dP}{dz}, \quad \frac{dh_l}{dz} = C_{P_l}\frac{dT}{dz} - C_{J_l}C_{P_l}\frac{dP}{dz}. \tag{8.8}$$

Combining Equation (8.7) and (8.8), thus:

$$\frac{dh_m}{dz} = \frac{W_g}{W_m}\left[C_{P_g}\frac{dT}{dz} - C_{J_g}C_{P_g}\frac{dP}{dz}\right] + \frac{W_l}{W_m}\left[C_{P_l}\frac{dT}{dz} - C_{J_l}C_{P_l}\frac{dP}{dz}\right]. \tag{8.9}$$

Therefore,

$$\frac{dh_m}{dz} = \left[\frac{W_g}{W_m}C_{P_g} + \frac{W_l}{W_m}C_{P_l}\right]\frac{dT}{dz} + \left[-\frac{W_g}{W_m}C_{J_g}C_{P_g} - \frac{W_l}{W_m}C_{J_l}C_{P_l}\right]\frac{dP}{dz}. \tag{8.10}$$

If the compression of liquid is ignored, then the J-T coefficient can be represented as:

$$C_{J_l} = -\frac{1}{C_{P_l}\cdot\rho_l}.$$

Thus,

$$\frac{dh_m}{dz} = \left[\frac{W_g}{W_m}C_{P_g} + \frac{W_l}{W_m}C_{P_l}\right]\frac{dT}{dz} + \left[-\frac{W_g}{W_m}C_{J_g}C_{P_g} + \frac{W_l}{W_m}\frac{1}{\rho_l}\right]\frac{dP}{dz}. \tag{8.11}$$

If let:

$$C_{P_m} = \frac{W_g}{W_m}C_{P_g} + \frac{W_l}{W_m}C_{P_l}, \quad C_{J_m} = -\frac{W_g}{W_m}C_{J_g}C_{P_g} + \frac{W_l}{W_m}\frac{1}{\rho_l},$$

then C_{P_m}, C_{J_m} is the specific heat capacity (SHC) and the J-T coefficient of the mixture of gas and liquid, respectively. Therefore, (8.11) can be written as:

$$\frac{dh_m}{dz} = C_{P_m}\frac{dT}{dz} + C_{J_m}\frac{dP}{dz}. \tag{8.12}$$

Ramey (1962) and Willhite (1967) discussed the radial transfer of heat between the fluid and the earth in detail. Over the infinitesimal dz shown in Figure 8.2, the radial transfer of heat from the fluid to the cement/earth interface can be described by:

$$dq_m = \frac{2\pi r_{to}U_{to}}{W}(T - T_k)dz. \tag{8.13}$$

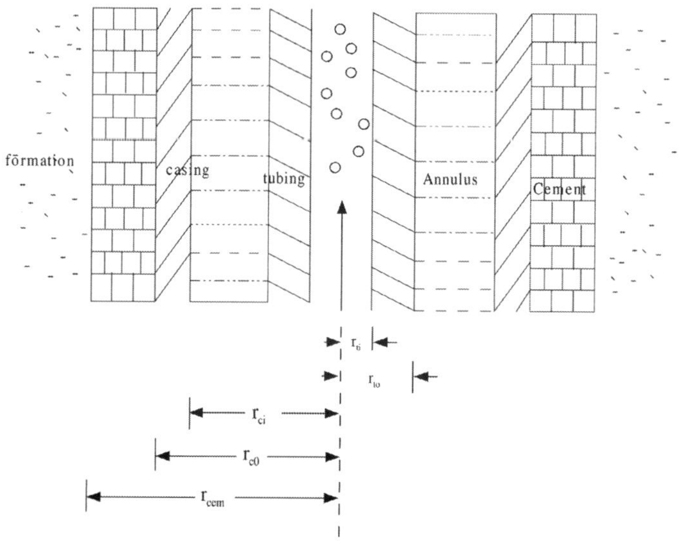

Figure 8.2 The radial transfer of heat.

The radial transfer of heat from the cement/earth interface to the surrounding earth is:

$$dq_m = \frac{2\pi K_e(T_k - T_e)}{Wf(t_D)}dz. \tag{8.14}$$

Combining Equation (8.14) and (8.13) gives the equation for the radial transfer of heat between the fluid and the surrounding earth is given by:

$$\frac{dq_m}{dz} = \frac{2\pi r_{to}U_{to}K_e}{W[k_e + f(t_D)r_{to}U_{to}]}(T - T_e). \tag{8.15}$$

Let:

$$a = \frac{2\pi r_{to}U_{to}K_e}{W[k_e + f(t_D)r_{to}U_{to}]},$$

then:

$$\frac{dq_m}{dz} = a(T - T_e). \tag{8.16}$$

From (8.6), we have:

$$a(T - T_e) + \frac{dh_m}{dz} + v_m\frac{dv_m}{dz} - g\cos\theta = 0.$$

Thus,

$$a(T - T_e) + C_{P_m}\frac{dT}{dz} + C_{J_m}\frac{dP}{dz} + v_m\frac{dv_m}{dz} - g\cos\theta = 0. \tag{8.17}$$

From the stated equation for the mixture of gas and liquid, $\rho_m = \frac{MP}{RZ_mT}$, we have:

$$T\frac{d\rho_m}{dz} + \rho_m\frac{dT}{dz} = \frac{M}{RZ_m}\frac{dP}{dz}. \tag{8.18}$$

Combining (8.1) and (8.18),

$$\frac{dv_m}{dz} = -\frac{v_m}{\rho_m}\frac{M}{TRZ_m}\frac{dP}{dz} + \frac{v_m}{T}\frac{dT}{dz}. \tag{8.19}$$

Therefore, (8.2) can be written as:

$$\frac{dP}{dz} = \rho_m g\cos\theta - \frac{\rho_m f v_m^2}{2d} + v_m^2\left[\frac{M}{TRZ_m}\frac{dP}{dz} - \frac{\rho_m}{T}\frac{dT}{dz}\right]. \tag{8.20}$$

Thus,

$$\left[1 - v_m^2\frac{M}{TRZ_m}\right]\frac{dP}{dz} = \rho_m g\cos\theta - \frac{\rho_m f v_m^2}{2d} - \frac{v_m^2\rho_m}{T}\frac{dT}{dz}$$

i.e.,

$$\frac{dP}{dz} = \frac{\rho_m g\cos\theta - \frac{\rho_m f v_m^2}{2d} - \frac{v_m^2\rho_m}{T}\frac{dT}{dz}}{1 - v_m^2\frac{M}{TRZ_m}}. \tag{8.21}$$

Using the state equation for the mixture of gas and liquid again, we have:

$$\frac{dP}{dz}=\frac{\rho_m g\cos\theta-\frac{\rho_m f v_m^2}{2d}-\frac{v_m^2\rho_m}{T}\frac{dT}{dz}}{1-\frac{\rho_m v_m^2}{P}}. \tag{8.22}$$

Combining (8.17) and (8.20), we have:

$$\left[C_{P_m}+\frac{v_m^2}{T}\right]\frac{dT}{dz}=-a(T-T_e)+\left[\frac{v_m^2}{\rho_m}\frac{M}{RZ_mT}-C_{J_m}C_{P_m}\right]\frac{dP}{dz}+g\cos\theta \tag{8.23}$$

i.e.,

$$\frac{dT}{dz}=\frac{-a(T-T_e)+\left[\frac{v_m^2}{\rho_m}\frac{M}{RZ_mT}-C_{J_m}C_{P_m}\right]\frac{dP}{dz}+g\cos\theta}{C_{P_m}+\frac{v_m^2}{T}}.$$

Using the state equation for the mixture of gas and liquid again, we have:

$$\frac{dT}{dz}=\frac{-a(T-T_e)+\left[\frac{v_m^2}{P}-C_{J_m}C_{P_m}\right]\frac{dP}{dz}+g\cos\theta}{C_{P_m}+\frac{v_m^2}{T}}. \tag{8.24}$$

Thus, we can obtain the coupled system model of differential equations on temperature and pressure as follows:

$$\begin{cases}\dfrac{dP}{dz}=\dfrac{\rho_m g\cos\theta-\frac{\rho_m f v_m^2}{2d}-\frac{v_m^2\rho_m}{T}\frac{dT}{dz}}{1-\frac{\rho_m v_m^2}{P}}\\[2ex] \dfrac{dT}{dz}=\dfrac{-a(T-T_e)+\left[\frac{v_m^2}{P}-C_{J_m}C_{P_m}\right]\frac{dP}{dz}+g\cos\theta}{C_{P_m}+\frac{v_m^2}{T}}\end{cases}. \tag{8.25}$$

The primal conditions are: pressure of the well bottom $P(z_0)=P_0$, temperature of the earth $T(z_0)=T_0$, and so on.

8.25 can be rewritten as:

$$\begin{cases}\dfrac{dP}{dz}=\dfrac{(C_{P_m}T+V_m^2)\left(\rho_m g\cos\theta-\frac{\rho_m f v_m^2}{2d}\right)-\rho_m v_m^2 T[a(T-T_e)+g\cos\theta]}{(C_{P_m}T+V_m^2)\left(1-\frac{\rho_m v_m^2}{P}\right)+\rho_m v_m^2\left[\frac{v_m^2}{P}-C_{J_m}C_{P_m}\right]}\\[2ex] \dfrac{dT}{dz}=\dfrac{-a(T-T_e)+\left[\frac{v_m^2}{P}-C_{J_m}C_{P_m}\right]\frac{dP}{dz}+g\cos\theta}{C_{P_m}+\frac{v_m^2}{T}}\\[2ex] P(z_0)=P_0,\quad T(z_0)=T_0\end{cases}. \tag{8.26}$$

8.2.3 **Model solution**

Let:

$$\frac{dP}{dz}=f_3(z;P,T),\qquad \frac{dT}{dz}=f_4(z;P,T).$$

Denote:

$$F=(f_1,f_2)^T,\quad y=(P,T)^T,\quad y(0)=y(z_0)=(P(z_0),T(z_0))^T,$$

then, the system of ordinary differential equations can be written as:

$$y'=F(z;y),\quad y(0)=y(z_0).$$

The norm of vector f is given as follows:

$$\|F\| = \max(|f_1|, |f_2|).$$

For f_1, f_2, there is:

$$|f_1| = \frac{|(C_{P_m}T + V_m^2)\left(\rho_m g \cos\theta - \frac{\rho_m f v_m^2}{2d}\right) + \rho_m v_m^2[a(T - T_e) - g\cos\theta]|}{\left|(C_{P_m}T + V_m^2)\left(1 - \frac{\rho_m v_m^2}{P}\right) + \rho_m v_m^2\left[\frac{v_m^2}{P} - C_{J_m}C_{P_m}\right]\right|}$$

$$\leq \frac{|(C_{P_m}T + V_m^2)|\left|\rho_m g \cos\theta - \frac{\rho_m f v_m^2}{2d}\right| + |\rho_m v_m^2||[a(T - T_e) - g\cos\theta]|}{\left|(C_{P_m}T + V_m^2)\left(1 - \frac{\rho_m v_m^2}{P}\right) + \rho_m v_m^2\left[\frac{v_m^2}{P} - C_{J_m}C_{P_m}\right]\right|}$$

$$\leq \frac{|\rho_m|(|K_1||K_2| + V_m^2|K_3|)}{|K_4|},$$

where $|(C_{P_m}T + V_m^2)| \leq |C_{P_m}T| + V_m^2 = |K_1|$, $\left|g\cos\theta - \frac{f v_m^2}{2d}\right| \leq |g\cos\theta| + \left|\frac{f v_m^2}{2d}\right| \leq g + \left|\frac{f v_m^2}{2d}\right| = |K_2|$, $|a(T - T_e) - g\cos\theta| \leq |a(T - T_e)| + g = |K_3|$, $|K_4| = \left|(C_{P_m}T + V_m^2)\left(1 - \frac{\rho_m v_m^2}{P}\right) + \rho_m v_m^2\left[\frac{v_m^2}{P} - C_{J_m}C_{P_m}\right]\right|$.

Since all parameters are bounded quantities, also $|K_1|, |K_2|, |K_3|, |\rho_m|, |K_4|$ are bounded. Let:

$$N_1 = \sup\left\{\frac{|\rho_m|(|K_1||K_2| + V_m^2|K_3|)}{|K_4|}\right\},$$

then,

$$|f_1| \leq N_1.$$

Similarly,

$$|f_2| = \left|\frac{-a(T - T_e) + \left[\frac{v_m^2}{P} - C_{J_m}C_{P_m}\right]\frac{dP}{dz} + g\cos\theta}{C_{P_m} + \frac{v_m^2}{T}}\right|$$

$$\leq \frac{|a(T - T_e)| + \left(\left|\frac{v_m^2}{P}\right| + |C_{J_m}C_{P_m}|\right)N_1 + g}{\left|C_{P_m} + \frac{v_m^2}{T}\right|}.$$

Let:

$$N_2 = \sup\left\{\frac{|a(T - T_e)| + \left(\left|\frac{v_m^2}{P}\right| + |C_{J_m}C_{P_m}|\right)N_1 + g}{\left|C_{P_m} + \frac{v_m^2}{T}\right|}\right\}$$

then, there is:

$$|f_2| \leq N_2.$$

Therefore,

$$\|F\| \leq \max\{N_1, N_2\}.$$

The partial differential of f_1, f_2 about P, T are bounded is discussed as follows:

$$\begin{cases} \dfrac{\partial f_1}{\partial P} = \dfrac{\frac{\rho_m v_m^2}{P^2}(C_{P_m}T + v_m^2)\left[\frac{f\rho_m v_m^2}{2d} + \frac{f\rho_m v_m^2 f_2}{T} - \rho_m g\cos\theta\right] + \frac{\rho_m v_m^2}{P^2}\left(1 - \frac{\rho_m v_m^2}{P}\right)v_m^2 f_1}{C_{P_m}T + v_m^2 + \left(1 - \frac{\rho_m v_m^2}{P}\right)\left(\frac{v_m^2}{P} - C_{J_m}C_{P_m}\right)} \\ \dfrac{\partial f_1}{\partial T} = \dfrac{\frac{\rho_m v_m^2}{P}(C_{P_m}T + v_m^2)\left[C_{P_m}\rho_m g\cos\theta - \frac{C_{P_m} f\rho_m v_m^2}{2d} - a\rho_m v_m^2(v_m^2 - PC_{J_m}C_{P_m})\right] + C_{P_m}\rho_m v_m^2[a(T - T_e) + g\cos\theta]}{\left[(C_{P_m}T + v_m^2)\left(1 - \frac{\rho_m v_m^2}{P}\right) + \rho_m v_m^2\left(\frac{v_m^2}{P} - C_{J_m}C_{P_m}\right)\right]^2} \end{cases}. \tag{8.27}$$

From the Boundedness of all parameters, we have:

$$\begin{cases} \left|\dfrac{\partial f_1}{\partial P}\right| \le \dfrac{\frac{|\rho_m v_m^2|}{P^2}|K_1|\left[\left|\frac{f\rho_m v_m^2}{2d}\right| + \left|\frac{f\rho_m v_m^2}{T}\right|N_2 + \rho_m g\right] + \frac{|\rho_m v_m^2|}{P^2}\left(1 + \left|\frac{\rho_m v_m^2}{P}\right|\right)v_m^2 N_1}{\left|C_{P_m}T + v_m^2 + \left(1 - \frac{\rho_m v_m^2}{P}\right)\left(\frac{v_m^2}{P} - C_{J_m}C_{P_m}\right)\right|} \\ \left|\dfrac{\partial f_1}{\partial T}\right| \le \dfrac{\left|\frac{\rho_m v_m^2}{P}\right|K_1[|C_{P_m}\rho_m|g + \left|\frac{C_{P_m} f\rho_m v_m^2}{2d}\right| + |a|\rho_m v_m^2(v_m^2 + |PC_{J_m}C_{P_m}|)] + |C_{P_m}\rho_m|v_m^2|K_3|}{|K_4|^2} \end{cases}. \tag{8.28}$$

Let:

$$M_{11} = \sup\left\{\frac{\frac{|\rho_m v_m^2|}{P^2}K_1\left[\left|\frac{f\rho_m v_m^2}{2d}\right| + \left|\frac{f\rho_m v_m^2}{T}\right|N_2 + \rho_m g\right] + \frac{|\rho_m v_m^2|}{P^2}\left(1 + \left|\frac{\rho_m v_m^2}{P}\right|\right)v_m^2 N_1}{\left|C_{P_m}T + v_m^2 + \left(1 - \frac{\rho_m v_m^2}{P}\right)\left(\frac{v_m^2}{P} - C_{J_m}C_{P_m}\right)\right|}\right\},$$

then,

$$\left|\frac{\partial f_1}{\partial P}\right| \le M_{11}$$

$$M_{12} = \frac{\left|\frac{\rho_m v_m^2}{P}\right|K_1[|C_{P_m}\rho_m|g + \left|\frac{C_{P_m} f\rho_m v_m^2}{2d}\right| + |a|\rho_m v_m^2(v_m^2 + |PC_{J_m}C_{P_m}|)] + |C_{P_m}\rho_m|v_m^2|K_3|}{|K_4|^2}$$

thus,

$$\left|\frac{\partial f_1}{\partial T}\right| \le M_{12}.$$

Similarly,

$$\begin{cases} \dfrac{\partial f_2}{\partial P} = \dfrac{\frac{v_m^2}{P^2}f_1 + \left[\frac{v_m^2}{P} - C_{J_m}C_{P_m}\right]\frac{\partial f_1}{\partial P}}{C_{P_m} + \frac{v_m^2}{T}} \\ \dfrac{\partial f_2}{\partial T} = \dfrac{\left[-a + \left(\frac{v_m^2}{P} - C_{J_m}C_{P_m}\right)\frac{\partial f_1}{\partial P}\right]\left[C_{P_m} + \frac{v_m^2}{T}\right] + \frac{v_m^2}{T^2}[-a(T - T_e) + \left(\frac{v_m^2}{P} - C_{J_m}C_{P_m}\right)f_1 + g\cos\theta]}{\left[C_{P_m} + \frac{v_m^2}{T}\right]^2} \end{cases}. \tag{8.29}$$

From the Boundedness of all parameters, we have:

$$\begin{cases} \left|\dfrac{\partial f_2}{\partial P}\right| \leq \dfrac{\frac{v_m^2}{P^2}N_1 + \left[\frac{v_m^2}{|P|} + |C_{J_m}C_{P_m}|\right]M_{11}}{\left|C_{P_m} + \frac{v_m^2}{T}\right|} \\ \left|\dfrac{\partial f_2}{\partial T}\right| \leq \dfrac{\left[|a| + \left(\frac{v_m^2}{|P|} + |C_{J_m}C_{P_m}|\right)M_{12}\right]\left[|C_{P_m}| + \frac{v_m^2}{|T|}\right] + \frac{v_m^2}{T^2}\left[|a(T-T_e)| + \left(\frac{v_m^2}{|P|} + |C_{J_m}C_{P_m}|\right)N_1 + g\right]}{\left|C_{P_m} + \frac{v_m^2}{T}\right|^2} \end{cases} \tag{8.30}$$

Let:

$$M_{21} = \sup\left\{\frac{\frac{v_m^2}{P^2}N_1 + \left[\frac{v_m^2}{|P|} + |C_{J_m}C_{P_m}|\right]M_{11}}{|C_{P_m} + \frac{v_m^2}{T}|}\right\},$$

$$M_{22} = \sup\left\{\frac{\begin{array}{c}\left[|a| + \left(\frac{v_m^2}{|P|} + |C_{J_m}C_{P_m}|\right)M_{12}\right]\left[|C_{P_m}| + \frac{v_m^2}{|T|}\right] \\ + \frac{v_m^2}{T^2}\left[|a(T-T_e)| + \left(\frac{v_m^2}{|P|} + |C_{J_m}C_{P_m}|\right)N_1 + g\right]\end{array}}{\left|C_{P_m} + \frac{v_m^2}{T}\right|^2}\right\}.$$

Thus,

$$\left|\frac{\partial f_2}{\partial P}\right| \leq M_{21}, \quad \left|\frac{\partial f_2}{\partial T}\right| \leq M_{22}.$$

The Lipschitz condition is very important when discussing the solution to the system of differential equations, thus we first consider the Lipschitz condition of $F(z; y)$. We write the primal problem again as follows:

$$\frac{dP}{dz} = f_1(z; P, T), \quad \frac{dT}{dz} = f_2(z; P, T).$$

This can be written as:

$$P' = f_1(z; P, T), \quad T' = f_4(z; \rho, v, P, T).$$

The primal condition is: $P(z_0) = P_0$, $T(z_0) = T_0$.

Using the Euler method, for $i = 0, 1, 2, \ldots$, we have:

$$P_{i+1} = P_i + (z_{i+1} - z_i)f_1(z_i; P_i, T_i), \quad T_{i+1} = T_i + (z_{i+1} - z_i)f_2(z_i; P_i, T_i).$$

Here P_i, T_i are intended to approximate $P(z_i)$, $T(z_i)$, where $z_0 < z_1 < z_2 \cdots$ is subdivision of the interval of integration. Let $y_i = (P_i, T_i)^T$, then:

$$y_{i+1} = y_i + (z_{i+1} - z_i)F(z_i; y_i), \quad i = 0, 1, 2, \ldots, n-1.$$

If set $h_i = z_{i+1} - z_i$, then for the subdivision above can be written as:

$$h = (h_0, h_1, \ldots, h_{n-1}).$$

If we connect y_0 and y_1, y_1 and $y_2, \ldots$ by straight lines we obtain the Euler polygon:

$$y_h(z) = y_i + (z - z_i)f(z_i; y_i) \quad \text{for } z_i \leq z \leq z_{i+1}.$$

Theorem 8.1. For $\| F(z; y)\| \leq N = \max\{N_1, N_2\}$, then there is the estimate:

$$\|y_i - y_0\| \leq N|z_i - z_0|,$$

for P_i, T_i defined by above way, where $y_i = (P_i, T_i)^T$.

For $|\frac{\partial f_k}{\partial P}| \leq M_{k1}$, $|\frac{\partial f_k}{\partial T}| \leq M_{k2}$, then

$$\|F(z;y) - F(z;\hat{y})\| \leq L\|y - \hat{y}\|,$$

where $k = 1, 2$; $L = \max_k \left(\sum_{i=1}^{2} M_{ki}\right)$.

Proof. (1) From $P_{i+1} = P_i + (z_{i+1} - z_i) f_1(z_i; P_i, T_i)$ and definition of $\|F(z; P, T)\|$, we have:

$$|P_{i+1} - P_i| = |z_{i+1} - z_i||f_1(z_i; P_i, T_i)| \leq N(z_{i+1} - z_i).$$

Therefore,

$$|P_i - P_{i-1}| \leq N(z_i - z_{i-1}), \quad \ldots, \quad |P_2 - P_1| \leq N(z_2 - z_1), \quad |P_1 - P_0| \leq N(z_1 - z_0).$$

Thus,

$$|P_i - P_{i-1}| + \cdots + |P_2 - P_1| + |P_1 - P_0| \leq N(z_i - z_0).$$

Since

$$|P_i - P_{i-1} + \cdots + P_2 - P_1 + P_1 - P_0| \leq |P_i - P_{i-1}| + \cdots + |P_2 - P_1| + |P_1 - P_0|,$$

so,

$$|P_i - P_0| \leq N(z_i - z_0).$$

Similarly,

$$|T_i - T_0| \leq N(z_i - z_0).$$

From definition of $\|y_i - y_0\|$, there is:

$$\|y_i - y_0\| \leq N(z_i - z_0).$$

(2) For $f_1(z;y)$, $f_2(z;y)$, $y = (P, T)^T$,

$$f_1(z;\hat{y}) - f_1(z;y) = \frac{\partial f_1}{\partial P}(\hat{P} - P) + \frac{\partial f_1}{\partial T}(\hat{T} - T).$$

Thus,

$$|f_1(z;\hat{y}) - f_1(z;y)| \leq \left|\frac{\partial f_1}{\partial P}\right| |\hat{P} - P| + \left|\frac{\partial f_1}{\partial T}\right| |\hat{T} - T|.$$

Let $\Delta y = \max\{|\hat{P} - P|, |\hat{T} - T|\}$, then:

$$|f_1(z;\hat{y}) - f_1(z;y)| \leq \left(\left|\frac{\partial f_1}{\partial P}\right| + \left|\frac{\partial f_1}{\partial T}\right|\right) \Delta y.$$

Similarly,

$$|f_2(z;\hat{y}) - f_2(z;y)| \leq \left(\left|\frac{\partial f_2}{\partial P}\right| + \left|\frac{\partial f_2}{\partial T}\right|\right) \Delta y.$$

From definition of norm, there is:

$$\|F(z;\hat{y}) - F(z;y)\| = \max\{|f_1(z;\hat{y}) - f_1(z;y)|, |f_2(z;\hat{y}) - f_2(z;y)|\}.$$

Let $L = \max_k \left(\sum_{i=1}^{2} M_{ki}\right)$ and $\|\hat{y} - y\| = \Delta y$, then:

$$\|F(z;\hat{y}) - F(z;y)\| \leq L\|\hat{y} - y\|.$$

We consider a subdivision of the interval of integration:

$$z_0, z_1, \ldots, z_{n-1}, z_n = Z.$$

Theorem 8.2. For a fixed subdivision h, let $y_h(x)$ and $\hat{y}_h(x)$ are the Euler polygons corresponding to the initial values y_0 and $\hat{y}_0$. respectively, for $F(z;y)$ satisfies the Lipschitz condition of Theorem 8.1. in a convex region which contains $(z;y_h(z))$ and $(z;\hat{y}_h(x))$ for $z_0 \le z \le Z$, then:

$$\|y_h(z) - \hat{y}_h(z)\| \le e^{L(z-z_0)}\|y_0 - \hat{y}_0\|.$$

Proof. From $y_{i+1} = y_i + (z_{i+1} - z_i)F(z_i;y_i)$, we have:

$$y_1 - y_0 = (z_1 - z_0)F(z_0;y_0), \quad \hat{y}_1 - \hat{y}_0 = (z_1 - z_0)F(z_0;\hat{y}_0).$$

Thus, we get:

$$y_1 - \hat{y}_1 = (y_0 - \hat{y}_0)(z_1 - z_0)[F(z_0;y_0) - F(z_0;\hat{y}_0)].$$

From Theorem 8.1, we have:

$$\|F(z;y) - F(z;\hat{y}\| \le L\|y - \hat{y}\|.$$

Therefore,

$$\|F(z_0;y_0) - F(z_0;\hat{y}_0\| \le L\|y_0 - \hat{y}_0\|.$$

Thus,

$$\|y_1 - \hat{y}_1\| \le (1 + L(z_1 - z_0))\|y_0 - \hat{y}_0\|.$$

From $(1 + L(z_1 - z_0)) \le e^{L(z_1-z_0)}$, we get:

$$\|y_1 - \hat{y}_1\| \le e^{L(z_1-z_0)}\|y_0 - \hat{y}_0\|.$$

If we repeat the same argument for $y_2 - \hat{y}_2$, and so on, we finally obtain:

$$\|y_h(z) - (\hat{y})_h(z)\| \le e^{L(z-z_0)}\|y_0 - \hat{y}_0\|.$$

Theorem 8.3. Let $F(z;y)$ be continuous, and $\|F(z;y)\| \le N$ and satisfy the Lipschitz condition of Theorem 8.1 on $D = \{(z;y) \mid z_0 \le z \le Z, \|y - y_0\| \le b\}$. If $Z - z_0 \le \frac{b}{N}$, then we have:

(1) For $|h| = \max_{i=0,1,2,\ldots,n-1}(|z_{i+1} - z_i|) \to 0$, the Euler polygons $y_{|h|}(z) = (\rho_{|h|}(z), v_{|h|}(z), P_{|h|}(z), T_{|h|}(z))^T$ converge uniformly to a continuous vector function $\phi(z)$.
(2) $\phi(z)$ is continuously differential and is a solution of primal problem on $z_0 \le z \le Z$.
(3) There exists no other solution of primal problem on $z_0 \le z \le Z$.

Proof. (1) Take an $\varepsilon > 0$. Since F is uniformly continuous on the compact D, there exists a $\delta > 0$, such that:

$$|z_2 - z_1| \le \delta \quad \text{and} \quad \|y_1 - y_2\| \le N\delta,$$

imply:

$$\|F(z_2;y_2) - F(z_1;y_1)\| \le \varepsilon.$$

Suppose that the subdivision h satisfies:

$$|z_{i+1} - z_i| \leq \delta, \quad \text{i.e. } |h| \leq \delta.$$

We study first the effects of adding new mesh-points. In a first step, we consider a subdivision $h(1)$, which is obtained by adding new points only to the first subinterval. We have the estimate $\|y_{h(1)}(z_1) - y_h(z_1)\| \leq \varepsilon|z_1 - z_0|$. Since subdivision h and $h(1)$ are identical on $z_1 \leq z \leq Z$, we can obtain by Theorem 8.2:

$$\|y_{h(1)}(z) - y_h(z)\| \leq e^{L(z-z_1)}\|y_{h(1)}(z_1) - y_h(z_1)\|.$$

Thus,

$$\|y_{h(1)}(z) - y_h(z)\| \leq e^{L(z-z_1)}|z_1 - z_0|\varepsilon, \quad \text{for } z_1 \leq z \leq Z.$$

We next add further points to the subinterval (z_1, z_2) and denote the new subdivision by $h(2)$. In the same way as above this lead to:

$$\|y_{h(2)}(z_1) - y_{h(1)}(z_1)\| \leq \varepsilon|z_2 - z_1|,$$

and

$$\|y_{h(2)}(z) - y_{h(1)}(z)\| \leq e^{L(z-z_2)}|z_2 - z_1|\varepsilon, \quad \text{for } z_2 \leq z \leq Z.$$

If we denote by $\hat{h}$ the final refinement, we obtain for $z_2 \leq z \leq z_{i+1}$:

$$\begin{aligned}
&\|y_{\hat{h}}(z) - y_h(z)\| \\
&\quad \leq \|y_{\hat{h}}(z) - y_{h(i-1)}(z)\| + \|y_{h(i-1)}(z) - y_{h(i-2)}(z)\| + \cdots + \|y_{h(2)}(z) - y_{h(1)}(z)\| \\
&\quad + \|y_{h(1)}(z) - y_h(z)\| \leq \varepsilon[e^{L(z-z_1)}|z_1 - z_0| + \cdots + e^{L(z-z_i)}|z_i - z_{i+1}|] \qquad (8.31) \\
&\leq \varepsilon\int_{z_0}^{z} e^{L(z-s)}ds = \frac{\varepsilon}{L}(e^{L(z-z_0)} - 1).
\end{aligned}$$

If we have different subdivisions h and $\acute{h}$, which both satisfy $|z_2 - z_1| \leq \delta$ and $\|y_1 - y_2\| \leq N\delta$, we introduce a third subdivision $\hat{h}$ which is a refinement of both subdivisions. Applying (8.31) to $\hat{h}$ and $\acute{h}$, we have:

$$\|y_h(z) - y_{\acute{h}}(z)\| \leq \|y_h(z) - y_{\hat{h}}(z)\| + \|y_{\hat{h}}(z) - y_{\acute{h}}(z)\| \leq 2\frac{\varepsilon}{L}(e^{L(z-z_0)} - 1).$$

For $\varepsilon > 0$ small enough, this becomes arbitrarily small and shows the uniform convergence of the Euler polygons to a continuous vector function $\phi(z)$.

(2) Let:

$$\varepsilon(\delta) := \sup\{\|F(z_1; y_1) - F(z_2; y_2)\|;\ |z_1 - z_2| \leq \delta;\ \|y_1 - y_2\| \leq N\delta,\ (z_i; y_i) \in D\}.$$

If z belongs to the subdivision h, then we can obtain:

$$\|y_h(z + \delta) - y_h(z) - \delta F(z; y_h(z))\| \leq \varepsilon(\delta)\delta.$$

By the limit $|h| \to 0$,

$$\|\phi(z + \delta) - \phi(z) - \delta F(z; \phi(z))\| \leq \varepsilon(\delta)\delta.$$

Since $\varepsilon(\delta) \to 0$ for $\delta \to 0$, this proves the differentiability of $\phi(z)$ and $\phi'(z) = F(z; \phi(z))$.

(3) From (2), the solution of the primal problem exists. Let $\phi(z)$ and $\varphi(z)$ be solutions of the primal problem, then:

$$\phi(z) = y_0 + \int_{z_0}^{z} F(x; \phi(x))dx \qquad (8.32)$$

and

$$\varphi(z) = y_0 + \int_{z_0}^{z} F(x;\varphi(x))dx. \tag{8.33}$$

From (8.32), (8.33) and Lipschitz condition of Theorem 8.1, we have:

$$\|\phi(z) - \varphi(z)\| \leq L\left|\int_{z_0}^{z} \|\phi(x) - \varphi(x)\|dx\right|. \tag{8.34}$$

Let:

$$g(z) = \int_{z_0}^{z} \|\phi(x) - \varphi(x)\|dx, \quad z \geq z_0.$$

Thus,

$$g'(z) = \|\phi(z) - \varphi(z)\|.$$

Therefore, (8.34) can be revised as:

$$g'(z) \leq Lg(z).$$

So,

$$(e^{-L(z-z_0)}g(z))' \leq 0,$$

thus,

$$e^{-L(z-z_0)}g(z) \leq g(z_0) = 0.$$

When $z \geq z_0$, $g(z) \geq 0$, thus:

$$g(z) \equiv 0, \quad z \geq z_0.$$

Therefore,

$$\phi(z) = \varphi(z).$$

8.2.4 Solving the model

To simplify the calculation, we divide the well into several short segments of the same length h. The length of a segment varies with variations in wall thickness, hole diameter, the fluid density inside and outside the pipe and well geometry. The model begins with a calculation at the bottom of the pipe. Gas pressure and temperature calculations are then performed for each successive "segment" of the pipe up to the top. We use the 4 order Runge-Kutta method to solve the model. Based on the afore-mentioned discussion, an algorithm is designed as follows.

Step 1. Obtain each point's inclination

$$\theta_j = \theta_{j-1} + \frac{\theta_k - \theta_{k-1}}{\Delta s_k}\Delta s_j,$$

where j represents a calculated segment point, s_k represents the measurement of the depth of the inclination angle θ_k and θ_{k-1}, Δs_j the step length of calculation.

Step 2. Calculate the gas condensation parameter Z_g:

If ($P < 35$ MPa)

$$Z_g = 1 + \left(0.31506 - \frac{1.0467}{T_{pr}} - \frac{0.5783}{T_{pr}^3}\right)\rho_{pr} + \left(0.053 - \frac{0.6123}{T_{pr}}\right)\rho_{pr}^2 + 0.6815\frac{\rho_{pr}^2}{T_{pr}^3},$$

where, $\rho_{pr} = 0.27\frac{P_{pr}}{T_{pr}}$, $T_{pr} = \frac{T}{T_{pc}}$, $P_{pr} = \frac{P}{P_{pc}}$.

Else

$$Z_g = (90.7x - 242.2x^2 + 42.4x^3)y^{(1.18+2.82x)} - (14.76x - 9.76x^2 + 4.58x^3)y + \frac{1+y+y^2+y^3}{(1-y)^3},$$

where,

$$\begin{aligned} F(y) &= -0.06125P_{pr}xe^{-1.2(1-x)^2} + (90.7x - 242.2x^2 + 42.4x^3)y^{2.18+2.82x} \\ &\quad + \frac{y+y^2+y^3-y^4}{(1-y^3)} - (14.76x - 9.76x^2 + 4.58x^3)y^2 \\ &= 0 \end{aligned}$$

$$x = \frac{1}{T_{pr}}.$$

Step 3. Calculate the dryness fraction of gas x:

$$x = \frac{1.205v_g}{\frac{1000\gamma_l}{GWR} + 1.205\gamma_g},$$

where GWR is a gas-liquid ratio, γ_g and γ_l are of relative density of the gas and liquid respectively.

Step 4. Calculate gas density ρ_m:

$$\rho_m = \frac{M_t}{V_t},$$

where,

$$\begin{aligned} M_t &= 1000\gamma_l + 1.205GWR \cdot \gamma_g, \\ v_t &= 1 + 0.0003458 \cdot \frac{ZT}{P} \cdot GWR. \end{aligned}$$

Step 5. Obtain gas velocity v_m:

$$v_m = \frac{q_t}{A},$$

where,

$$q_t = \left(1 + 0.0003458 \cdot \frac{ZT}{P}\right) \cdot \frac{Q_{gsc}}{GWR}$$

A is the area of pipe, Q_{gsc} is the quantity of gas.

Step 6. Calculate the gas heat ratio C_{P_g}:

$$C_{P_g} = 1697.5107P^{0.0661}T^{0.0776}.$$

Step 7. Calculate the gas-liquid mixture heat ratio C_{P_m}:

$$C_{P_m} = C_{P_g}x + C_{P_l}(1-x),$$

where, C_{pl} is the liquid heat ratio,

$$C_{pl} = 4.2\,\text{kJ}/(\text{kg} \cdot {}^\circ\text{C}).$$

Step 8. Calculate the gas-liquid mixture J-T parameter C_{J_m}:

$$C_{J_m} = -\frac{xC_{P_g}}{C_{P_m}}C_{J_l} + \frac{1-x}{\rho_l C_{P_m}},$$

where,

$$C_{J_l} = -\frac{1}{C_{P_l}\rho_l}.$$

Step 9. Let the right parts of the coupled differential equations to be functions F_i, where $(i = 1, 2)$. Then we can obtain a system of coupled functions as follows:

$$\begin{cases} F_1 = \dfrac{\rho_m g \cos\theta - \frac{f_m \rho_m v_m^2}{2d} - \frac{\rho_m v_m^2}{T} F_2}{1 - \frac{\rho_m v_m^2}{P}} \\ F_2 = \dfrac{g\cos\theta + \left(\frac{v_m^2}{P} - C_{J_m} C_{P_m}\right) F_1 - \alpha(T - T_e)}{C_{P_m} + \frac{v_m^2}{T}} \end{cases}$$

where,

$$\alpha = \frac{2\pi r_{to} U_{to} Ke}{r_{to} U_{to} f(t_D) + Ke},$$

$$U_{ti}^{-1} = \frac{1}{hc + hr} + \frac{r_{ti} \ln\left(\frac{r_{cem}}{r_{co}}\right)}{k_{cem}} + \frac{r_{ti} \ln\left(\frac{r_{ci}}{r_{to}}\right)}{k_{ang}}$$

and

$$C_J = \frac{R}{C_P} \frac{(2r_A - r_B T - 2r_B BT)Z - (2r_A B + r_B AT)}{3Z^2 - 2Z + A - B - B^2)T}$$

$$A = \frac{r_A P}{T}, \quad B = \frac{r_B P}{T}, \quad r_A = \frac{0.42747\alpha T_{pc}^2}{P_{pc}}, \quad r_B = \frac{0.08664 C_b T_{pc}}{P_{pc}},$$

$$\alpha_i = [1 + m(1 - T_{pr}^{0.5})]^2, \quad m = 0.48 + 1.574w - 0.176w^2$$

and

$$\begin{cases} f(t_D) = 1.1281\sqrt{t_D}\,(1 - 0.3\sqrt{t_D}) \quad (t_D \le 1.5) \\ f(t_D) = \left(1 + \dfrac{0.6}{t_D}\right)[0.4063 + 0.5\ln(t_D)] \quad (t_D > 1.5) \end{cases}$$

$$t_D = \frac{t\alpha}{r_{wb}^2}$$

and

$$\frac{1}{\sqrt{f_m}} = 1.14 - 2\lg\left(\frac{0.00001524}{r_{ti}} + \frac{21.25}{Re^{0.9}}\right).$$

Step 10. Assume that P, T to be y_i $(i = 1, 2)$, respectively. Then we can obtain some basic parameters as follows:

$$\begin{cases} a_i = F_i[y_1, y_2, y_3, y_4] \\ b_i = F_i\left[y_1 + \dfrac{h}{2}a_1, y_2 + \dfrac{h}{2}a_2, y_3 + \dfrac{h}{2}a_3, y_4 + \dfrac{h}{2}a_4\right] \end{cases}$$

Step 11. Calculate the gas-liquid mixture's dry degree, pressure and temperature at point $(j + 1)$:

$$y_i^{(j+1)} = y_i^j + \frac{h}{6}(a_i + 2b_i + 2c_i + d_i) \quad (i = 1, 2, 3, 4) \quad (j = 1, 2, \ldots, n).$$

Step 12. Repeat steps 1 to 11 until y_i^n is calculated.

Table 8.1 Pipe parameters.

Diameter [mm]	Thickness [mm]	Weight [kg]	Expansion coefficient [m/m·K]	Young's modulus [GPa]	Poisson's ratio [–]	Using length [m]
88.9	12.95	23.791	0.0000115	215	0.3	700
88.9	9.53	18.28	0.0000115	215	0.3	2850
88.9	7.34	15.043	0.0000115	215	0.3	1430
88.9	6.45	13.582	0.0000115	215	0.3	600
88.9	6.45	13.582	0.0000115	215	0.3	350
73	5.51	9.493	0.0000115	215	0.3	185

Table 8.2 Well parameters.

Measured depth [m]	Internal diameter [mm]	External diameter [mm]
3301.7	154.78	177.8
5936.83	152.5	177.8
6115	108.62	127

8.2.5 **Numerical simulation**

As was described previously, the algorithm begins with a calculation for the pipe at the bottom of the pipe. The calculations are performed for each successive "segment" of the pipe from the bottom to the top.

8.2.5.1 *Parameters*

In this simulation, we study a pipe in X well, which is in Sichuan province, China. All the needed parameters are given as follows:

Depth of the well = 6115 m
The well Bottom pressure = 40 MPa
Critical pressure = 4.968 MPa
Gas specific weight = 0.6434 kg/m^3
Ground thermal conductivity parameter = 2.06 W/m·K
Gas-liquid ratio = 100000
Geothermal gradient = 2.18 °C/100 m
Gas outputs = 500000 m^3/day
Length of one segment = 1 m

The parameters of the pipes, the inclined well, the inclination, the azimuth and vertical depth are as follows.

8.2.5.2 *Main results*

After calculation, we obtain a series of results for this well. The detailed source code can be seen in **Appendix 4**. Gas pressures and temperatures as shown in Table 8.4.

At the same time, we obtain a comparative result with a measurement value.

The relative error is as shown in Table 8.7.

8.2.6 **Sensitivity analysis**

To study how different gas-liquid ratios, ground thermal conductivity parameters, geothermal gradients and gas outputs influence the gas pressure and the temperature, different gas-liquid

Table 8.3 Azimuth, inclination and vertical depth.

Number	Measured depth [m]	Inclination [°]	Azimuth [°]	Vertical depth [m]
1	1000	2.82	240.84	999.88
2	1200	2.28	237.69	1199.53
3	1300	1.13	213.69	1299.49
4	2800	1.19	26.21	2799.41
5	3000	1.74	44.39	2999.25
6	3400	1.92	190.95	3399.21
7	3900	1.98	268.9	3899.14
8	4000	2.00	297.38	3999.11
9	4100	4.68	324.34	4098.96
10	4200	1.97	302.88	4198.74
11	4300	1.03	204.57	4298.7
12	4400	1.54	164.16	4398.68
13	4500	2.37	195.11	4498.61
14	4600	2.12	214.67	4598.54
15	4700	1.96	216.31	4698.47
16	4800	3.04	229.14	4798.38
17	4900	3.59	243.86	4898.23
18	5000	5.79	366.45	4997.87
19	5100	8.14	258.61	5097.01
20	5200	7.01	236.71	5196.1
21	5300	5.78	239.1	5295.5
22	5400	5.05	244.42	5395.04
23	5500	3.92	228.03	5494.72
24	5600	4.44	233.71	5594.49
25	5700	5.03	234.87	5694.17
26	5800	5.13	233.21	5793.77
27	5900	4.53	234.82	5893.44
28	6000	3.67	232.4	5993.2
29	6115	4.94	233.11	6107.88

ratios, ground thermal conductivity parameters, geothermal gradients and gas outputs were used. Using the algorithm and simulation, a series of results was obtained.

Firstly, we used three different gas-liquid ratios; 5000, 50000 and 100000; and the other parameters remained the same. The gas pressure and temperature distribution figures were shown in Figures 8.3 and 8.4.

From Figures 8.3 and 8.4, it can be seen that the gas temperature increases from the top to the bottom of the pipe. At the same pipe depth, if the gas-liquid ratio increases, the gas temperatures increase. Gas pressure also increases from the top to the bottom of the pipe. At the same pipe depth, if the gas-liquid ratio increases, pressure also increases.

Secondly, three different ground thermal conductivity parameters were used; 1.7, 2.06 and 2.3 W/m·K; while the other parameters remained the same. Then the gas pressure and temperature distribution figures were determined as shown in Figures 8.5 and 8.6.

From Figures 8.5 and 8.6, it can be seen that the gas temperature increases from the top to the bottom of the pipe. At the same pipe depth, if the ground thermal conductivity parameter increases, then the gas temperature decreases. The gas pressure can be seen to decrease from the top to the bottom of the pipe. At the same pipe depth, if the gas-liquid ratio increases, the pressure also increases.

Then, three different geothermal gradients were used; 2.18, 2.4 and 2.6 °C/100 m, with the other parameters remaining the same. The gas pressure and the temperature distribution figures were as shown in Figures 8.7 and 8.8.

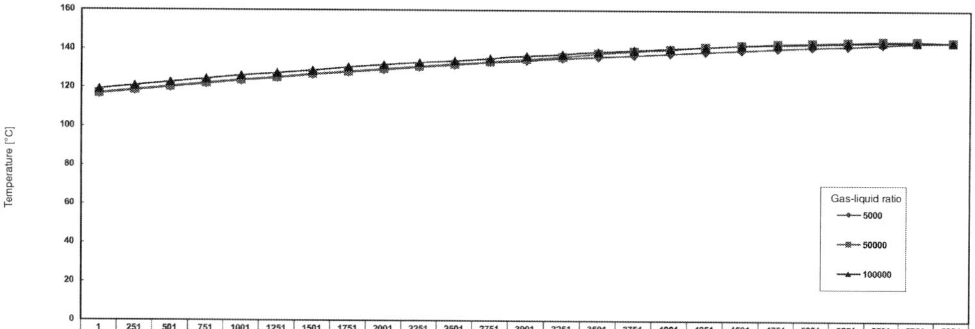

Gas-liquid ratio	1	251	501	751	1001	1251	1501	1751	2001	2251	2501	2751
5000	116.534	118.341	120.088	121.771	123.376	124.915	126.39	127.8	129.149	130.437	131.668	132.847
50000	116.959	118.75	120.514	122.24	123.8	125.302	126.75	128.146	129.497	130.809	132.091	133.361
100000	119.088	120.886	122.626	124.303	125.894	127.42	128.881	130.278	131.614	132.891	134.115	135.294

Gas-liquid ratio	3001	3251	3501	3751	4001	4251	4501	4751	5001	5251	5501	5751	5904
5000	133.98	135.076	136.154	137.095	137.963	138.797	139.606	140.409	141.224	141.929	142.666	143.496	144.106
50000	134.646	136.017	137.792	139.197	140.195	141.151	142.058	142.899	143.627	143.985	144.17	144.182	144.106
100000	136.438	137.568	138.722	139.659	140.484	141.256	141.976	142.645	143.25	143.643	143.925	144.08	144.106

Depth [m]

Figure 8.3 Gas temperature distribution.

Table 8.4 Results of gas temperature and pressure.

Number	Measured depth [m]	Temperature [°C]	Pressure [MPa]
1	1	119.0880	37.7800
2	251	120.8860	37.8770
3	501	122.6260	37.9740
4	751	124.3030	38.0710
5	1001	125.8940	38.1680
6	1251	127.4200	38.2640
7	1501	128.8810	38.3610
8	1751	130.2780	38.4580
9	2001	131.6140	38.5540
10	2251	132.8910	38.6490
11	2501	134.1150	38.7440
12	2751	135.2940	38.8390
13	3001	136.4380	38.9320
14	3251	137.5680	39.0240
15	3501	138.7220	39.1150
16	3751	139.6590	39.2080
17	4001	140.4840	39.3010
18	4251	141.2560	39.3940
19	4501	141.9760	39.4850
20	4751	142.6450	39.5770
21	5001	143.2500	39.6670
22	5251	143.6430	39.7590
23	5501	143.9250	39.8510
24	5751	144.0800	39.9430
25	5904	144.1060	40.0000

Table 8.5 Calculation results.

Location	Temperature [°C]	Pressure [MPa]
Well head	119.0880	37.7800
Well bottom	144.1060	40.0000

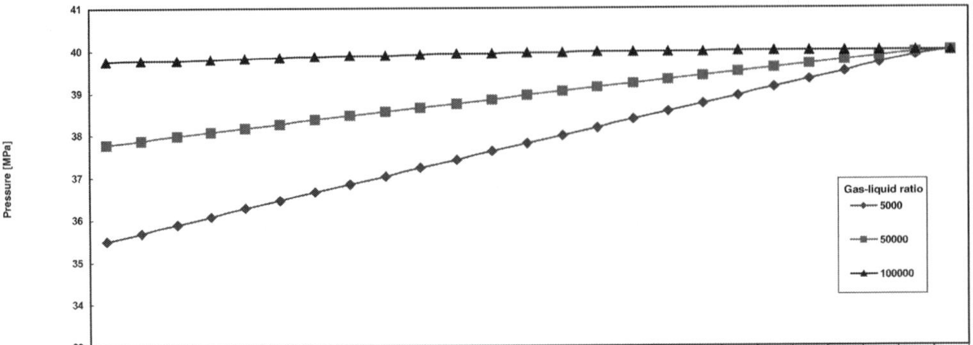

Depth [m]	1	251	501	751	1001	1251	1501	1751	2001	2251	2501	2751
5000	35.492	35.685	35.877	36.069	36.262	36.454	36.647	36.839	37.032	37.224	37.415	37.606
50000	37.78	37.877	37.974	38.071	38.168	38.264	38.361	38.458	38.554	38.649	38.744	38.839
100000	39.738	39.758	39.777	39.796	39.815	39.834	39.852	39.87	39.887	39.904	39.919	39.933

Depth [m]	3001	3251	3501	3751	4001	4251	4501	4751	5001	5251	5501	5751	5904
5000	37.797	37.986	38.173	38.361	38.552	38.741	38.931	39.121	39.31	39.5	39.691	39.883	40
50000	38.932	39.024	39.115	39.208	39.301	39.394	39.485	39.577	39.667	39.759	39.851	39.943	40
100000	39.946	39.957	39.965	39.972	39.976	39.982	39.985	39.992	39.993	39.997	39.998	39.998	40

Depth [m]

Figure 8.4 Gas pressure distribution.

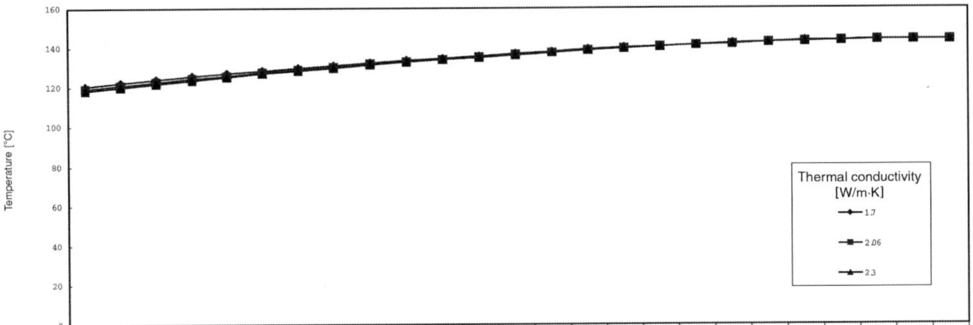

Depth [m]	1	251	501	751	1001	1251	1501	1751	2001	2251	2501	2751
1.7	120.684	122.354	123.971	125.529	126.999	128.408	129.758	131.05	132.287	133.474	134.616	135.722
2.06	119.088	120.886	122.626	124.303	125.894	127.42	128.881	130.278	131.614	132.891	134.115	135.294
2.3	118.194	120.063	121.871	123.614	125.274	126.865	128.388	129.845	131.236	132.565	133.837	135.058

Depth [m]	3001	3251	3501	3751	4001	4251	4501	4751	5001	5251	5501	5751	5904
1.7	136.805	137.886	139.015	139.91	140.688	141.422	142.11	142.755	143.342	143.718	143.989	144.142	144.171
2.06	136.438	137.568	138.722	139.659	140.484	141.256	141.976	142.645	143.25	143.643	143.925	144.08	144.106
2.3	136.24	137.4	138.575	139.541	140.396	141.195	141.939	142.632	143.258	143.672	143.971	144.139	144.171

Depth [m]

Figure 8.5 Gas temperature distribution.

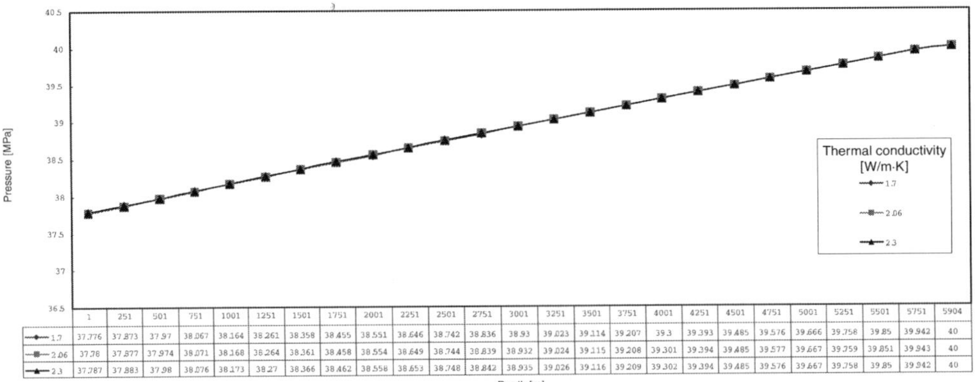

Depth [m]	1	251	501	751	1001	1251	1501	1751	2001	2251	2501	2751
1.7	37.776	37.873	37.97	38.067	38.164	38.261	38.358	38.455	38.551	38.646	38.742	38.836
2.06	37.78	37.877	37.974	38.071	38.168	38.264	38.361	38.458	38.554	38.649	38.744	38.839
2.3	37.787	37.883	37.98	38.076	38.173	38.27	38.366	38.462	38.558	38.653	38.748	38.842

Depth [m]	3001	3251	3501	3751	4001	4251	4501	4751	5001	5251	5501	5751	5904
1.7	38.93	39.023	39.114	39.207	39.3	39.393	39.485	39.576	39.666	39.758	39.85	39.942	40
2.06	38.932	39.024	39.115	39.208	39.301	39.394	39.485	39.577	39.667	39.759	39.851	39.943	40
2.3	38.935	39.026	39.116	39.209	39.302	39.394	39.485	39.576	39.667	39.758	39.85	39.942	40

Depth [m]

Figure 8.6 Gas pressure distribution.

Table 8.6 Measurement results.

Location	Temperature [°C]	Pressure [MPa]
Well head	116.416	36.254
Well bottom	144.1060	40.0000

Table 8.7 Relative error.

Temperature [°C]	Pressure [MPa]
0.022952	0.042092

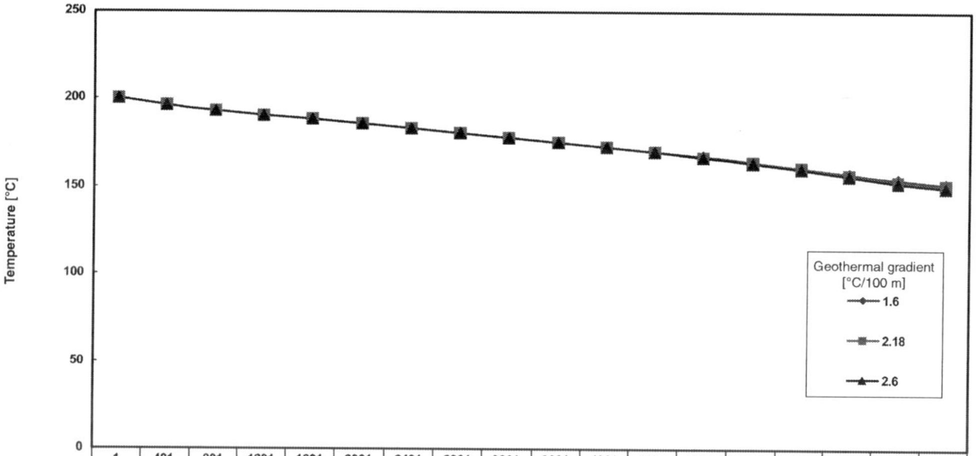

	1	401	801	1201	1601	2001	2401	2801	3201	3601	4001	4401	4801	5201	5601	6001	6401	6801
1.6	200	196.254	193.35	190.723	188.218	185.761	183.309	180.832	178.306	175.712	173.031	170.244	167.331	164.27	161.035	157.599	154.14	152.096
2.18	200	196.246	193.335	190.701	188.192	185.734	183.281	180.804	178.276	175.674	172.977	170.158	167.192	164.044	160.675	157.04	153.312	151.082
2.6	200	196.241	193.324	190.686	188.175	185.715	183.263	180.785	178.254	175.644	172.93	170.082	167.063	163.831	160.331	156.494	152.487	150.052

Depth [m]

Figure 8.7 Gas temperature distribution.

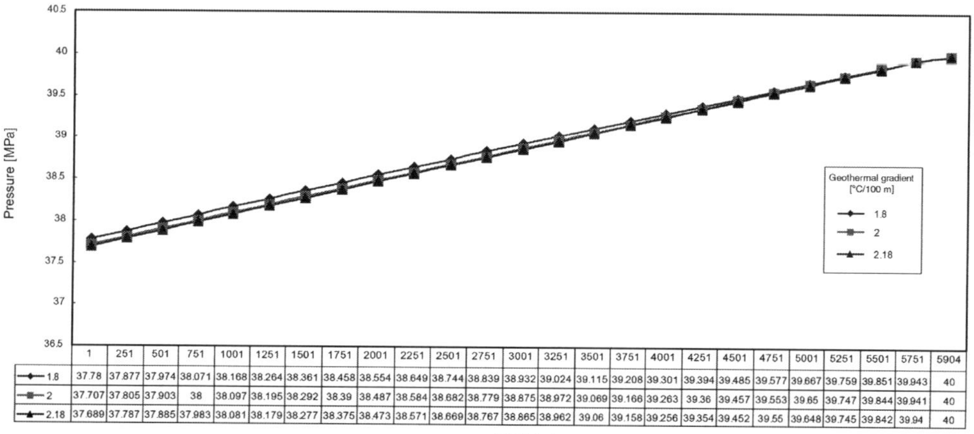

	1	251	501	751	1001	1251	1501	1751	2001	2251	2501	2751	3001	3251	3501	3751	4001	4251	4501	4751	5001	5251	5501	5751	5904
1.8	37.78	37.877	37.974	38.071	38.168	38.264	38.361	38.458	38.554	38.649	38.744	38.839	38.932	39.024	39.115	39.208	39.301	39.394	39.485	39.577	39.667	39.759	39.851	39.943	40
2	37.707	37.805	37.903	38	38.097	38.195	38.292	38.39	38.487	38.584	38.682	38.779	38.875	38.972	39.069	39.166	39.263	39.36	39.457	39.553	39.65	39.747	39.844	39.941	40
2.18	37.689	37.787	37.885	37.983	38.081	38.179	38.277	38.375	38.473	38.571	38.669	38.767	38.865	38.962	39.06	39.158	39.256	39.354	39.452	39.55	39.648	39.745	39.842	39.94	40

Depth [m]

Figure 8.8 Gas pressure distribution.

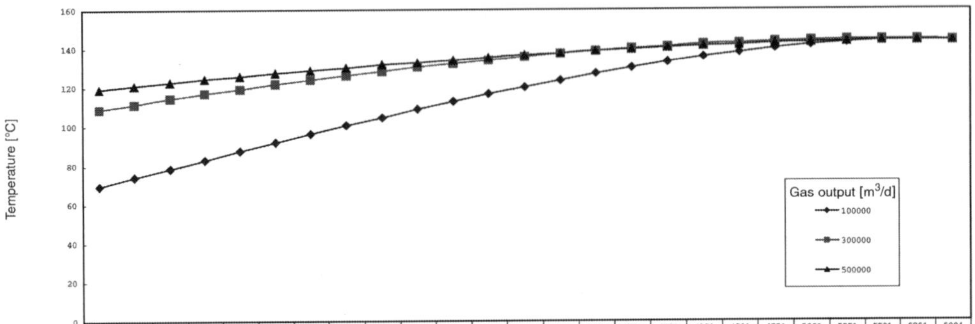

Gas output [m³/d]	1	251	501	751	1001	1251	1501	1751	2001	2251	2501
100000	69.35	73.952	78.515	83.031	87.494	91.897	96.231	100.486	104.654	108.723	112.679
300000	108.846	111.596	114.27	116.861	119.367	121.783	124.104	126.327	128.446	130.458	132.356
500000	119.088	120.886	122.626	124.303	125.894	127.42	128.881	130.278	131.614	132.891	134.115

Gas output [m³/d]	2751	3001	3251	3501	3751	4001	4251	4501	4751	5001	5251	5501	5751	5904
100000	116.507	120.193	123.717	127.057	130.191	133.095	135.74	138.096	140.126	141.793	143.059	143.874	144.187	144.106
300000	134.137	135.796	137.325	138.72	139.974	141.083	142.04	142.839	143.474	143.94	144.23	144.337	144.254	144.106
500000	135.294	136.438	137.568	138.722	139.659	140.484	141.256	141.976	142.645	143.25	143.643	143.925	144.08	144.106

Depth [m]

Figure 8.9 Gas temperature distribution.

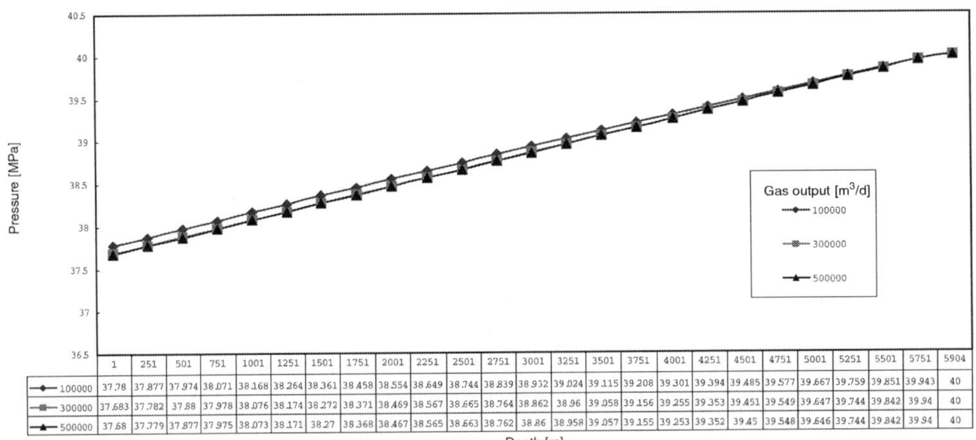

Gas output [m³/d]	1	251	501	751	1001	1251	1501	1751	2001	2251	2501
100000	37.78	37.877	37.974	38.071	38.168	38.264	38.361	38.458	38.554	38.649	38.744
300000	37.683	37.782	37.88	37.978	38.076	38.174	38.272	38.371	38.469	38.567	38.665
500000	37.68	37.779	37.877	37.975	38.073	38.171	38.27	38.368	38.467	38.565	38.663

Gas output [m³/d]	2751	3001	3251	3501	3751	4001	4251	4501	4751	5001	5251	5501	5751	5904
100000	38.839	38.932	39.024	39.115	39.208	39.301	39.394	39.485	39.577	39.667	39.759	39.851	39.943	40
300000	38.764	38.862	38.96	39.058	39.156	39.255	39.353	39.451	39.549	39.647	39.744	39.842	39.94	40
500000	38.762	38.86	38.958	39.057	39.155	39.253	39.352	39.45	39.548	39.646	39.744	39.842	39.94	40

Depth [m]

Figure 8.10 Gas pressure's distribution.

From Figures 8.7 and 8.8, it can be seen that the gas temperature increases from the top to the bottom of the pipe. At the same pipe depth, if the geothermal gradient increases, then the gas temperature also increases. The gas pressure increases from the top to the bottom of the pipe. At the same pipe depth, if the gas-liquid ratio increases, the pressure decreases.

Lastly, three different gas outputs were used: 100000, 300000 and 500000 m^3/day; and the other parameters remained the same. The gas pressure and temperature distribution figures were determined as shown in Figures 8.9 and 8.10.

From Figures 8.9 and 8.10, it can be seen that the gas temperature increases from the top to the bottom of the pipe. At the same pipe depth, if the gas output increases, temperature also increases. The gas pressure increases from the top to the bottom of the pipe and at the same pipe depth, if the gas-liquid ratio increases, the pressure decreases.

8.3 PTPTV-GW

For the coupled system of differential equations model (ρ, V, P, T), we set the following assumptions:

1. There is a steady one dimensional in the tubing, and all featured parameters are homochromatic at any transverse cross section.
2. Heat transfer in one dimension is steady from the tubing to the second interface, but the heat transfer in one dimension is unsteady from the second interface to the surrounding formation.
3. The heat losses from the tubing and the surrounding earth are radial.
4. The vertical formation temperature is considered to be linearly distributed based on known geothermal gradients.

8.3.1 The coupled system differential equations model

Consider the flow system depicted in Figure 8.1.

A straight cylindrical flow tube with an inclination angle θ, a constant cross-sectional flow area A, a hydraulic diameter, and a total length Z. Through this tubing gas flows from the bottom to the top with a mass flow rate w. The distance co-ordinate in the flow direction along the tubing is denoted z.

Mass, momentum, and energy balances, along with the density, pressure, volume, temperature (P, ρ, V, T) relation of the gas, are used to generate the constitutive equations.

Material balance: The amount of gas in a given control volume is the product of volume and density. Volume is given by the product of the area (or average area, when the area varies with the length) and the gas velocity. Thus, under steady-state conditions, the mass balance equation for a differential depth, dz, of the well in terms of gas density ρ, and velocity v is written as:

$$\frac{d(\rho A v)}{dz} = 0$$

i.e.,

$$\rho \frac{dv}{dz} + v \frac{d\rho}{dz} = 0. \tag{8.35}$$

Momentum balance: For a steady flow system, the pressure gradient $\frac{dP}{dz}$, is balanced by the static head, $\rho g \cos\theta$, the friction head, $\frac{\rho f v^2}{2d}$, and the kinetic head, $(\rho v)\left(\frac{dv}{dz}\right)$. Thus,

$$\frac{dP}{dz} = -\rho v \frac{dv}{dz} + \rho g \cos\theta - \frac{\rho f v^2}{2d}$$

i.e.,

$$\frac{1}{\rho}\frac{dP}{dz} = -v \frac{dv}{dz} + g \cos\theta - \frac{f v^2}{2d}. \tag{8.36}$$

Energy balance: We take the bottom of well as the origin of the coordinate axis and the vertical as the positive direction. We take the distance element of length dz of the tubing, then fluid containing energy flowing into the distance element of the tubing: the inner energy $E(z)$, the kinetic energy $\frac{1}{2}mv^2(z)$, potential energy $mgz\cos\theta$ (θ is inclination angle), and pressure energy $P(z)V(z)$. The sum of inner energy and pressure energy is a flux of fluid, $H(z) = E(z) + P(z)V(z)$. When fluid flows out from distance element, then energy contains: inner energy $E(z + dz)$, the pressure gradient, $\frac{dP}{dz}$, the kinetic energy $\frac{1}{2}mv^2(z + dz)$, potential energy $mg(z + dz)\cos\theta$, and pressure energy $P(z + dz)V(z + dz)$. Let dQ be the radial transfer of heat of tubing.

According to the rule of energy balance: the energy of fluid flowing into the infinitesimal equals the energy sum of losses and fluid flowing out from the infinitesimal.

In particular, for flowing gas, we have:

$$H(z)+\frac{1}{2}mv^2(z)-mgz\cos\theta=H(z+dz)+\frac{1}{2}mv^2(z+dz)-mg(z+dz)\cos\theta+dQ. \quad (8.37)$$

From (8.37), we get:

$$-\frac{dQ}{dz}=\frac{dH}{dz}+mv\frac{dv}{dz}-mg\cos\theta. \quad (8.38)$$

(8.38) divided by m, we have:

$$\frac{dq}{dz}=-\frac{dh}{dz}-v\frac{dv}{dz}+g\cos\theta. \quad (8.39)$$

Ramey (1962) and Willhite (1967) discussed the radial transfer of heat between fluid and earth in detail. Over the differential element dz shown in Figure 8.2 the radial transfer of heat from fluid to the cement/earth interface can be described by:

$$dq=\frac{2\pi r_{to}U_{to}}{w}(T-T_k)dz, \quad (8.40)$$

the radial transfer of heat from the cement/earth interface to the surrounding earth is:

$$dq=\frac{2\pi K_e(T_k-T_e)}{wf(t_D)}dz. \quad (8.41)$$

Combining Equation (8.40) and (8.41) gives the equation for the radial transfer of heat between the fluid and the surrounding earth:

$$\frac{dq}{dz}=\frac{2\pi r_{to}U_{to}K_e}{w[k_e+f(t_D)r_{to}U_{to}]}(T-T_e). \quad (8.42)$$

Using basic thermodynamic principles, we can obtain the enthalpy term of Equation (8.26) to:

$$\frac{dh}{dz}=\left(\frac{\partial h}{\partial P}\right)_T\frac{dP}{dz}+\left(\frac{\partial h}{\partial T}\right)_P\frac{dT}{dz}, \quad (8.43)$$

where:

$$\left(\frac{\partial h}{\partial T}\right)_P=C_P. \quad (8.44)$$

From the definition of the J-T coefficient, we have:

$$C_J=\left(\frac{\partial T}{\partial P}\right)_T=-\frac{(\partial h/\partial P)_T}{(\partial h/\partial T)_P},$$

thus,

$$\left(\frac{\partial h}{\partial P}\right)_T=-C_JC_P. \quad (8.45)$$

Therefore, the overall enthalpy change in a flowing fluid is:

$$\frac{dh}{dz}=-C_JC_P\frac{dP}{dz}+C_P\frac{dT}{dz}. \quad (8.46)$$

Combining Equations (8.39), (8.42) and (8.46), we can obtain:

$$\frac{2\pi r_{to}U_{to}K_e}{w[k_e+f(t)r_{to}U_{to}]}(T-T_e)-C_JC_P\frac{dP}{dz}+C_P\frac{dT}{dz}+v\frac{dv}{dz}-g\cos\theta=0. \quad (8.47)$$

Let $a = \frac{2\pi r_{to} U_{to} K_e}{w[k_e + f(t) r_{to} U_{to}]}$, thus Equation (8.47) can be written as:

$$a(T - T_e) - C_J C_P \frac{dP}{dz} + C_P \frac{dT}{dz} + v\frac{dv}{dz} - g\cos\theta = 0. \tag{8.48}$$

From state equation of gas, $\rho = \frac{MP}{RZ_g T}$, we have:

$$T\frac{d\rho}{dz} + \rho\frac{dT}{dz} = \frac{M}{RZ_g}\frac{dP}{dz}. \tag{8.49}$$

Combining (8.35), (8.36), (8.48) and (8.49), we can obtain the coupled system of differential equations for gas density, gas velocity, gas pressure and temperature:

$$\begin{cases} \dfrac{d\rho}{dz} = \dfrac{\left(C_J\rho - \frac{M}{RZ_g}\right)\left[\rho g\cos\theta + \frac{f\rho v^2}{2d}\right] + \frac{\rho a(T-T_e) - \rho g\cos\theta}{C_P}}{T + v^2\left[\frac{1}{C_p} + C_J\rho - \frac{M}{RZ_g}\right]} \\ \dfrac{dv}{dz} = -\dfrac{v}{\rho}\dfrac{d\rho}{dz} \\ \dfrac{dP}{dz} = v^2\dfrac{d\rho}{dz} - \rho g\cos\theta - \dfrac{f\rho v^2}{2d} \\ \dfrac{dT}{dz} = C_J\dfrac{dP}{dz} + \dfrac{\left[\frac{v^2}{\rho}\frac{d\rho}{dz} + g\cos\theta - a(T - T_e)\right]}{C_P} \end{cases}. \tag{8.50}$$

The primal conditions being as follows:

$$P(z_0) = P_0, \quad T(z_0) = T_0, \quad \rho(z_0) = \frac{MP_0}{RZ_g}, \quad v(z_0) = \frac{w}{A\rho_0}.$$

In the following section, the solution of the coupled system model will be discussed by the Euler approximate method and three theorems will be produced.

8.3.2 Solution of the model

Let:

$$\frac{d\rho}{dz} = f_1(z;\rho,v,P,T), \quad \frac{dv}{dz} = f_2(z;\rho,v,P,T), \quad \frac{dP}{dz} = f_3(z;\rho,v,P,T), \quad \frac{dT}{dz} = f_4(z;\rho,v,P,T).$$

Denote:

$$F = (f_1, f_2, f_3, f_4)^T, \quad y = (\rho, v, P, T)^T, \quad y(0) = y(z_0) = (\rho(z_0), v(z_0), P(z_0), T(z_0))^T,$$

then, the systems of ordinary differential equations can be written as:

$$y' = F(z;y), \quad y(0) = y(z_0).$$

The norm of vector f is given as follows,

$$\|F\| = \max(|f_1|, |f_2|, |f_3|, |f_4|).$$

For f_1, f_2, f_3, f_4, there is:

$$|f_1| = \frac{\left|\left(C_J\rho - \frac{M}{RZ_g}\right)\left(\rho g\cos\theta - \frac{f\rho v^2}{2d}\right) + \frac{\rho a(T-T_e) - \rho g\cos\theta}{C_P}\right|}{\left|T + v^2\left(\frac{1}{C_P} + C_J\rho - \frac{M}{RZ_g}\right)\right|} \leq \frac{|\rho||K_1||K_2| + |\frac{\rho}{C_P}||K_3|}{|K_4|}$$

$$= \frac{|\rho|(|K_1||K_2| + |\frac{1}{C_P}||K_3|)}{|K_4|},$$

where, $|K_1|=\left|C_J\rho-\frac{M}{RZ_g}\right|\leq|C_J\rho|+\left|\frac{M}{RZ_g}\right|$, $|K_2|=\left|g\cos\theta-\frac{fv^2}{2d}\right|\leq|g\cos\theta|+\left|\frac{fv^2}{2d}\right|\leq g+\left|\frac{fv^2}{2d}\right|$, $|K_3|=|a(T-T_e)-g\cos\theta|\leq|a(T-T_e)|+g$, $|K_4|=\left|T+v^2\left(\frac{1}{C_P}+C_J\rho-\frac{M}{RZ_g}\right)\right|$.

Since all parameter are bounded quantities, $|K_1|, |K_2|, |K_3|, |\rho|, |\frac{1}{\rho}|, |K_4|$ are bounded.

Let:

$$N_1=\sup\left\{\frac{|\rho||K_1||K_2|+\left|\frac{\rho}{C_P}\right||K_3|}{|K_4|}\right\},$$

then,

$$|f_1|\leq N_1.$$

Similarly,

$$|f_2|=\left|-\frac{v}{\rho}f_1\right|=\left|\frac{v}{\rho}\right||f_1|\leq\left|\frac{v}{\rho}\right|N_1.$$

Let:

$$N_2=\sup\left\{\left|\frac{v}{\rho}\right|N_1\right\},$$

then, there is:

$$|f_2|\leq N_2.$$

Similarly,

$$|f_3|=\left|v^2f_1-\rho g\cos\theta-\frac{f\rho v^2}{2d}\right|\leq|v^2||f_1|+|\rho g\cos\theta|+\left|\frac{f\rho v^2}{2d}\right|\leq|v^2||f_1|+|\rho g|+\left|\frac{f\rho v^2}{2d}\right|.$$

Let:

$$N_3=\sup\left\{|v^2||f_1|+|\rho g|+\left|\frac{f\rho v^2}{2d}\right|\right\},$$

then,

$$|f_3|\leq N_3.$$

Similarly,

$$\begin{aligned}|f_4| &= \left|C_Jf_3+\frac{\frac{v^2}{\rho}f_1+g\cos\theta-a(T-T_e)}{C_P}\right|\leq|C_Jf_3|+\frac{\left|\frac{v^2}{\rho}f_1\right|+|g\cos\theta|+|a(T-T_e)|}{|C_P|}\\ &\leq |C_J|N_3+\frac{\left|\frac{v^2}{\rho}\right|N_1+g+|a(T-T_e)|}{|C_P|}.\end{aligned}$$

Let:

$$N_4=\sup\left\{|C_J|N_3+\frac{\left|\frac{v^2}{\rho}\right|N_1+g+|a(T-T_e)|}{|C_P|}\right\},$$

then, there is

$$|f_4|\leq N_4.$$

Therefore,

$$\|F\|\leq\max\{N_1,N_2,N_3,N_4\}.$$

The partial differential of f_1, f_2, f_3, f_4 about ρ, v, P, T are bounded is discussed as follows:

$$\frac{\partial f_1}{\partial \rho} = \frac{\left(2C_J\rho - \frac{M}{RZ_g}\right)\left(g\cos\theta - \frac{fv^2}{2d}\right) + \frac{a(T-T_e)-g\cos\theta}{C_P}}{T + v^2\left(\frac{1}{C_P} + C_J\rho - \frac{M}{RZ_g}\right)} - \frac{C_J\left[\left(C_J\rho - \frac{M}{RZ_g}\right)\left(\rho g\cos\theta - \frac{f\rho v^2}{2d}\right) + \frac{\rho a(T-T_e)-\rho g\cos\theta}{C_P}\right]}{\left[T + v^2\left(\frac{1}{C_P} + C_J\rho - \frac{M}{RZ_g}\right)\right]^2}$$

$$\frac{\partial f_1}{\partial v} = \frac{-\frac{f\rho v}{d}\left(C_J\rho - \frac{M}{RZ_g}\right)}{T + v^2\left(\frac{1}{C_P} + C_J\rho - \frac{M}{RZ_g}\right)} - \frac{2v\left(\frac{1}{C_P} + C_J\rho - \frac{M}{RZ_g}\right)\left[\left(C_J\rho - \frac{M}{RZ_g}\right)\left(\rho g\cos\theta - \frac{f\rho v^2}{2d}\right) + \frac{\rho a(T-T_e)-\rho g\cos\theta}{C_P}\right]}{\left[T + v^2\left(\frac{1}{C_P} + C_J\rho - \frac{M}{RZ_g}\right)\right]^2}$$

$$\frac{\partial f_1}{\partial P} = \frac{\rho\frac{\partial C_J}{\partial P}\left(\rho g\cos\theta - \frac{f\rho v^2}{2d}\right) - \frac{\rho a(T-T_e)-\rho g\cos\theta}{C_P^2}\frac{\partial C_P}{\partial P}}{T + v^2\left(\frac{1}{C_P} + C_J\rho - \frac{M}{RZ_g}\right)} - \frac{v^2\left(-\frac{1}{C_P^2}\frac{\partial C_P}{\partial P} + \rho\frac{\partial C_J}{\partial P}\right)\left[\left(C_J\rho - \frac{M}{RZ_g}\right)\left(\rho g\cos\theta - \frac{f\rho v^2}{2d}\right) + \frac{\rho a(T-T_e)-\rho g\cos\theta}{C_P}\right]}{\left[T + v^2\left(\frac{1}{C_P} + C_J\rho - \frac{M}{RZ_g}\right)\right]^2}$$

$$\frac{\partial f_1}{\partial T} = \frac{\rho\frac{\partial C_J}{\partial T}\left(\rho g\cos\theta - \frac{f\rho v^2}{2d}\right) + \frac{\rho a C_P-(\rho a(T-T_e)-\rho g\cos\theta)\frac{\partial C_P}{\partial T}}{C_P^2}}{T + v^2\left(\frac{1}{C_P} + C_J\rho - \frac{M}{RZ_g}\right)} - \frac{\left[1 + v^2\left(\frac{-1}{C_P^2}\frac{\partial C_P}{\partial T} + \rho\frac{\partial C_J}{\partial T}\right)\right]\left[\left(C_J\rho - \frac{M}{RZ_g}\right)(\rho g\cos\theta - \frac{f\rho v^2}{2d}) + \frac{\rho a(T-T_e)-\rho g\cos\theta}{C_P}\right]}{\left[T + v^2\left(\frac{1}{C_P} + C_J\rho - \frac{M}{RZ_g}\right)\right]^2}.$$

Then,

$$\left|\frac{\partial f_1}{\partial \rho}\right| = \left|\frac{(C_J\rho + K_1)K_2 + \frac{K_3}{C_P}}{K_4} - \frac{C_J(K_1K_2 + K_3)}{K_4^2}\right| \leq \frac{|C_J\rho||K_1| + |K_1||K_2| + \frac{|K_3|}{|C_P|}}{|K_4|} + \frac{|C_J|(|K_1||K_2| + |K_3|)}{K_4^2}.$$

Let:

$$M_{11} = \sup\left\{\frac{|C_J\rho||K_1| + |K_1||K_2| + \frac{|K_3|}{|C_P|}}{|K_4|} + \frac{|C_J|(|K_1||K_2| + |K_3|)}{K_4^2}\right\},$$

then,

$$\left|\frac{\partial f_1}{\partial \rho}\right| \leq M_{11}.$$

Similarly,

$$\left|\frac{\partial f_1}{\partial v}\right| = \left|-\frac{2\rho v\left(\frac{1}{C_P}+K_1\right)(K_1K_2+K_3)}{K_4^2} - \frac{f\rho v K_1}{dK_4}\right|$$

$$\leq \frac{|f\rho v||K_1|}{|dK_4|} + \frac{2|\rho v|\left(\left|\frac{1}{C_P}\right|+|K_1|\right)(|K_1||K_2|+|K_3|)}{K_4^2}.$$

Let:

$$M_{12} = \sup\left\{\frac{|f\rho v||K_1|}{|dK_4|} + \frac{2|\rho v|\left(\left|\frac{1}{C_P}\right|+|K_1|\right)(|K_1||K_2|+|K_3|)}{K_4^2}\right\},$$

thus:

$$\left|\frac{\partial f_1}{\partial v}\right| \leq M_{12}.$$

Similarly, there have:

$$\left|\frac{\partial f_1}{\partial P}\right| = \left|\frac{\rho^2\frac{\partial C_J}{\partial P}K_2 - \frac{K_3}{C_P}\frac{\partial C_P}{\partial P}}{K_4} - \frac{v\left(-\frac{1}{C_P^2}\frac{\partial C_P}{\partial P}+\rho\frac{\partial C_J}{\partial P}\right)(K_1K_2+K_3)}{K_4^2}\right|$$

$$\leq \frac{\left|(\rho)^2\left|\frac{\partial C_J}{\partial P}\right||K_2| + \left|\frac{K_3}{C_P}\right|\frac{\partial C_P}{\partial P}\right|}{|K_4|} + \frac{|v|\left(\frac{1}{C_P^2}\left|\frac{\partial C_P}{\partial P}\right|+|\rho|\left|\frac{\partial C_J}{\partial P}\right|\right)(|K_1||K_2|+|K_3|)}{K_4^2}.$$

Let:

$$M_{13} = \sup\left\{\frac{\left|(\rho)^2\left|\frac{\partial C_J}{\partial P}\right||K_2| + \left|\frac{K_3}{C_P}\right|\frac{\partial C_P}{\partial P}\right|}{|K_4|} + \frac{|v|\left(\frac{1}{C_P^2}\left|\frac{\partial C_P}{\partial P}\right|+|\rho|\left|\frac{\partial C_J}{\partial P}\right|\right)(|K_1||K_2|+|K_3|)}{K_4^2}\right\},$$

then,

$$\left|\frac{\partial f_1}{\partial P}\right| \leq M_{13}.$$

Similarly,

$$\left|\frac{\partial f_1}{\partial T}\right| = \left|\frac{\rho^2\frac{\partial C_J}{\partial T}K_2 + \frac{\rho a C_P - \frac{\partial C_P}{\partial T}K_3}{C_P^2}}{K_4} - \frac{1+v^2\left(-\frac{1}{C_P^2}\frac{\partial C_P}{\partial T}+\rho\frac{\partial C_J}{\partial T}(K_1K_2+K_3)\right)}{K_4^2}\right|$$

$$\leq \frac{(\rho)^2\left|\frac{\partial C_J}{\partial T}\right||K_2| + \frac{|\rho||a||C_P|+\left|\frac{\partial C_P}{\partial T}\right||K_3|}{C_P^2}}{|K_4|} + \frac{1+v^2\left(\frac{1}{C_P^2}\left|\frac{\partial C_P}{\partial T}\right|+|\rho|\left|\frac{\partial C_J}{\partial T}\right|(|K_1||K_2|+|K_3|)\right)}{K_4^2}.$$

Let:

$$M_{14} = \sup\left\{\frac{(\rho)^2\left|\frac{\partial C_J}{\partial T}\right||K_2| + \frac{|\rho||a||C_P|+\left|\frac{\partial C_P}{\partial T}\right||K_3|}{C_P^2}}{|K_4|} + \frac{1+v^2\left(\frac{1}{C_P^2}\left|\frac{\partial C_P}{\partial T}\right|+|\rho|\left|\frac{\partial C_J}{\partial T}\right|(|K_1||K_2|+|K_3|)\right)}{K_4^2}\right\},$$

thus,

$$\left|\frac{\partial f_1}{\partial T}\right| \leq M_{14},$$

where, $\left|C_J\rho - \frac{M}{RZ_g}\right| \leq |C_J\rho| + \left|\frac{M}{RZ_g}\right| = |K_1|$, $\left|g\cos\theta - \frac{fv^2}{2d}\right| \leq |g\cos\theta| + \left|\frac{fv^2}{2d}\right| \leq g + \left|\frac{fv^2}{2d}\right| = |K_2|$, $|a(T-T_e) - g\cos\theta| \leq |a(T-T_e)| + g = |K_3|$, $\left|T + v^2\left(\frac{1}{C_P} + C_J\rho - \frac{M}{RZ_g}\right)\right| = |K_4|$.

According to similar method, there are:

$$\begin{cases} \dfrac{\partial f_2}{\partial \rho} = -\dfrac{v}{\rho}^2 f_1 - \dfrac{v}{\rho}\dfrac{\partial f_1}{\partial \rho} \\ \dfrac{\partial f_2}{\partial v} = -\dfrac{1}{\rho} f_1 - \dfrac{v}{\rho}\dfrac{\partial f_1}{\partial v} \\ \dfrac{\partial f_2}{\partial P} = -\dfrac{v}{\rho}\dfrac{\partial f_1}{\partial P} \\ \dfrac{\partial f_2}{\partial T} = -\dfrac{v}{\rho}\dfrac{\partial f_1}{\partial T} \end{cases}$$

Thus,

$$\left|\frac{\partial f_2}{\partial \rho}\right| \leq \frac{|v|^2}{\rho}|f_1| + \left|\frac{v}{\rho}\right|\left|\frac{\partial f_1}{\partial \rho}\right| \leq \frac{|v|^2}{\rho} N_1 + \left|\frac{v}{\rho}\right| M_{11}.$$

Let:

$$M_{21} = \sup\left\{\frac{|v|^2}{\rho} N_1 + \left|\frac{v}{\rho}\right| M_{11}\right\},$$

thus,

$$\left|\frac{\partial f_2}{\partial \rho}\right| \leq M_{21}.$$

Similarly,

$$\frac{\partial f_2}{\partial v} = -\frac{1}{\rho} f_1 - \frac{v}{\rho}\frac{\partial f_1}{\partial v} \leq \frac{1}{|\rho|}|f_1| + \left|\frac{v}{\rho}\right|\left|\frac{\partial f_1}{\partial v}\right| \leq \frac{1}{|\rho|} N_1 + \left|\frac{v}{\rho}\right| M_{12}.$$

Let:

$$M_{22} = \sup\left\{\frac{1}{|\rho|} N_1 + \left|\frac{v}{\rho}\right| M_{12}\right\},$$

then,

$$\left|\frac{\partial f_2}{\partial v}\right| \leq M_{22}.$$

Similarly,

$$\frac{\partial f_2}{\partial P} = -\frac{v}{\rho}\frac{\partial f_1}{\partial P}$$

$$\left|\frac{\partial f_2}{\partial P}\right| = \left|\frac{v}{\rho}\right|\left|\frac{\partial f_1}{\partial P}\right| \leq \left|\frac{v}{\rho}\right| M_{13}.$$

Let:

$$M_{23} = \sup\left\{\left|\frac{v}{\rho}\right| M_{13}\right\},$$

then,

$$\left|\frac{\partial f_2}{\partial P}\right| \leq M_{23}.$$

Similarly,

$$\frac{\partial f_2}{\partial T} = -\frac{v}{\rho}\frac{\partial f_1}{\partial T}$$

$$\left|\frac{\partial f_2}{\partial T}\right| = \left|\frac{v}{\rho}\right|\left|\frac{\partial f_1}{\partial T}\right| \leq \left|\frac{v}{\rho}\right| M_{14}.$$

Let:

$$M_{24} = \sup\left\{\left|\frac{v}{\rho}\right| M_{14}\right\},$$

then,

$$\left|\frac{\partial f_2}{\partial T}\right| \leq M_{24}.$$

The partial differential of f_3 about ρ, v, P, T may written as follows:

$$\begin{cases} \dfrac{\partial f_3}{\partial \rho} &= v^2\dfrac{\partial f_1}{\partial \rho} + g\cos\theta - \dfrac{fv^2}{2d} \\ \dfrac{\partial f_3}{\partial v} &= 2vf_1 + v^2\dfrac{\partial f_1}{\partial v} - \dfrac{f\rho v}{d} \\ \dfrac{\partial f_3}{\partial P} &= v^2\dfrac{\partial f_1}{\partial P} \\ \dfrac{\partial f_3}{\partial T} &= v^2\dfrac{\partial f_1}{\partial T} \end{cases}$$

and

$$\begin{cases} \dfrac{\partial f_4}{\partial \rho} &= C_J\dfrac{\partial f_3}{\partial \rho} + \dfrac{1}{C_P}\left(-\dfrac{v^2}{\rho^2}f_1 + \dfrac{v^2}{\rho}\dfrac{\partial f_1}{\partial \rho}\right) \\ \dfrac{\partial f_4}{\partial v} &= C_J\dfrac{\partial f_3}{\partial v} + \dfrac{1}{C_P}\left(\dfrac{2v}{\rho}f_1 + \dfrac{v^2}{\rho}\dfrac{\partial f_1}{\partial v}\right) \\ \dfrac{\partial f_4}{\partial P} &= C_J\dfrac{\partial f_3}{\partial p} + \dfrac{\partial C_J}{\partial p}f_3 + \dfrac{\frac{v^2}{\rho}\frac{\partial f_1}{\partial p}C_P - \left[\frac{v^2}{\rho}f_1 + g\cos\theta - a(T-T_e)\right]\frac{\partial C_P}{\partial p}}{C_P^2} \\ \dfrac{\partial f_4}{\partial T} &= C_J\dfrac{\partial f_3}{\partial T} + \dfrac{\partial C_J}{\partial T}f_3 + \dfrac{\left(\frac{v^2}{\rho}\frac{\partial f_1}{\partial T} - a\right)C_P - \left[\frac{v^2}{\rho}f_1 + g\cos\theta - a(T-T_e)\right]\frac{\partial C_P}{\partial T}}{C_P^2} \end{cases}$$

Repeating the above method, we can get:

$$\left|\frac{\partial f_3}{\partial \rho}\right| \leq M_{31}, \quad \left|\frac{\partial f_3}{\partial v}\right| \leq M_{32}, \quad \left|\frac{\partial f_3}{\partial P}\right| \leq M_{33}, \quad \left|\frac{\partial f_3}{\partial T}\right| \leq M_{34},$$

$$\left|\frac{\partial f_4}{\partial \rho}\right| \leq M_{41}, \quad \left|\frac{\partial f_4}{\partial v}\right| \leq M_{42}, \quad \left|\frac{\partial f_4}{\partial P}\right| \leq M_{43}, \quad \left|\frac{\partial f_4}{\partial T}\right| \leq M_{44}.$$

The Lipschitz condition is very important in discussing the solution of the system of differential equations, thus we firstly consider the Lipschitz condition of $F(z;y)$. We write the primal problem again as follows:

$$\frac{d\rho}{dz} = f_1(z;\rho,v,P,T), \quad \frac{dv}{dz} = f_2(z;\rho,v,P,T), \quad \frac{dP}{dz} = f_3(z;\rho,v,P,T), \quad \frac{dT}{dz} = f_4(z;\rho,v,P,T).$$

This can be written as:

$$\rho' = f_1(z; \rho, v, P, T), \quad v' = f_2(z; \rho, v, P, T), \quad P' = f_3(z; \rho, v, P, T), \quad T' = f_4(z; \rho, v, P, T).$$

The primal condition is: $\rho(z_0) = \rho_0$, $v(z_0) = v_0$, $P(z_0) = P_0$, $T(z_0) = T_0$.
Using the Euler method, for $i = 0, 1, 2, \ldots$, we have:

$$\rho_{i+1} = \rho_i + (z_{i+1} - z_i) f_1(z_i; \rho_i, v_i, P_i, T_i), \quad v_{i+1} = v_i + (z_{i+1} - z_i) f_2(z_i; \rho_i, v_i, P_i, T_i),$$

$$P_{i+1} = P_i + (z_{i+1} - z_i) f_3(z_i; \rho_i, v_i, P_i, T_i), \quad T_{i+1} = T_i + (z_{i+1} - z_i) f_4(z_i; \rho_i, v_i, P_i, T_i).$$

Here ρ_i, v_i, P_i, T_i are intended to approximate $\rho(z_i), v(z_i), P(z_i), T(z_i)$, where $z_0 < z_1 < z_2 \cdots$ is subdivision of the interval of integration. Let $y_i = (\rho_i, v_i, P_i, T_i)^T$, then:

$$y_{i+1} = y_i + (z_{i+1} - z_i) F(z_i; y_i), \quad i = 0, 1, 2, \ldots, n-1.$$

If set $h_i = z_{i+1} - z_i$, then for the subdivision above can be written as:

$$h = (h_0, h_1, \ldots, h_{n-1}).$$

If we connect y_0 and y_1, y_1 and $y_2, \ldots$ by straight lines we obtain the Euler polygon then we obtain

$$y_h(z) = y_i + (z - z_i) f(z_i; y_i) \quad \text{for } z_i \leq z \leq z_{i+1}.$$

Theorem 8.4. For $\|F(z; y)\| \leq N = \max\{N_1, N_2, N_3, N_4\}$, then there is for ρ_i, v_i, P_i, T_i defined by above way the estimate:

$$\|y_i - y_0\| \leq N|z_i - z_0|,$$

where $y_i = (\rho_i, v_i, P_i, T_i)^T$.

For $\left|\frac{\partial f_k}{\partial \rho}\right| \leq M_{k1}$, $\left|\frac{\partial f_k}{\partial v}\right| \leq M_{k2}$, $\left|\frac{\partial f_k}{\partial P}\right| \leq M_{k3}$, $\left|\frac{\partial f_k}{\partial T}\right| \leq M_{k4}$, then

$$\|F(z; y) - F(z; \hat{y})\| \leq L\|y - \hat{y}\|,$$

where $k = 1, 2, 3, 4$; $L = \max_k(\sum_{i=1}^{4} M_{ki})$.

Proof. (1) From $\rho_{i+1} = \rho_i + (z_{i+1} - z_i) f_1(z_i; \rho_i, v_i, P_i, T_i)$ and definition of $\|F(z; \rho, v, P, T)\|$, we have:

$$|\rho_{i+1} - \rho_i| = |z_{i+1} - z_i||f_1(z_i; \rho_i, v_i, P_i, T_i)| \leq N(z_{i+1} - z_i).$$

Therefore,

$$|\rho_i - \rho_{i-1}| \leq N(z_i - z_{i-1}), \quad \ldots, \quad |\rho_2 - \rho_1| \leq N(z_2 - z_1), \quad |\rho_1 - \rho_0| \leq N(z_1 - z_0).$$

Thus,

$$|\rho_i - \rho_{i-1}| + \cdots + |\rho_2 - \rho_1| + |\rho_1 - \rho_0| \leq N(z_i - z_0).$$

Since:

$$|\rho_i - \rho_{i-1} + \cdots + \rho_2 - \rho_1 + \rho_1 - \rho_0| \leq |\rho_i - \rho_{i-1}| + \cdots + |\rho_2 - \rho_1| + |\rho_1 - \rho_0|,$$

so,

$$|\rho_i - \rho_0| \leq N(z_i - z_0).$$

Similarly,

$$|v_i - v_0| \leq N(z_i - z_0),$$
$$|P_i - P_0| \leq N(z_i - z_0),$$
$$|T_i - T_0| \leq N(z_i - z_0).$$

From definition of $\|y_i - y_0\|$, there is:

$$\|y_i - y_0\| \leq N(z_i - z_0).$$

(2) For $f_1(z;y), f_2(z;y), f_3(z;y), f_4(z;y), y = (\rho, v, P, T)^T$,

$$f_1(z;\hat{y}) - f_1(z;y) = \frac{\partial f_1}{\partial \rho}(\hat{\rho} - \rho) + \frac{\partial f_1}{\partial v}(\hat{v} - v) + \frac{\partial f_1}{\partial P}(\hat{P} - P) + \frac{\partial f_1}{\partial T}(\hat{T} - T).$$

Thus,

$$|f_1(z;\hat{y}) - f_1(z;y)| \leq \left|\frac{\partial f_1}{\partial \rho}\right| |\hat{\rho} - \rho| + \left|\frac{\partial f_1}{\partial v}\right| |\hat{v} - v| + \left|\frac{\partial f_1}{\partial P}\right| |\hat{P} - P| + \left|\frac{\partial f_1}{\partial T}\right| |\hat{T} - T|.$$

Let $\Delta y = \max\{|\hat{\rho} - \rho|, |\hat{v} - v|, |\hat{P} - P|, |\hat{T} - T|\}$, then:

$$|f_1(z;\hat{y}) - f_1(z;y)| \leq \left(\left|\frac{\partial f_1}{\partial \rho}\right| + \left|\frac{\partial f_1}{\partial v}\right| + \left|\frac{\partial f_1}{\partial P}\right| + \left|\frac{\partial f_1}{\partial T}\right|\right) \Delta y.$$

Similarly,

$$|f_2(z;\hat{y}) - f_2(z;y)| \leq \left(\left|\frac{\partial f_2}{\partial \rho}\right| + \left|\frac{\partial f_2}{\partial v}\right| + \left|\frac{\partial f_2}{\partial P}\right| + \left|\frac{\partial f_2}{\partial T}\right|\right) \Delta y,$$

$$|f_3(z;\hat{y}) - f_3(z;y)| \leq \left(\left|\frac{\partial f_3}{\partial \rho}\right| + \left|\frac{\partial f_3}{\partial v}\right| + \left|\frac{\partial f_3}{\partial P}\right| + \left|\frac{\partial f_3}{\partial T}\right|\right) \Delta y,$$

$$|f_4(z;\hat{y}) - f_4(z;y)| \leq \left(\left|\frac{\partial f_4}{\partial \rho}\right| + \left|\frac{\partial f_4}{\partial v}\right| + \left|\frac{\partial f_4}{\partial P}\right| + \left|\frac{\partial f_4}{\partial T}\right|\right) \Delta y.$$

From definition of norm, there is:

$$\|F(z;\hat{y}) - F(z;y)\| = \max\{|f_1(z;\hat{y}) - f_1(z;y)|, |f_2(z;\hat{y}) - f_2(z;y)|, |f_3(z;\hat{y}) - f_3(z;y)|, |f_4(z;\hat{y}) - f_4(z;y)|\}.$$

Let $L = \max_k \left(\sum_{i=1}^4 M_{ki}\right)$ and $\|\hat{y} - y\| = \Delta y$, then:

$$\|F(z;\hat{y}) - F(z;y)\| \leq L\|\hat{y} - y\|.$$

We consider a subdivision of the interval of integration

$$z_0, z_1, \ldots, z_{n-1}, z_n = Z.$$

Theorem 8.5. For a fixed subdivision h, let $y_h(x)$ and $\hat{y}_h(x)$ are the Euler polygons corresponding to the initial values y_0 and $\hat{y}_0$. respectively, for $F(z;y)$ satisfies the Lipschitz condition of Theorem 8.4. in a convex region which contains $(z;y_h(z))$ and $(z;\hat{y}_h(x))$ for $z_0 \leq z \leq Z$, then:

$$\|y_h(z) - \hat{y}_h(z)\| \leq e^{L(z-z_0)}\|y_0 - \hat{y}_0\|.$$

Proof. From $y_{i+1} = y_i + (z_{i+1} - z_i)F(z_i;y_i)$, we have:

$$y_1 - y_0 = (z_1 - z_0)F(z_0;y_0), \quad \hat{y}_1 - \hat{y}_0 = (z_1 - z_0)F(z_0;\hat{y}_0).$$

Thus, we get:

$$y_1 - \hat{y}_1 = (y_0 - \hat{y}_0)(z_1 - z_0)[F(z_0;y_0) - F(z_0;\hat{y}_0)].$$

From Theorem 8.4, we have:

$$\|F(z;y) - F(z;\hat{y}\| \leq L\|y - \hat{y}\|.$$

Therefore,

$$\|F(z_0;y_0) - F(z_0;\hat{y}_0\| \leq L\|y_0 - \hat{y}_0\|.$$

Thus,

$$\|y_1 - \hat{y}_1\| \leq (1 + L(z_1 - z_0))\|y_0 - \hat{y}_0\|.$$

From $(1 + L(z_1 - z_0)) \leq e^{L(z_1 - z_0)}$, we get:

$$\|y_1 - \hat{y}_1\| \leq e^{L(z_1 - z_0)}\|y_0 - \hat{y}_0\|.$$

If we repeat the same argument for $y_2 - \hat{y}_2$, $y_3 - \hat{y}_3$, and so on, we finally obtain:

$$\|y_h(z) - (\hat{y})_h(z)\| \leq e^{L(z - z_0)}\|y_0 - \hat{y}_0\|.$$

Theorem 8.6. Let $F(z;y)$ be continuous, and $\|F(z;y)\| \leq N$ and satisfy the Lipschitz condition of Theorem 8.4 on $D = \{(z;y)|z_0 \leq z \leq Z, \|y - y_0\| \leq b\}$. If $Z - z_0 \leq \frac{b}{N}$, then we have:

(1) For $|h| = \max_{i=0,1,2,\ldots,n-1}(|z_{i+1} - z_i|) \to 0$, the Euler polygons $y_{|h|}(z) = (P_{|h|}(z), T_{|h|}(z))^T$ converge uniformly to a continuous vector function $\phi(z)$.
(2) $\phi(z)$ is continuously differential and is a solution of primal problem on $z_0 \leq z \leq Z$.
(3) There exists no other solution of primal problem on $z_0 \leq z \leq Z$.

Proof. (1) Take an $\varepsilon > 0$. Since F is uniformly continuous on the compact D, there exists a $\delta > 0$, such that:

$$|z_2 - z_1| \leq \delta \quad \text{and} \quad \|y_1 - y_2\| \leq N\delta,$$

imply

$$\|F(z_2;y_2) - F(z_1;y_1)\| \leq \varepsilon.$$

Suppose that the subdivision h satisfies:

$$|z_{i+1} - z_i| \leq \delta, \quad \text{i.e. } |h| \leq \delta.$$

We study first the effects of adding new mesh-points. In a first step, we consider a subdivision $h(1)$, which is obtained by adding new points only to the first subinterval. We have the estimate $\|y_{h(1)}(z_1) - y_h(z_1)\| \leq \varepsilon|z_1 - z_0|$. Since subdivision h and $h(1)$ are identical on $z_1 \leq z \leq Z$, we can obtain by Theorem 8.5:

$$\|y_{h(1)}(z) - y_h(z)\| \leq e^{L(z - z_1)}\|y_{h(1)}(z_1) - y_h(z_1)\|.$$

Thus,

$$\|y_{h(1)}(z) - y_h(z)\| \leq e^{L(z - z_1)}|z_1 - z_0|\varepsilon, \quad \text{for } z_1 \leq z \leq Z.$$

We next add further points to the subinterval (z_1, z_2) and denote the new subdivision by $h(2)$. In the same way as above this leads to:

$$\|y_{h(2)}(z_1) - y_{h(1)}(z_1)\| \leq \varepsilon|z_2 - z_1|,$$

and

$$\|y_{h(2)}(z) - y_{h(1)}(z)\| \leq e^{L(z - z_2)}|z_2 - z_1|\varepsilon, \quad \text{for } z_2 \leq z \leq Z.$$

If we denote by $\hat{h}$ the final refinement, we obtain for $z_2 \le z \le z_{i+1}$:

$$\begin{aligned}\|y_{\hat{h}}(z) - y_h(z)\| &\le \|y_{\hat{h}}(z) - y_{h(i-1)}(z)\| + \cdots + \|y_{h(2)}(z) - y_{h(1)}(z)\| + \|y_{h(1)}(z) - y_h(z)\| \\ &\le \varepsilon[e^{L(z-z_1)}|z_1 - z_0| + \cdots + e^{L(z-z_i)}|z_i - z_{i+1}|] \\ &\le \varepsilon \int_{z_0}^{Z} e^{L(z-s)} ds = \frac{\varepsilon}{L}(e^{L(z-z_0)} - 1).\end{aligned}$$

If we now have different subdivisions h and $\acute{h}$, which both satisfy $|z_2 - z_1| \le \delta$ and $\|y_1 - y_2\| \le N\delta$, we introduce a third subdivision $\hat{h}$ which is a refinement of both subdivisions. And apply the above inequality to $\hat{h}$ and $\acute{h}$, we have:

$$\|y_h(z) - y_{\acute{h}}(z)\| \le \|y_h(z) - y_{\hat{h}}(z)\| + \|y_{\hat{h}}(z) - y_{\acute{h}}(z)\| \le 2\frac{\varepsilon}{L}(e^{L(z-z_0)} - 1).$$

For $\varepsilon > 0$ small enough, this becomes arbitrarily small and shows the uniform convergence of the Euler polygons to a continuous vector function $\phi(z)$. (2) Let:

$$\varepsilon(\delta) := \sup\{\|F(z_1; y_1) - F(z_2; y_2)\|;\ |z_1 - z_2| \le \delta;\ \|y_1 - y_2\| \le N\delta,\ (z_i; y_i) \in D\}.$$

If z belongs to the subdivision h, then we can determine

$$\|y_h(z + \delta) - y_h(z) - \delta F(z; y_h(z))\| \le \varepsilon(\delta)\delta.$$

By the limit $|h| \to 0$,

$$\|\phi(z + \delta) - \phi(z) - \delta F(z; \phi(z))\| \le \varepsilon(\delta)\delta.$$

Since $\varepsilon(\delta) \to 0$ for $\delta \to 0$, this proves the differentiability of $\phi(z)$ and $\phi'(z) = F(z; \phi(z))$.

(3) From (2), the solution of the primal problem exists. Let $\phi(z)$ and $\varphi(z)$ be solutions of the primal problem, then:

$$\phi(z) = y_0 + \int_{z_0}^{z} F(x; \phi(x))dx \tag{8.51a}$$

and

$$\varphi(z) = y_0 + \int_{z_0}^{z} F(x; \varphi(x))dx. \tag{8.51b}$$

From (8.51a), (8.51b) and Lipschitz condition of Theorem 8.5, we have:

$$\|\phi(z) - \varphi(z)\| \le L\left|\int_{z_0}^{z} \|\phi(x) - \varphi(x)\|dx\right|. \tag{8.51c}$$

Let:

$$g(z) = \int_{z_0}^{z} \|\phi(x) - \varphi(x)\|dx, \quad z \ge z_0.$$

Thus,

$$g'(z) = \|\phi(z) - \varphi(z)\|.$$

Therefore, (8.51) can be revised as:

$$g'(z) \le Lg(z).$$

So,

$$(e^{-L(z-z_0)}g(z))' \le 0,$$

thus,

$$e^{-L(z-z_0)}g(z) \le g(z_0) = 0.$$

When $z \geq z_0$, $g(z) \geq 0$, thus

$$g(z) \equiv 0, \quad z \geq z_0.$$

Therefore,

$$\phi(z) = \varphi(z).$$

8.3.3 Solving the model

To simplify the calculation, the well is divided into several short segments of the same length. The length of a segment varies with variations in wall thickness, hole diameter, the fluid density inside and outside the pipe and well geometry. The model begins with a calculation at the bottom of the pipe. Then the gas density, velocity, pressure and temperature calculations are performed for each successive "segment" of the pipe up to the surface. By means of the afore-mentioned discussion, the 4 order Runge-Kutta method is used to solve the model. Therefore, the algorithm is designed as follows.

Step 1. Obtain each point's inclination:

$$\theta_j = \theta_{j-1} + \frac{\theta_k - \theta_{k-1}}{\Delta s_k} \Delta s_j,$$

where j represents segment point of calculation, Δs_k represents measurement depth of inclination angle θ_k and θ_{k-1}, Δs_j the step length of calculation.

Step 2. Calculate the gas condensing parameter Z_g:

If ($P < 35$ MPa)

$$Z_g = 1 + \left(0.31506 - \frac{1.0467}{T_{pr}} - \frac{0.5783}{T_{pr}^3}\right)\rho_{pr} + \left(0.053 - \frac{0.6123}{T_{pr}}\right)\rho_{pr}^2 + 0.6815\frac{\rho_{pr}^2}{T_{pr}^3},$$

where, $\rho_{pr} = 0.27\frac{P_{pr}}{T_{pr}}$, $T_{pr} = \frac{T}{T_{pc}}$, $P_{pr} = \frac{P}{P_{pc}}$.

Else

$$Z_g = (90.7x - 242x^2 + 42.4x^3)y^{(1.18+2.82x)} - (14.76x - 9.76x^2 + 4.58t^3)y + \frac{1 + y + y^2 + y^3}{(1-y)^3},$$

where,

$$\begin{aligned} F(y) &= -0.06125P_{pr}xe^{-1.2(1-x)^2} + (90.7x - 242.2x^2 + 42.4x^3)y^{2.18+2.82x} \\ &\quad + \frac{y + y^2 + y^3 - y^4}{(1-y^3)} - (14.76x - 9.76x^2 + 4.58x^3)y^2 \\ &= 0 \end{aligned}$$

$$x = \frac{1}{T_{pr}}.$$

Step 3. Calculate the original conditions of the gas density ρ and velocity v at point j. Assume the gas pressure and temperature at dot j are already known as P_j and T_j, then ρ_j and v_j can be obtained by the following equations:

$$\rho_j = 0.000001 \times 3484.48\gamma_g \frac{P_j}{ZT_j},$$

$$v_j = \frac{101000 \times 300000T_j}{293 \times 86400P_jA}.$$

Step 4. Let the right parts of the coupled differential equations to be functions f_i, where ($i = 1, 2, 3, 4$). Then we can obtain a system of coupled functions as follows:

$$\begin{cases} f_1 = \dfrac{\left(C_J\rho - \frac{M}{RZ_g}\right)\left[\rho g\cos\theta + \frac{f\rho v^2}{2d}\right] + \frac{\rho a(T-T_e)-\rho g\cos\theta}{C_P}}{T + v^2\left[\frac{1}{C_p} + C_J\rho - \frac{M}{RZ_g}\right]} \\ f_2 = -\dfrac{v}{\rho}\dfrac{d\rho}{dz} \\ f_3 = v^2\dfrac{d\rho}{dz} - \rho g\cos\theta - \dfrac{f\rho v^2}{2d} \\ f_4 = C_J\dfrac{dP}{dz} + \dfrac{\left[\frac{v^2}{\rho}\frac{d\rho}{dz} + g\cos\theta - a(T-T_e)\right]}{C_P} \end{cases},$$

where,

$$a = \frac{2\pi r_{to}U_{to}K_e}{w[k_e + f(t_D)r_{to}U_{to}]}$$

$$U_{ti}^{-1} = \frac{1}{hc+hr} + \frac{r_{ti}\ln\left(\frac{r_{\text{cem}}}{r_{co}}\right)}{k_{\text{cem}}} + \frac{r_{ti}\ln\left(\frac{r_{ci}}{r_{to}}\right)}{k_{\text{ang}}}$$

$$C_P = 1243 + 3.14T + 7.931\times10^{-4}T^2 - 6.881\times10^{-7}T^3$$

and

$$C_J = \frac{R}{C_P}\frac{(2r_A - r_BT - 2r_BBT)Z - (2r_AB + r_BAT)}{3Z^2 - 2Z + A - B - B^2)T}$$

$$A = \frac{r_AP}{T},\quad B = \frac{r_BP}{T},\quad r_A = \frac{0.42747\alpha T_{pc}^2}{P_{pc}},\quad r_B = \frac{0.08664C_bT_{pc}}{P_{pc}},$$

$$\alpha_i = [1 + m(1 - T_{pr}^{0.5})]^2,\quad m = 0.48 + 1.574w - 0.176w^2$$

and

$$\begin{cases} f(t_D) = 1.1281\sqrt{t_D}\left(1 - 0.3\sqrt{t_D}\right) & (t_D \le 1.5) \\ f(t_D) = \left(1 + \dfrac{0.6}{t_D}\right)[0.4063 + 0.5\ln(t_D)] & (t_D > 1.5) \end{cases}$$

$$t_D = \frac{t\alpha}{r_{wb}^2}$$

and

$$\frac{1}{\sqrt{f}} = 1.14 - 2\lg\left(\frac{0.00001524}{r_{ti}} + \frac{21.25}{Re^{0.9}}\right).$$

Step 5. Assume P, T, v, ρ to be y_i ($i = 1, 2, 3, 4$), respectively. Then we can obtain some basic parameters as follows:

$$\begin{cases} a_i = f_i\,[y_1, y_2, y_3, y_4] \\ b_i = f_i\left[y_1 + \dfrac{h}{2}a_1, y_2 + \dfrac{h}{2}a_2, y_3 + \dfrac{h}{2}a_3, y_4 + \dfrac{h}{2}a_4\right] \\ c_i = f_i\left[y_1 + \dfrac{h}{2}b_1, y_2 + \dfrac{h}{2}b_2, y_3 + \dfrac{h}{2}b_3, y_4 + \dfrac{h}{2}b_4\right] \\ d_i = f_i\,[y_1 + hc_1, y_2 + hc_2, y_3 + hc_3, y_4 + hc_4] \end{cases}$$

Step 6. Calculate the gas density, gas velocity, gas pressure and temperature at point $(j+1)$:

$$y_i^{(j+1)} = y_i^j + \frac{h}{6}(a_i + 2b_i + 2c_i + d_i) \quad (i = 1, 2, 3, 4) \quad (j = 1, 2, \ldots, n).$$

Step 7. Repeat the second to sixth step until calculating y_i^n.

8.3.4 Numerical simulation

As was described previously, the algorithm begins with a calculation at the bottom of the pipe. The calculations are performed for each successive "segment" of the pipe to the surface.

8.3.4.1 *Parameters*

In this simulation, we study a pipe at the *X* well, which is in Sichuan Province, China. All the needed parameters are as follows:

Internal fluid density = 1000 kg/m^3
External fluid density = 1000 kg/m^3
Depth of the well = 7100 m
Friction coefficient = 1.2
Ground temperature = 16°C
Geothermal gradient = 2.18 °C/100 m
Length of one segment = 1 m

Parameters of pipes, inclined well, inclination, azimuth and vertical depth are given in Tables 8.1, 8.2 and 8.3.

8.3.5 Results and analysis

To study the different gas outputs, geothermal gradients and thermal conductivity parameters that influence the gas density, velocity, pressure and temperature, we use different gas outputs; 300000 m^3/d, 500000 m^3/d and 700000 m^3/d; geothermal gradients and thermal conductivity parameters. To study the difference between the model with the J-T coefficient and without the J-T coefficient, we present results allowing for both conditions. Using an algorithm and a simulation, a series of results for the temperature, pressure, velocity and density with variations in the products, geothermal gradients and thermal conductivity parameters were determined and displayed in tables and figures. In the following we present an analysis of the results.

8.3.5.1 *Temperature*

In Table 8.8 and Figure 8.11, temperatures for the different gas outputs with the J-T coefficient are shown. In Table 8.9 and Figure 8.12, the temperatures of the different gas outputs without the J-T coefficient are shown. It is clear from these results that

(1) Temperature is a nonlinear distribution.
(2) At the same depth, the temperature increases with an increase in production. This is because the increase in production results in an increase in velocity, thus the heat loss decreases.
(3) The J-T coefficient weakly affects the temperature. Because the change in the tubular diameter is small, the J-T coefficient is also small. When considering the J-T effect, the temperature reduces. Since this is a gas well, the J-T cooling is much more significant than the J-T heating.

Table 8.8 Temperature [°C] for different gas outputs with J-T coefficient.

Depth [m]	300000 m^3/d	500000 m^3/d	700000 m^3/d	Depth [m]	300000 m^3/d	500000 m^3/d	700000 m^3/d
0	73.223	114.134	124.580	4000	149.401	153.954	155.452
400	83.923	119.091	128.287	4400	153.484	156.675	157.763
800	93.896	123.849	131.849	4800	157.604	159.194	159.940
1200	103.118	128.356	135.258	5200	160.189	161.512	161.987
1600	111.746	132.684	138.356	5600	162.906	163.640	163.909
2000	119.749	136.819	141.733	6000	165.256	165.588	165.710
2400	127.048	140.726	144.772	6400	167.261	167.351	167.385
2800	133.639	144.393	147.665	6800	168.907	168.920	168.925
3200	139.529	147.814	150.407	7100	169.510	169.510	169.513
3600	144.766	150.995	153.001				

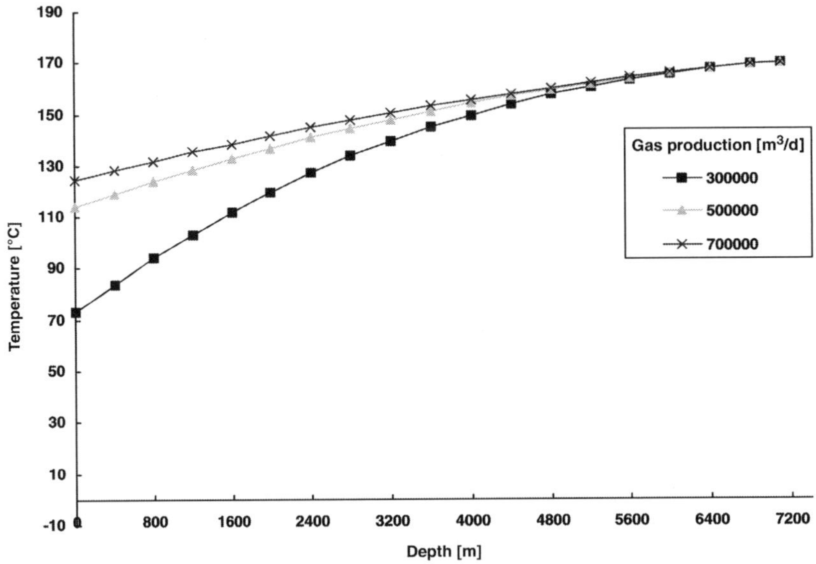

Figure 8.11 Temperature for different gas outputs with J-T coefficient.

Table 8.9 Temperature [°C] for different gas output without J-T coefficient.

Depth [m]	300000 m^3/d	500000 m^3/d	700000 m^3/d	Depth [m]	300000 m^3/d	500000 m^3/d	700000 m^3/d
0	73.763	114.546	125.092	4000	149.510	154.056	155.591
400	84.400	119.476	128.746	4400	153.571	156.767	157.880
800	94.318	124.135	132.259	4800	157.132	159.268	160.035
1200	103.486	128.656	135.624	5200	160.240	161.570	162.063
1600	112.065	132.951	138.880	5600	162.943	163.683	163.967
2000	120.023	137.055	142.021	6000	165.281	165.619	165.752
2400	127.281	140.932	145.025	6400	167.267	167.370	167.411
2800	133.835	144.573	147.886	6800	168.910	168.925	168.935
3200	139.692	147.968	150.599	7100	169.512	169.512	169.512
3600	144.901	151.127	153.166				

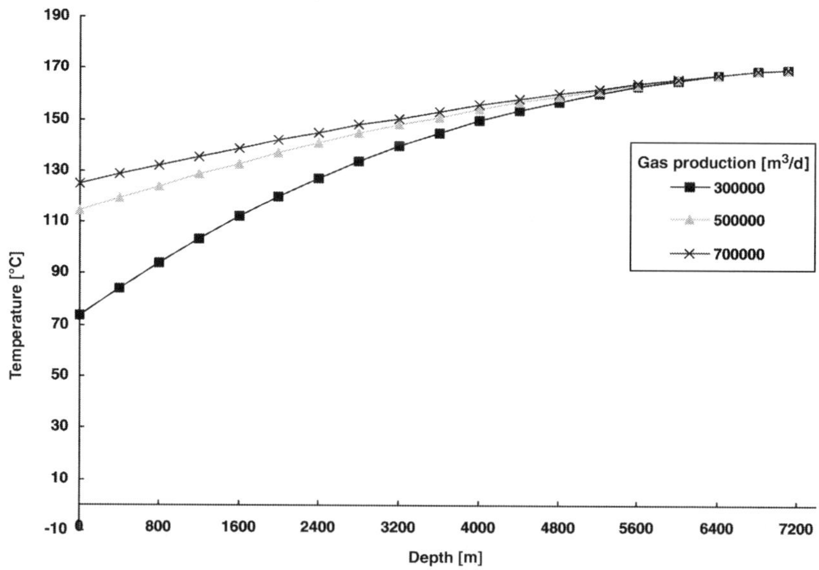

Figure 8.12 Temperature for different gas outputs without J-T coefficient.

Table 8.10 Temperature [°C] for different geothermal gradients with J-T coefficient.

Depth [m]	2.18°C/100 m	2.4°C/100 m	2.5°C/100 m	Depth [m]	2.18°C/100 m	2.4°C/100 m	2.5°C/100 m
0	73.223	81.615	85.291	4000	149.401	164.638	171.567
400	83.923	93.613	97.921	4400	153.484	168.870	175.867
800	93.896	104.719	109.573	4800	157.604	172.552	179.594
1200	103.118	114.911	120.23	5200	160.189	175.741	182.811
1600	111.746	124.375	130.091	5600	162.906	178.491	185.575
2000	119.749	133.084	139.133	6000	165.256	180.851	187.940
2400	127.048	140.963	147.285	6400	167.261	182.839	189.924
2800	133.639	148.020	154.558	6800	168.907	184.489	191.543
3200	139.529	154.276	160.983	7100	169.510	185.064	192.134
3600	144.766	159.794	166.629				

In Table 8.10 and Figure 8.13, the temperatures of the different geothermal gradients with the J-T coefficient are shown. In Table 8.11 and Figure 8.14, the temperatures of the different thermal conductivity parameters with the J-T coefficient are shown. It is clear that:

(1) The temperature increases with an increase in the geothermal gradients.
(2) The J-T coefficient weakly affects the temperature. When considering the J-T effect, the temperature decreases.

In Table 8.12 and Figure 8.15, the temperatures of the different thermal conductivity parameters with the J-T coefficient are shown. In Table 8.13 and Figure 8.16, the temperatures of the different thermal conductivity parameters without the J-T coefficient are shown. It is clear that:

(1) The temperature decreases as the thermal conductivity parameter increases.
(2) At the same time, the J-T coefficient weakly affects the temperature. When considering the J-T effect, the temperature decreases.

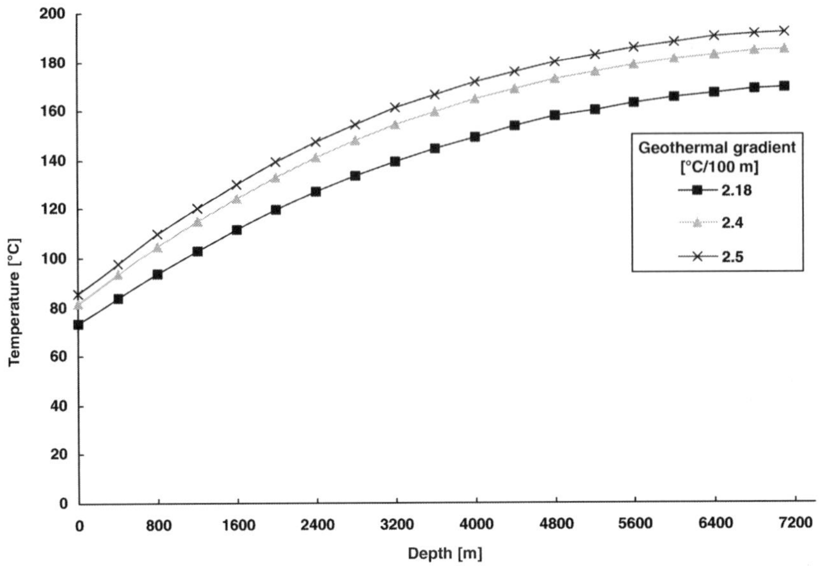

Figure 8.13 Temperature for different geothermal gradients with J-T coefficient.

Table 8.11 Temperature [°C] for different geothermal gradients without J-T coefficient.

Depth [m]	2.18°C/100 m	2.4°C/100 m	2.5°C/100 m	Depth [m]	2.18°C/100 m	2.4°C/100 m	2.5°C/100 m
0	73.763	82.199	85.896	4000	149.510	164.749	171.680
400	84.400	94.128	98.454	4400	153.571	168.959	175.957
800	94.318	104.719	110.039	4800	157.132	172.621	179.664
1200	103.486	115.304	120.635	5200	160.240	175.793	182.863
1600	112.065	124.713	130.430	5600	162.943	178.528	185.613
2000	120.023	133.373	139.429	6000	165.281	180.876	187.965
2400	127.281	141.207	147.534	6400	167.267	182.854	189.938
2800	133.835	148.225	154.767	6800	168.910	184.475	191.549
3200	139.692	154.445	161.155	7100	169.512	185.067	192.137
3600	144.901	159.933	166.769				

8.3.5.2 *Pressure*

In Table 8.14 and Figure 8.17, the pressure of the different gas outputs with the J-T coefficient is shown. In Table 8.15 and Figure 8.18, the pressure of the different gas outputs without the J-T coefficient is shown. It is clear that:

(1) At the same depth, the pressure decreases with an increase in production. This is because the flow velocity increases with an increase in output resulting in a drop in increased frictional pressure.
(2) The J-T coefficient has no effect on the pressure.

In Table 8.16 and Figure 8.19, the pressure of the different geothermal gradients with the J-T coefficient is shown. In Table 8.17 and Figure 8.20, the pressure of the different geothermal gradients without the J-T coefficient is shown. It is clear that:

(1) The pressure increases with an increase in the geothermal gradients.
(2) The J-T coefficient has no effect on pressure.

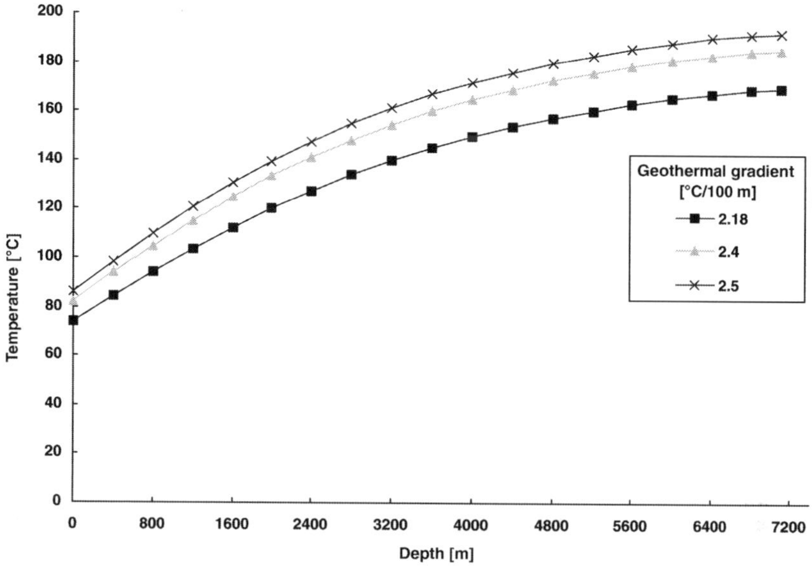

Figure 8.14 Temperature for different geothermal gradients without J-T coefficient.

Table 8.12 Temperature [°C] for different thermal conductivities with J-T coefficient.

Depth [m]	0.52 W/m·K	0.57 W/m·K	2.42 W/m·K	Depth [m]	0.52 W/m·K	0.57 W/m·K	2.42 W/m·K
0	73.223	72.896	70.151	4000	149.401	149.359	148.998
400	83.923	83.639	81.258	4400	153.484	153.453	153.182
800	93.896	93.654	91.611	4800	157.604	157.041	156.842
1200	103.118	102.912	101.178	5200	160.189	160.173	160.029
1600	111.746	111.574	110.123	5600	162.906	162.894	162.791
2000	119.749	119.608	118.413	6000	165.256	165.248	165.175
2400	127.048	126.934	125.964	6400	167.261	167.247	167.206
2800	133.639	133.548	132.772	6800	168.907	168.903	168.896
3200	139.529	139.457	138.846	7100	169.510	169.51	169.508
3600	144.766	144.711	144.237				

In Table 8.18 and Figure 8.21, the pressure of the different thermal conductivity parameters with the J-T coefficient is shown. In Table 8.19 and Figure 8.22, the pressure in the different thermal conductivity parameters without the J-T coefficient is shown. It is clear that:

(1) The thermal conductivity parameters have little effect on the pressure.
(2) The J-T coefficient has no effect on the pressure.

In order to elaborate on when and when not to include the J-T effect in modelling temperature and pressure using our model, the following explanation is provided:

(1) For $C_J = \frac{\partial P}{\partial T}$, if the variation in P is very small relative to T, i.e., $C_J \to 0$, the J-T affect should not be considered.
(2) In fact, $C_J \to 0$, cannot be assured in advance. On the other hand, the model provides two independent models to compute the four parameters: P, T, V, ρ. Thus, it not only verifies whether the J-T effect exists, but determines its influence.

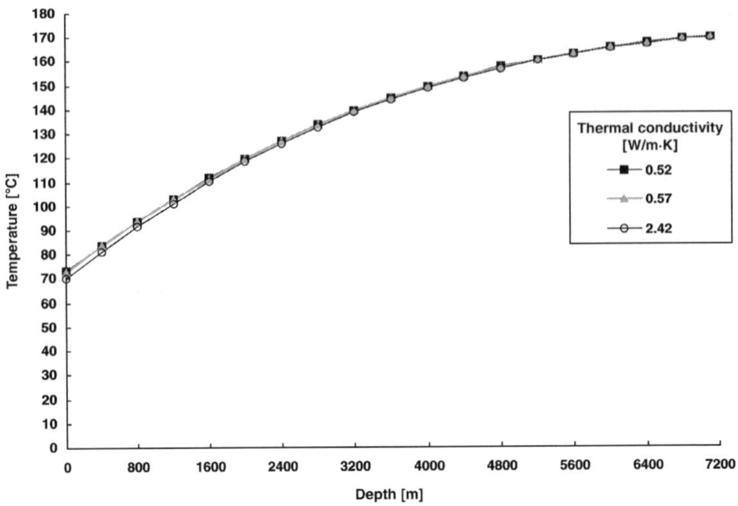

Figure 8.15 Temperature for different thermal conductivities with J-T coefficient.

Table 8.13 Temperature [°C] for different thermal conductivities without J-T coefficient.

Depth [m]	0.52 W/m·K	0.57 W/m·K	2.42 W/m·K	Depth [m]	0.52 W/m·K	0.57 W/m·K	2.42 W/m·K
0	73.763	73.438	70.705	4000	149.510	149.469	149.109
400	84.400	84.119	81.749	4400	153.571	153.54	153.27
800	94.318	94.077	92.044	4800	157.132	157.109	156.911
1200	103.486	103.282	101.556	5200	160.240	160.224	160.081
1600	112.065	111.894	110.451	5600	162.943	162.931	162.829
2000	120.023	119.882	118.694	6000	165.281	165.273	165.2
2400	127.281	127.167	126.202	6400	167.267	167.262	167.221
2800	133.835	133.744	132.973	6800	168.910	168.909	168.902
3200	139.692	139.621	139.013	7100	169.512	169.512	169.511
3600	144.901	144.846	144.374				

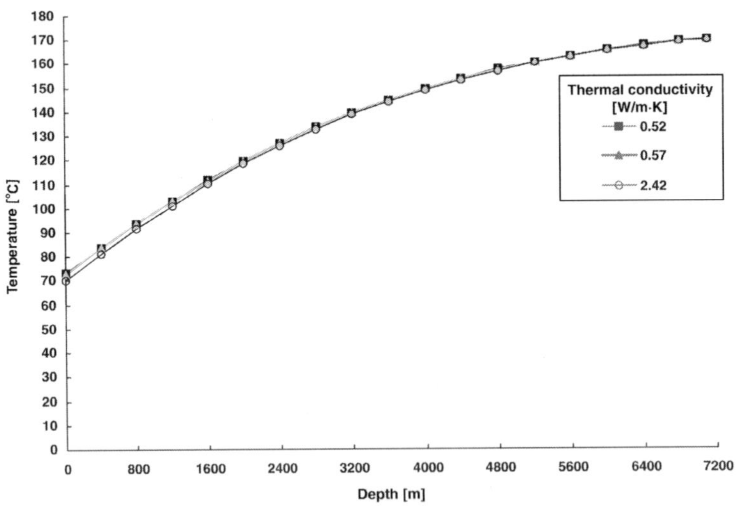

Figure 8.16 Temperature for different thermal conductivities without J-T coefficient.

Table 8.14 Pressure [MPa] for different gas outputs with J-T coefficient.

Depth [m]	300000 m³/d	500000 m³/d	700000 m³/d	Depth [m]	300000 m³/d	500000 m³/d	700000 m³/d
0	26.676	23.618	17.027	4000	34.322	32.733	30.143
400	27.491	24.545	18.483	4400	35.049	33.637	31.367
800	28.294	25.474	19.905	4800	35.774	34.542	32.588
1200	29.084	26.404	21.299	5200	36.497	35.449	33.806
1600	29.862	27.322	22.638	5600	37.220	36.358	35.022
2000	30.623	28.229	23.929	6000	37.939	37.268	36.236
2400	31.375	29.130	25.194	6400	38.665	38.196	37.480
2800	32.120	30.029	26.441	6800	39.429	39.233	38.936
3200	32.858	30.929	27.681	7100	39.731	39.624	39.508
3600	33.592	31.830	28.914				

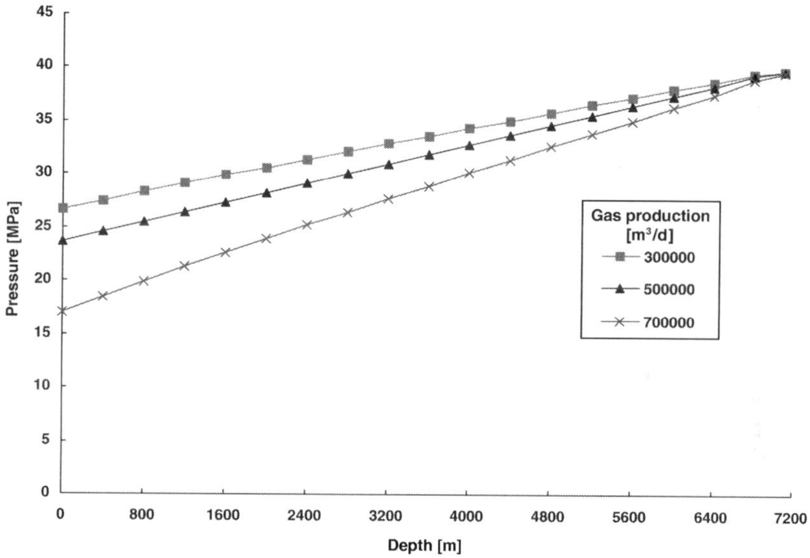

Figure 8.17 Pressure for different gas outputs with J-T coefficient.

Table 8.15 Pressure [MPa] for different gas outputs without J-T coefficient.

Depth [m]	300000 m³/d	500000 m³/d	700000 m³/d	Depth [m]	300000 m³/d	500000 m³/d	700000 m³/d
0	26.678	23.619	17.026	4000	34.322	32.733	30.143
400	27.493	24.545	18.481	4400	35.049	33.637	31.367
800	28.295	25.474	19.904	4800	35.774	34.542	32.588
1200	29.085	26.404	21.298	5200	36.497	35.449	33.806
1600	29.861	27.322	22.637	5600	37.218	36.358	35.022
2000	30.623	28.229	23.928	6000	37.939	37.268	36.236
2400	31.376	29.130	25.193	6400	38.665	38.196	37.480
2800	32.120	30.029	26.441	6800	39.429	39.233	38.936
3200	32.858	30.929	27.681	7100	39.731	39.624	39.508
3600	33.592	31.830	28.914				

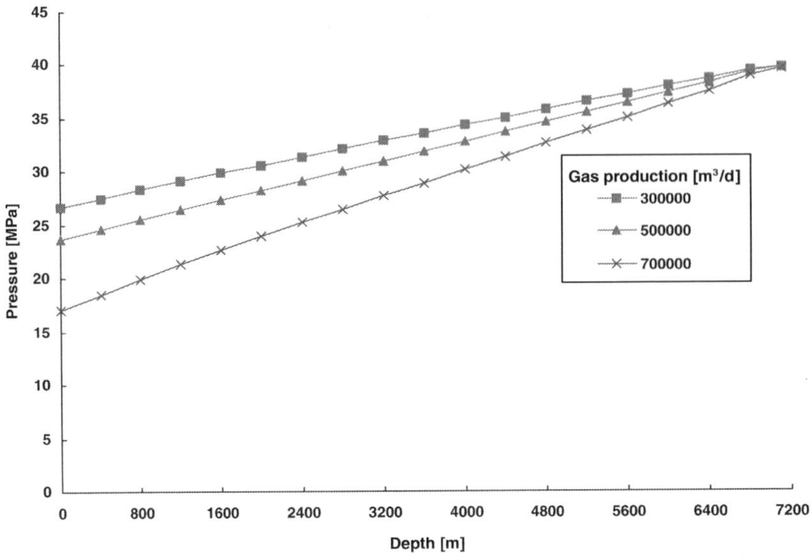

Figure 8.18 Pressure for different gas outputs without J-T coefficient.

Table 8.16 Pressure [MPa] for different geothermal gradients with J-T coefficient.

Depth [m]	2.18°C/100 m	2.4°C/100 m	2.5°C/100 m	Depth [m]	2.18°C/100 m	2.4°C/100 m	2.5°C/100 m
0	26.676	26.979	27.108	4000	34.322	34.450	34.504
400	27.491	27.778	27.9	4400	35.049	35.160	35.207
800	28.294	28.564	28.679	4800	35.774	35.867	35.907
1200	29.084	29.337	29.445	5200	36.497	36.573	36.605
1600	29.862	30.096	30.196	5600	37.220	37.278	37.303
2000	30.623	30.840	30.933	6000	37.939	37.982	38.000
2400	31.375	31.575	31.659	6400	38.665	38.691	38.702
2800	32.120	32.301	32.378	6800	39.429	39.441	39.445
3200	32.858	33.022	33.091	7100	39.731	39.737	39.739
3600	33.592	33.738	33.799				

8.3.5.3 *Gas velocity*

It is clear that:

(1) At the same depth, the gas velocity increases with an increase in the production.
(2) When the production is constant, the gas velocity varies as the depth of the well decreases.
(3) The J-T coefficient has a weak effect on gas velocity. Therefore, the main influence on gas velocity is gas production.

8.3.5.4 *Gas density*

It is clear that:

(1) At the same depth, gas density decreases with an increase in production.
(2) When the production is constant, gas density varies as the depth of the well decreases.
(3) The J-T coefficient has an effect on gas density, but it is weak. Therefore the main influence on gas density is gas production.

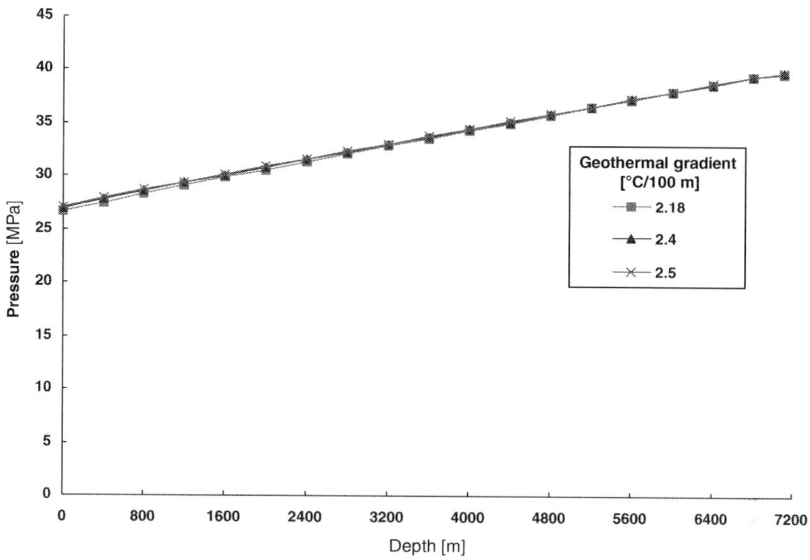

Figure 8.19 Pressure for different geothermal gradients without the J-T coefficient.

Table 8.17 Pressure [MPa] for different geothermal gradients without J-T coefficient.

Depth [m]	2.18°C/100 m	2.4°C/100 m	2.5°C/100 m	Depth [m]	2.18°C/100 m	2.4°C/100 m	2.5°C/100 m
0	26.678	26.981	27.110	4000	34.322	34.450	34.504
400	27.493	27.780	27.902	4400	35.049	35.160	35.207
800	28.295	28.566	28.681	4800	35.774	35.867	35.907
1200	29.085	29.338	29.446	5200	36.497	36.573	36.605
1600	29.861	30.096	30.196	5600	37.218	37.278	37.303
2000	30.623	30.841	30.933	6000	37.939	37.982	38.000
2400	31.376	31.575	31.659	6400	38.665	38.691	38.702
2800	32.120	32.301	32.378	6800	39.429	39.441	39.445
3200	32.858	33.022	33.091	7100	39.731	39.737	39.739
3600	33.592	33.738	33.799				

8.3.6 **Error analysis**

In this section, we compare the results of our calculation with actual measured results in a wellhead, when the gas output is 500000 m^3/day with the J-T coefficient, the results of which are shown in Table 8.24.

We use the method of Cullender and Smith (1956) to calculate the pressure of the well. The detailed mathematical model is shown in the following.

Considering the flowing gas column, we have:

$$\int_{p_{tf}}^{p_{wf}} \frac{\frac{p}{TZ}dp}{(\frac{p}{TZ})^2 + F^2} = \int_0^H 0.03415\gamma_g \, dH,$$

where,

$$F^2 = \frac{1.324 \times 10^{-18} f Q_{sc}^2}{d^5}.$$

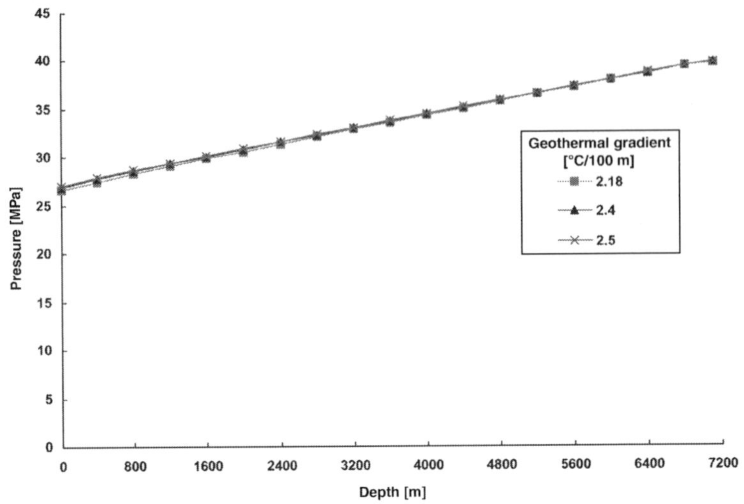

Figure 8.20 Pressure for different geothermal gradients without J-T coefficient.

Table 8.18 Pressure [MPa] for different thermal conductivities with J-T coefficient.

Depth [m]	0.52 W/m·K	0.57 W/m·K	2.42 W/m·K	Depth [m]	0.52 W/m·K	0.57 W/m·K	2.42 W/m·K
0	26.676	26.673	26.652	4000	34.322	34.322	34.321
400	27.491	27.489	27.473	4400	35.049	35.049	35.049
800	28.294	28.293	28.279	4800	35.774	35.774	35.773
1200	29.084	29.083	29.073	5200	36.497	36.497	36.496
1600	29.862	29.86	29.852	5600	37.220	37.22	37.218
2000	30.623	30.622	30.616	6000	37.939	37.939	37.939
2400	31.375	31.575	31.371	6400	38.665	38.665	38.665
2800	32.120	32.119	32.116	6800	39.429	39.429	39.429
3200	32.858	32.858	32.856	7100	39.731	39.731	39.731
3600	33.592	33.592	33.591				

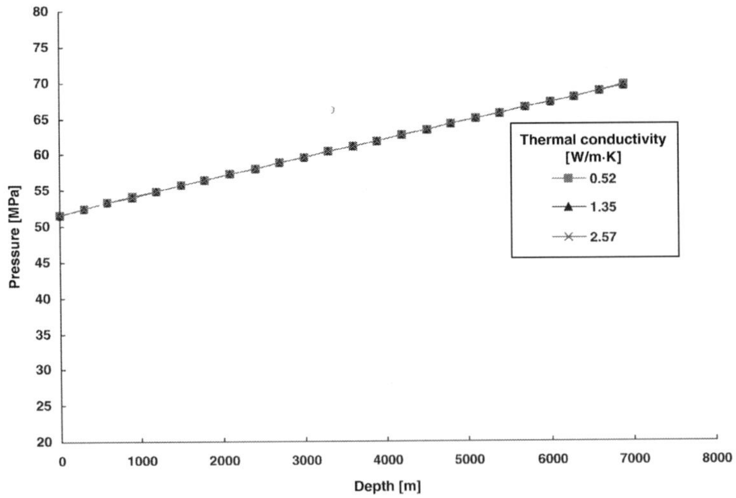

Figure 8.21 Pressure for different thermal conductivities with J-T coefficient.

Table 8.19 Pressure [MPa] for different thermal conductivities without J-T coefficient.

Depth [m]	0.52 W/m·K	0.57 W/m·K	2.42 W/m·K	Depth [m]	0.52 W/m·K	0.57 W/m·K	2.42 W/m·K
0	26.678	26.675	26.654	4000	34.322	34.322	34.321
400	27.493	27.491	27.474	4400	35.049	35.049	35.049
800	28.295	28.294	28.281	4800	35.774	35.774	35.773
1200	29.085	29.084	29.074	5200	36.497	36.497	36.496
1600	29.861	29.86	29.852	5600	37.218	37.22	37.218
2000	30.623	30.623	30.617	6000	37.939	37.939	37.939
2400	31.376	31.375	31.371	6400	38.665	38.665	38.665
2800	32.120	32.12	32.116	6800	39.429	39.429	39.429
3200	32.858	32.858	32.856	7100	39.731	39.731	39.731
3600	33.592	33.592	33.591				

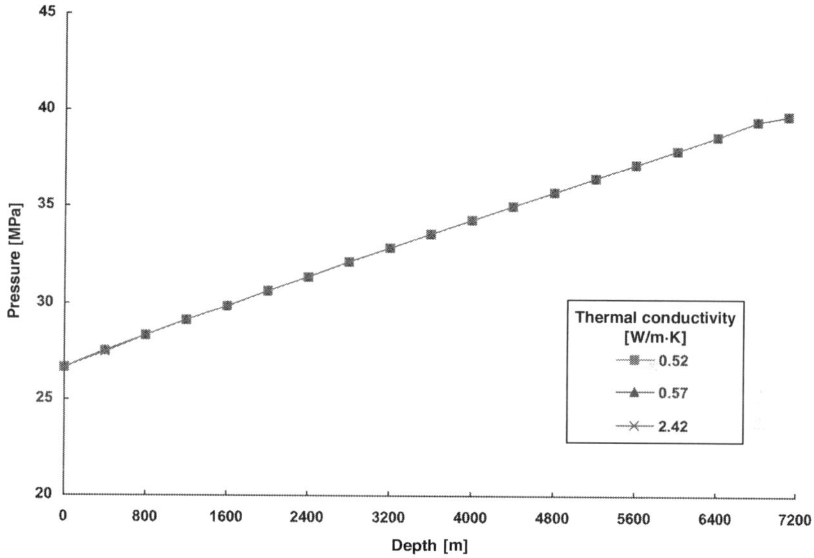

Figure 8.22 Pressure for different thermal conductivities without J-T coefficient.

Table 8.20 Velocity [m/s] for different gas outputs with J-T coefficient.

Depth [m]	300000 m^3/d	500000 m^3/d	700000 m^3/d	Depth [m]	300000 m^3/d	500000 m^3/d	700000 m^3/d
0	4.396	7.347	10.330	4000	4.247	7.843	12.850
400	4.404	7.425	10.580	4400	4.199	7.877	13.210
800	4.406	7.491	10.810	4800	4.144	7.907	13.600
1200	4.403	7.541	11.010	5200	4.080	7.937	14.050
1600	4.396	7.590	11.220	5600	5.123	7.965	14.570
2000	4.384	7.637	11.450	6000	5.236	7.999	15.220
2400	4.369	7.683	11.690	6400	5.326	8.030	15.970
2800	4.347	7.727	11.950	6800	5.350	8.059	16.870
3200	4.320	7.768	12.220	7100	5.360	8.069	17.280
3600	4.287	7.807	12.520				

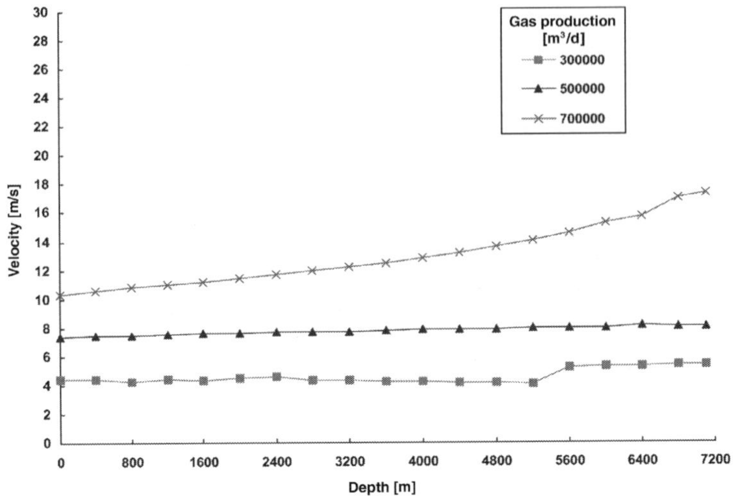

Figure 8.23 Gas velocity for different gas outputs with J-T coefficient.

Table 8.21 Velocity [m/s] for different gas outputs without J-T coefficient.

Depth [m]	300000 m^3/d	500000 m^3/d	700000 m^3/d	Depth [m]	300000 m^3/d	500000 m^3/d	700000 m^3/d
0	4.416	7.367	10.330	4000	4.237	7.853	12.850
400	4.406	7.423	10.58	4400	4.119	7.876	13.210
800	4.426	7.493	10.810	4800	4.134	7.906	13.600
1200	4.408	7.546	11.01	5200	4.060	7.938	14.050
1600	4.356	7.640	11.220	5600	5.223	7.955	14.580
2000	4.484	7.667	11.45	6000	5.256	7.969	15.220
2400	4.569	7.685	11.690	6400	5.316	8.130	15.670
2800	4.344	7.737	11.950	6800	5.360	8.069	16.970
3200	4.310	7.762	12.22	7100	5.380	8.089	17.380
3600	4.227	7.802	12.520				

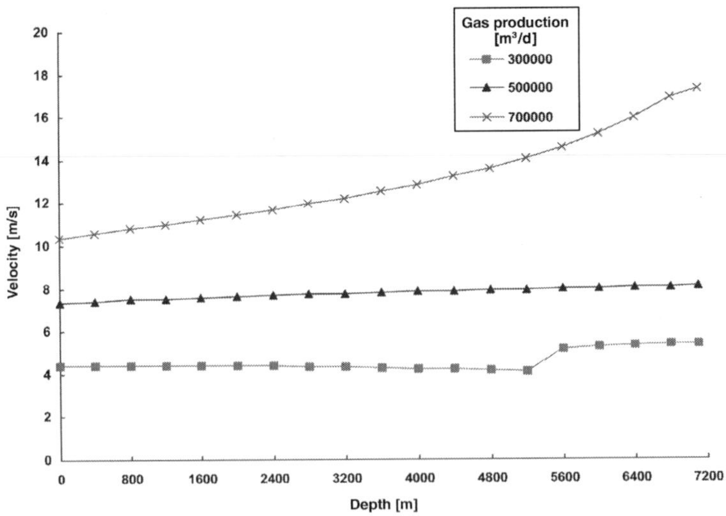

Figure 8.24 Gas velocity for different gas outputs without J-T coefficient.

Table 8.22 Density [kg/m^3] for different gas outputs with J-T coefficient.

Depth [m]	300000 m^3/d	500000 m^3/d	700000 m^3/d	Depth [m]	300000 m^3/d	500000 m^3/d	700000 m^3/d
0	155.220	140.943	92.122	4000	160.674	146.41	130.252
400	154.952	141.126	94.356	4400	162.499	147.19	133.267
800	154.863	141.629	99.696	4800	164.677	148.019	136.195
1200	154.976	142.179	104.642	5200	158.200	148.91	139.06
1600	155.238	142.787	109.265	5600	157.936	149.845	141.86
2000	155.631	143.288	113.287	6000	157.562	150.808	144.582
2400	156.195	143.825	117.06	6400	156.892	151.829	147.261
2800	156.963	144.387	120.559	6800	155.200	153.167	150.48
3200	157.947	145.001	123.908	7100	154.789	154.789	154.78
3600	159.171	145.678	127.135				

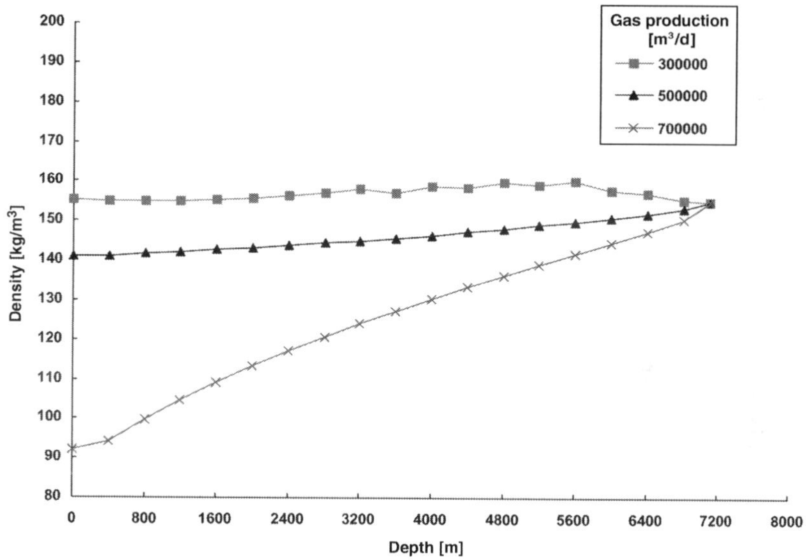

Figure 8.25 Gas density for different gas productions with J-T coefficient.

Let:

$$I = \frac{\frac{p}{TZ}}{\left(\frac{p}{TZ}\right)^2 + F^2}.$$

Thus,

$$p_{mf} = p_{tf} + \frac{0.03415\gamma_g H}{I_{mf} + I_{tf}}, \quad p_{wf} = p_{mf} + \frac{0.03415\gamma_g H}{I_{wf} + I_{mf}}.$$

We calculate the temperature using the average temperature method:

$$\bar{T} = \frac{T_{tf} + T_{wf}}{2}.$$

As was described previously, the algorithm begins with a calculation at the bottom of the pipe. Figures 8.28 and 8.27.

When the wellhead temperature and the bottom temperature are the same as our methods, the temperature is a linear distribution using the average temperature method. However, using our method the result is a nonlinear distribution, which is more reasonable and realistic.

Table 8.23 Density [kg/m^3] for different gas outputs without J-T coefficient.

Depth [m]	300000 m^3/d	500000 m^3/d	700000 m^3/d	Depth [m]	300000 m^3/d	500000 m^3/d	700000 m^3/d
0	155.220	140.963	92.222	4000	158.653	146.51	130.352
400	154.954	141.226	94.366	4400	158.469	147.29	133.367
800	154.865	141.649	99.796	4800	159.634	148.119	136.295
1200	154.979	142.199	104.652	5200	159.178	148.96	139.16
1600	155.240	142.767	109.365	5600	160.145	149.865	141.89
2000	155.633	143.388	113.288	6000	157.582	150.828	144.584
2400	156.195	143.835	117.16	6400	156.922	151.839	147.263
2800	156.960	144.487	120.659	6800	155.300	153.187	150.46
3200	157.940	145.101	123.928	7100	154.689	154.889	154.889
3600	157.158	145.688	127.235				

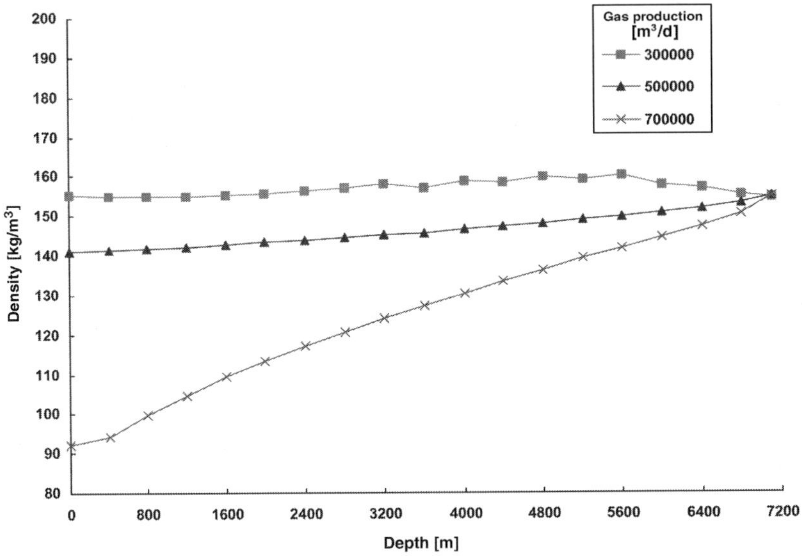

Figure 8.26 Gas density for different gas outputs without J-T coefficient.

Table 8.24 Comparative results.

Wellhead	Temperature [°C]	Pressure [MPa]	Velocity [m/s]	Density [kg/m^3]
Measurement results	110.243	22.638	7.122	137.878
Calculation results	114.134	23.618	7.347	140.943
Relative error	4.28%	4.32%	3.29%	2.22%

When the bottom pressure is the same as in our method, the algorithm begins with a calculation at the bottom of the pipe. To calculate each pressure using the Cullender and Smith (C-S) method, we determine the temperature using the average temperature method. From the wellhead pressure comparison with the measured result, our calculation errors are less than when using the C-S method, at less than 5%. The results are shown in Table 8.25. The C-S method only

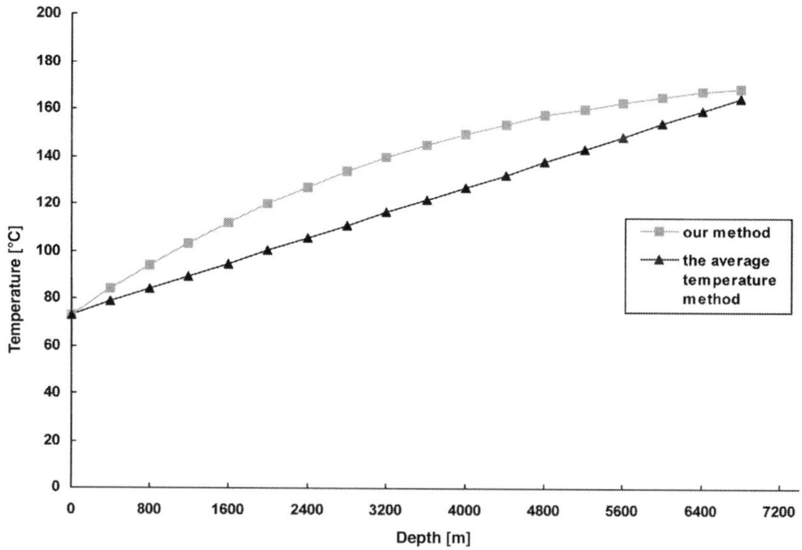

Figure 8.27 Temperature comparison.

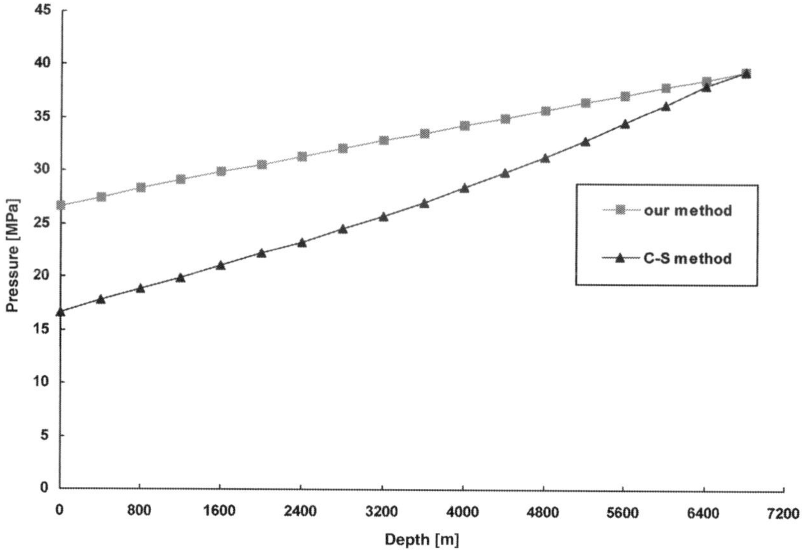

Figure 8.28 Pressure comparison.

Table 8.25 Comparative results of C-S method.

Wellhead	Pressure [MPa]
Measurement results	22.638
C-S results	16.591
Relative error	21.91%

calculates pressure, but is unable to calculate temperature. Thus it ignores the fact that pressure and temperature are interdependent. Therefore, our method is more feasible and effective.

8.4 PDTPVD-GLTPTF

Considering the partial differential equation model for pressure P, temperature T, velocity v, and density ρ, we assume the following:

1. The gas liquid two-phase flow in the tubing transient is in one dimension of the flow direction.
2. The pressure at all points is equal in the transverse cross-section.
3. When the gas-liquid two-phase reaches thermodynamic equilibrium, the temperature at all points is equal in the transverse cross-section.
4. There is no mass transfer between the gas and liquid.

8.4.1 **Prediction model**

Consider the flow system depicted in Figure 8.1, a straight cylindrical flow tube with an inclination angle θ, a constant cross-sectional flow area A, a hydraulic diameter d, and a total length Z. Through this tubing, the fluid flows from the bottom to the top with a mass flow rate w. The distance co-ordinate in the flow direction along the tubing is denoted z. Mass, momentum, and energy balances, along with the pressure, temperature, velocity and density, relative to the transient gas liquid two-phase flow, as well as the stated equation, are used to generate the constitutive equations.

Mass balance: We take the bottom of the well as the origin of the coordinate axis and the vertical up as a positive direction. Let dz and dt denote the differential depth and differential time respectively. It follows from mass balance law that:

$$\rho_m v_m A dt - \left[\rho_m v_m A + \frac{\partial}{\partial z}(\rho_m v_m A)\right] dt = \frac{\partial}{\partial t}(\rho_m A dz) dt,$$

which can be transformed as:

$$\frac{\partial \rho_m}{\partial t} + v_m \frac{\partial \rho_m}{\partial z} + \rho_m \frac{\partial v_m}{\partial z} = 0. \tag{8.49}$$

Momentum balance: The forces on dz includes pressure on the cross section, the gravity and friction, which can be expressed as $\frac{\partial P}{\partial z} A dz$, $\rho_m g A \cos\theta dz$ and $\frac{f \rho_m v_m^2}{2d} A dz$ respectively. During the time dt, the momentum flowing into dz is $\rho_m v_m^2 A$ and momentum flowing out from dz is $\rho_m v_m^2 A + \frac{\partial}{\partial z}(\rho_m v_m^2 A) dz$. In addition, the increment momentum during time dt is $\frac{\partial}{\partial t}(\rho_m v_m A dz)$. It follows that:

$$\frac{\partial(\rho_m v_m)}{\partial t} + \frac{\partial P}{\partial z} + \frac{\partial(\rho_m v_m^2)}{\partial z} + \rho_m g \cos\theta + \frac{f \rho_m v_m^2}{2d} = 0. \tag{8.50}$$

Energy balance: The types of energy considered in this paper include inner energy, pressure energy, kinetic energy and potential energy, where the inner energy and pressure energy are collectively referred to as enthalpy. The energy flowing into dz can be given by:

$$H_m(z) + \frac{1}{2}(\rho_m v_m A) v_m^2 + (\rho_m v_m A) gz cos\theta.$$

The first item refers to enthalpy, the second item refers to kinetic energy and the third item refers to potential energy. Similarly, the energy flowing out of dz can be given by:

$$\left[H_m(z) + \frac{\partial H_m}{\partial z}\right] dz + \frac{1}{2}\left[\rho_m v_m^3 + \frac{\partial}{\partial z}(\rho_m v_m^3) dz\right] A + (\rho_m v_m A) g(z + dz)\cos\theta.$$

Ramey (1962) and Willhite (1967) discussed the radial transfer of heat between the fluid and the earth in detail. The differential element dz was shown in Figure 8.2. During time dt, the radial heat transfer from the fluid to the cement/earth interface can be described by $2\pi r_{to}U_{to}(T - T_k)dz$. The radial heat transfer from the cement/earth interface to the surrounding earth during time dt is $2\pi K_e(T_k - T_e)/f(t_D)dz$. Thus the radial heat transfer between the gas and the surrounding earth tube is:

$$\frac{2\pi r_{to}U_{to}K_e}{r_{to}U_{to}f(t_D) + K_e}(T - T_e)dz.$$

The heat generated by the friction with tubing is:

$$\frac{f\rho_m v_m^3}{dz}Adz.$$

In addition, the variable quantity of the energy during time dt is:

$$\left[\frac{\partial}{\partial t}\rho_m C_{pm}T\right]Adt,$$

where $C_{pm} = (W_g/W_m)C_{pg} + (W_l/W_m)C_{pl}$. Let $a = \frac{2\pi r_{to}U_{to}K_e}{w[k_e + f(t)r_{to}U_{to}]}$, by the energy conservation law, we have:

$$\frac{\partial}{\partial t}(P_m C_{pm}T)A + \frac{\partial}{\partial z}\left[H_m + \frac{1}{2}\rho_m v_m^3 A + \rho_m g v_m z\cos\theta\right] + a(T - T_e) + \frac{f\rho_m v_m^2 A}{2d} = 0. \quad (8.51)$$

From $C_{pm} = (W_g/W_m)C_{pg} + (W_l/W_m)C_{pl}$, we get:

$$\frac{\partial H_m}{\partial z} = \frac{W_g}{W_m}\frac{\partial H_g}{\partial z} + \frac{W_l}{W_m}\frac{\partial H_l}{\partial z}. \quad (8.52)$$

For general enthalpy, $\frac{\partial H}{\partial z} = C_p\frac{\partial T}{\partial z} - C_J C_p\frac{\partial P}{\partial z}$, then:

$$\frac{\partial H_g}{\partial z} = C_{pg}\frac{\partial T}{\partial z} - C_{Jg}C_{pg}\frac{\partial P}{\partial z}, \quad \frac{\partial H_l}{\partial z} = C_{pl}\frac{\partial T}{\partial z} - C_{Jl}C_{pl}\frac{\partial P}{\partial z}. \quad (8.53)$$

Therefore, we have:

$$\frac{\partial H_m}{\partial z} = \frac{W_g}{W_m}\left[C_{pg}\frac{\partial T}{\partial z} - C_{Jg}C_{pg}\frac{\partial P}{\partial z}\right] + \frac{W_l}{W_m}\left[C_{pl}\frac{\partial T}{\partial z} - C_{Jl}C_{pl}\frac{\partial P}{\partial z}\right].$$

Let:

$$C_{pm} = \left(\frac{W_g}{W_m}\right)C_{pg} + \left(\frac{W_l}{W_m}\right)C_{pl}, \quad C_{Jm} = -\left[\left(\frac{W_g}{W_m}\right)C_{Jg}C_{pg} + \left(\frac{W_l}{W_m}\right)C_{Jl}C_{pl}\right],$$

then C_{pm}, C_{Jm} is the Specific Heat Capacity (SHC) and the J-T coefficient of the mixture of gas and liquid, respectively. Therefore,

$$\frac{\partial H_m}{\partial z} = C_{pm}\frac{\partial T}{\partial z} + C_{Jm}C_{pm}\frac{\partial P}{\partial z}. \quad (8.54)$$

Thus, the above equation can be written as:

$$C_{pm}\left(T\frac{\partial\rho_m}{\partial z} + \rho_m\frac{\partial T}{\partial z}\right)A + C_{pm}\frac{\partial T}{\partial z} + C_{Jm}C_{pm}\frac{\partial P}{\partial z} + \frac{1}{2}\left(v_m^3\frac{\partial\rho_m}{\partial z} + 3\rho_m v_m^2\frac{\partial v_m}{\partial z}\right)$$
$$+ \left(v_m z\frac{\partial\rho_m}{\partial z} + \rho_m z\frac{\partial v_m}{\partial z} + \rho_m v_m\right)g\cos\theta + a(T - T_e) + \frac{f\rho_m v_m^3 A}{2d} = 0. \quad (8.55)$$

From initial condition and boundary condition, we can obtain the system of partial differential equations on velocity, pressure and temperature as follows:

$$\begin{cases} \dfrac{\partial v_m}{\partial z} = -\dfrac{1}{\rho_m}\left(\dfrac{\partial \rho_m}{\partial t} + v_m \dfrac{\partial \rho_m}{\partial z}\right) \\ \dfrac{\partial P}{\partial z} = -\rho_m \dfrac{\partial v_m}{\partial t} - v_m \dfrac{\partial \rho_m}{\partial t} - v_m^2 \dfrac{\partial \rho_m}{\partial z} - 2\rho_m v_m \dfrac{\partial v_m}{\partial z} - \rho_m g \cos\theta \\ \dfrac{\partial T}{\partial z} = -\dfrac{1}{C_{pm}}\left[\left(T\dfrac{\partial \rho_m}{\partial z} + \rho_m \dfrac{\partial T}{\partial z}\right)A + C_{pm}\dfrac{\partial T}{\partial z} + C_{Jm}C_{pm}\dfrac{\partial P}{\partial z} + \dfrac{1}{2}\left(v_m^3 \dfrac{\partial \rho_m}{\partial z} + 3\rho_m v_m^2 \dfrac{\partial v_m}{\partial z}\right)\right. \\ \quad \left. + \left(v_m z \dfrac{\partial \rho_m}{\partial z} + \rho_m z \dfrac{\partial v_m}{\partial z} + \rho_m v_m\right) g\cos\theta + a(T - T_e) + \dfrac{f\rho_m v_m^3 A}{2d}\right] \\ P(z,0) = \Phi_P(z), \quad T(z,0) = \Phi_T(z), \quad v_m(z,0) = \Phi_{vm}(z) \\ P(0,t) = \zeta_P(t), \quad T(0,t) = \zeta_T(t), \quad v_m(0,t) = \zeta_{vm}(t) \end{cases} \tag{8.56}$$

Observe that there are three equations but four unknown numbers P, T, ρ_m, v_m, so it is necessary to supplement the stated equation of the mixture $\rho_m = \frac{MP}{RZ_mT}$. Then the above system and the stated equation constitute the mathematical model of predicting velocity, pressure and temperature and density.

8.4.2 **Model solution**

To simplify the calculation, we divide the wells into several short segments of the same length h and divide the testing time into several short segments of the same length τ. The model begins with a calculation at the bottom of the pipe and the initial time. Then the density, velocity, pressure and temperature calculations are performed for each successive "segment" of the pipe up to the surface as well as for each successive "segment" of time up to the end point of the test time.

8.4.2.1 *Finite difference scheme*

We adopt the finite difference method to develop the numerical model of the problem as it is suitable for programming and can achieve high accuracy. The first step in the finite difference method is to determine the finite difference scheme, which requires that the partial derivatives are substituted with a different quotient. Then the partial differential equations are transformed into algebraic equations. The approximate solutions to the original partial differential equations can then be calculated by solving the algebraic equations using a method based on the Taylor series expansion. The construction of the finite difference scheme is as follows.

For system and the stated equation, there are three types of formula needing to be discretized, i.e. $\frac{\partial U}{\partial z}$, $\frac{\partial U}{\partial t}$ and U. Divide the $z-t$ plane sets of equal rectangle of sides h, τ, by equally spaced grid lines parallel ot, defined by $z_j = jh, j = 1, 2, \ldots, m$, and equally spaced grid lines parallel to oz, defined by $t_k = k\tau, k = 1, 2, \ldots, n$, (see Figure 8.29).

For convenience, let F_j^k denotes $F(z_j, t_k)$. Discretize the formulas at point $(z_{j+\frac{1}{2}}, t_k)$, and then we have three difference forms as follows:

$$\left(\frac{\partial U}{\partial t}\right)_{j+\frac{1}{2}}^{k+1} = \frac{U_{j+1}^{k+1} + U_j^{k+1} - U_{j+1}^k - U_j^k}{2\tau}, \quad \left(\frac{\partial U}{\partial z}\right)_{j+\frac{1}{2}}^{k+1} = \frac{U_{j+1}^{k+1} + U_j^{k+1}}{h}, \quad U_{j+\frac{1}{2}}^{k+1} = \frac{U_{j+1}^{k+1} + U_j^{k+1}}{2}.$$

Then the finite difference scheme can be obtained as follows:

$$(v_m)_{j+1}^{k+1} = (v_m)_j^{k+1} - \Delta_1, \quad P_{j+1}^{K+1} = P_j^{k+1} - \Delta_2,$$

$$T_{j+1}^{k+1} = \Psi T_j^{k+1}, T_{j+1}^k, T_j^k, \rho_{j+1}^{k+1}, \rho_j^{k+1}, v_{j+1}^{k+1}, v_j^{k+1}),$$

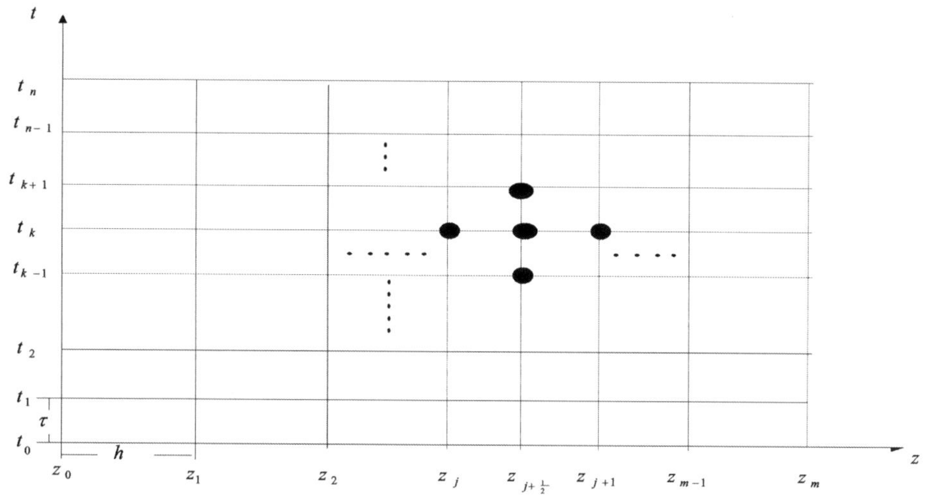

Figure 8.29 Discretization of the finite solution region.

$$(\rho_m)_{j+1}^{k+1} = \frac{MP_{j+1}^{k+1}}{RZ_m T_{j+1}^{k+1}},$$

where:

$$\Delta_1 = h\left[\frac{(\rho_m)_{j+1}^{k+1} + (\rho_m)_j^{k+1}}{2}\right]^{-1}\left[\frac{(\rho_m)_{j+1}^{k+1} + (\rho_m)_j^{k+1} - (\rho_m)_{j+1}^{k} - (\rho_m)_j^{k}}{2\tau}\right.$$
$$\left. + \frac{(v_m)_{j+1}^{k+1} + (v_m)_j^{k+1}}{2} \cdot \frac{(v_m)_{j+1}^{k+1} - (v_m)_j^{k+1}}{h}\right],$$

$$\Delta_2 = h\left\{\frac{(\rho_m)_{j+1}^{k+1} + (\rho_m)_j^{k+1}}{2} \cdot \frac{(v_m)_{j+1}^{k+1} + (v_m)_j^{k+1} - (v_m)_{j+1}^{k} - (v_m)_j^{k}}{2\tau}\right.$$
$$+ \frac{(v_m)_{j+1}^{k+1} + (v_m)_j^{k+1}}{2} \cdot \frac{(\rho_m)_j^{k+1} - (\rho_m)_{j+1}^{k} - (\rho_m)_j^{k}}{2\tau}$$
$$+ \left(\frac{(v_m)_{j+1}^{k+1} + (v_m)_j^{k+1}}{2}\right)^2 \cdot \frac{(\rho_m)_{j+1}^{k+1} - (\rho_m)_j^{k+1}}{h}$$
$$+ \frac{[(v_m)_{j+1}^{k+1} + (v_m)_j^{k+1}][(\rho_m)_{j+1}^{k+1} + (\rho_m)_j^{k+1}]}{2} \cdot \frac{(v_m)_{j+1}^{k+1} - (v_m)_j^{k+1}}{h}$$
$$+ \frac{[(\rho_m)_{j+1}^{k+1} + (\rho_m)_j^{k+1}](\cos\theta_j + \cos\theta_{j+1})}{4}$$
$$\left. + \frac{(f_{j+1}^{k+1} + f_j^{k+1})[(\rho_m)_{j+1}^{k+1} + (\rho_m)_j^{k+1}][(v_m)_{j+1}^{k+1} + (v_m)_j^{k+1}]^2}{16}\right\}.$$

After replacing $\frac{\partial T}{\partial z}, \frac{\partial T}{\partial t}, \frac{\partial P}{\partial z}, \frac{\partial T}{\partial z}, \frac{\partial v}{\partial z}, T, P, \rho_m, v_m$, by their difference scheme, we obtain:

$$T_{j+1}^{k+1} = \Psi(T_j^{k+1}, T_{j+1}^{k}, T_j^{k}, (\rho_m)_{j+1}^{k+1}, (\rho_m)_j^{k+1}, (v_m)_{j+1}^{k+1}.$$

8.4.3 Calculation of some parameters

In this section, we will give the calculating method of some parameters:

1. Each point's inclination: $\theta_j = \theta_{j-1} + (\theta_k - \theta_{k-1}\Delta s_j/\Delta s_k)$, where j represents segment point of calculation, s_k represents measurement depth of inclination angle θ_k, θ_{k-1}, Δs_k are the step length of calculation.
2. Gas condensing parameter Z_g:
   If ($p < 35$ MPa)

$$Z_g = 1 + \left(0.31506 - \frac{1.0467}{T_{pr}} - \frac{0.5783}{T_{pr}^3}\right)\rho_{pr} + \left(0.053 - \frac{0.6123}{T_{pr}}\right)\rho_{pr}^2 + 0.6815\frac{\rho_{pr}^2}{T_{pr}^3},$$

   where, $\rho_{pr} = 0.27\frac{P_{pr}}{T_{pr}}$, $T_{pr} = \frac{T}{T_{pc}}$, $P_{pr} = \frac{P}{P_{pc}}$.
   Else

$$\begin{aligned} Z_g &= (90.7x - 242x^2 + 42.4x^3)y^{(1.18+2.82x)} - (14.76x - 9.76x^2 + 4.58t^3)y \\ &\quad + \frac{1+y+y^2+y^3}{(1-y)^3}, \end{aligned}$$

   where,

$$\begin{aligned} F(y) &= -0.06125P_{pr}xe^{-1.2(1-x)^2} + (90.7x - 242.2x^2 + 42.4x^3)y^{2.18+2.82x} \\ &\quad + \frac{y+y^2+y^3-y^4}{(1-y^3)} - (14.76x - 9.76x^2 + 4.58x^3)y^2 \\ &= 0 \end{aligned}$$

$$x = \frac{1}{T_{pr}}.$$

3. Transient heat transfer function (Hasan and Kabir, 1991):

$$\begin{cases} f(t_D) = 1.1281\sqrt{t_D}\left(1 - 0.3\sqrt{t_D}\right) & (t_D \le 1.5) \\ f(t_D) = \left(1 + \dfrac{0.6}{t_D}\right)[0.4063 + 0.5\ln(t_D)] & (t_D > 1.5) \end{cases}$$

$$t_D = \frac{t\alpha}{r_{wb}^2}.$$

4. Friction factor f: $1/\sqrt{f} = 1.14 - 2\log(e/d + 2.125/Re^{0.9}$
5. Dryness fraction of gas λ: $\lambda = 1.205\gamma_g/(1000\gamma_l/GWR + 1.205\gamma_g)$ where GWR is a gas-liquid ratio γ_g and γ_l are of relative density of the gas and liquid, respectively.
6. Density of mixture ρ_m: $\rho_m = M_t/V_t$, $M_t = 1000\gamma_l + 1 \cdot 205GWR.\gamma_l$, $V_t = 1 + 0.0003458ZT/p \cdot GWR$
7. Velocity of mixture v_m: $v_m = q_t/A$, where $q_t = (1 + 0.0003458ZT/p)Q_{gsc}/GWR$, Q_{gsc} is the quantity of gas.
8. Mixture heat ratio $C_{pm} = \lambda C_{pg} + (1-\lambda)C_{pl}$, $C_{pg} = 1697.5107P^{0.0661}T^{0.0776}$, $C_{pl} =$ 4.2 kJ/(kg · °C)
9. Mixture J-T parameter $C_{Jm} = -(\lambda C_{pg}/C_{pm})C_{Jl} + (1-\lambda)/(\rho_l C_{pm})$, where $C_{pl} = -1/C_{pl}\rho_l$.

8.4.3.1 *Definite conditions*

To solve the prediction model, some definite conditions, such as the initial condition and the boundary condition are needed. The initial conditions are determined using the static gas column method. The initial conditions are made pressure, temperature and the density distribution along the well at the initial time. The bottom-hole pressure is calculated using the following formula: $P_0^k = \sqrt{(p_0)^2 + \frac{\mu q_{sc} P_{sc} T_{bh} Z_g}{2\pi\beta H_T T_{sc}} \ln \frac{4\beta t_k}{r t_{ri}^2 \Psi \mu C_t}}$.

8.4.3.2 *Solution process*
After the finite difference scheme is determined, the density, velocity, pressure and temperature can be calculated for each successive "segment" of length and time as follows.

Step 1. Set step length of depth and time. In addition, we denote the relatively tolerant error by ε. The smaller h, τ, ε are, the more accurate the results are. We set $h = 0.5$ (m), $\tau = 60$ (s) and $\varepsilon = 5\%$.
Step 2. Set $j = 0$, $k = 0$.
Step 3. Generate $(\tilde{\rho}_m)_{j+1}^{k+1}$ from interval I randomly, the I is determined by the history data and experts' experience.
Step 4. Calculate $(v_m)_{j+1}^{k+1}$.
Step 5. Calculate $(P)_{j+1}^{k+1}$.
Step 6. Calculate $(T)_{j+1}^{k+1}$.
Step 7. Calculate $(\rho)_{j+1}^{k+1}$.
Step 8. If $\left|\frac{[(\tilde{\rho}_m)_{j+1}^{k+1} - (\rho_m)_{j+1}^{k+1}]}{(\tilde{\rho}_m)_{j+1}^{k+1}}\right| \leq \varepsilon$, let $k = k + 1$ and return to step 3, otherwise return to step 3.
Step 9. Repeat from the third step to the eighth step until $j = m$, where m is the number of depth segments.
Step 10. Repeat from the third step to the ninth step until $k = n$, where n is the number of depth segments.

8.4.4 Example calculation

The calculations are performed for each successive segment of the pipe up to the surface as well as the time up to the end point of testing time.

8.4.4.1 *Parameters*
In this simulation, we study a pipe in X well, which is located in China. The needed parameters are given as following:

Depth of the well = 6115 m
Well bottom pressure = 40 MPa
Critical pressure = 4.968 MPa
Gas specific weight = 0.6434 kg/m^3
Ground thermal conductivity parameter = 2.06 W/m·K
Gas-liquid ratio = 100000
Ground temperature = 16°C
Geothermal gradient = 2.18°C/100 m
Gas outputs = 500000 m^3/d
Porosity = 0.2
Roughness of the well inner surface = 0.000015
Comprehensive compression coefficient = 0.03
One segment time = 60 s
Testing time = 1 day

The parameters of the pipes, the inclined well, the inclination, the azimuth and the vertical depth are as in Tables 8.1, 8.2 and 8.3.

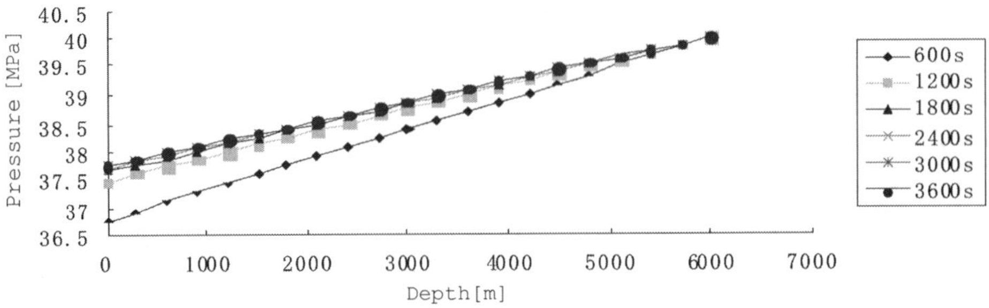

Figure 8.30 Pressure distribution at different times.

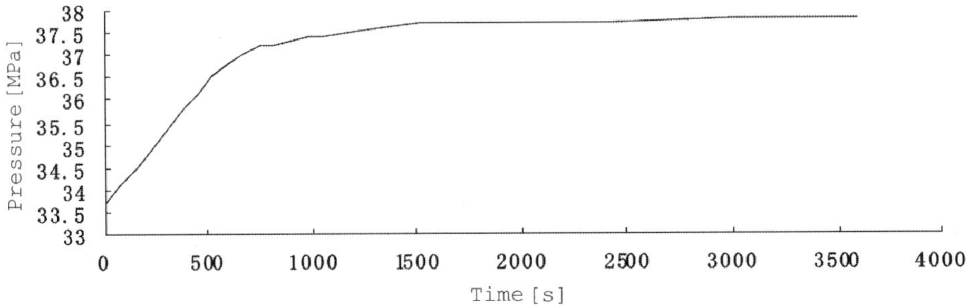

Figure 8.31 Pressure at well head.

8.4.4.2 *Results analysis*

To study how the temperature, pressure, velocity and density vary with time and depth, we show the six result curves as graphs, which represent the distribution along the well at 600 s, 1200 s, 1800 s, 2400 s, 3000 s and 3600 s respectively. In addition, in order to compare these results with the results from the steady-state model, the relative error is calculated.

Pressure: The pressure distributions along the well at different times are shown by Figure 8.30.

From Figure 8.30, it can be seen that when the time is fixed, the pressure increases with increasing depth. This is because the flow velocity increases with an increase in outputs resulting in an increase in the frictional pressure drop. When the depth is fixed, the pressure increases with time. This is because as time increases, the flow increases and the resulting frictional heat leads to an increase in the well head pressure, while the pressure difference between the well head and the bottom decreases. In addition, as can be seen from Figure 8.30, the pressure changes quickly in the early stages, but stabilizes over time. To test how the pressure of a fixed depth varies with time, we looked at the pressure at the well head , the results for which are shown in Figure 8.31.

As shown in Figure 8.31, the pressure of the well head changes quickly in the early stages, but basically stabilizes from 1200 s onwards. At 3600 s, the relative error between the pressure at the well head and the result calculated using the steady-state model is less than 1%. This shows that the pressure calculated using the transient flow model is consistent with the steady-state flow model as the flow stabilizes.

Temperature: Temperature distributions along the well at different times are shown in Figure 8.32.

From Figure 8.32, it can be seen that when the time is fixed, the temperature increases with increasing depth. When the depth is fixed, the temperature increases with the increasing time. This is because as time increases, the flow increases and the additional frictional heat leads to an increase in the well head temperature, while the temperature difference between the well head and bottom decreases. In addition, it can be seen from Figure 8.32, the temperature changes quickly in the early stages, but stabilizes over time.

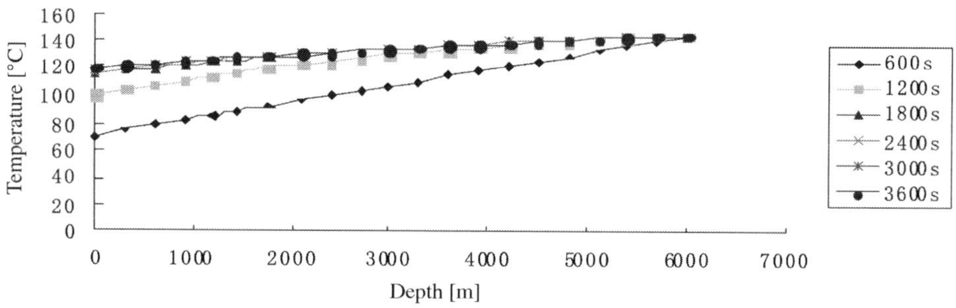

Figure 8.32 Temperature distribution at different times.

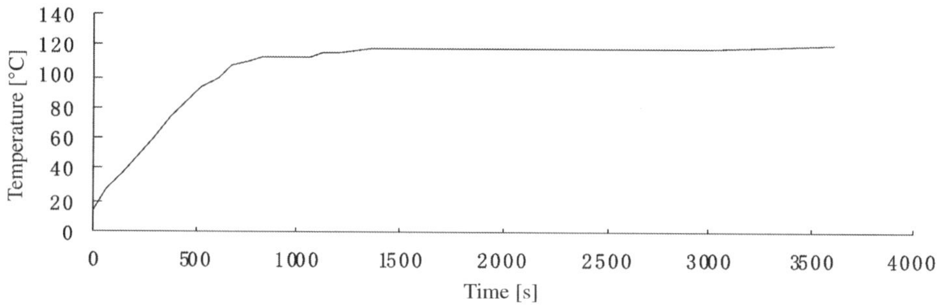

Figure 8.33 Temperature at well head.

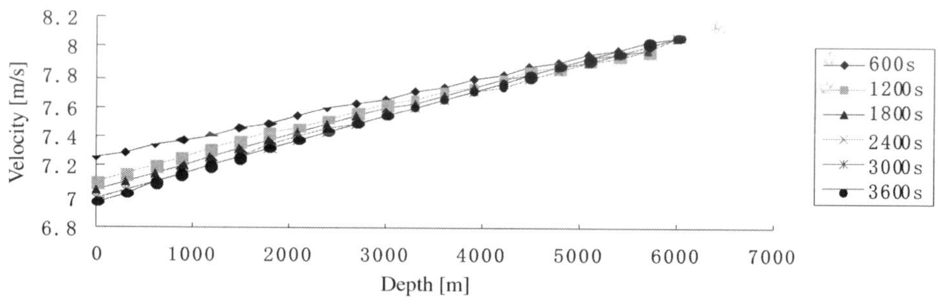

Figure 8.34 Gas phase velocity distribution at different times.

In order to test how the temperature of a fixed depth varies with time, we chose the well head and the results are shown in Figure 8.33.

As shown in Figure 8.33, the temperature at the well head changes quickly in the early stages, but basically stabilizes from 1200 s on wards. At 3600 s, the relative error between the temperature at the well head and the results calculated using the steady-state model is less than 1%. This shows that the temperature calculated using the transient flow model is consistent with the steady-state flow model when the flow stabilizes.

Velocity: The gas phase velocity distributions along the well at different times are shown in Figure 8.34.

From Figure 8.34, it can be seen that when the time is fixed, the gas phase velocity increases with increasing depth. In order to test how the velocity of a fixed depth varies with time, we chose to measure at the well head and the results are shown in Figure 8.35.

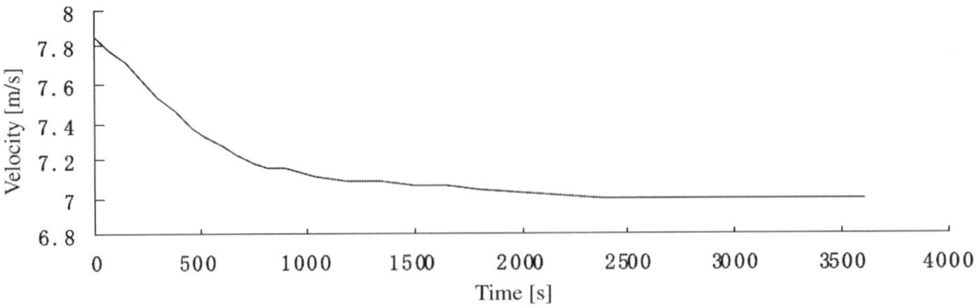

Figure 8.35 Gas phase velocity at well head.

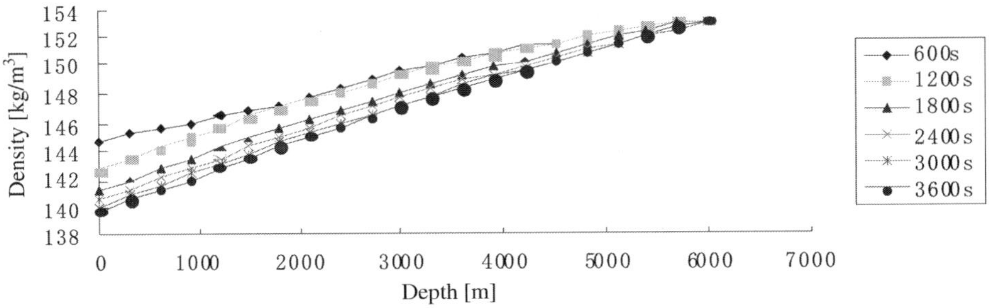

Figure 8.36 Gas phase density distribution at different times.

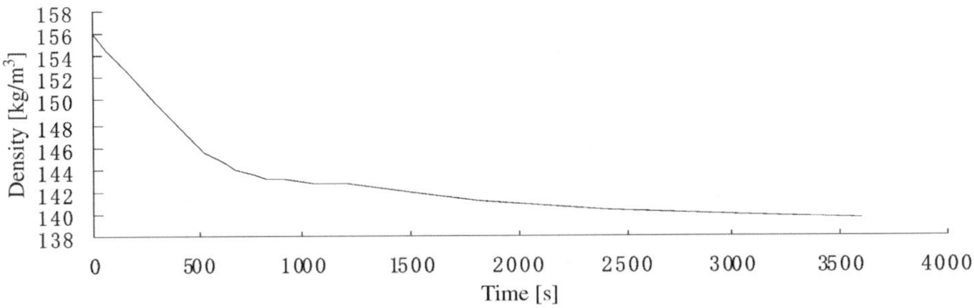

Figure 8.37 Gas phase density at well head.

As shown in Figure 8.35, the gas phase density at the well head changes quickly in the early stages, but basically stabilizes from 1800 s onwards. At 3600 s, the relative error between the gas phase velocity at the well head and the result calculated using the steady-state model is less than 1.5%, which demonstrates that the gas phase velocity calculated using the transient flow model is consistent with steady-state flow model as the flow stabilizes.

Density: Gas phase density distributions along the well at different times are shown in Figure 8.36.

From Figure 8.36, it can be seen that when the time is fixed, the gas phase density increases with increasing depth. To test how the density of a fixed depth varies with time, we chose to measure at the well head and the results are shown in Figure 8.37.

As shown in Figure 8.37, the gas phase density at the well head changes quickly in the early stages, but basically stabilizes from 1800 s onwards. At 3600 s, the relative error between the gas phase density at the well head and the result calculated using the steady-state model is less than 1.8%, which demonstrates that the gas phase density calculated using the transient flow model is consistent with the steady-state flow model as the flow stabilizes.

8.5 NMSOGW-TTBF

Considering the differential equation model for P, T, v, we have the following assumptions:

1. The flow is a one-dimensional fully developed three-phase flow with negligible heat or mass transfer between the phases.
2. When the oil/gas/water three-phase reaches a thermodynamic equilibrium, the temperatures at all points are equal in the transverse cross-section.
3. The gas-wall and the oil-wall friction shear stresses are ignored and there is a mixture-wall friction at the place of the water-wall friction shear stress.
4. When the gas-liquid two-phase reaches thermodynamic equilibrium, the temperatures at all points are equal at the transverse cross-section.
5. Oil and water are assumed to be incompressible phases since their compressibility is small.
6. The phase pressures are assumed to be equal at a given axial location, while the liquid phase is taken as a continuous phase.

8.5.1 **The coupled system model**

Within the tubing, the gas flow takes place under turbulent flow conditions. Consider the flow system depicted in Figure 8.1. Mass, momentum, and energy balances, along with the pressure, temperature and velocity, relative to the transient gas-liquid-liquid three-phase flow, as well as the stated equation are used to generate the constitutive equation.

Material balance: Taking the bottom of well as the origin of the coordinate axis and vertical up as a positive direction. Let dz and dt denote the differential depth and differential time respectively. First, we use a unified mass equation for each phase. Thus, during the time dt, the mass flowing into dz is $\rho_k V_k \alpha_k A dt$, the mass flowing out of dz is $[\rho_k V_k \alpha_k A + \frac{\partial}{\partial z}(\rho_k V_k \alpha_k A)]dt$, and the mass increment in dz caused by deformation of tube and compression of fluid is $\frac{\partial}{\partial t}(\rho_k \alpha_k A dz)dt$. It follows from mass balance law that:

$$\rho_k V_k \alpha_k A dt - \left[\rho_k V_k \alpha_k A + \frac{\partial}{\partial z}(\rho_k V_k \alpha_k A)\right] dt = \frac{\partial}{\partial t}(\rho_k \alpha_k A dz)dt \tag{8.57}$$

which can be transformed as:

$$\frac{\partial(\rho_k \alpha_k)}{\partial t} + \frac{\partial(\rho_k \alpha_k V_k)}{\partial z} = 0. \tag{8.58}$$

Thus, for each phase, we have the mass equation as follows:

Liquid flows:

$$\frac{\partial(\rho_l \alpha_l)}{\partial t} + \frac{\partial(\rho_l \alpha_l V_l)}{\partial z} = 0. \tag{8.59}$$

Gas flows:

$$\frac{\partial(\rho_g \alpha_g)}{\partial t} + \frac{\partial(\rho_g \alpha_g V_g)}{\partial z} = 0. \tag{8.60}$$

Momentum equation: A unified momentum equation can be represented as follows:

$$\underbrace{\text{Sum vector of the forces}} = \underbrace{\text{the momentum flowing out from } dz}$$
$$- \underbrace{\text{the momentum flowing into } dz + \text{the change of momentum during time } dt}.$$

The force on dz include as follows:

The pressure on cross section: $[\alpha_k P_k - (\alpha_k P_k + \frac{\partial(\alpha_k P_k)}{\partial z}]A = -\frac{\partial(\alpha_k P_k)}{\partial z} Adz$
The gravity which flows in the opposite direction: $\rho_k g \cos\theta Adz$
The friction which flows in the opposite direction: $\tau_{kb} S_{kb} dz$
The interfacial shear stress which flows in the opposite direction: $\tau_{kj} S_{kj} dz$
The momentum flowing out from dz: $\rho_k \alpha_k v_k^2 A + \frac{\partial(\rho_k \alpha_k v_k^2)}{\partial} A$
The momentum flowing into dz: $\rho_k \alpha_k v_k^2 A$
The change of momentum during time dt: $\frac{\partial(\rho_k \alpha_k v_k)}{\partial t} Adz$

The subscript k represents gas (g), liquid (l) and implies a phase that is different from phase k. From the law of momentum conservation, there have momentum equation for each phase as follows:

Liquid flows:

$$\frac{\partial(\rho_l \alpha_l v_l)}{\partial t} + \frac{\partial(\rho_l \alpha_l v_l^2)}{\partial z} + \alpha_l \frac{\partial P_l}{\partial z} = -\rho_l g \alpha_l \cos\theta - \frac{\tau_{lb} S_{lb}}{A} + \frac{\tau_{lg} S_{lg}}{A}. \tag{8.61}$$

Gas flows:

$$\frac{\partial(\rho_g \alpha_g v_g)}{\partial t} + \frac{\partial(\rho_g \alpha_g v_g^2)}{\partial z} + \alpha_g \frac{\partial P_g}{\partial z} = -\rho_g g \alpha_g \cos\theta - \frac{\tau_{gb} S_{gb}}{A} - \frac{\tau_{lg} S_{lg}}{A}. \tag{8.62}$$

Here, $\tau_{lb} S_{lb}$, $\tau_{gb} S_{gb}$ mean friction shear stress about liquid-wall and gas-wall respectively. $\tau_{lg} S_{lg}$ is the shear force between two phase. In this work, we discuss the flow is the bubbly flow ($\tau_{gb} S_{gb} = 0$). The interfacial shear stress and interfacial mass transfer among phases are neglect ($\tau_{lg} S_{lg} = 0$).

Therefore (8.61), (8.62) can be written as:

Liquid flows:

$$\frac{\partial(\rho_l \alpha_l v_l)}{\partial t} + \frac{\partial(\rho_l \alpha_l v_l^2)}{\partial z} + \alpha_l \frac{\partial P_l}{\partial z} = -\rho_l g \alpha_l \cos\theta - \frac{\tau_{lb} S_{lb}}{A}. \tag{8.63}$$

Gas flows:

$$\frac{\partial(\rho_g \alpha_g v_g)}{\partial t} + \frac{\partial(\rho_g \alpha_g v_g^2)}{\partial z} + \alpha_g \frac{\partial P_g}{\partial z} = -\rho_g g \alpha_g \cos\theta. \tag{8.64}$$

Energy equation: A unified momentum equation can be represented as follows:

$$\underbrace{\text{the translate energy to the second net}} = \underbrace{\text{the energy flowing out from } dz}$$
$$- \underbrace{\text{the energy flowing into } dz + \text{the change of energy during time } dt}.$$

The types of energy considered include; inner energy, pressure energy, kinetic energy and potential energy, where the inner energy and pressure energy are collectively referred to as enthalpy. There is no mass interchange due to phase change or a chemical reaction among the two immiscible liquids and air. So, the unified energy equation can be expressed as in Cazarez *et al.* (2010):

$$\frac{\partial(\rho_k \alpha_k e_k)}{\partial t} + \frac{\partial(\rho_k \alpha_k e_k v_k)}{\partial z} = \alpha_k \frac{\partial P_k}{\partial t} + \rho_k g \alpha_k \rho_k \cos\theta + dQ. \tag{8.65}$$

Combined mass equation, we have:

$$\rho_k\alpha_k\left(\frac{\partial e_k}{\partial t}+v_k\frac{\partial e_k}{\partial z}\right)=\alpha_k\frac{\partial P_k}{\partial t}+\rho_k g\alpha_k\rho_k\cos\theta+dQ, \tag{8.66}$$

where, $e_k=h_k+\frac{v_k^2}{2}$, h_k, which satisfies following relation, denotes specific enthalpy:

$$\begin{cases}\dfrac{\partial h_k}{\partial z}=C_{p_k}\dfrac{\partial t_k}{\partial z}-\eta_k\dfrac{\partial P_k}{\partial z}\\ \dfrac{\partial h_k}{\partial t}=C_{p_k}\dfrac{\partial t_k}{\partial t}-\eta_k\dfrac{\partial P_k}{\partial t}\end{cases} \tag{8.67}$$

η_k is J-T coefficient defined as:

$$\begin{cases}\eta_k=0, \quad k=g\\ \eta_k=-\dfrac{1}{C_{p_k}l}, \quad k=l\end{cases} \tag{8.68}$$

In Equation (8.65), dQ denotes the radial heat transfer between the gas and the surrounding earth tube. Ramey (1962) and Willhite (1967) discussed the radial transfer of heat between the fluid and the earth in detail. As in (8.2), the radial heat transfer from the fluid to the cement/earth interface can be described by:

$$dQ=\alpha_k\frac{2\pi r_{to}U_{to}K_e}{r_{to}U_{to}f(t_D)+K_e}(T-T_e)dz. \tag{8.69}$$

Let $a=\frac{2\pi r_{to}U_{to}K_e}{r_{to}U_{to}f(t_D)+K_e}$, using the relationship of Equations (8.65), (8.66), (8.67), (8.68), (8.69), we have the gas energy equation is:

$$C_{P_g}\left(\frac{\partial T_g}{\partial t}+v_g\frac{\partial T_g}{\partial z}\right)+v_g\left(\frac{\partial v_g}{\partial t}+v_g\frac{\partial v_g}{\partial z}\right)-\frac{1}{\rho_g}\frac{\partial P_g}{\partial t}+v_g g\cos\theta-\frac{a\alpha_g(T-T_e)}{\rho_g}=0. \tag{8.70}$$

Similarly, the liquid flow energy equation can be written as:

$$\alpha_l\rho_l C_{P_l}\left(\frac{\partial T_l}{\partial t}+v_l\frac{\partial T_l}{\partial z}\right)+\alpha_l\rho_l v_l\left(\frac{\partial v_l}{\partial t}+v_l\frac{\partial v_l}{\partial z}\right)-\alpha_l v_l\frac{\partial P_l}{\partial t}+\alpha_l\rho_l v_l g\cos\theta-a\alpha_l(T-T_e)=0. \tag{8.71}$$

By combining Equations (8.59), (8.60), (8.63), (8.64), (8.70), (8.71), we can determine the coupled differential equations system model as follows:

$$\begin{cases}\dfrac{\partial(\rho_l\alpha_l)}{\partial t}+\dfrac{\partial(\rho_l\alpha_l V_l)}{\partial z}=0\\ \dfrac{\partial(\rho_g\alpha_g)}{\partial t}+\dfrac{\partial(\rho_g\alpha_g V_g)}{\partial z}=0\\ \dfrac{\partial(\rho_l\alpha_l v_l)}{\partial t}+\dfrac{\partial(\rho_l\alpha_l v_l^2)}{\partial z}+\alpha_l\dfrac{\partial P_l}{\partial z}=-\rho_l g\alpha_l\cos\theta-\dfrac{\tau_{lb}S_{lb}}{A}\\ \dfrac{\partial(\rho_g\alpha_g v_g)}{\partial t}+\dfrac{\partial(\rho_g\alpha_g v_g^2)}{\partial z}+\alpha_g\dfrac{\partial P_g}{\partial z}=-\rho_g g\alpha_g\cos\theta\\ C_{P_g}\left(\dfrac{\partial T_g}{\partial t}+v_g\dfrac{\partial T_g}{\partial z}\right)+v_g\left(\dfrac{\partial v_g}{\partial t}+v_g\dfrac{\partial v_g}{\partial z}\right)-\dfrac{1}{\rho_g}\dfrac{\partial P_g}{\partial t}+v_g g\cos\theta-\dfrac{a\alpha_g(T-T_e)}{\rho_g}=0\\ \alpha_l\rho_l C_{P_l}\left(\dfrac{\partial T_l}{\partial t}+v_l\dfrac{\partial T_l}{\partial z}\right)+\alpha_l\rho_l v_l\left(\dfrac{\partial v_l}{\partial t}+v_l\dfrac{\partial v_l}{\partial z}\right)-\alpha_l v_l\dfrac{\partial P_l}{\partial t}+\alpha_l\rho_l v_l g\cos\theta\\ \quad -a\alpha_l(T-T_e)=0,\end{cases} \tag{8.72}$$

where, $\alpha_l = \alpha_w + \alpha_o$, $\rho_l = \frac{\alpha_w \rho_w + \alpha_o \rho_o}{\alpha_l}$, $P_l = P_g = P$, $\alpha_l + \alpha_g = 1$, $C_{P_l} = \frac{\rho_o \alpha_o C_{P_o} + \rho_w \alpha_w C_{P_w}}{\rho_l \alpha_l}$, $T_g = T_l = T$.

8.5.2 Model analysis

A two-fluid model system is obviously much more complex to compute than a one-fluid system. For one thing, the number of differential equations for a two-fluid model is twice that of a one-fluid model. Also, the non-linearity and coupling of the equations are much stronger owing to the interactions between the two-fluids. The equations are coupled not only within those for the same phase, but also among those for different phases. However, the solution to the coupled equations is highly sensitive to the initial values. Numerical efficiency is also strongly affected by the methods used for discretization and iteration, and relaxation. Because of the coupling of the equations, to solve the equations directly is extremely complicated. Here, we put forward a simplified solution model.

First, we need a pressure equation: $P = \rho_g g v_g A$. So,

$$\alpha_g \frac{\partial P}{\partial t} + v_g \alpha_g \frac{\partial P}{\partial z} + P \frac{\partial v_g}{\partial z} = -P v_g \frac{\partial \alpha_g}{\partial z} - P \frac{\partial \alpha_g}{\partial t} \tag{8.73}$$

$$-(1-\alpha_g) \frac{\partial v_l}{\partial z} = -\frac{\partial \alpha_g}{\partial z} - v_l \frac{\partial \alpha_g}{\partial z}. \tag{8.74}$$

Then, to eliminate $\frac{\partial \alpha_g}{\partial z}$, we can get a new equation as:

$$\frac{\alpha_g}{P} \frac{\partial P}{\partial t} + \frac{v_g \alpha_g}{P} \frac{\partial P}{\partial z} + (v_g - v_l) \frac{\partial \alpha_g}{\partial z} + \alpha_g \frac{\partial v_g}{\partial z} + (1-\alpha_g) \frac{\partial v_l}{\partial z} = 0. \tag{8.75}$$

Substituting the Equations (8.59), (8.60) into the Equations (8.63), (8.64), the following equations are obtained:

$$\frac{\partial v_l}{\partial t} + v_l \frac{\partial v_l}{\partial z} + \frac{1}{\rho_l} \frac{\partial P}{\partial z} - \frac{P}{\rho_l (1-\alpha_g)} \frac{\partial \alpha_g}{\partial z} = \frac{\tau_{lb} S_{lb} - \rho_l g (1-\alpha_g) A^2 \cos\theta}{\rho_l (1-\alpha_g) A^2} \tag{8.76}$$

$$\frac{\partial v_g}{\partial t} + v_g \frac{\partial v_g}{\partial z} + v_g \frac{\partial v_g}{\partial z} + \frac{1}{\rho_g} \frac{\partial P}{\partial z} + \frac{P}{\rho_g \alpha_g} \frac{\partial \alpha_g}{\partial z} = g \cos\theta. \tag{8.77}$$

These modified equations can be written in a compact form:

$$A \frac{\partial U}{\partial t} + B \frac{\partial U}{\partial z} = D, \tag{8.78}$$

where A and B are coefficient matrices, D is a vector containing all algebraic terms and U is the solution vector:

$$A = \begin{bmatrix} \frac{\alpha_g}{P} & 0 & 0 \\ 0 & \rho_g \alpha_g A & 0 \\ 0 & 0 & \rho_l (1-\alpha_l) A \end{bmatrix}$$

$$B = \begin{bmatrix} \dfrac{v_g\alpha_g}{P} & \alpha_g & 1-\alpha_g \\ A\alpha_g & \rho_g\alpha_g A v_g & 0 \\ A(1-\alpha_g) & 0 & \rho_l(1-\alpha_l)v_l A \end{bmatrix}$$

$$D = \begin{bmatrix} -(v_g - v_l)\dfrac{\partial\alpha_g}{\partial z} \\ -\tau_{lb}S_{lb} - \rho_g\alpha_g gA\cos\theta \\ -\rho_l g(1-\alpha_g)A\cos\theta + AP\dfrac{\partial\alpha_g}{\partial z} \end{bmatrix}$$

$$U = \begin{bmatrix} P & v_g & v_l \end{bmatrix}$$

By solving the above equations, we can obtain the value of vector U. We are then able to determine the temperature at every point of the well using the value of U in the energy equation. In this way, we can reduce the complexity and the coupling of the system.

8.5.3 **Numerical solution**

There are several numerical methods for solving partial differential equations, such as finite difference methods, finite element methods and finite volume methods. The finite difference technique is far more widely used than any other technique. This technique is implemented by replacing all derivatives with difference quotients. When the geometry is not complicated as in the case of pipes, the finite difference method is easier and faster than other methods. Generally, we often use an implicit scheme method. The two methods both have advantages and disadvantages. The implicit scheme method has good stability but poor computing time, whereas the characteristic line difference method has a rapid calculation but some time step limits. Here, we use the implicit scheme difference method.

The grid arrangement is staggered as described by Harlow and Welch (1965). The staggered grid arrangement for a one-dimensional domain is shown in Figure 8.3. The advantage of the grid is that it can avoid unreasonable pressure differences.

The equations are hyperbolic systems of the first-order, so in this case we usually use a difference scheme, such as the Lax-Friedrichs form (Xu *et al.*, 2013b), or the Lax-Wendroff form (Gourlay, 1968) and the upstream form. The systems discretization was determined by applying a first order upstream implicit scheme to the spatial derivatives and a first completely implicit scheme to the time derivatives. The model 8.78 can be written as a matrix in a discretized form as:

$$E_j(u_j^0)u_j^{t+\delta t} = F_j(u_j^0, u_j^t, u_{j-1}^{t+\delta t}), \tag{8.79}$$

where the superscripts t and $t+\delta t$ indicate that the dependent variables are calculated at the old and new times, respectively, and are also the cell number where the variable is calculated. In (8.79), the variables with the subscript $j-1$ and the superscript t are known since these are the inlet variables and the initial condition, respectively. Also, in these equations the superscript o represents the dummy variables for the iterative method and u is the column vector for the dependent variables which are given by: $u = [\rho \quad v_g \quad v_l]^T$.

The Equations (8.79) are linear equations and there are many solutions. After considering the many large and sometimes ill-posed linear systems, we adopt the Generalized Minimum Residual Method for the solution. Then, we use the same difference scheme as in the energy equations by putting in the pressure and velocity values, we can easily derive the temperature value.

8.5.4 Calculation of some parameters

In this section, we will give the calculating method of some parameters.

Step 1. Obtain each point's inclination:

$$\theta_j = \theta_{j-1} + \frac{\theta_k - \theta_{k-1}}{\Delta s_k}\Delta s_j,$$

where j represents segment point of calculation, Δs_k represents measurement depth of inclination angle θ_k and θ_{k-1}, Δs_j step length of calculation.

Step 2. Transient heat transfer function (Hasan and Kabir (1991)):

$$\begin{cases} f(t_D) = 1.1281\sqrt{t_D}\left(1 - 0.3\sqrt{t_D}\right) & (t_D \le 1.5) \\ f(t_D) = \left(1 + \dfrac{0.6}{t_D}\right)[0.4063 + 0.5\ln(t_D)] & (t_D > 1.5) \end{cases}$$

$$t_D = \frac{t\alpha}{r_{wb}^2}.$$

Step 3. Friction between liquid and wall:

$$\frac{\tau_{lb}S_{lb}}{A} = \frac{f\rho_l v_l^2}{2d}.$$

Step 4. Friction factor f:

$$\frac{1}{\sqrt{f}} = 1.14 - 2\lg\left(\frac{0.00001524}{r_{ti}} + \frac{21.25}{Re^{0.9}}\right).$$

Step 5. Gas condensing parameter Z_g:

If ($P < 35$ MPa)

$$Z_g = 1 + \left(0.31506 - \frac{1.0467}{T_{pr}} - \frac{0.5783}{T_{pr}^3}\right)\rho_{pr} + \left(0.053 - \frac{0.6123}{T_{pr}}\right)\rho_{pr}^2 + 0.6815\frac{\rho_{pr}^2}{T_{pr}^3},$$

where, $\rho_{pr} = 0.27\frac{P_{pr}}{T_{pr}}$, $T_{pr} = \frac{T}{T_{pc}}$, $P_{pr} = \frac{P}{P_{pc}}$.

Else

$$Z_g = (90.7x - 242x^2 + 42.4x^3)y^{(2.18+2.82x)} - (14.76x - 9.76x^2 + 4.58t^3)y + \frac{1 + y + y^2 + y^3}{(1-y)^3},$$

where,

$$\begin{aligned} F(y) &= -0.06125P_{pr}xe^{-1.2(1-x)^2} + (90.7x - 242.2x^2 + 42.4x^3)y^{2.18+2.82x} \\ &\quad + \frac{y + y^2 + y^3 - y^4}{(1-y^3)} - (14.76x - 9.76x^2 + 4.58x^3)y^2 \\ &= 0 \end{aligned}$$

$$x = \frac{1}{T_{pr}}.$$

8.5.5 Initial condition and boundary condition

To solve model 8.79, the initial conditions and the boundary conditions need to be added. The initial conditions are made up of the pressure temperature and velocity distribution along the well at the initial time or at the steady-state. Here, we adopt the value at the initial time using actual measurements. The boundary conditions set are the gas and liquid flow at the bottom and the pressure at the top.

8.5.5.1 *Solution*

Based on the afore-mentioned discussion, the algorithm is designed as follows.

Step 1. Set the step length for the depth and time. Compute each point's inclination.
Step 2. Given the initial conditions and boundary conditions.
Step 3. Compute the parameters for matrix E and F under the initial conditions or the last times variables.
Step 4. Calculate the linear equations to determine the value of vector u.
Step 5. Calculate the mass formulas under the new velocity value.
Step 6. Calculate the temperature T using the energy formulas under new pressure and velocity values. However, the equations in this step are redundant. Therefore, because an temperature value in each phase has been obtained, we adopt a average value in this step: $T = \frac{T_g + T_l}{2}$.
Step 7. Calculate the gas density, because a temperature value in each phase has been density using the gas state equation: $\rho_g = \frac{MP}{RZ_g T}$.
Step 8. Repeat Steps 3 to 6 until the time step is reached.

8.5.6 Example calculation

As described previously, the algorithm begins with a calculation for the pipe at the bottom of the pipe at initial time. The calculations are performed for each successive segment of the pipe up to the surface as well as the time up to the testing time end point.

8.5.6.1 *Parameters*

In this simulation, we study a pipe in X well, which is located in China. The parameters are as follows:

Depth of the well = 7100 m
Well bottom pressure = 40 MPa
Critical pressure = 4.968 MPa
Gas specific weight = 0.6434 kg/m^3
Ground thermal conductivity parameter = 2.06 W/m·K
Gas-liquid ratio = 100000
Ground temperature = 16°C
Geothermal gradient = 2.18°C/100 m
Gas outputs = 500000 m^3/d
Roughness of the inner surface of the well = 0.000015
Gas compression coefficient = 0.03
Time of one segment = 60 s
Testing time = 1 day

The parameters for the pipes, inclined well, inclination, azimuth and vertical depth are given in Tables 8.1, 8.2 and 8.3.

8.5.6.2 *Results analysis*

The pressure distributions along the well at different times are shown in Figure 8.38. From Figure 8.38, it can be seen that when the time is fixed, the pressure increases with increasing

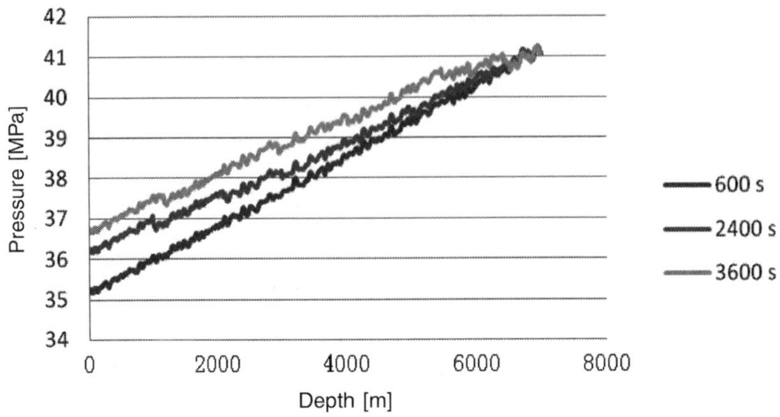

Figure 8.38 Pressure distribution at different times.

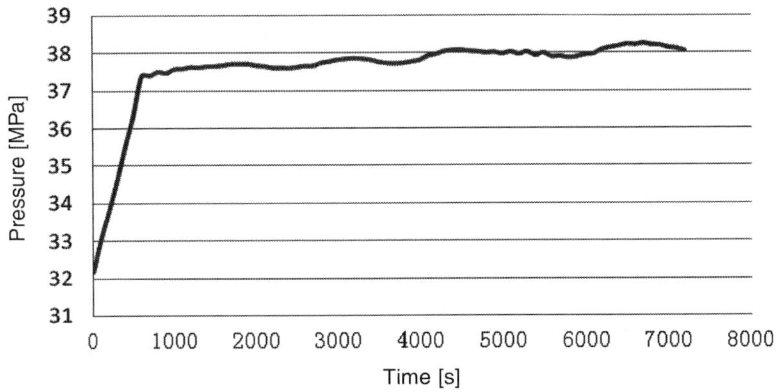

Figure 8.39 Pressures at well head.

depth. When the depth is fixed, the pressure increases as the time increases. In addition, as can be seen from Figure 8.39, the pressure changes quickly in the early stages, but stabilizes over time.

The gas phase velocity distributions along the well at different times are shown in Figure 8.40. From Figure 8.40, it can be seen that when the time is fixed, the gas phase velocity increases with increasing depth. As shown in Figure 8.41, the gas phase velocity at the well head changes quickly in the early stages, but basically stabilizes from 1000 s onwards. However, from Figure 8.42, it can be seen that when the time is fixed, the liquid phase velocity fluctuates from 4 to 7 m/s and has a very weak mounting tendency. The liquid phase velocity at the well head also has a little change in form as shown in Figure 8.43. As was expected the gas velocity is much greater than that of the liquid because of its smaller density.

The temperature profiles are shown in Figure 8.44. From Figure 8.44, it can be seen that when the time is fixed, the temperature increases along with increasing depth. This is because as the

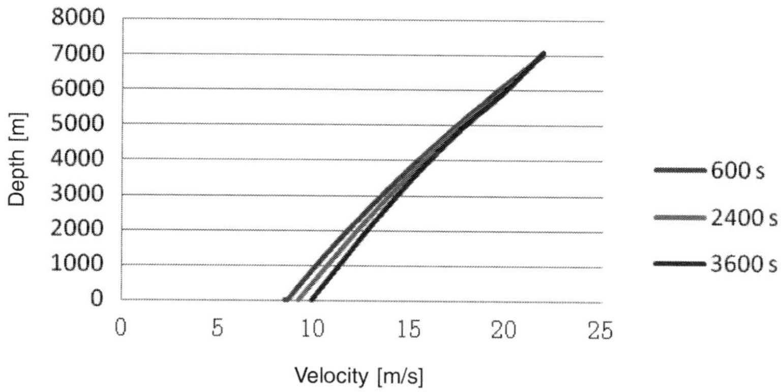

Figure 8.40 Gas phase velocity distribution at different times.

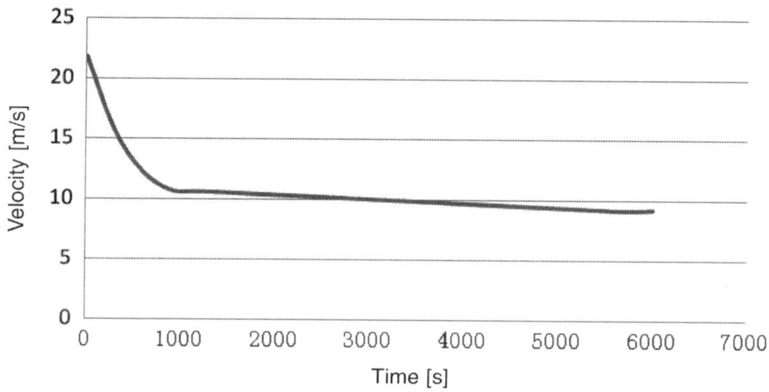

Figure 8.41 Gas phase velocity at well head.

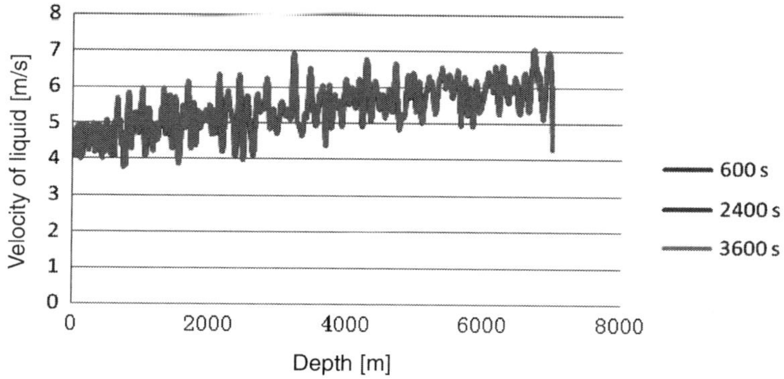

Figure 8.42 Liquid phase velocity distribution at different times.

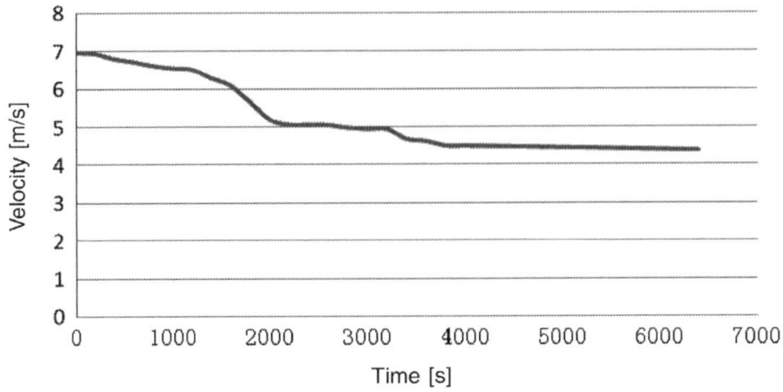

Figure 8.43 Liquid phase velocity at well head.

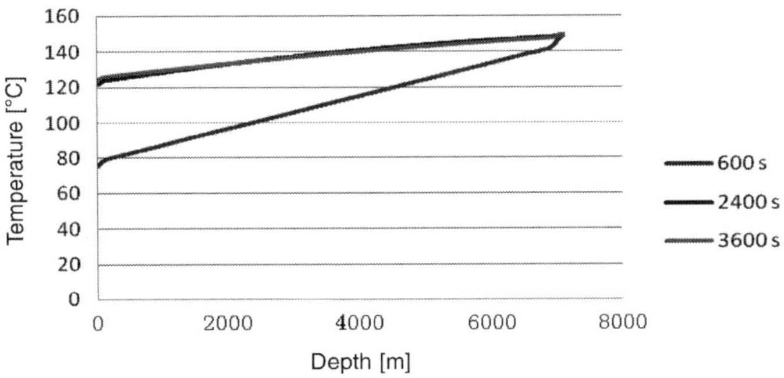

Figure 8.44 Temperature distribution at different times.

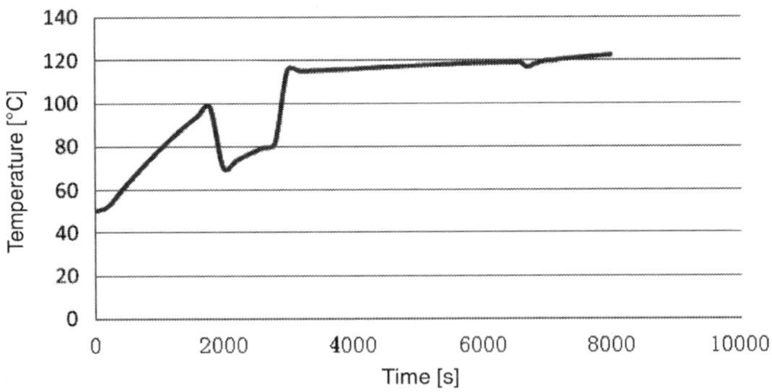

Figure 8.45 Temperatures at well head.

time increases, the flow increases and this increase in frictional heat leads to an increase in the well head temperature, while the temperature difference between the well head and the bottom decreases. In addition, as can be seen from Figure 8.45, the temperature changes quickly in the early stages, but stabilizes over time.

8.6 PDPTVD-TBF

Considering the differential equation model for P, T, v, a, we set the following assumptions:

1. The gas-liquid-liquid flow in the tubing is in one dimension of the flow direction with negligible heat or mass transfer between the phases.
2. When the oil/gas/water three-phase reaches a thermodynamic equilibrium, the temperature at all points is equal at the transverse cross-section.
3. The gas-wall and the oil-wall friction shear stresses, such as the mixture-wall friction at the point of the water-wall friction shear stress, are ignored.
4. Oil and water are assumed to be incompressible phases as their compressibility is small.
5. Phase pressures are assumed to be equal at a given axial location, where the liquid phase is taken as a continuous phase.

8.6.1 **The coupled system model**

Within the tubing, the gas flow takes place under turbulent flow conditions. Consider the flow system depicted in Figure 8.1, a straight cylindrical flow tube with an inclination angle θ, a constant cross-sectional flow area A, a hydraulic diameter d, and a total length Z. Through this tubing the gas flows from the bottom to the top with a mass flow rate W. The distance co-ordinate in the flow direction along the tubing is denoted z.

The mass, momentum, and energy balances, along with the gas dryness fraction, pressure and the temperature in relation to the gas, are used to generate the constitutive equations. The mass, momentum, and energy balances, along with the pressure, temperature and velocity, relative to the steady gas-liquid-liquid three-phase flow, as well as the stated equation for gas are used to generate the constitutive equation.

Material balance: In this work, the interfacial mass transfer among the phases was neglected, since a phase change (condensation or evaporation) did not occur. We take the top of the well as the origin of the coordinate axis and vertical up as a positive direction. Let dz denote the differential depth. Thus, under a steady-state condition, the equations solved for each phase follow the law of fluid dynamics, as in (Bonizzi, 2003):

Liquid phase (mixture) continuity equation:

$$\frac{d(\rho_l \alpha_l v_l A)}{dz} = 0. \tag{8.80}$$

Gas continuity equation:

$$\frac{d(\rho_g \alpha_g v_g A)}{dz} = 0. \tag{8.81}$$

Momentum equation: The unified momentum equation as follows:

$$\underbrace{\text{Sum vector of the forces}} = \underbrace{\text{the momentum flowing out from } dz} - \underbrace{\text{the momentum flowing into } dz}$$

The force on dz include as follows:

(1) The pressure on the cross section:

$$\left[\alpha_k P_k - \left(\alpha_k P_k + \frac{d(\alpha_k P)}{dz}\right)\right] A = -\frac{d(\alpha_k P)A}{dz}$$

(2) The gravity which flows in the opposite direction:

$$\rho_k g \cos\theta A\, dz$$

(3) The friction which flows in the opposite direction:

$$\tau_{kb} S_{kb}\, dz$$

(4) The interfacial shear stress which flows in the opposite direction:

$$\tau_{kj} S_{kj}\, dz$$

(5) The momentum flowing out from dz:

$$\rho_k \alpha_k v_k^2 A + A \frac{d(\rho_k \alpha_k v_k^2)}{dz}$$

(6) The momentum flowing into dz:

$$\rho_k \alpha_k v_k^2 A$$

The subscript k represents the gas (g), and liquid (l) and implies a phase that is different from phase k. From the law of momentum conservation, we have a momentum equation for each phase as follows:

Liquid phase (mixture) momentum equation:

$$\frac{d(\rho_l \alpha_l v_l^2)}{dz} + \frac{d(\alpha_l P)}{dz} = -\rho_l g \alpha_l \cos\theta - \frac{\tau_{lb} S_{lb}}{A} + \frac{\tau_{lg} S_{lg}}{dz} \tag{8.82}$$

Gas momentum equation:

$$\frac{d(\rho_g \alpha_g v_g^2)}{dz} + \frac{d(\alpha_g P)}{dz} = -\rho_g g \alpha_g \cos\theta - \frac{\tau_{gb} S_{gb}}{A} + \frac{\tau_{gl} S_{gl}}{dz} \tag{8.83}$$

Here, $\tau_{lb} S_{lb}$, $\tau_{gb} S_{gb}$ mean friction shear stress about gas-wall and liquid-wall respectively. $\tau_{lg} S_{lg}$ is the shear force between two phase.

Energy equation: The unified momentum equation is as follows:

$$\underbrace{\text{the translate energy the second net}} = \underbrace{\text{the energy flowing out from } dz}$$
$$- \underbrace{\text{the energy flowing into } dz}.$$

The types of energy considered in this paper include inner energy, pressure energy, kinetic energy and potential energy. The energy flowing out from dz can be given by:

(1) The internal energy: $U_k(z)$
(2) The kinetic energy: $\frac{m_k v_k^2(z)}{2}$
(3) The potential energy: $m_k g z \cos\theta$
(4) The pressure energy: $P_k(z) v_k(z)$

The energy flowing out from dz can be given by:

(1) The internal energy: $U_k(z + \Delta z)$
(2) The kinetic energy: $\frac{m_k v_k^2(z+\Delta z)}{2}$
(3) The potential energy: $m_k g(z + \Delta z)\cos\theta$
(4) The pressure energy: $P_k(z + \Delta z) v_k(z + \Delta z)$.

where, the inner energy and pressure energy are collectively referred to as enthalpy, and dQ represents the tubing heat radial.

According to the rule of the energy balance: the energy of the fluid flowing into the infinitesimal equals the energy sum of losses and fluid flowing out from the infinitesimal:

$$H_k(z)+\frac{1}{2}m_k v_k^2(z)-m_k gz\cos\theta=H_k(z+\Delta z)+\frac{1}{2}m_k v_k^2(z+\Delta z)-m_k g(z+\Delta z)\cos\theta+dQ. \tag{8.84}$$

Equation (8.84) can be written as:

$$\frac{dH_k}{dz}+m_k v_k\frac{dv_k}{dz}-m_k g\cos\theta+\frac{dQ}{dz}. \tag{8.85}$$

Divided by m_k:

$$\frac{dq_k}{dz}+v_k\frac{dv_k}{dz}-g\cos\theta+\frac{dh_k}{dz} \tag{8.86}$$

h_k, which satisfies the following relation, denotes specific enthalpy:

$$\frac{dH_k}{dz}=C_{Pk}\frac{dT_k}{dz}-\eta_k C_{Pk}\frac{dP_k}{dz} \tag{8.87}$$

η_k is the J-T coefficient which is defined as:

$$\begin{cases}\eta_k=0, & \text{if } k=g\\ \eta_k=-\dfrac{1}{C_{Pk}\rho_k}, & \text{if } k=l\end{cases}. \tag{8.88}$$

In equation (8.86), dq_k denotes the radial heat transfer between the gas and the surrounding earth tube. Ramey (1962) and Willhite (1967) discussed the radial transfer of heat between the fluid and the earth in detail. As in Figure 8.2, the radial heat transfer from the fluid to the cement/earth interface can be described by:

$$dq_k=\frac{2\pi r_{to}U_{to}(T-T_k)}{W}dz. \tag{8.89}$$

And the radial heat transfer from the cement/earth interface to the surrounding earth is:

$$dq_k=\frac{2\pi K_e(T_k-T_e)}{Wf(t_D)}dz. \tag{8.90}$$

Thus the radial heat transfer between the gas and the surrounding earth tube is:

$$dq_k=\frac{2\pi r_{to}U_{to}K_e(T-T_e)}{W(r_{to}U_{to}f(t_D)+K_e)}dz. \tag{8.91}$$

Let $a=\frac{2\pi r_{to}U_{to}K_e}{W(r_{to}U_{to}f(t_D)+K_e)}$, substituting 8.87, 8.88 and 8.91 into 8.86, we have energy equation for gas phase as follows:

$$C_{Pg}\rho_g v_g\frac{dT}{dz}+\rho_g v_g^2\frac{dv_g}{dz}+v_g\rho_g g\cos\theta-a_g(T-T_e)=0. \tag{8.92}$$

Similarly, the liquid phase (mixture) energy equation can be written as:

$$C_{Pl}\rho_l v_l\frac{dT}{dz}+\rho_l v_l^2\frac{dv_l}{dz}+v_l\rho_l g\cos\theta-a_l(T-T_e)=0. \tag{8.93}$$

We can also add a gas state equation:

$$\rho_g=\frac{MP}{RZ_gT}. \tag{8.94}$$

Combining (8.80), (8.81), (8.82), (8.83), (8.92), (8.93) and (8.94), the coupled system model of differential equations can be presented as follows:

$$\begin{cases} \dfrac{d(\rho_l\alpha_l v_l A)}{dz}=0 \\ \dfrac{d(\rho_g\alpha_g v_g A)}{dz}=0 \\ \dfrac{d(\rho_l\alpha_l v_l^2)}{dz}+\dfrac{d(\alpha_l P)}{dz}=-\rho_l g\alpha_l\cos\theta-\dfrac{\tau_{lb}S_{lb}}{A}+\dfrac{\tau_{lg}S_{lg}}{dz} \\ \dfrac{d(\rho_g\alpha_g v_g^2)}{dz}+\dfrac{d(\alpha_g P)}{dz}=-\rho_g g\alpha_g\cos\theta-\dfrac{\tau_{gb}S_{gb}}{A}+\dfrac{\tau_{gl}S_{gl}}{dz} \\ C_{Pg}\rho_g v_g\dfrac{dT}{dz}+\rho_g v_g^2\dfrac{dv_g}{dz}+v_g\rho_g g\cos\theta-a_g(T-T_e)=0 \\ C_{Pl}\rho_l v_l\dfrac{dT}{dz}+\rho_l v_l^2\dfrac{dv_l}{dz}+v_l\rho_l g\cos\theta-a_l(T-T_e)=0 \\ \rho_g=\dfrac{MP}{RZ_gT} \\ P_l=P_g=P,\quad \alpha_l+\alpha_g=1,\quad T_l=T_g=T. \end{cases} \tag{8.95}$$

The primal conditions being as follows:

$$P(z_0)=P_0,\quad T(z_0)=T_0,\quad \rho_g(z_0)=\frac{MP_0}{RZ_g},\quad v(z_0)=\frac{w}{A\rho_0}.$$

8.6.2 Model analysis

A two-fluid model system is obviously more complex to compute than a one-fluid system. For one thing, the number of differential equations for a two-fluid model is twice that of a one-fluid model. Also, the non-linearity and coupling of the equations are much stronger owing to the interactions between the two-fluids. The equations are coupled not only within those for the same phase, but also among those for different phases.

To solve Equations (8.83), we can also use the four-order Runge-Kutta method. In the solution process, we found that the coupled equations were highly sensitive to gas holdup and its value usually exceeds the bounds in the actual calculation as its range (0 1) is small, so this result was considered unreasonable. From this is can be concluded that gas holdup should not be a variable. We applied a simplified model to eliminate this factor and to determine the other value of the variable. The pressure equation is defined as:

$$P=\rho_g g v_g A. \tag{8.96}$$

Substituting the (8.96) into the continuity equations for the gas and liquid phase, the gas and liquid material balance equation can be expressed as:

$$v_g\alpha_g\frac{dP}{dz}+Pv_g\frac{d\alpha_g}{dz}+P\alpha_g\frac{dv_g}{dz}=0 \tag{8.97}$$

$$-(1-\alpha_g)\frac{dv_l}{dz}=-v_l\frac{d\alpha_g}{dz}. \tag{8.98}$$

Substituting the Equations (8.97) and (8.98) into the momentum equation, the following equations can be obtained:

$$\rho_g\alpha_g v_g\frac{dv_g}{dz}+\alpha_g\frac{dP}{dz}+P\frac{d\alpha_g}{dz}=-\rho_l g\alpha_l\cos\theta-\frac{\tau_{lb}S_{lb}}{A}+\frac{\tau_{lg}S_{lg}}{dz} \tag{8.99}$$

$$\rho_l\alpha_l v_l\frac{dv_l}{dz}+\alpha_l\frac{dP}{dz}+P\frac{d\alpha_l}{dz}=-\rho_g g\alpha_g\cos\theta-\frac{\tau_{gl}S_{gj}}{dz}. \tag{8.100}$$

Substituting the Equations (8.98) and (8.99) into eliminate the shear force between the two phases, the equation is:

$$\rho_g\alpha_g v_g\frac{dv_g}{dz}+\rho_l\alpha_l v_l\frac{dv_l}{dz}+\frac{dP}{dz}=-\rho_l g\alpha_l\cos\theta-\rho_g g\alpha_g\cos\theta-\frac{\tau_{lb}S_{lb}}{A}. \tag{8.101}$$

Combining (8.92), (8.93), (8.97), (8.98) and (8.99), a new coupled differential equations system for pressure, gas velocity and liquid velocity is determined:

$$\begin{cases}\dfrac{dv_l}{dz}=\dfrac{v_l}{1-\alpha_g}\dfrac{d\alpha_g}{dz}\\[2ex] \dfrac{dv_g}{dz}=\dfrac{Pv_g\frac{d\alpha_g}{dz}-\alpha_g v_g\left[g\cos\theta(\alpha_g\rho_g+\alpha_l\rho_l)+\frac{\tau_{lb}S_{lb}}{A}+\alpha_l\rho_l v_l\frac{dv_l}{dz}\right]}{\rho_g\alpha_g^2v_g^2-P\alpha_g}\\[2ex] \dfrac{dP}{dz}=\dfrac{-P\alpha_g\frac{dv_g}{dz}-Pv_g\frac{d\alpha_g}{dz}}{\alpha_g v_g}\\[2ex] \dfrac{dT_l}{dz}=-\dfrac{v_l}{C_{Pl}}\dfrac{dv_l}{dz}-\dfrac{1}{\rho_l C_{Pl}}\dfrac{dP}{dz}-\dfrac{g\cos\theta}{C_{Pl}}-\dfrac{\alpha_l(T_{T_e})}{\rho_l v_l C_{Pl}}\\[2ex] \dfrac{dT_g}{dz}=-\dfrac{v_g}{C_{Pg}}\dfrac{dv_g}{dz}-\dfrac{g\cos\theta}{C_{Pg}}-\dfrac{\alpha_g(T_{T_e})}{\rho_g v_g C_{Pg}}\\[2ex] P(z_0)=P_0,\quad v_g(z_0)=v_{g0},\quad v_l(z_0)=v_{l0},\quad T(z_0)=T_0,\quad \rho_{g0}=\dfrac{MP_0}{RZ_gT_0}.\end{cases} \tag{8.102}$$

To determine the numerical solution, we start with a guess value of α_g at the bottom of the pipe using an experience value for the well (generally a low value). Then we determine the value of the pressure, gas velocity and liquid velocity at the same point. The continuity equations are tested and if they are not satisfied, a new guess value (a higher value) is used and this step is repeated. At the next computation point, we use the value of α_g at the last point. We could also obtain the temperature value using the value of the pressure, gas velocity and liquid velocity at every point using energy equations. It is clear that we would get two temperature values using the gas and liquid phase energy equations. Because of the redundancy of the equations, we use an average value for the real value.

8.6.3 Numerical solution

To simplify the calculation, we divided the well into several short segments of the same length h. The length of a segment varies depending on variations in well thickness, hole diameter, the fluid density inside and outside the pipe and well geometry. Then the velocity, pressure and temperature calculations are performed for each successive "segment" of the pipe up to the surface by means of the afore-mentioned discussion, we use the four order Runge-Kutta method to simulate the model.

8.6.3.1 *Solution process*

From the above the analysis, an algorithm is designed as follows.

Step 1. Set the depth step length and denote the relative tolerance error using ε. The smaller h, ε is, the more accurate the results. However, this would lead to rapid increase in the calculation time. We set $h=1$ (m) and $\varepsilon=5\%$.

Step 2. Compute each point's inclination.

Step 3. Given the initial conditions.
Step 4. Compute the parameters under the initial conditions or the last time variables.
Step 5. Let the right parts of the coupled differential equations be functions f_i, where $(i = 1,2,3,4,5)$. From this we can then obtain a coupled functions system as follows:

$$\begin{cases} f_1 = \dfrac{v_l}{1-\alpha_g}\dfrac{d\alpha_g}{dz} \\ f_2 = \dfrac{Pv_g\frac{d\alpha_g}{dz} - \alpha_g v_g[g\cos\theta(\alpha_g\rho_g + \alpha_l\rho_l) + \frac{\tau_{lb}S_{lb}}{A} + \alpha_l\rho_l v_l\frac{dv_l}{dz}]}{\rho_g\alpha_g^2 v_g^2 - P\alpha_g} \\ f_3 = \dfrac{-P\alpha_g\frac{dv_g}{dz} - Pv_g\frac{d\alpha_g}{dz}}{\alpha_g v_g} \\ f_4 = -\dfrac{v_l}{C_{Pl}}\dfrac{dv_l}{dz} - \dfrac{1}{\rho_l C_{Pl}}\dfrac{dP}{dz} - \dfrac{g\cos\theta}{C_{Pl}} - \dfrac{\alpha_l(T_{T_e})}{\rho_l v_l C_{Pl}} \\ f_5 = -\dfrac{v_g}{C_{Pg}}\dfrac{dv_g}{dz} - \dfrac{g\cos\theta}{C_{Pg}} - \dfrac{\alpha_g(T_{T_e})}{\rho_g v_g C_{Pg}} \\ P(z_0) = P_0, \quad v_g(z_0) = v_{g0}, \quad v_l(z_0) = v_{l0}, \quad T(z_0) = T_0, \quad \rho_{g0} = \dfrac{MP_0}{RZ_g T_0}. \end{cases}$$

Step 6. Assume P, v_g, v_l, T_g, T_l, to be y_i $(i = 1,2,3,4,5)$, respectively. Some parameters can be written as follows:

$$\begin{cases} a_{1i} = f_i(y_1, y_2, y_3, y_4) \\ b_{1i} = f_i(y_1 + ha_1/2, y_2 + ha_2/2, y_3 + ha_3/2, y_4 + ha_4/2) \\ c_{1i} = f_i(y_1 + hb_1/2, y_2 + hb_2/2, y_3 + hb_3/2, y_4 + hb_4/2) \\ d_{1i} = f_i(y_1 + hc_1, y_2 + hc_2, y_3 + hc_3, y_4 + hc_4) \end{cases}$$

and

$$\begin{cases} a_{2i} = f_i(y_1, y_2, y_3, y_5) \\ b_{2i} = f_i(y_1 + ha_1/2, y_2 + ha_2/2, y_3 + ha_3/2, y_5 + ha_4/2) \\ c_{2i} = f_i(y_1 + hb_1/2, y_2 + hb_2/2, y_3 + hb_3/2, y_5 + hb_4/2) \\ d_{2i} = f_i(y_1 + hc_1, y_2 + hc_2, y_3 + hc_3, y_5 + hc_4). \end{cases}$$

Step 7. Calculate the gas velocity, liquid velocity, pressure and temperature at point $(j+1)$:

$$y_{ki}^{(j+1)} = y_{ki}^j + h(a_{ki} + 2b_{ki} + 2c_{ki} + d_{ki})/6, \quad i = 1,2,3,4,5; \; j = 1,2,\ldots,n; \quad k = 1,2.$$

Step 8. Compute the gas density:

$$\rho_{kg}^j = \frac{MP_k^j}{RZ_g T_k^j}, \quad j = 1,2,\ldots,n; \; k = 1,2.$$

Step 9. Repeat the second to sixth steps until y_{ki}^n is calculated.
Step 10. Compute the average value of the gas velocity, liquid velocity, pressure, temperature and gas density at every point:

$$y_i^j = (y_{1i}^j + y_{2i}^j)/2, \quad i = 1,2,3,4,5; \qquad \rho_g^j = (\rho_{1g}^j + \rho_{2g}^j)/2, \quad j = 1,2,\ldots,n.$$

8.6.3.2 *Initial condition and calculation of some parameters*
To solve the model, some definite conditions, i.e. the initial conditions, need to be added. The initial conditions are; pressure, temperature and velocity distribution along the well at the well top. In this section, we give the calculation method for some of the parameters:

1. Obtain each point's inclination:

$$\theta_j = \theta_{j-1} + \frac{\theta_k - \theta_{k-1}}{\Delta s_k}\Delta s_j,$$

where j represents the segment point of calculation, and Δs_k represents the measurement depth of the inclination angle θ_k and θ_{k-1}, Δs_j, and the step length of calculation.

2. Transient heat transfer function (Hasan and Kabir, 1991):

$$\begin{cases} f(t_D) = 1.1281\sqrt{t_D}\left(1 - 0.3\sqrt{t_D}\right) & (t_D \le 1.5) \\ f(t_D) = \left(1 + \dfrac{0.6}{t_D}\right)[0.4063 + 0.5\ln(t_D)] & (t_D > 1.5) \end{cases}$$

$$t_D = \frac{t\alpha}{r_{wb}^2}.$$

3. The frictional force between the liquid phase and the pipe wall. In the present model, the gas-wall friction and the oil-wall friction are neglected, and the friction force is represented by the mixture-wall interaction as given Taitel *et al.* (1989):

$$\frac{\tau_{lb} S_{lb}}{A} = \frac{f \rho_l v_l^2}{2d}.$$

It was thought that the friction factor was affected by the Reynolds numbers for fluid in a pipeline and therefore the Blasius equation (Abbasbandy, 2007; Wang, 2004) should be adopted. Many researchers such as Brauner (Brauner, 2001; Brauner and Ullmann, 2002; Brauner *et al.*, 1998) have used the equation to compute the frictional force between the oil or water and the pipe wall:

$$f_e = \begin{cases} \dfrac{64}{Re}, & Re \le 2000 \\ \dfrac{0.3164}{Re^{0.25}}, & Re > 2000. \end{cases}$$

$$Re_l = \frac{\rho_l v_l d}{\mu_l}$$

$$\mu_l = e^{1.003 + 0.01479(1.8T+32) + 0.00001982(1.8T+32)^5}.$$

4. Gas condensing parameter Z_g:
   If ($p < 35$ MPa)

$$Z_g = 1 + \left(0.31506 - \frac{1.0467}{T_{pr}} - \frac{0.5783}{T_{pr}^3}\right)\rho_{pr} + \left(0.053 - \frac{0.6123}{T_{pr}}\right)\rho_{pr}^2 + 0.6815\frac{\rho_{pr}^2}{T_{pr}^3},$$

where, $\rho_{pr} = 0.27\frac{P_{pr}}{T_{pr}}$, $T_{pr} = \frac{T}{T_{pc}}$, $P_{pr} = \frac{P}{P_{pc}}$.
   Else

$$Z_g = (90.7x - 242x^2 + 42.4x^3)y^{(1.18+2.82x)} - (14.76x - 9.76x^2 + 4.58t^3)y + \frac{1 + y + y^2 + y^3}{(1-y)^3},$$

where,

$$F(y) = -0.06125P_{pr}xe^{-1.2(1-x)^2} + (90.7x - 242.2x^2 + 42.4x^3)y^{2.18+2.82x} + \frac{y+y^2+y^3-y^4}{(1-y^3)} - (14.76x - 9.76x^2 + 4.58x^3)y^2 = 0$$

$$x = \frac{1}{T_{pr}}.$$

5. Calculate the original conditions of the gas density, gas velocity and liquid velocity at the initial point. Assume that the pressure and temperature at the first computation point are already known:

$$\alpha_l = \alpha_w + \alpha_o, \quad \rho_l = \frac{\alpha_w \rho_w + \alpha_o \rho_o}{\alpha_l}, \quad C_{Pl} = \frac{\rho_o \alpha_o C_{Po} + \rho_w \alpha_w C_{Pw}}{\rho_l \alpha_l}, \quad \rho_{go} = \frac{MP_o}{RZ_g T_o}.$$

8.6.4 **Numerical simulation**

As described previously, the algorithm begins with a calculation for the pipe at the bottom of the pipe. The calculations are performed for each successive segment of the pipe up to the surface.

8.6.4.1 *Parameters*

In this simulation, we study a pipe in *X* well, located in China. The parameters are as follows:

Depth of the well = 7100 m
Well bottom pressure = 40 MPa
Critical pressure = 4.968 MPa
Gas specific weight = 0.6434 kg/m^3
Ground thermal conductivity parameter = 2.06 W/m·K
Gas-liquid ratio = 100000
Ground temperature = 16°C
Geothermal gradient = 2.18°C/100 m
Gas outputs = 500000 m^3/d
Roughness of the inner surface of the well = 0.000015

The parameters for the pipes, the inclined well, the inclination, the azimuth and the vertical depth are given in Tables 8.1, 8.2 and 8.3.

8.6.4.2 *Main results*

Using the algorithm designed above, we obtained a series of results for this well. The value of the gas velocity, liquid velocity, pressure, temperature and gas density in the well-bore are shown in Table 8.26.

We also obtained a comparative result using the measurement value in Table 8.23.

8.6.5 **Sensitivity analysis**

To study how different gas outputs, ground thermal conductivity parameters and geothermal gradients influence the result, we used different gas outputs and geothermal gradients. Using the algorithm and simulation, we obtained a further series of results.

From Figures 8.46 to 8.55, we can see how different gas outputs and geothermal gradients influence the gas velocity, liquid velocity, pressure, temperature and gas density. Three different geothermal gradients; 1.8, 2 and 2.18 °C/100 m were used to test the influence, with the other parameters remaining the same. At the same pipe depth, if the geothermal gradient increases, then

Table 8.26 Gas velocity, liquid velocity, pressure, temperature, and gas density.

Number	Depth [m]	Gas velocity [m/s]	Liquid velocity [m/s]	Gas density [kg/m^3]	Temperature [°C]	Pressure [MPa]
1	1	24.0923	13.65832	152.0939	160	26.8543
2	251	23.09781	13.15157	154.2652	160.182	27.2861
3	501	22.01219	12.75175	157.3852	160.4397	27.39698
4	751	21.10253	12.3748	159.8005	160.6453	27.96778
5	1001	20.33891	12.0186	161.5814	160.8033	28.41054
6	1251	19.66564	11.68565	163.0457	160.9374	28.73772
7	1501	18.5231	11.36803	168.4707	161.3991	29.00721
8	1751	18.07647	11.06624	168.7879	161.441	30.00459
9	2001	17.64079	10.78163	169.1489	161.4859	30.06404
10	2251	16.89795	10.51192	172.3621	161.7789	30.13154
11	2501	16.18378	10.25763	175.7631	162.0926	30.72493
12	2751	15.49544	10.01706	179.2547	162.4189	31.35411
13	3001	15.1085	9.779836	180.1091	162.5075	32.00134
14	3251	14.55478	9.557113	182.9218	162.7802	32.16054
15	3501	14.11463	9.345681	184.6904	162.957	32.68357
16	3751	13.52555	9.138729	188.5386	163.3389	33.01321
17	4001	13.1031	8.938986	190.778	163.5656	33.73109
18	4251	12.72154	8.75197	192.7167	163.7649	34.14978
19	4501	12.3284	8.574998	194.9974	164.0033	34.51284
20	4751	11.84029	8.398595	199.0358	164.4266	34.94068
21	5001	11.41695	8.233741	202.4863	164.793	35.69942
22	5251	11.02317	8.073987	205.7895	165.1492	36.34922
23	5501	10.63466	7.922223	209.3397	165.5365	36.97275
24	5751	10.35072	7.777391	211.3627	165.7614	37.64436
25	6001	9.952495	7.636786	215.7401	166.2541	38.02792
26	6251	9.794312	7.501457	215.9121	166.263	38.85974
27	6501	9.472653	7.369309	219.3738	166.663	38.89148
28	6751	9.209321	7.241372	221.9558	166.9616	39.5515
29	7001	9.043878	7.118207	222.687	167.0377	40
30	7100	9	7	223.024	167.0892	40.04453

Table 8.27 Comparative results.

Well-at 250 m depth	Pressure [MPa]	Temperature [°C]	Gas velocity [m/s]	Liquid velocity [m/s]	Gas density [kg/m^3]
Calculation results	27.2861	160.182	23.09781	13.15157	154.2652
Measurement results	28.3042	157.352	22.1002	12.9875	156.3842
Relative error	0.0360	0.018	0.045	0.0126	0.0136

the temperature also increases but at a decreasing rate from the top to the bottom of the pipe. At the same pipe depth, if the geothermal gradient increases or decreases, the pressure, gas velocity, liquid velocity and gas density show little change.

Three different gas outputs; 300000, 500000 and 700000 m^3/d; were used to test the influence with the other parameters remaining the same. At the same pipe depth, if the gas output increases, the gas velocity, liquid velocity, pressure and temperature all increase. However, the gas density decreases.

In this section, a two-fluid model for a three-phase bubbly flow in vertical pipes has been presented. Due to the gas holdup sensitivity, a new simple model, which can be easily applied to calculate pressure drop, temperature, gas velocity, liquid velocity and gas density in HTHP wells,

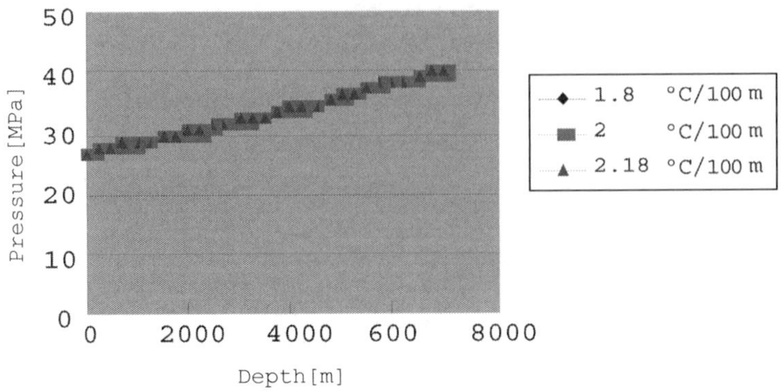

Figure 8.46 Pressure distribution at different geothermal gradients.

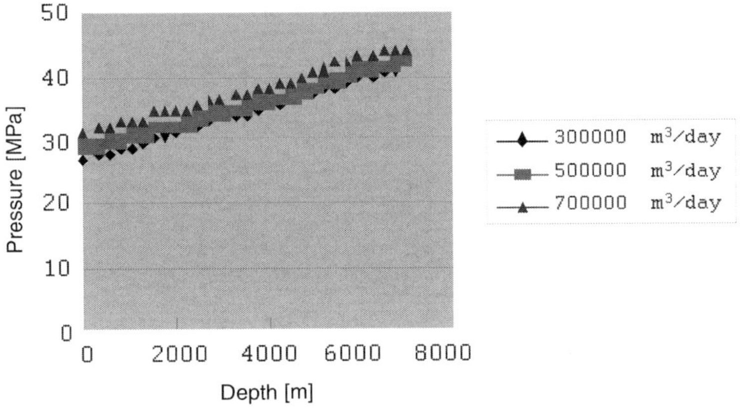

Figure 8.47 Pressure distribution at different gas outputs.

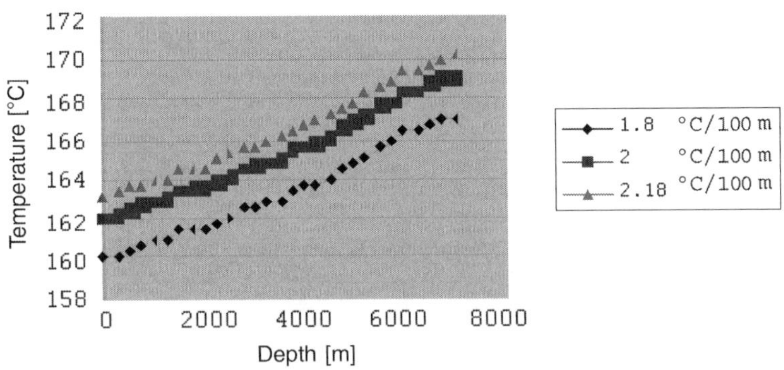

Figure 8.48 Temperature distribution at different geothermal gradients.

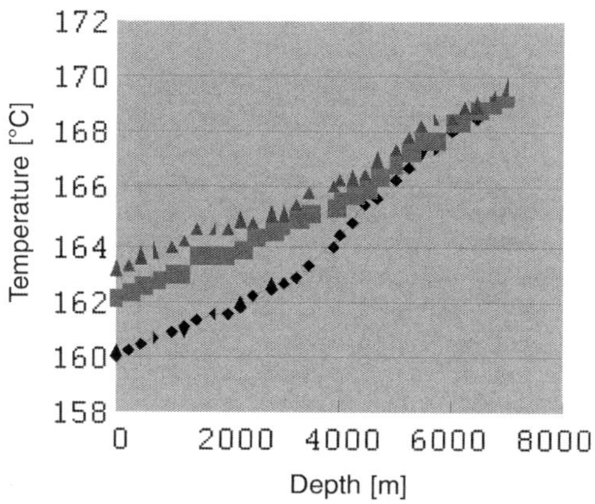

Figure 8.49 Temperature distribution at different gas outputs.

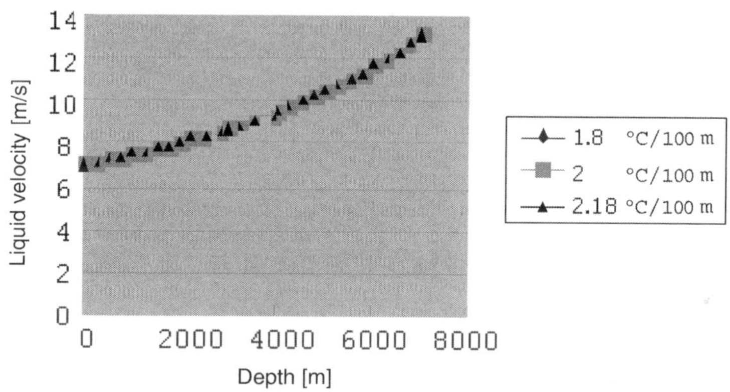

Figure 8.50 Liquid velocity distribution at different geothermal gradients.

is constructed and an algorithm with the four order Runge-Kutta method is proposed to solve the model. The accuracy and reliability of the proposed model was validated by a case at the *X* Well (HTHP well), at 7100 m depth in China. The model predictions were in very good agreement with the field data. A sensitivity analysis was conducted for the model. The curves-for the pressure, temperature, gas velocity, liquid velocity and gas density along the depth of the well were plotted with different gas outputs and geothermal gradients, intuitively reflecting gas flow law and the characteristics of the heat transfer in formations. There is a high degree of confidence in the practicality of the method to provide meaningful prediction and the results provide a technical reliability for the process design of well testing in HTHP gas wells and a dynamic analysis for injection.

8.7 PPTHVD-STF

Considering the differential equation model for *P*, *T*, *v*, *a*, we set the following assumptions:

1. The gas-liquid-liquid flow in the tubing is in one dimension of the flow direction with negligible heat or mass transfer between phases.

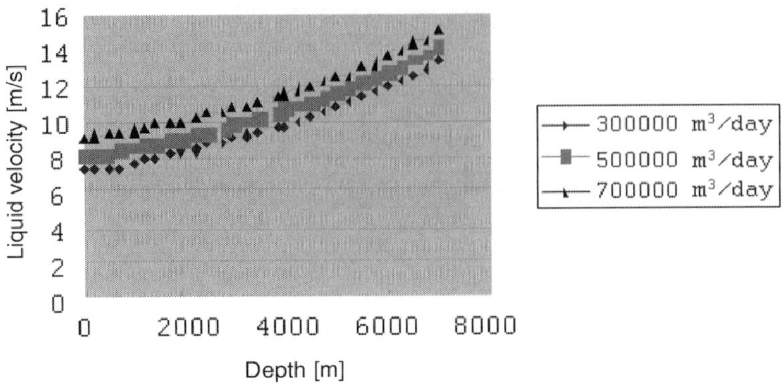

Figure 8.51 Liquid velocity distribution at different gas outputs.

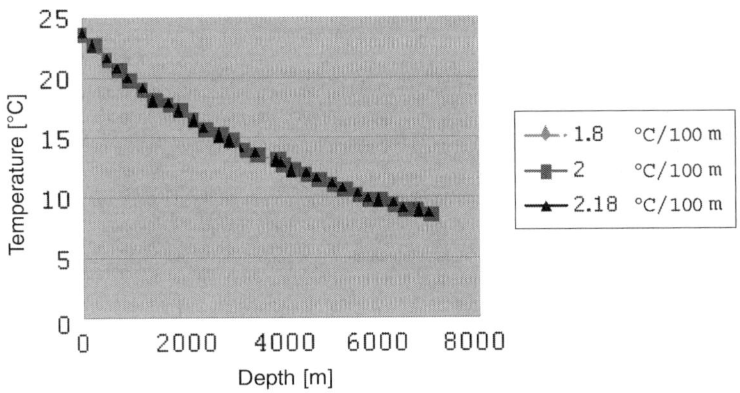

Figure 8.52 Gas velocity distribution at different geothermal gradients.

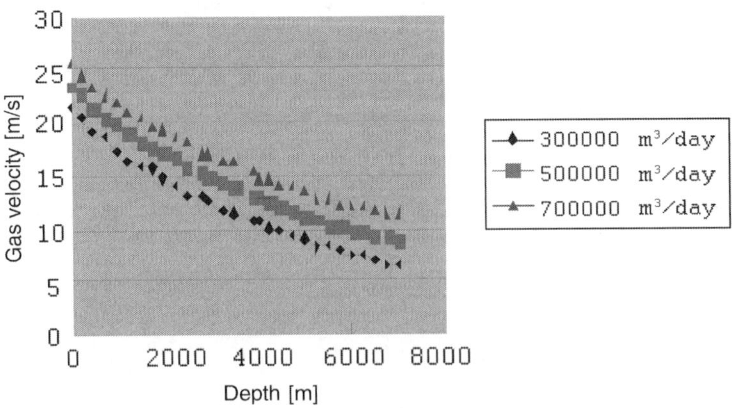

Figure 8.53 Gas velocity distribution at different gas outputs.

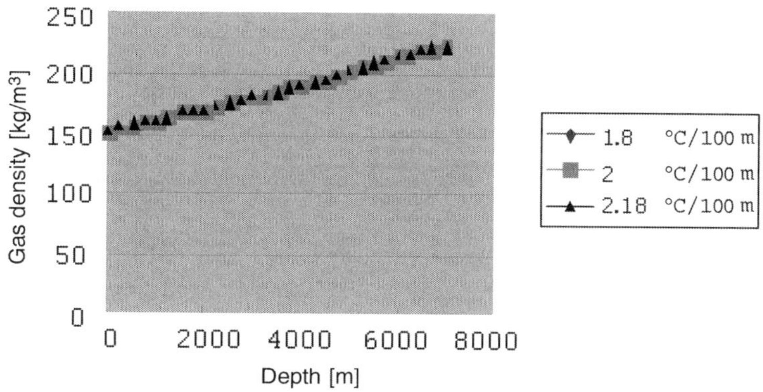

Figure 8.54 Gas density distribution at different geothermal gradients.

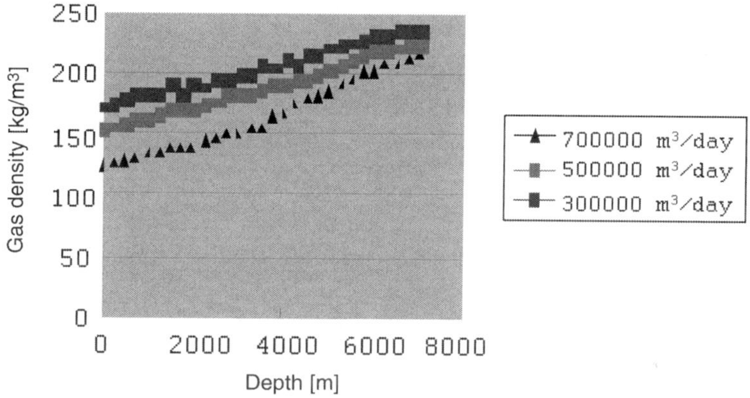

Figure 8.55 Gas density distribution at different gas outputs.

2. When the oil/gas/water three-phase reaches thermodynamic equilibrium, the temperature at all points is equal in the transverse cross-section.
3. The oil-water two phase flow has a flat interface.
4. Oil and water are assumed to be incompressible phases since their compressibility is small.
5. Phase pressures are assumed to be equal at a given axial location.
6. The gas flow is considered an approximate ideal gas.

8.7.1 The coupled system model

A three-phase in an infinitesimal section is considered as shown in Figure 8.56. Referring to Figure 8.56, the flow of the three fluids is considered. It is assumed that water is heavier than oil and flows at the bottom, while the oil flows on the top.

8.7.1.1 *Material balance*

The top of the well is considered the origin of the coordinate axis and the vertical up is the positive direction. Let dz denote the differential depth. Thus, under a steady-state condition for each phase, it follows the law of fluid dynamics:

Water flows continuity equation:

$$\frac{d(\rho_\omega \alpha_\omega v_\omega A)}{dz} = 0. \tag{8.103}$$

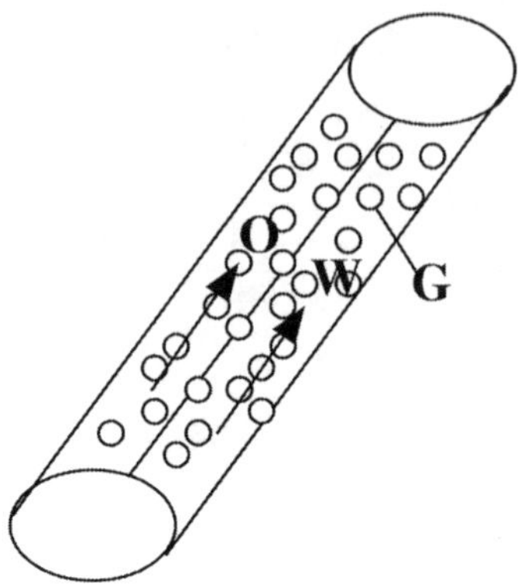

Figure 8.56 Bubbly gas three phase flow

Oil flows continuity equation:

$$\frac{d(\rho_o \alpha_o v_o A)}{dz} = 0. \tag{8.104}$$

Gas flows continuity equation:

$$\frac{d(\rho_g \alpha_g v_g A)}{dz} = 0. \tag{8.105}$$

Momentum equation: We can put a unified momentum equation as follows:

$$\underbrace{\text{Sum vector of the forces}} = \underbrace{\text{the momentum flowing out from } dz} - \underbrace{\text{the momentum flowing into } dz}.$$

The force on dz include as follows:

The pressure on cross section:

$$\left[\alpha_k P_k - \left(\alpha_k P_k + \frac{d(\alpha_k P_k)}{dz}\right) A\right] = -\frac{d(\alpha_k P_k)}{dz} A.$$

The gravity which flows in the opposite direction:

$$\rho_k g \cos\theta A\, dz.$$

The friction which flows in the opposite direction:

$$\tau_{kb} S_{kb}\, dz.$$

The interfacial shear stress which flows in the opposite direction:

$$\tau_{kj} S_{kj}\, dz.$$

The momentum flowing out from dz:

$$\rho_k \alpha_k \upsilon_k^2 A + A \frac{d(\rho_k \alpha_k \upsilon_k^2)}{dz}.$$

The momentum flowing into dz:

$$\rho_k \alpha_k \upsilon_k^2 A.$$

The subscript k represents gas (g), oil (o), water (w) and implies a phase that is different from phase k. From the law of momentum conservation, we have momentum equation for each phase as follows:

Water flows momentum equation:

$$\frac{d(\rho_\omega \alpha_\omega \upsilon_\omega^2)}{dz} + \frac{d(\alpha_\omega P)}{dz} = -\rho_\omega \alpha_\omega g \cos\theta - \frac{\tau_{wb} S_{wb}}{A} - \frac{\tau_{wo} S_{wo}}{A}. \tag{8.106}$$

Oil flows momentum equation:

$$\frac{d(\rho_o \alpha_o \upsilon_o^2)}{dz} + \frac{d(\alpha_o P)}{dz} = -\rho_o \alpha_o g \cos\theta - \frac{\tau_{ob} S_{ob}}{A} - \frac{\tau_{ow} S_{ow}}{A}. \tag{8.107}$$

Gas flows momentum equation:

$$\frac{d(\rho_g \alpha_g \upsilon_g^2)}{dz} + \frac{d(\alpha_g P)}{dz} = -\rho_g \alpha_g g \cos\theta. \tag{8.108}$$

Here, $\tau_{wb} S_{wb}$, $\tau_{ob} S_{ob}$ mean friction shear stress about water-wall and oil-wall respectively. $\tau_{wo} S_{wo} = \tau_{ow} S_{ow}$ is the shear force between oil phase and water phase.

Energy equation: The unified energy equation is:

$$\underbrace{\text{The translate energy the second net}} = \underbrace{\text{The energy flowing out from } dz} - \underbrace{\text{The energy flowing into } dz}.$$

The types of energy considered in this paper include inner energy, pressure energy, kinetic energy and potential energy.

The energy flowing into dz includes follow types:

The internal energy: $H_k(z)$;
The kinetic energy: $\frac{m_k \upsilon_k^2(z)}{2}$;
The potential energy: $m_k g z \cos\theta$;
The pressure energy: $P_k(z)\upsilon_k(z)$.

The energy flowing out from dz includes follow types:

1. The internal energy: $H_k(z + \Delta z)$;
2. The kinetic energy: $\frac{m_k \upsilon_k^2(z+\Delta z)}{2}$;
3. The potential energy: $m_k g(z + \Delta z)\cos\theta$;
4. The pressure energy: $P_k(z + \Delta z)\upsilon_k(z + \Delta z)$.

Where, the inner energy and pressure energy are collectively referred to as enthalpy, dQ represents the radial of the heat of the tubing.

According to the rule of the energy balance: the energy of fluid flowing into the distance element equals the energy sum of losses and fluid flowing out from the infinitesimal:

$$H_k(z) + \frac{m_k \upsilon_k^2(z)}{2} - m_k g z \cos\theta = H_k(z + \Delta z) + \frac{m_k \upsilon_k^2(z + \Delta z)}{2} - m_k g(z + \Delta z)\cos\theta + dQ \tag{8.109}$$

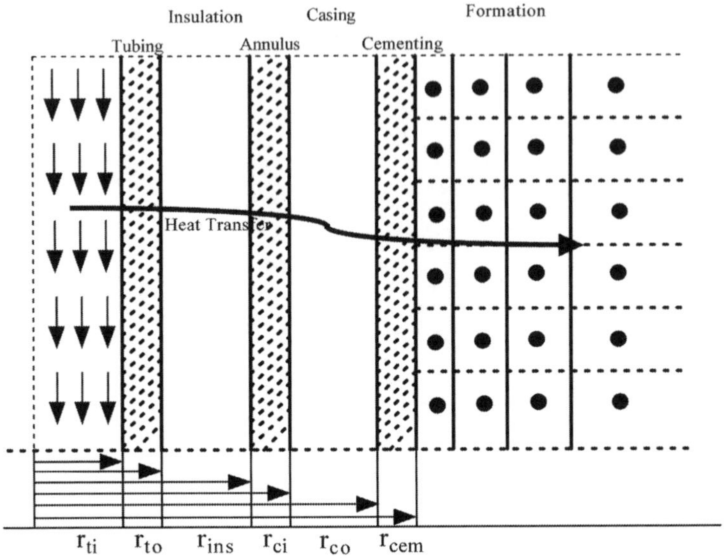

Figure 8.57 The radial transfer of heat.

(8.109) can be written as:

$$\frac{dH_k}{dz} + m_k \upsilon_k \frac{d\upsilon_k}{dz} - m_k g \cos\theta + \frac{dQ}{dz} = 0. \tag{8.110}$$

Divided by m_k yields:

$$\frac{dq_k}{dz} + \upsilon_k \frac{d\upsilon_k}{dz} - g\cos\theta + \frac{dh_k}{dz} = 0 \tag{8.111}$$

h_k, which satisfies following relation, denotes specific enthalpy.

$$\frac{dh_k}{dz} = C_{pk}\frac{dT_k}{dz} - \eta_k C_{pk}\frac{dP_k}{dz}, \tag{8.112}$$

where η_k is J-T coefficient defined as:

$$\begin{cases} \eta_k = 0, & k = g \\ \eta_k = -\dfrac{1}{C_{pk}\rho_k}, & k = 0, \omega \end{cases} \tag{8.113}$$

In Equation (8.109), dq_k denotes the radial heat transfer between the gas and the surrounding earth tube. Ramey (1962) and Willhite (1967) discussed the radial transfer of heat between the fluid and the earth in detail. As Figure 8.57, the radial heat transfer from the fluid to the cement/earth interface can be described by:

$$dq_{fe} = \frac{2\pi r_{t0} U_{t0}}{w}(T - T_k)dz. \tag{8.114}$$

And the radial heat transfer from the cement/earth interface to the surrounding earth is:

$$dq_{es} = \frac{2\pi K_e}{wf(t_D)}(T_k - T_e)dz. \tag{8.115}$$

Thus the radial heat transfer between the gas and the surrounding earth tube is:

$$dq = \frac{2\pi r_{t0} U_{t0} K_e}{w(r_{t0} U_{t0} f(t_D) + K_e)} (T - T_0) dz. \quad (8.116)$$

Let $a = \frac{2\pi r_{t0} U_{t0} K_e}{w(r_{t0} U_{t0} f(t_D) + K_e)}$, substituting (8.112), (8.113) and (8.114) into (8.111), we have energy equation for water phase as follows:

$$\rho_w C_{pw} \frac{dT_w}{dz} + \rho_w \upsilon_w \frac{d\upsilon_w}{dz} + \frac{dP_w}{dz} - \rho_w g \cos\theta + a(T - T_e) = 0. \quad (8.117)$$

Similarly, the oil flow energy equation can be written as:

$$\rho_o C_{po} \frac{dT_o}{dz} + \rho_o \upsilon_o \frac{d\upsilon_o}{dz} + \frac{dP_o}{dz} - \rho_o g \cos\theta + a(T - T_e) = 0. \quad (8.118)$$

From Figure 8.57, we suppose the gases are fully dispersed in oil-water two phase flows. Therefore, we do not consider the gas energy equation.

Finally, we can also add a gas state equation:

$$\rho_g = \frac{MP_g}{RZ_g T}. \quad (8.119)$$

Combined (8.103), (8.104), (8.105), (8.106), (8.107), (8.108), (8.109), (8.110), (8.111), the coupled system model of differential equations is as follows:

$$\begin{cases}
\frac{d(\rho_\omega \alpha_\omega \nu_\omega A)}{dz} = 0 \\
\frac{d(\rho_o \alpha_o \nu_o A)}{dz} = 0 \\
\frac{d(\rho_g \alpha_g \nu_g A)}{dz} = 0 \\
\frac{d(\rho_\omega \alpha_\omega \upsilon_\omega^2)}{dz} + \frac{d(\alpha_\omega P_w)}{dz} = -\rho_\omega \alpha_\omega g \cos\theta - \frac{\tau_{wb} S_{wb}}{A} - \frac{\tau_{wo} S_{wo}}{A} \\
\frac{d(\rho_o \alpha_o \upsilon_o^2)}{dz} + \frac{d(\alpha_o P_o)}{dz} - -\rho_o \alpha_o g \cos\theta - \frac{\tau_{ob} S_{ob}}{A} - \frac{\tau_{ow} S_{ow}}{A} \\
\frac{d(\rho_g \alpha_g \upsilon_g^2)}{dz} + \frac{d(\alpha_g P_g)}{dz} = -\rho_g \alpha_g g \cos\theta \\
\rho_w C_{pw} \frac{dT_w}{dz} + \rho_w \upsilon_w \frac{d\upsilon_w}{dz} + \frac{dP_w}{dz} - \rho_w g \cos\theta + a(T_w - T_e) = 0 \\
\rho_o C_{po} \frac{dT_o}{dz} + \rho_o \upsilon_o \frac{d\upsilon_o}{dz} + \frac{dP_o}{dz} - \rho_o g \cos\theta + a(T_o - T_e) = 0 \\
\rho_g = \frac{MP_g}{RZ_g T} \\
P_w = P_o = P_g = P, \quad \alpha_w + \alpha_o + \alpha_g = 1, \quad T_w = T_o = T_g = T, \quad y(z_0) = \varphi(z_0).
\end{cases} \quad (8.120)$$

8.7.2 Model analysis

A three-fluid model system is obviously more complex to compute than a one-fluid or two-fluid system. For one thing, the number of differential equations for a three-fluid model is twice that of a two-fluid model. Also, the non-linearity and coupling of the equations are much stronger owing to the interactions amongst the three-fluids. The equations are coupled not only within those of the same phase, but also among those of different phases. Because of the coupling of the equations, solving the equations directly is extremely complicated. Here, we apply a simplified model.

First, considering that the gas density is only a function of pressure $\rho_g = f(P_g)$, then:

$$\frac{d\rho_g}{dx} = \frac{d\rho_g}{dP}\frac{dP}{dx} = \frac{1}{c^2}\frac{dP}{dx}, \tag{8.121}$$

where c is the sound velocity, which is calculated using the following equation:

$$c = \sqrt{\frac{M}{ZRT}}. \tag{8.122}$$

So, the (8.103), (8.104), (8.105) can be written as:

$$\alpha_w\frac{dv_w}{dz} + v_w\frac{d\alpha_w}{dz} = 0 \tag{8.123}$$

$$\alpha_o\frac{dv_o}{dz} + v_o\frac{d\alpha_o}{dz} = 0 \tag{8.124}$$

$$v_g(1-\alpha_w-\alpha_o)\frac{dP}{dz} - Pv_g\frac{d\alpha_o}{dz} - Pv_g\frac{d\alpha_w}{dz} + P(1-\alpha_w-\alpha_o)\frac{dv_g}{dz} = 0. \tag{8.125}$$

Substituting the Equations (8.104), (8.105), (8.125) into the momentum equation, respectively, the following equations can be obtained:

$$\rho_w\alpha_w v_w\frac{dv_w}{dz} + \alpha_w\frac{dP}{dz} + P\frac{d\alpha_w}{dz} = -\rho_w g\alpha_w\cos\theta - \frac{\tau_{wb}S_{wb}}{A} - \frac{\tau_{wo}S_{wo}}{A} \tag{8.126}$$

$$\rho_o\alpha_o v_o\frac{dv_o}{dz} + \alpha_o\frac{dP}{dz} + P\frac{d\alpha_o}{dz} = -\rho_o g\alpha_o\cos\theta - \frac{\tau_{ob}S_{ob}}{A} - \frac{\tau_{ow}S_{ow}}{A} \tag{8.127}$$

$$\rho_g(1-\alpha_w-\alpha_o)v_g\frac{dv_g}{dz} + (1-\alpha_w-\alpha_o)\frac{dP}{dz} - P\frac{\alpha_o}{dz} - P\frac{\alpha_w}{dz} = -\rho_g g(1-\alpha_w-\alpha_o)\cos\theta. \tag{8.128}$$

Sum the (8.126), (8.127), to eliminate the shear force between two phases, the equations can be acquired:

$$\rho_w\alpha_w v_w\frac{dv_w}{dz} + \rho_o\alpha_o v_o\frac{dv_o}{dz} + (\alpha_w+\alpha_o)\frac{dP}{dz} + P\left(\frac{d\alpha_w}{dz} + \frac{d\alpha_o}{dz}\right)$$
$$= -g\cos\theta(\rho_w\alpha_w + \rho_o\alpha_o) - \frac{\tau_{wb}S_{wb}}{A} - \frac{\tau_{ob}S_{ob}}{A} \tag{8.129}$$

Combining (8.117), (8.118), (8.123), (8.124), (8.125) and (8.129), the coupled differential equations system on pressure, velocity and temperature in mathematical form is obtained and the well-posedness of the model is analyzed as follows:

$$\begin{cases} \alpha_w \dfrac{d\upsilon_w}{dz} + \upsilon_w \dfrac{d\alpha_w}{dz} = 0 \\ \alpha_o \dfrac{d\upsilon_o}{dz} + \upsilon_o \dfrac{d\alpha_o}{dz} = 0 \\ \upsilon_g(1-\alpha_w-\alpha_o)\dfrac{dP}{dz} - P\upsilon_g \dfrac{d\alpha_o}{dz} - P\upsilon_g \dfrac{d\alpha_w}{dz} + P(1-\alpha_w-\alpha_o)\dfrac{d\upsilon_g}{dz} = 0 \\ \rho_w\alpha_w\upsilon_w \dfrac{d\upsilon_w}{dz} + \rho_o\alpha_o\upsilon_o \dfrac{d\upsilon_o}{dz} + (\alpha_w+\alpha_o)\dfrac{dP}{dz} + P\left(\dfrac{d\alpha_w}{dz} + \dfrac{d\alpha_o}{dz}\right) + g\cos\theta(\rho_w\alpha_w + \rho_o\alpha_o) \\ \quad + \dfrac{\tau_{wb}S_{wb} + \tau_{ob}S_{ob}}{A} = 0 \\ \dfrac{d(\rho_g\alpha_g\upsilon_g^2)}{dz} + \dfrac{d(\alpha_g P)}{dz} = -\rho_g\alpha_g g\cos\theta \\ \rho_w C_{pw}\dfrac{dT}{dz} + \rho_w\upsilon_w \dfrac{d\upsilon_w}{dz} + \dfrac{dP}{dz} - \rho_w g\cos\theta + a(T-T_e) = 0 \\ \rho_o C_{po}\dfrac{dT}{dz} + \rho_o\upsilon_o \dfrac{d\upsilon_o}{dz} + \dfrac{dP}{dz} - \rho_o g\cos\theta + a(T-T_e) = 0 \\ \rho_g = \dfrac{MP}{RZ_gT} \\ P(z_0) = P_0, \quad \upsilon_g(z_0) = \upsilon_{g0}, \quad \upsilon_o(z_0) = \upsilon_{o0}, \upsilon_w(z_0) = \upsilon_{w0} \\ \alpha_w(z_0) = \alpha_{w0}, \quad \alpha_o(z_0) = \alpha_{o0}, \quad T_w(z_0) = T_0. \end{cases} \tag{8.130}$$

8.7.2.1 *A well-posedness analysis of the system*

The task here is to check the nature of the proposed model, and to establish whether the system is stable. As in Cazarez-Candia and Vásquez-Cruz (2005), the equations for the two liquids (oil and water) are combined to obtain the mixture as in Cazarez *et al.* (2010). The modified model is given as follows:

$$\begin{cases} \dfrac{d\alpha_g}{dz} = \dfrac{-\alpha_g(\rho_g^2 + \rho_g\upsilon_g^2 P - 2P\upsilon_g^2)\left(\frac{f\rho_l\upsilon_l^2}{2d} + \rho_l g\cos\theta(1-\alpha_g)\right) + \alpha_g(1-\alpha_g)\rho_g^3 g\cos\theta}{2\rho_g^2\upsilon_g^2 - \rho_g^3\upsilon_g^2 - 2\alpha_g\rho_g^2\upsilon_g^2 + \alpha_g\rho_g^3\upsilon_g^2 + 2\alpha_g\rho_g^2\rho_l\upsilon_l^2 - 4P\alpha_g\upsilon_g^2\rho_l\upsilon_l^2 + 2P\alpha_g\rho_g\rho_l\upsilon_g^2\upsilon_l^2} \\ \dfrac{d\upsilon_l}{dz} = \dfrac{\upsilon_l\frac{d\alpha_g}{dz}}{1-\alpha_g} \\ \dfrac{dP}{dz} = \dfrac{-\rho_l g\cos\theta(1-\alpha_g) - \frac{f\rho_l\upsilon_l^2}{2d} - 2\rho_l(1-\alpha_g)\upsilon_l\frac{d\upsilon_l}{dz}}{1-\alpha_g} \\ \dfrac{d\upsilon_g}{dz} = -\dfrac{\alpha_g\upsilon_g P\frac{dP}{dz} + \rho_g^2\upsilon_g\frac{d\alpha_g}{dz}}{\rho_g^2\alpha_g} \\ \dfrac{dT}{dz} = \dfrac{\rho_l g\cos\theta - \alpha(T-T_e) - \frac{dP}{dz} - \rho_l\upsilon_l\frac{d\upsilon_l}{dz}}{\rho_l C_{Pl}}, \end{cases}$$

where,

$$\alpha_l = \alpha_o + \alpha_w, \quad \rho_l = \frac{\rho_w\alpha_w + \rho_o\alpha_o}{\alpha_l}, \quad \upsilon_l = \upsilon_o = \upsilon_w, \quad C_{Pl} = \frac{\rho_w C_{Pw} + \rho_o C_{Po}}{\alpha_l \rho_l}, \quad T_l = T_o = T_w.$$

Let:

$$\frac{d\alpha_g}{dz} = f_1(z;\alpha_g,\upsilon_l,P,\upsilon_g,T), \quad \frac{d\upsilon_l}{dz} = f_2(z;\alpha_g,\upsilon_l,P,\upsilon_g,T), \quad \frac{dP}{dz} = f_3(z;\alpha_g,\upsilon_l,P,\upsilon_g,T),$$

$$\frac{d\upsilon_g}{dz} = f_4(z;\alpha_g,\upsilon_l,P,\upsilon_g,T), \quad \frac{dT}{dz} = f_5(z;\alpha_g,\upsilon_l,P,\upsilon_g,T).$$

Denote:

$$F = (f_1, f_2, f_3, f_4, f_5)^T, \quad y = (\alpha_g, \upsilon_l, P, \upsilon_g, T)^T,$$

$$y(0) = y(z_0) = (\alpha_g(z_0), \upsilon_l(z_0), P(z_0), \upsilon_g(z_0), T(z_0))^T,$$

then, the systems of ordinary differential equations can be written as:

$$y' = F(z;y), \quad y(0) = y(z_0).$$

The norm of vector f was given as follows:

$$\|F\| = \max(|f_1|, |f_2|, |f_3|, |f_4|, |f_5|).$$

For f_1, f_2, f_3, f_4, f_5,

$$\begin{aligned} |f_1| &= \frac{|-\alpha_g(\rho_g^2 + \rho_g\upsilon_g^2 P - 2P\upsilon_g^2)(\frac{f\rho_l\upsilon_l^2}{2d} + \rho_l g\cos\theta(1-\alpha_g)) + \alpha_g(1-\alpha_g)\rho_g^3 g\cos\theta|}{|2\rho_g^2\upsilon_g^2 - \rho_g^3\upsilon_g^2 - 2\alpha_g\rho_g^2\upsilon_g^2 + \alpha_g\rho_g^3\upsilon_g^2 + 2\alpha_g\rho_g^2\rho_l\upsilon_l^2 - 4P\alpha_g\upsilon_g^2\rho_l\upsilon_l^2 + 2P\alpha_g\rho_g\rho_l\upsilon_g^2\upsilon_l^2|} \\ &\le \frac{|K_1||K_2| + |K_3|}{|K_4|}, \end{aligned}$$

where,

$$\begin{aligned} |K_1| &= |-\alpha_g(\rho_g^2 + \rho_g\upsilon_g^2 P - 2P\upsilon_g^2)| \le |\rho_g^2| + |\rho_g||\upsilon_g^2||P| + 2|P||\upsilon_g^2|, \\ |K_2| &= \left|\frac{f\rho_l\upsilon_l^2}{2d} + \rho_l g\cos\theta(1-\alpha_g)\right| \le 2\frac{|f||\rho_l||\upsilon_l^2|}{|d|} + |\rho_l||g|, \\ |K_3| &= |\alpha_g(1-\alpha_g)\rho_g^3 g\cos\theta| \le |\rho_g^3||g|, \\ |K_4| &= |2\rho_g^2\upsilon_g^2 - \rho_g^3\upsilon_g^2 - 2\alpha_g\rho_g^2\upsilon_g^2 + \alpha_g\rho_g^3\upsilon_g^2 + 2\alpha_g\rho_g^2\rho_l\upsilon_l^2 - 4P\alpha_g\upsilon_g^2\rho_l\upsilon_l^2 + 2P\alpha_g\rho_g\rho_l\upsilon_g^2\upsilon_l^2|. \end{aligned}$$

Since all parameters are bounded quantities, therefore $|K_1|, |K_2|, |K_3|, |K_4|$ are bounded. Let $N_1 = \sup\left\{\frac{|K_1||K_2|+|K_3|}{|K_4|}\right\}$, then, $|f_1| \le N_1$.

Similarly, $|f_2| = \left|\frac{\upsilon_l}{1-\alpha_g} f_1\right| \le \left|\frac{\upsilon_l}{1-\alpha_g}\right| N_1$. Let $N_2 = \sup\left\{\left|\frac{\upsilon_l}{1-\alpha_g}\right| N_1\right\}$, then, $|f_2| \le N_2$.

$|f_3| \le \frac{|\rho_l||g| + \left|\frac{|f||\rho_l||\upsilon_l^2|}{2|d|}\right| + 2|\rho_l||\upsilon_l|N_2}{|1-\alpha_g|}$. Let $N_3 = \sup\left\{\frac{|\rho_l||g| + |\frac{|f||\rho_l||\upsilon_l^2|}{2|d|}| + 2|\rho_l||\upsilon_l|N_2}{|1-\alpha_g|}\right\}$, then, $|f_3| \le N_3$.

$|f_4| \le \frac{|\upsilon_g||P|N_3 + |\rho_g^2||\upsilon_g|N_1}{|\rho_g^2|}$. Let $N_4 = \sup\left\{\frac{|\upsilon_g||P|N_3 + |\rho_g^2||\upsilon_g|N_1}{|\rho_g^2|}\right\}$, then, $|f_4| \le N_4$.

$|f_5| \le \frac{|g|}{|C_{Pl}|} + \frac{|a||T-T_e|}{|\rho_l||C_{Pl}|} + \frac{|f_3|}{|\rho_l||C_{Pl}|} + \frac{|\upsilon_l||f_2|}{|C_{Pl}|}$. Let $N_5 = \sup\left\{\frac{|g|}{|C_{Pl}|} + \frac{|a||T-T_e|}{|\rho_l||C_{Pl}|} + \frac{|f_3|}{|\rho_l||C_{Pl}|} + \frac{|\upsilon_l||f_2|}{|C_{Pl}|}\right\}$, then, $|f_5| \le N_5$.

Therefore,

$$\|F\| \le \max\{N_1, N_2, N_3, N_4, N_5\}.$$

The partial differential of f_1, f_2, f_3, f_4, f_5 about $\alpha_g, \upsilon_l, P, \upsilon_g, T$ are bounded discussed as follows:

$$\frac{\partial f_1}{\partial \alpha_g} = \frac{-K_2(\rho_g^2 + \rho_g \upsilon_g^2 P - 2\upsilon_g^2) + (1-2\alpha_g)\rho_g^3 g\cos\theta}{K_4} - \frac{(-2\rho_g^2\upsilon_g^2 + 2\rho_g^2\rho_l\upsilon_l^2 - 4\upsilon_g^2\rho_l\upsilon_l^2 + 2P\rho_g\rho_l\upsilon_g^2\upsilon_l^2)(K_1K_2+K_3)}{K_4^2}$$

$$\frac{\partial f_1}{\partial \upsilon_l} = \frac{\frac{f\rho_l\upsilon_l}{d}K_1}{K_4} - \frac{(4\alpha_g\rho_g^2\rho_l\upsilon_l - 8\alpha_g\upsilon_g^2\rho_l\upsilon_l + 4P\alpha_g\rho_g\rho_l\upsilon_g^2\upsilon_l)(K_1K_2+K_3)}{K_4^2}$$

$$\frac{\partial f_1}{\partial P} = \frac{K_1 + \alpha_g\rho_g g\cos\theta(1-\alpha_g)}{K_4} - \frac{(\upsilon_g^2 - \rho_g\upsilon_g^2 + \alpha_g\rho_l\upsilon_l^2)(K_1K_2+K_3)}{K_4^2}$$

$$\frac{\partial f_1}{\partial \upsilon_g} = \frac{-\alpha_g\rho_g\upsilon_g^2K_1}{K_4} - \frac{(4\rho_g^2\upsilon_g - 2\rho_g^3\upsilon_g - 4\alpha_g\rho_g^2\upsilon_g - 8\alpha_g\upsilon_g\rho_l\upsilon_l^2 + 4P\alpha_g\rho_g\rho_l\upsilon_g\upsilon_l)(K_1K_2+K_3)}{K_4^2}$$

$$\frac{\partial f_1}{\partial T} = 0.$$

Then,

$$\left|\frac{\partial f_1}{\partial \alpha_g}\right| = \left|\frac{-K_2(\rho_g^2 + \rho_g \upsilon_g^2 P - 2\upsilon_g^2) + (1-2\alpha_g)\rho_g^3 g\cos\theta}{K_4} - \frac{(-2\rho_g^2\upsilon_g^2 + 2\rho_g^2\rho_l\upsilon_l^2 - 4\upsilon_g^2\rho_l\upsilon_l^2 + 2P\rho_g\rho_l\upsilon_g^2\upsilon_l^2)(K_1K_2+K_3)}{K_4^2}\right|$$

$$\leq \frac{(|\rho_g^2| + |\rho_g||\upsilon_g^2||P| + 2|\upsilon_g^2|)|K_2| + |\rho_g^3||g|}{|K_4|} + \frac{(2|\rho_g^2||\upsilon_g^2| + 2|\rho_g^2||\rho_l||\upsilon_l^2| + 4|\upsilon_g^2||\rho_l||\upsilon_l^2| + 2|P||\rho_g||\rho_l||\upsilon_g^2||\upsilon_l^2|)(|K_1||K_2| + |K_3|)}{|K_4^2|}.$$

Let:

$$M_{11} = \sup\left\{\frac{(|\rho_g^2| + |\rho_g||\upsilon_g^2||P| + 2|\upsilon_g^2|)|K_2| + |\rho_g^3||g|}{|K_4|} + \frac{(2|\rho_g^2||\upsilon_g^2| + 2|\rho_g^2||\rho_l||\upsilon_l^2| + 4|\upsilon_g^2||\rho_l||\upsilon_l^2| + 2|P||\rho_g||\rho_l||\upsilon_g^2||\upsilon_l^2|)(|K_1||K_2| + |K_3|)}{|K_4^2|}\right\},$$

then,

$$\left|\frac{\partial f_1}{\partial \alpha_g}\right| \leq M_{11}.$$

$$\left|\frac{\partial f_1}{\partial \upsilon_l}\right| = \left|\frac{\frac{f\rho_l\upsilon_l}{d}K_1}{K_4} - \frac{(4\alpha_g\rho_g^2\rho_l\upsilon_l - 8\alpha_g\upsilon_g^2\rho_l\upsilon_l + 4P\alpha_g\rho_g\rho_l\upsilon_g^2\upsilon_l)(K_1K_2+K_3)}{K_4^2}\right|$$

$$\leq \frac{|K_1| + |\rho_g||g|}{|K_4|} + \frac{(4|\rho_g^2||\rho_l||\upsilon_l| + 8|\upsilon_g^2||\rho_l||\upsilon_l| + 4|P||\rho_g||\rho_l||\upsilon_g^2||\upsilon_l|)(|K_1||K_2| + |K_3|)}{|K_4^2|}.$$

Let:

$$M_{12} = \sup\left\{\frac{|K_1| + |\rho_g||g|}{|K_4|} + \frac{(4|\rho_g^2||\rho_l||\upsilon_l| + 8|\upsilon_g^2||\rho_l||\upsilon_l| + 4|P||\rho_g||\rho_l||\upsilon_g^2||\upsilon_l|)(|K_1||K_2| + |K_3|)}{|K_4^2|}\right\},$$

then,

$$\left|\frac{\partial f_1}{\partial \upsilon_l}\right| \leq M_{12}.$$

$$\begin{aligned}\left|\frac{\partial f_1}{\partial P}\right| &= \left|\frac{K_1+\alpha_g\rho_g g\cos\theta(1-\alpha_g)}{K_4} - \frac{(\upsilon_g^2-\rho_g\upsilon_g^2+\alpha_g\rho_l\upsilon_l^2)(K_1K_2+K_3)}{K_4^2}\right| \\ &\leq \frac{|K_1|+|\rho_g||g|}{|K_4|} + \frac{(|\upsilon_g^2|+|\rho_g||\upsilon_g^2|+|\rho_l||\upsilon_l^2|)(|K_1||K_2|+|K_3|)}{|K_4^2|}.\end{aligned}$$

Let:

$$M_{13} = \sup\left\{\frac{|K_1|+|\rho_g||g|}{|K_4|} + \frac{(|\upsilon_g^2|+|\rho_g||\upsilon_g^2|+|\rho_l||\upsilon_l^2|)(|K_1||K_2|+|K_3|)}{|K_4^2|}\right\},$$

then,

$$\left|\frac{\partial f_1}{\partial P}\right| \leq M_{13}.$$

$$\begin{aligned}\left|\frac{\partial f_1}{\partial \upsilon_g}\right| &= \left(\left|\frac{-\alpha_g\rho_g\upsilon_g^2K_1}{K_4} - \frac{(4\rho_g^2\upsilon_g-2\rho_g^3\upsilon_g-4\alpha_g\rho_g^2\upsilon_g-8\alpha_g\upsilon_g\rho_l\upsilon_l^2+4P\alpha_g\rho_g\rho_l\upsilon_g\upsilon_l)(K_1K_2+K_3)}{K_4^2}\right|\right) \\ &\leq \frac{|\rho_g||\upsilon_g^2||K_1|}{|K_4|} + \frac{(4|\rho_g^2||\upsilon_g|+2|\rho_g^3||\upsilon_g|+4|\rho_g^2||\upsilon_g|+8|\upsilon_g||\rho_l||\upsilon_l^2|+4|P||\rho_g||\rho_l||\upsilon_g||\upsilon_l|)(|K_1||K_2|+|K_3|)}{|K_4^2|}\end{aligned}$$

Let:

$$\begin{aligned}M_{14} &= \sup\left\{\frac{|\rho_g||\upsilon_g^2||K_1|}{|K_4|}\right. \\ &+ \left.\frac{(4|\rho_g^2||\upsilon_g|+2|\rho_g^3||\upsilon_g|+4|\rho_g^2||\upsilon_g|+8|\upsilon_g||\rho_l||\upsilon_l^2|+4|P||\rho_g||\rho_l||\upsilon_g||\upsilon_l|)(|K_1||K_2|+|K_3|)}{|K_4^2|}\right\},\end{aligned}$$

then,

$$\left|\frac{\partial f_1}{\partial \upsilon_g}\right| \leq M_{14}, \quad \left|\frac{\partial f_1}{\partial T}\right| = 0 \leq M_{15}.$$

According to similar method, the system was obtained as follows:

$$\begin{cases}\dfrac{\partial f_2}{\partial \alpha_g} = \dfrac{\upsilon_l}{1-\alpha_g}\dfrac{\partial f_1}{\partial \alpha_g} + \dfrac{\upsilon_l f_1}{(1-\alpha_g)^2} \\ \dfrac{\partial f_2}{\partial \upsilon_l} = \dfrac{f_1+\upsilon_l\frac{\partial f_1}{\partial \upsilon_l}}{1-\alpha_g} \\ \dfrac{\partial f_2}{\partial P} = \dfrac{\upsilon_l}{1-\alpha_g}\dfrac{\partial f_1}{\partial P} \\ \dfrac{\partial f_2}{\partial \upsilon_g} = \dfrac{\upsilon_l}{1-\alpha_g}\dfrac{\partial f_1}{\partial \upsilon_g} \\ \dfrac{\partial f_2}{\partial T} = 0.\end{cases}$$

Thus, $\left|\frac{\partial f_2}{\partial \alpha_g}\right| \le \frac{|\upsilon_l|}{|1-\alpha_g|}M_{11} + \frac{|\upsilon_l|N_1}{|(1-\alpha_g)^2|}$. Let $M_{21} = \sup\left\{\frac{|\upsilon_l|}{|1-\alpha_g|}M_{11} + \frac{|\upsilon_l|N_1}{|(1-\alpha_g)^2|}\right\}$, then, $\left|\frac{\partial f_2}{\partial \alpha_g}\right| \le M_{21}$.

$\left|\frac{\partial f_2}{\partial \upsilon_l}\right| \le \frac{N_1+|\upsilon_l|M_{12}}{|1-\alpha_g|}$. Let $M_{22} = \sup\left\{\frac{N_1+|\upsilon_l|M_{12}}{|1-\alpha_g|}\right\}$, then, $\left|\frac{\partial f_2}{\partial \upsilon_l}\right| \le M_{22}$.

$\left|\frac{\partial f_2}{\partial P}\right| \le \frac{|\upsilon_l|M_{13}}{|1-\alpha_g|}$. Let $M_{23} = \sup\left\{\frac{|\upsilon_l|M_{13}}{|1-\alpha_g|}\right\}$, then, $\left|\frac{\partial f_2}{\partial P}\right| \le M_{23}$.

$\left|\frac{\partial f_2}{\partial \upsilon_g}\right| \le \frac{|\upsilon_l|M_{14}}{|1-\alpha_g|}$. Let $M_{24} = \sup\left\{\frac{|\upsilon_l|M_{14}}{|1-\alpha_g|}\right\}$, then, $\left|\frac{\partial f_2}{\partial \upsilon_g}\right| \le M_{24}$ and $\left|\frac{\partial f_2}{\partial T}\right| = 0 \le M_{25}$.

The partial differential of f_3, f_4, f_5 about $\alpha_g, \upsilon_l, P, \upsilon_g, T$ may be written as follows, respectively:

$$\begin{cases}
\dfrac{\partial f_3}{\partial \alpha_g} = \dfrac{\rho_l g\cos\theta + 2\rho_l \upsilon_l f_2 - 2\rho_l(1-\alpha_g)\upsilon_l \frac{\partial f_2}{\partial \alpha_g}}{1-\alpha_g} + \dfrac{\rho_l g\cos\theta(1-\alpha_g) + \frac{f\rho_l \upsilon_l^2}{2d} + 2\rho_l(1-\alpha_g)\upsilon_l f_2}{(1-\alpha_g)^2} \\
\dfrac{\partial f_3}{\partial \upsilon_l} = \dfrac{\frac{-f\rho_l \upsilon_l}{d} - 2\rho_l(1-\alpha_g)f_2 - 2\rho_l(1-\alpha_g)\upsilon_l \frac{\partial f_2}{\partial \upsilon_l}}{1-\alpha_g} \\
\dfrac{\partial f_3}{\partial P} = \dfrac{-2\rho_l(1-\alpha_g)\upsilon_l}{1-\alpha_g}\dfrac{\partial f_2}{\partial P} \\
\dfrac{\partial f_3}{\partial \upsilon_g} = \dfrac{-2\rho_l(1-\alpha_g)\upsilon_l}{1-\alpha_g}\dfrac{\partial f_2}{\partial \upsilon_g} \\
\dfrac{\partial f_3}{\partial T} = 0
\end{cases}$$

$$\begin{cases}
\dfrac{\partial f_4}{\partial \alpha_g} = \dfrac{\upsilon_g P f_3 + \alpha_g \upsilon_g P \frac{\partial f_3}{\partial \alpha_g} + \rho_g^2 \upsilon_g \frac{\partial f_1}{\partial \alpha_g}}{\rho_g^2 \alpha_g} - \dfrac{\rho_g^2(\alpha_g \upsilon_g P f_3 + \rho_g^2 \upsilon_g f_1)}{\rho_g^4 \alpha_g^2} \\
\dfrac{\partial f_4}{\partial \upsilon_l} = \dfrac{\alpha_g \upsilon_g P \frac{\partial f_3}{\partial \upsilon_l} + \rho_g^2 \upsilon_g \frac{\partial f_1}{\partial \upsilon_l}}{\rho_g^2 \alpha_g} \\
\dfrac{\partial f_4}{\partial P} = \dfrac{\alpha_g P f_3 + \alpha_g \upsilon_g P \frac{\partial f_3}{\partial \upsilon_g} + \rho_g^2 f_1 + \rho_g^2 \upsilon_g \frac{\partial f_1}{\partial \upsilon_g}}{\rho_g^2 \alpha_g} \\
\dfrac{\partial f_4}{\partial \upsilon_g} = \dfrac{\alpha_g \upsilon_g f_3 + \alpha_g \upsilon_g P \frac{\partial f_3}{\partial P} + \rho_g^2 \upsilon_g \frac{\partial f_1}{\partial P}}{\rho_g^2 \alpha_g} \\
\dfrac{\partial f_4}{\partial T} = 0
\end{cases}$$

and

$$\begin{cases}
\dfrac{\partial f_5}{\partial \alpha_g} = \dfrac{-\frac{\partial f_3}{\partial \alpha_g} - \rho_l \upsilon_l \frac{\partial f_2}{\partial \alpha_g}}{\rho_l C_{Pl}} \\
\dfrac{\partial f_5}{\partial \upsilon_l} = \dfrac{-\frac{\partial f_3}{\partial \upsilon_l} - \rho_l f_2 - \rho_l \upsilon_l \frac{\partial f_2}{\partial \upsilon_l}}{\rho_l C_{Pl}} \\
\dfrac{\partial f_5}{\partial P} = \dfrac{-\frac{\partial f_3}{\partial P} - \rho_l \upsilon_l \frac{\partial f_2}{\partial P}}{\rho_l C_{Pl}} - \dfrac{\left[\rho_l \frac{\partial C_{Pl}}{\partial P}\right]\left[\rho_l g\cos\theta - a(T-T_e) - f_3 - \rho_l \upsilon_l f_2\right]}{\rho_l^2 C_{Pl}^2} \\
\dfrac{\partial f_5}{\partial \upsilon_g} = \dfrac{-\frac{\partial f_3}{\partial \upsilon_g} - \rho_l \upsilon_l \frac{\partial f_2}{\partial \upsilon_g}}{\rho_l C_{Pl}} \\
\dfrac{\partial f_5}{\partial T} = \dfrac{a - \frac{\partial f_3}{\partial T} - \rho_l \upsilon_l \frac{\partial f_2}{\partial T}}{\rho_l C_{Pl}} - \dfrac{\rho_l \frac{\partial C_{Pl}}{\partial T}\left(\rho_l g\cos\theta - \alpha(T-T_e) - \frac{dP}{dz} - \rho_l \upsilon_l \frac{d\upsilon_l}{dz}\right)}{\rho_l^2 C_{Pl}^2}
\end{cases}.$$

Repeating the above method, the follow conditions can be obtained:

$$\left|\frac{\partial f_3}{\partial \alpha_g}\right| \le M_{31}, \quad \left|\frac{\partial f_3}{\partial \upsilon_l}\right| \le M_{32}, \quad \left|\frac{\partial f_3}{\partial P}\right| \le M_{33}, \quad \left|\frac{\partial f_3}{\partial \upsilon_g}\right| \le M_{34}, \quad \left|\frac{\partial f_3}{\partial T}\right| \le M_{35}$$

$$\left|\frac{\partial f_4}{\partial \alpha_g}\right| \le M_{41}, \quad \left|\frac{\partial f_4}{\partial \upsilon_l}\right| \le M_{42}, \quad \left|\frac{\partial f_4}{\partial P}\right| \le M_{43}, \quad \left|\frac{\partial f_4}{\partial \upsilon_g}\right| \le M_{44}, \quad \left|\frac{\partial f_4}{\partial T}\right| \le M_{45}$$

$$\left|\frac{\partial f_5}{\partial \alpha_g}\right| \le M_{51}, \quad \left|\frac{\partial f_5}{\partial \upsilon_l}\right| \le M_{52}, \quad \left|\frac{\partial f_5}{\partial P}\right| \le M_{53}, \quad \left|\frac{\partial f_5}{\partial \upsilon_g}\right| \le M_{54}, \quad \left|\frac{\partial f_5}{\partial T}\right| \le M_{55}.$$

The Lipschitz condition is very important in discussing the solution of the system of differential equations, thus, the Lipschitz condition of $F(z;y)$ was considered firstly. The primal problem was written as follows:

$$\begin{aligned} \frac{d\alpha_g}{dz} &= f_1(z;\alpha_g,\upsilon_l,P,\upsilon_g,T), \quad \frac{d\upsilon_l}{dz} = f_2(z;\alpha_g,\upsilon_l,P,\upsilon_g,T), \quad \frac{dP}{dz} = f_3(z;\alpha_g,\upsilon_l,P,\upsilon_g,T), \\ \frac{d\upsilon_g}{dz} &= f_4(z;\alpha_g,\upsilon_l,P,\upsilon_g,T), \quad \frac{dT}{dz} = f_5(z;\alpha_g,\upsilon_l,P,\upsilon_g,T). \end{aligned}$$

Its can be written as:

$$\begin{aligned} \alpha_g' &= f_1(z;\alpha_g,\upsilon_l,P,\upsilon_g,T), \quad \upsilon_l' = f_2(z;\alpha_g,\upsilon_l,P,\upsilon_g,T), \quad P' = f_3(z;\alpha_g,\upsilon_l,P,\upsilon_g,T), \\ \upsilon_g' &= f_4(z;\alpha_g,\upsilon_l,P,\upsilon_g,T), \quad T' = f_5(z;\alpha_g,\upsilon_l,P,\upsilon_g,T). \end{aligned}$$

The primal condition is:

$$\alpha_g(z_0) = \alpha_{g0}, \quad \upsilon_l(z_0) = \upsilon_{l0}, \quad P(z_0) = P_0, \quad \upsilon_g(z_0) = \upsilon_{g0}, \quad T(z_0) = T_0.$$

Using the Euler method, for $i = 0, 1, 2, \ldots$:

$$\begin{aligned} \alpha_{g(i+1)} &= \alpha_{gi} + (z_{i+1} - z_i) f_1(z_i;\alpha_{gi},\upsilon_{li},P_i,\upsilon_{gi},T_i), \quad \upsilon_{l(i+1)} = \upsilon_{li} + (z_{i+1} - z_i) f_2(z_i;\alpha_{gi},\upsilon_{li},P_i,\upsilon_{gi},T_i), \\ P_{i+1} &= P_i + (z_{i+1} - z_i) f_3(z_i;\alpha_{gi},\upsilon_{li},P_i,\upsilon_{gi},T_i), \quad \upsilon_{g(i+1)} = \upsilon_{gi} + (z_{i+1} - z_i) f_4(z_i;\alpha_{gi},\upsilon_{li},P_i,\upsilon_{gi},T_i), \\ T_{g(i+1)} &= T_i + (z_{i+1} - z_i) f_5(z_i;\alpha_{gi},\upsilon_{li},P_i,\upsilon_{gi},T_i). \end{aligned}$$

Here $\alpha_{gi}, \upsilon_{li}, P_i, \upsilon_{gi}, T_i$ are intended to approximate $\alpha_g(z_i), \upsilon_l(z_i), P(z_i), \upsilon_g(z_i), T(z_i)$, where $z_0 < z_1 < z_2 \ldots$ is subdivision of the interval of integration.

Let $y_i = (\alpha_{gi}, \upsilon_{li}, P_i, \upsilon_{gi}, T_i)^T$, then:

$$y_{i+1} = y_i + (z_{i+1} - z_i) F(z_i; y_i), \quad i = 0, 1, 2, \ldots, n-1.$$

If set $h_i = z_{i+1} - z_i$, then for the subdivision above can be written as:

$$h = (h_0, h_1, \ldots, h_{n-1}).$$

If we connect y_0 and y_1, y_1 and $y_2, \ldots$ etc. by straight lines we obtain the Euler polygon:

$$y_h(z) = y_i + (z - z_i) f(z_i; y_i) \quad \text{for } z_i \le z \le z_{i+1}.$$

Theorem 8.7. For $\|F(z;y)\| \le N = \max\{N_1, N_2, N_3, N_4, N_5\}$, then, for $\alpha_{gi}, \upsilon_{li}, P_i, \upsilon_{gi}, T_i$ defined by above way the estimate:

$$\|y_i - y_0\| \le N|z_i - z_0|,$$

where $y_i = (\alpha_{gi}, \upsilon_{li}, P_i, \upsilon_{gi}, T_i)^T$.

For $\left|\frac{\partial f_k}{\partial \alpha_g}\right| \leq M_{k1}, \left|\frac{\partial f_k}{\partial \upsilon_l}\right| \leq M_{k2}, \left|\frac{\partial f_k}{\partial P}\right| \leq M_{k3}, \left|\frac{\partial f_k}{\partial \upsilon_g}\right| \leq M_{k4}, \left|\frac{\partial f_k}{\partial T}\right| \leq M_{k5}$, then

$$\|F(z;y) - F(z;\hat{y})\| \leq L\|y - \hat{y}\|,$$

where $k = 1, 2, 3, 4, 5$; $L = \max_k(\sum_{i=1}^{5} M_{ki})$.

Proof. (1) From $\alpha_{g(i+1)} = \alpha_{gi} + (z_{i+1} - z_i) f_1(z_i; \alpha_{gi}, \upsilon_{li}, P_i, \upsilon_{gi}, T_i)$ and definition of $\|F(z; \alpha_g, \upsilon_l, P, \upsilon_g, T)\|$,

$$|\alpha_{g(i+1)} - \alpha_{gi}| = |z_{i+1} - z_i||f_1(z_i; \alpha_{gi}, \upsilon_{li}, P_i, \upsilon_{gi}, T_i)| \leq N(z_{i+1} - z_i).$$

Therefore,

$$|\alpha_{gi} - \alpha_{g(i-1)}| \leq N(z_i - z_{i-1}), \quad \ldots, \quad |\alpha_{g2} - \alpha_{g1}| \leq N(z_2 - z_1), \quad |\alpha_{g1} - \alpha_{g0}| \leq N(z_1 - z_0).$$

Thus,

$$|\alpha_{gi} - \alpha_{g(i-1)}| + \cdots + |\alpha_{g2} - \alpha_{g1}| + |\alpha_{g1} - \alpha_{g0}| \leq N(z_i - z_0).$$

Since:

$$|\alpha_{gi} - \alpha_{g(i-1)} + \cdots + \alpha_{g2} - \alpha_{g1} + \alpha_{g1} - \alpha_{g0}| \leq |\alpha_{gi} - \alpha_{g(i-1)}| + \cdots + |\alpha_{g2} - \alpha_{g1}| + |\alpha_{g1} - \alpha_{g0}|,$$

so,

$$|\alpha_{gi} - \alpha_{g0}| \leq N(z_i - z_0).$$

Similarly, $|\upsilon_{li} - \upsilon_{l0}| \leq N(z_i - z_0)$, $|P_i - P_0| \leq N(z_i - z_0)$, $|\upsilon_{gi} - \upsilon_{g0}| \leq N(z_i - z_0)$, $|T_i - T_0| \leq N(z_i - z_0)$.

From definition of $\|y_i - y_0\|$, we get:

$$\|y_i - y_0\| \leq N(z_i - z_0).$$

(2) For $f_1(z;y), f_2(z;y), f_3(z;y), f_4(z;y), f_5(z;y), y = (\alpha_g, \upsilon_l, P, \upsilon_g, T)^T$,

$$f_1(z;\hat{y}) - f_1(z;y) = \frac{\partial f_1}{\partial \alpha_g}(\hat{\alpha}_g - \alpha_g) + \frac{\partial f_1}{\partial \upsilon_l}(\hat{\upsilon}_l - \upsilon_l) + \frac{\partial f_1}{\partial P}(\hat{P} - P) + \frac{\partial f_1}{\partial \upsilon_g}(\hat{\upsilon}_g - \upsilon_g) + \frac{\partial f_1}{\partial T}(\hat{T} - T).$$

Thus,

$$|f_1(z;\hat{y}) - f_1(z;y)| \leq \left|\frac{\partial f_1}{\partial \alpha_g}\right| |\hat{\alpha}_g - \alpha_g| + \left|\frac{\partial f_1}{\partial \upsilon_l}\right| |\hat{\upsilon}_l - \upsilon_l| + \left|\frac{\partial f_1}{\partial P}\right| |\hat{P} - P| + \left|\frac{\partial f_1}{\partial \upsilon_g}\right| |\hat{\upsilon}_g - \upsilon_g| + \left|\frac{\partial f_1}{\partial T}\right| |\hat{T} - T|.$$

Let $\Delta y = \max\{|\hat{\alpha}_g - \alpha_g|, |\hat{\upsilon}_l - \upsilon_l|, |\hat{P} - P|, |\hat{\upsilon}_g - \upsilon_g|, |\hat{T} - T|\}$, then:

$$|f_1(z;\hat{y}) - f_1(z;y)| \leq \left(\left|\frac{\partial f_1}{\partial \alpha_g}\right| + \left|\frac{\partial f_1}{\partial \upsilon_l}\right| + \left|\frac{\partial f_1}{\partial P}\right| + \left|\frac{\partial f_1}{\partial \upsilon_g}\right| + \left|\frac{\partial f_1}{\partial T}\right|\right)\Delta y.$$

Similarly, we can get:

$$|f_2(z;\hat{y}) - f_2(z;y)| \leq \left(\left|\frac{\partial f_2}{\partial \alpha_g}\right| + \left|\frac{\partial f_2}{\partial v_l}\right| + \left|\frac{\partial f_2}{\partial P}\right| + \left|\frac{\partial f_2}{\partial v_g}\right| + \left|\frac{\partial f_2}{\partial T}\right|\right)\Delta y,$$

$$|f_3(z;\hat{y}) - f_3(z;y)| \leq \left(\left|\frac{\partial f_3}{\partial \alpha_g}\right| + \left|\frac{\partial f_3}{\partial v_l}\right| + \left|\frac{\partial f_3}{\partial P}\right| + \left|\frac{\partial f_3}{\partial v_g}\right| + \left|\frac{\partial f_3}{\partial T}\right|\right)\Delta y,$$

$$|f_4(z;\hat{y}) - f_4(z;y)| \leq \left(\left|\frac{\partial f_4}{\partial \alpha_g}\right| + \left|\frac{\partial f_4}{\partial v_l}\right| + \left|\frac{\partial f_4}{\partial P}\right| + \left|\frac{\partial f_4}{\partial v_g}\right| + \left|\frac{\partial f_4}{\partial T}\right|\right)\Delta y,$$

$$|f_5(z;\hat{y}) - f_5(z;y)| \leq \left(\left|\frac{\partial f_5}{\partial \alpha_g}\right| + \left|\frac{\partial f_5}{\partial v_l}\right| + \left|\frac{\partial f_5}{\partial P}\right| + \left|\frac{\partial f_5}{\partial v_g}\right| + \left|\frac{\partial f_5}{\partial T}\right|\right)\Delta y.$$

From definition of norm, there is:

$$\begin{aligned}\|F(z;\hat{y}) - F(z;y)\| = \max\{&|f_1(z;\hat{y}) - f_1(z;y)|, |f_2(z;\hat{y}) - f_2(z;y)|, |f_3(z;\hat{y}) - f_3(z;y)|,\\ &|f_4(z;\hat{y}) - f_4(z;y)|, |f_5(z;\hat{y}) - f_5(z;y)|\}.\end{aligned}$$

Let $L = \max_k(\sum_{i=1}^{5} M_{ki})$ and $\|\hat{y} - y\| = \Delta y$, then:

$$\|F(z;\hat{y}) - F(z;y)\| \leq L\|\hat{y} - y\|.$$

Therefore, the differential equations systems F is continuous and satisfies Lipschitz conditions. The solution is confirmed and gain the only from the original problem (Birkhoff and Rota, 1978).

The modified system of equations that can be written in a compact form as:

$$A\frac{dU}{dz} = B, \tag{8.131}$$

where A is coefficient matrices, B is a vector containing all algebraic terms and U is the solution vector:

$$A = \begin{bmatrix} v_g(1-\alpha_w-\alpha_o) & 0 & o & P(1-\alpha_w-\alpha_o) & -Pv_g & -Pv_g & 0 \\ 0 & \alpha_w & 0 & 0 & v_w & 0 & 0 \\ 0 & 0 & \alpha_o & 0 & 0 & v_o & 0 \\ \alpha_w+\alpha_o & \rho_w v_w \alpha_w & \rho_o v_o \alpha_o & 0 & P & P & 0 \\ 0 & 0 & \rho_g(1-\alpha_w-\alpha_o)v_g & -P & -P & 0 & \\ 1 & \rho_w v_w & 0 & 0 & 0 & 0 & \rho_w C_{pw} \\ 1 & 0 & \rho_o v_o & 0 & 0 & 0 & \rho_o C_{po} \end{bmatrix}$$

$$B = \begin{bmatrix} 0 \\ 0 \\ 0 \\ -\rho_w g \alpha_w \cos\theta - \rho_o g \alpha_o \cos\theta - \dfrac{\tau_{wb} S_{wb}}{A} - \dfrac{\tau_{ob} S_{ob}}{A} \\ -\rho_g g(1-\alpha_w-\alpha_w)\cos\theta \\ \rho_w g\cos\theta - a(T-T_e) \\ \rho_o g\cos\theta - a(T-T_e) \end{bmatrix} \quad U = \begin{bmatrix} P \\ v_w \\ v_o \\ v_g \\ \alpha_w \\ \alpha_o \\ T \end{bmatrix}$$

There are several numerical methods for solving the ordinary differential equations, such as the Runge-Kutta methods, linear multi-step methods and predictor-correcting methods. The Runge-Kutta technique is far more widely used than any other techniques for defined initial conditions. The stability of the numerical algorithm is given as follows.

The stability analysis of the fourth-order Runge-Kutta method

From the solution process above, at every computation point the model is a linear first-order ordinary differential equation, so cannot be solved analytically; however, in science and engineering, a numeric approximation to the solution is often suitable enough. There are many numerical methods, such as the Euler Method, the Trapezoidal Method, the θ Method, the Adams Method, and the Runge-Kutta Method. The Runge-Kutta method is an important method for the approximation of solutions from ordinary differential equations. The four-order explicit RK4 method has the following characteristics of generality; less calculation, high velocity, high efficiency, and high precision. However, it has strict conditions for stability. With this in mind, stability is discussed in this section.

For simplicity, equations for the liquid mixture were also used to analyse the stability. The system can be translated to the following form:

$$D\frac{dU}{dz} = FU + G,$$

where,

$$D = \begin{bmatrix} v_g \alpha_g P & \rho_g^2 \upsilon_g & \rho_g^2 \upsilon_g & 0 & 0 \\ 0 & -\upsilon_l & 0 & 1-\alpha_g & 0 \\ 1-\alpha_g & 0 & 0 & 2\alpha_l \rho_l \upsilon_l & 0 \\ \alpha_g(\upsilon_g^2 P + \rho_g) & \rho_g^2 \upsilon_g^2 & \rho_g^2 \upsilon_g^2 & 0 & 0 \\ 1 & 0 & 0 & \upsilon_l \rho_l & \rho_l C_{Pl} \end{bmatrix} \quad G = \begin{bmatrix} 0 \\ 0 \\ -g\cos\theta\,\rho_l \\ 0 \\ g\cos\theta + a\rho_l T_e \end{bmatrix}$$

$$F = \begin{bmatrix} 0 & 0 & 0 & 0 & 0 \\ 0 & 0 & 0 & 0 & 0 \\ 0 & g\cos\theta\,\rho_l & 0 & f\rho_l/(2d) & 0 \\ 0 & -g\rho_g^2\cos\theta & 0 & 0 & 0 \\ 0 & 0 & 0 & 0 & -a\rho_l \end{bmatrix} \quad U = \begin{bmatrix} P \\ \alpha_g \\ \upsilon_g \\ \upsilon_l \\ T \end{bmatrix}$$

Considering the characteristic equation, the mathematical characters for a set of ordinary differential equations can be found for the solving of the following eigenvalue system:

$$det[\lambda D - F] = 0,$$

$$\lambda_1 = \lambda_2 = \lambda_3 = 0, \quad \lambda_4 = -\frac{a}{C_{Pl}}$$

$$\lambda_5 = \frac{2\rho_g^2 g\cos\theta d(1-2\alpha_g) + 2\alpha_g^2 \rho_g g\cos\theta d(\rho_g - \rho_l) + \alpha_g \rho_g f \upsilon_l(\rho_g - P\upsilon_g^2)}{2d\rho_g[\rho_g \upsilon_g^2(\alpha_g - 1) - 2\alpha_g^2 \rho_l \upsilon_l^2] + 4P\rho_l d\alpha_g(\alpha_g - 1)\upsilon_g^2 \upsilon_l^2 + 4d\alpha_g \rho_g(\rho_l \upsilon_l^2 - \rho_g \upsilon_g^2)}$$
$$+ \frac{2\alpha_g \rho_l g\cos\theta d(\rho_g - P\upsilon_g^2 + P\alpha_g \upsilon_g^2)}{2d\rho_g[\rho_g \upsilon_g^2(\alpha_g - 1) - 2\alpha_g^2 \rho_l \upsilon_l^2] + 4P\rho_l d\alpha_g(\alpha_g - 1)\upsilon_g^2 \upsilon_l^2 + 4d\alpha_g \rho_g(\rho_l \upsilon_l^2 - \rho_g \upsilon_g^2)}.$$

In engineering practice, the value, $\lambda_4 < 0$ and $\lambda_5 > 0$ can be found. Therefore, the differential system solution tends towards divergence. However, the value of the positive eigenvalue is very small (from 0.009 to 0.1) which indicates that for a solution cycle, the approximate solution is weakly influenced. These models belong to a well-conditioned model and even though the absolute value of the negative characteristic parameter range is very small it has an important impact on the RK4 methods. For ordinary differential equations, if the numerical results are stable with a fixed grid size h, the product of λ and h must fall into an absolutely stable interval. The RK4 interval is $[-2.78, 0]$ (Birkhoff and Rota, 1978). Therefore, the system is stable at grid $h < 28$.

The Equations for (8.131) are large linear equations and there are many solution methods. Considering the large and sometimes ill-posed linear systems, we adopt the Singular Value Decomposition (SVD) Method, which stabilizes the solutions.

8.7.3 Numerical solution

To simplify the calculation, we divided the well into several short segments of the same length h. The length of a segment varies depending on variations in well thickness, hole diameter, the fluid density inside and outside the pipe and well geometry. Then, velocity, pressure and temperature calculations are performed for each successive "segment" of the pipe up to the surface By means of the afore-mentioned discussion, we use the four order Runge-Kutta method to simulate the model.

8.7.3.1 *Initial condition and calculation of some parameters*

In order to solve model 8.131, some definite conditions, i.e. the initial conditions, need to be added. The initial conditions are the distribution of the pressure, temperature and velocity along the well at the well top. In this section, we give the calculation method for some of parameters:

1. Each point's inclination:

$$\theta_j = \theta_{j-1} + (\theta_k - \theta_{k-1})\Delta s_j / \Delta s_k,$$

   where j represents segment point of calculation, Δs_k represents measurement depth of inclination angle θ_k, θ_{k-1}, Δs_j step length of calculation.
2. Transient heat transfer function:

$$f(t_D) = \begin{cases} 1.128\sqrt{t_D}(1 - 0.3\sqrt{t_D}), & t_D \leq 1.5 \\ (0.4063 + 0.5\ln t_D)(1 + 0.6/t_D), & t_D > 1.5. \end{cases}$$

3. Frictional force between liquid phase and pipe wall:
   In the present model, the gas-wall friction and the oil-wall friction are neglected, and the friction force is represented by the mixture-wall interaction given by Taitel *et al.*, 1995:

$$\frac{\tau_{kb} S_{kb}}{A} = \frac{f_k \rho_k v_k^2}{2d}$$

   They thought friction factor was affected by the Reynolds numbers of fluid in pipeline and the Blasius equation should be adopt. Many researchers such as Brauner and Ullmann (2002) and Brauner *et al.* (1998) had used the equation to compute the frictional force between oil or water and pipe wall:

$$f_k = \begin{cases} \dfrac{Re}{64}, & Re \leq 2000 \\ \dfrac{0.3164}{Re^{0.25}}, & Re > 1.5 \end{cases}, \quad Re_k = \frac{\rho_k v_k d}{\mu_k}.$$

4. Gas condensing parameter:

$$Z_g = \begin{cases} 1 + \left(0.31506 - \dfrac{1.0467}{T_{pr}} - \dfrac{0.5783}{T_{pr}^3}\right)\rho_{pr} + \left(0.053 - \dfrac{0.6123}{T_{pr}}\right)\rho_{pr}^2 + \dfrac{0.6815\rho_{pr}^2}{T_{pr}^3}, \\ \text{if } P < 35\,\text{MPa}; \\ (90.7x - 242x^2 + 42.4x^3)y^{1.18+2.82x} - (14.76x - 9.76x^2 + 4.58x^3)y + \dfrac{1 + y + y^2 + y^3}{(1-y)^3}, \\ \text{otherwise.} \end{cases}$$

$$\begin{aligned} F(y) &= -0.06125P_{pr}x\exp[-1.2(1-x)^2] + (90.7x - 242.2x^2 + 42.4x3)y^{1.18+2.82x} \\ &\quad + \frac{1 + y + y^2 + y^3}{(1-y)^3} - (14.76x - 9.76x^2 + 4.58x3)y^2 = 0, \\ x &= \frac{1}{T_{pr}}. \end{aligned}$$

5. Calculating the original conditions of the gas velocity, water velocity, oil velocity, oil-cut, water-cut, pressure and temperature at initial point. Assume that the pressure and temperature at the first compute point are already known:

$$\rho_{g0} = \frac{MP_0}{RZ_0 T_0}.$$

8.7.3.2 *Solution process*

The proposed algorithm procedure for solving the problem is designed as follows. The overall algorithm program flow diagram is presented in Figure 8.58.

Step 1. Setting step length of depth. In addition, the relatively tolerant error is denoted by ε. The smaller $\Delta\lambda$, ε are, the more accurate the results are. However, it will lead to rapid increasing calculating time. Here, set $\Delta h = 250$ (m), $\Delta\lambda = 1$, and $\varepsilon = 5\%$.

Step 2. Compute each point's inclination.

Step 3. Given the initial conditions and let $h = 0$.

Step 4. Compute the parameters under the initial conditions or the last time variables and let $\lambda = 0$.

Step 5. Let the differential equations to be functions f_i , where $(i = 1,2,3,4,5,6,7)$. Then a system of coupled functions is derived as follows:

$$\begin{cases} \upsilon_g(1-\alpha_w-\alpha_g)f_1 - P\upsilon_g f_6 - P\upsilon_g f_5 + P(1-\alpha_w-\alpha_g)f_4 = 0 \\ \alpha_w f_2 + \upsilon_w f_5 = 0 \\ \alpha_o f_3 + \upsilon_o f_6 = 0 \\ \rho_w\alpha_w\upsilon_w f_2 + \rho_o\alpha_o\upsilon_o f_3 + (\alpha_w+\alpha_o)f_1 + P(f_5+f_6) \\ \quad = -\rho_w g\alpha_w\cos\theta - \rho_o g\alpha_o\cos\theta - \dfrac{\tau_{wb}S_{wb}}{A} - \dfrac{\tau_{ob}S_{ob}}{A} \\ \rho_g(1-\alpha_w-\alpha_o)\upsilon_g f_4 + (1-\alpha_w-\alpha_o)f_1 - Pf_6 - Pf_5 = -\rho_g g((1-\alpha_w-\alpha_o)\cos\theta \\ \rho_w C_{pw} f_7 + \rho_w\upsilon_w f_2 + f_1 - \rho_w g\cos\theta + a(T-T_e) = 0 \\ \rho_o C_{po} f_7 + \rho_o\upsilon_o f_3 + f_1 - \rho_o g\cos\theta + a(T-T_e) = 0. \end{cases}$$

Step 6. Assume that $P, \upsilon_w, \upsilon_o, \upsilon_g, \alpha_w, \alpha_o, T$ to be y_i $(i = 1,2,3,4,5,6,7)$, respectively. Through solving the equations by SVD methods, some basic parameters can be gotten as follows:

$$\begin{cases} a_i = f_i(y_1, y_2, y_3, y_4, y_5, y_6, y_7) \\ b_i = f_i(y_1 + ha_1/2, y_2 + ha_2/2, y_3 + ha_3/2, y_4 + ha_4/2, y_5 + ha_5/2, y_6 + ha_6/2, y_7 + ha_7/2) \\ c_i = f_i(y_1 + hb_1/2, y_2 + hb_2/2, y_3 + hb_3/2, y_4 + hb_4/2, y_5 + hb_5/2, y_6 + hb_6/2, y_7 + hb_7/2) \\ d_i = f_i(y_1 + hc_1, y_2 + hc_2, y_3 + hc_3, y_4 + hc_4, y_5 + hc_5, y_6 + hc_6, y_7 + hc_7). \end{cases}$$

Step 7. Calculate the gas velocity, water velocity, oil velocity, oil-cut, water-cut, pressure and temperature at point $(j+1)$:

$$y_i^{j+1} = y_i^j + h(a_i + 2b_i + 2c_i + d_i)/6, \qquad i = 1,2,3,4,5,6,7 \quad j = 1,2,\ldots,n.$$

Step 8. Compute the gas density:

$$\rho_g^j = \frac{MP_k^j}{RZ_g T_k^j}.$$

Step 9. $\lambda = \lambda + \Delta\lambda$. Repeat steps 6 to 8 until $\lambda > \frac{\Delta h}{\Delta\lambda}$.

Step 10. $h = h + \Delta h$. Repeat steps 4 to 9 until $h < h_{\max}$ is calculated.

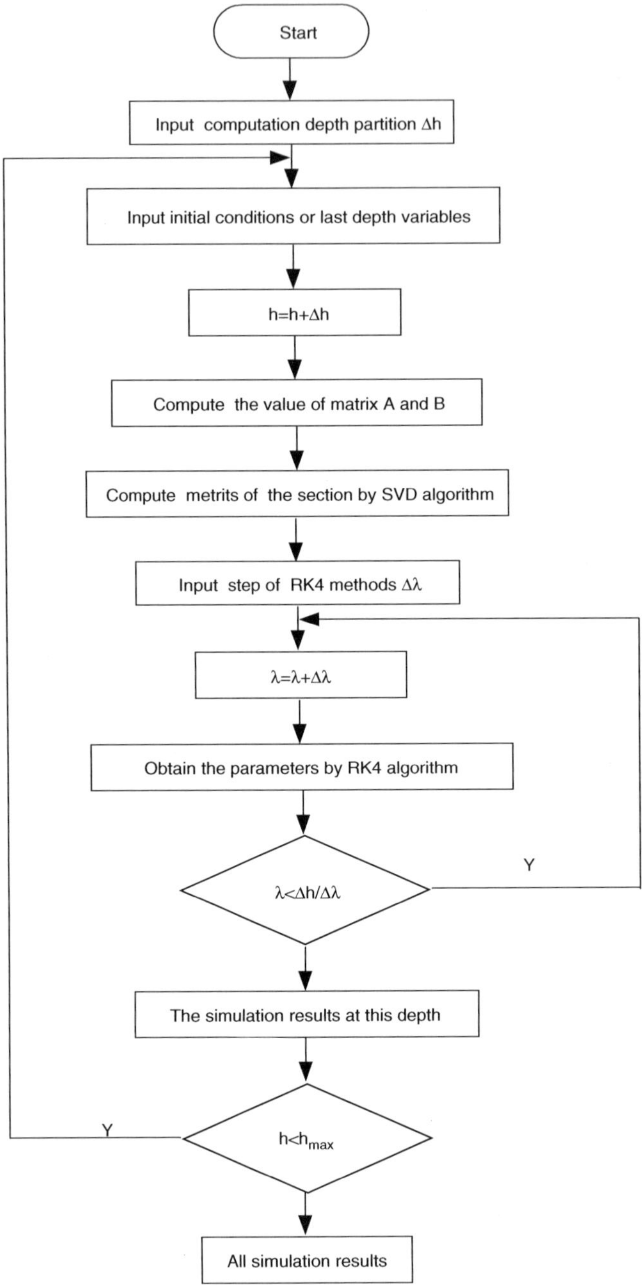

Figure 8.58 The overall procedure of the model.

8.7.4 Numerical simulation and results discussion

Some models were validated using the experimental data and it was found to bean effective method which needs to be more deeply analysed. To study the suitability of the model, some parameters were adjusted to reflect totally different operating conditions. However, as a result of the test conditions, it was not possible to totally simulate the real situation, so the error was large. A 4570 m-deep-well

Table 8.28 Parameters of pipes.

Diameter [mm]	Thickness [mm]	Weight [kg]	Expansion coefficient [m/m·K]	Young's modulus [GPa]	Poisson's ratio [–]	Using length [m]
88.9	9.53	18.9	0.0000115	215	0.3	1400
88.9	7.34	15.18	0.0000115	215	0.3	750
88.9	6.45	13.69	0.0000115	215	0.3	4200
73	7.82	12.8	0.0000115	215	0.3	600
73	5.51	9.52	0.0000115	215	0.3	150

Table 8.29 Well parameters.

Measured depth [m]	Internal diameter [mm]	External diameter [mm]
4325.69	168.56	193.7
6301.7	168.3	193.7
7100	121.42	146.1

is considered a deep well, and a 7100 m-deep-well is considered a super-deep well, so completion tests for deep wells has become an emerging new problem. In the applied basic theory for deep well testing research, tubular string mechanical analysis has been shown to be extremely complex as fluid temperature and tubing pressure heavily affect the force of the tubular string. Therefore, because in such an experiment it is difficult to model the real tubular string, here the real data is derived by comparing the consistency between the model and the real situation.

As was described previously, the algorithm begins with a calculation for the pipe at the bottom of the pipe. The calculations are performed for each successive segment of the pipe up to the surface.

8.7.4.1 *Parameters*

In this simulation, we study a pipe in X well, which is located in China. All the needed parameters are given as follows:

Depth of the well = 7100 m
Well bottom pressure = 40 MPa
Critical pressure = 4.968 MPa
Gas specific weight = 0.6434 kg/m^3
Ground thermal conductivity parameter = 2.06 W/m·K
Ground temperature = 16°C
Geothermal gradient = 2.18°C/100 m
Roughness of the inner surface of the well = 0.000015

The parameters for the pipes, the inclined well, the inclination, the azimuth and vertical depth are given in Tables 8.28, 8.29 and 8.30.

8.7.4.2 *Step analysis*

The effect of grid fineness was investigated. First, the adequacy of the number of grids in the $RK4$ method to obtain independent results is tested with the maximum steps being no less than 10. As shown in Figures 8.59 and 8.60, grids with 125, 250, 500 and 2500 steps were chosen at which point the pressure along the depth becomes practically grid-independent for the 250 and 500-grid solutions. An interesting phenomenon was found in that more precise results were not gained with a higher number of grids, mainly because errors tended to increase from accumulation.

8.7.4.3 *Main results*

Based on the algorithm designed above, we obtained a series of results for this well. The value of the gas velocity, water velocity, oil velocity, oil-cut, water-cut, pressure, temperature and gas density in the well are shown in Table 8.31.

Table 8.30 Parameters of azimuth, inclination and vertical depth.

Measured depth [m]	Inclination [°]	Azimuth [°]	Vertical depth [m]	Measured depth [m]	Inclination [°]	Azimuth [°]	Vertical depth [m]
0	0	120.33	0	3605	2.05	123.25	3603.36
303	1.97	121.2	302.87	3901	0.16	121.45	3899.22
600	1.93	120.28	599.73	4183	2.92	121.24	4181.09
899	0.75	126.57	898.59	4492	2.73	129.22	4489.95
1206	1.25	124.9	1205.45	4816.07	1.98	121.61	4813.87
1505	1.04	124.62	1504.32	5099.07	2.74	129.93	5096.74
1800	0.49	123.75	1799.18	5394.07	0.13	120.46	5391.61
2105	2.49	125.27	2104.04	5706.07	0.63	129.59	5703.47
2401	1.27	123.13	2399.91	5983.07	2.09	120.14	5980.34
2669	2.44	120.12	2667.79	6302.07	2.69	122.91	6299.19
3021	0.14	127.39	3019.63	6597.07	2.45	129.41	6594.06
3299	1.18	122.6	3297.5	6911.12	0.15	124.88	6907.96
7100	1.15	123.2	7085.88				

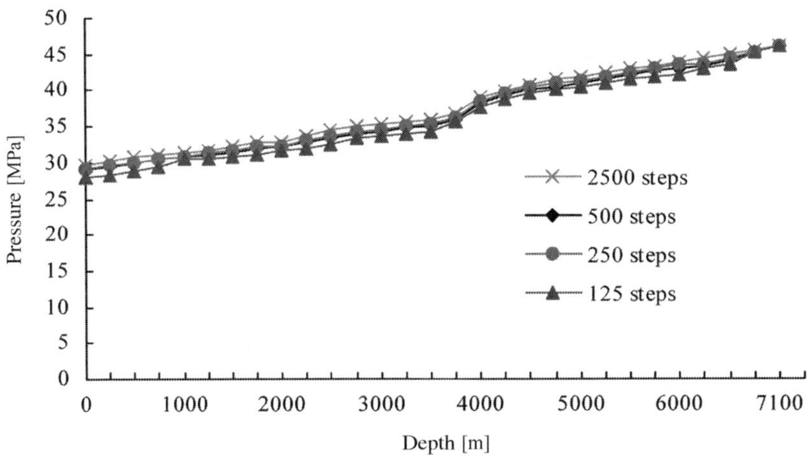

Figure 8.59 Step independence of the pressure simulation results.

At the same time, we obtained a comparative result with the measurement values as shown in Table 8.32.

8.7.4.4 *Trend analysis*

A trend analysis was carried out to examine whether the developed model is physically correct. To test the developed model, the distribution of the oil velocity, gas velocity, water velocity, water-cut, oil-cut, pressure, temperature and density in the well were determined.

Figure 8.59 shows the pressure profiles. Pressure distributions along the well at different depths are shown by Figure 8.59. From Figure 8.60, it can be seen that the pressure increases along with increasing depth. Similarly, we can also see the other variables trends in Figure 8.60.

8.7.5 **Sensitivity analysis**

Here, we consider the influence two parameters have on the result. The first is the gas outputs, with three different gas outputs used; 300000 m^3/d, 500000 m^3/d and 700000 m^3/d. In order to study the difference between a model with a radial transfer of heat from the tubing and one without, we present

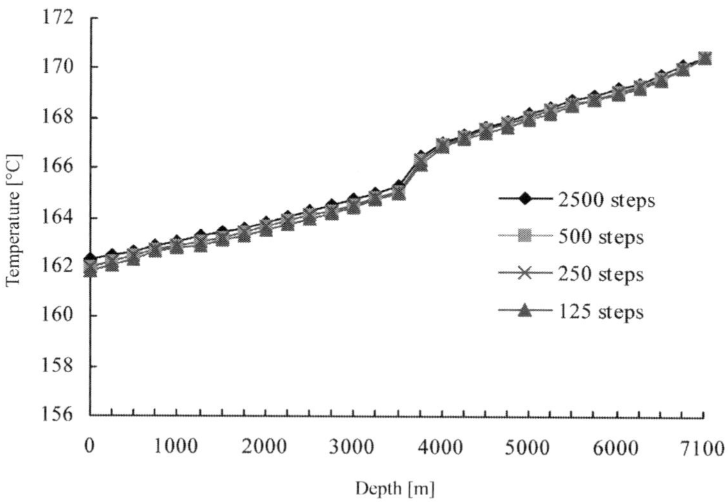

Figure 8.60 Step independence of the temperature simulation results.

Table 8.31 Gas velocity, liquid velocity, pressure, temperature, and gas density.

Number	Depth [m]	Gas velocity [m/s]	Water velocity [m/s]	Oil velocity [m/s]	Temperature [°C]	Water-cut [–]	Oil-cut [–]	Pressure [MPa]	Gas density [kg/m^3]
1	1	24.1580	20.1065	11.3866	160.0450	0.0365	0.3819	27.3532	152.0781
2	251	23.4568	19.5633	10.3332	160.2215	0.0374	0.3813	27.3913	154.2194
3	501	22.9864	19.0949	9.6912	160.4448	0.0391	0.3786	27.8886	156.9380
4	751	22.5498	18.8179	9.3431	160.6667	0.0406	0.3765	28.3775	159.6071
5	1001	22.1535	18.4321	8.8936	160.8360	0.0415	0.3758	28.7377	161.5689
6	1251	21.9875	18.0444	8.4803	160.9788	0.0429	0.3739	29.0325	163.1722
7	1501	21.8567	17.8474	8.2842	161.3102	0.0447	0.3710	29.7460	167.0533
8	1751	21.1578	17.6924	8.1363	161.4675	0.0475	0.3663	30.0672	168.7956
9	2001	20.9873	17.2551	7.7542	161.5575	0.0490	0.3641	30.2368	169.7118
10	2251	20.6874	17.2457	7.6630	161.6714	0.0515	0.3599	30.4553	170.8924
11	2501	20.5673	17.1416	7.6473	161.9712	0.0542	0.3554	31.0633	174.1827
12	2751	20.4854	17.1248	7.5411	162.3115	0.0553	0.3539	31.7473	177.8769
13	3001	19.2753	16.9819	7.6753	162.4883	0.0568	0.3515	32.0912	179.7291
14	3251	18.6250	16.7991	7.3057	162.6621	0.0588	0.3478	32.4226	181.5113
15	3501	17.5903	16.7058	7.2791	162.9649	0.0601	0.3448	33.0060	184.6470
16	3751	16.3395	16.5341	7.1135	163.0539	0.0608	0.3424	33.1643	185.4944
17	4001	14.6166	16.4638	7.0117	163.5062	0.0700	0.3335	34.0187	190.0730
18	4251	13.7848	16.4025	6.9447	163.7323	0.0729	0.3316	34.4344	192.2942
19	4501	12.4774	16.3553	6.8994	163.9050	0.0731	0.3304	34.7462	193.9573
20	4751	11.2480	16.2257	6.7723	164.1836	0.0736	0.3300	35.2459	196.6192
21	5001	10.7876	16.0956	6.6563	164.5867	0.0762	0.3273	35.9669	200.4536
22	5251	10.2104	16.0217	6.5959	164.8689	0.0871	0.3159	36.4658	203.1012
23	5501	9.5266	15.8677	6.4704	165.2564	0.0913	0.3114	37.1433	206.6885
24	5751	8.1742	15.7702	6.3960	165.5248	0.0971	0.3054	37.6100	209.1558
25	6001	7.8504	15.6993	6.3444	165.8080	0.1099	0.2920	38.0926	211.7005
26	6251	6.4680	15.6682	6.3244	166.2895	0.1253	0.2761	38.9053	215.9769
27	6501	5.7118	15.5916	6.2695	166.3224	0.1388	0.2623	38.9746	216.3457
28	6751	5.6114	15.4979	6.2030	166.6521	0.1874	0.2232	39.5203	219.2079
29	7001	5.2532	15.4663	6.0329	166.9846	0.2010	0.2034	40.0661	222.0656
30	7100	5.0415	14.9865	5.9872	167.0450	0.2020	0.1983	40.8555	225.0328

Table 8.32 Comparative results.

Well-at 250 m depth	Gas velocity [m/s]	Water velocity [m/s]	Oil velocity [m/s]	Temperature [°C]	Water-cut [–]	Oil-cut [–]	Pressure [MPa]	Gas density [kg/m^3]
Calculation results	23.324	19.535	10.132	160.202	0.0373	0.3818	27.421	154.204
Measurement results	22.1002	18.273	10.012	157.352	0.0451	0.378	28.3042	156.3842

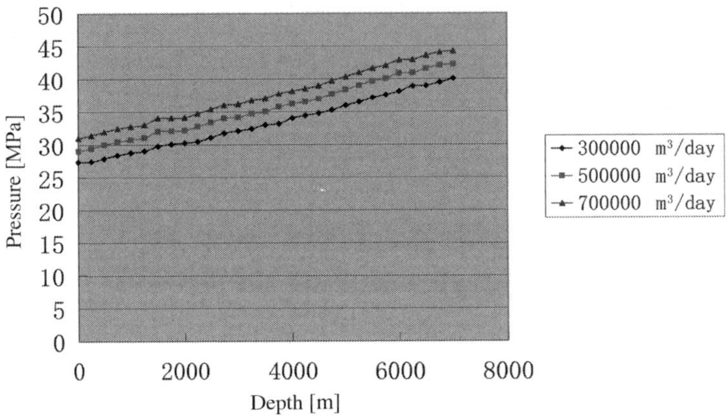

Figure 8.61 Pressure distribution.

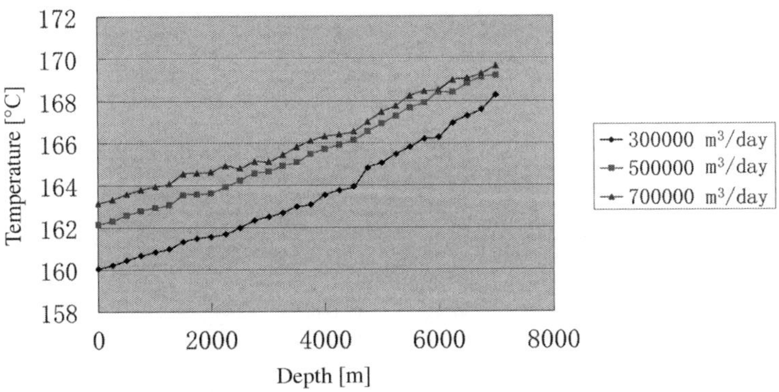

Figure 8.62 Temperature distribution.

results allowing for the two conditions. Using the algorithm and simulation, we obtained a series of results.

The values at the different gas outputs

For this purpose, synthetic sets were prepared, where in each set only the gas output parameter was changed, while the other parameters were kept constant. As Figures 8.61 to 8.68 show, if the gas output increases, then the gas velocity, water velocity, oil velocity, pressure and temperature all increase, but the gas density decreases. This investigation shows that the gas output has a strong influence on the temperature and gas velocity. Further, there is almost no change in the water-cut and oil-cut. Hence,

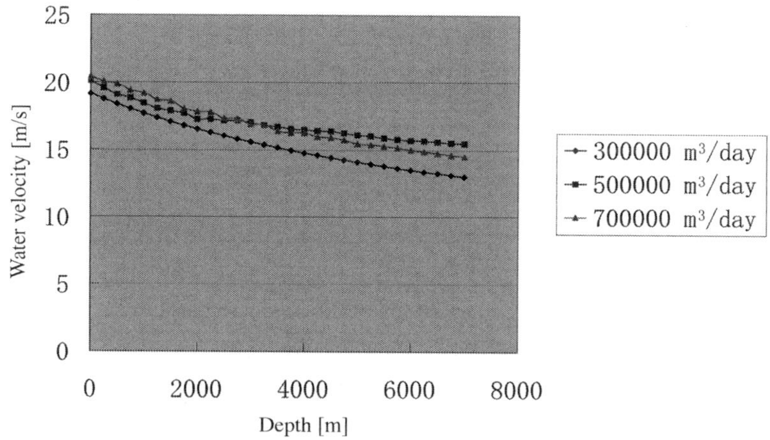

Figure 8.63 Water velocity distribution.

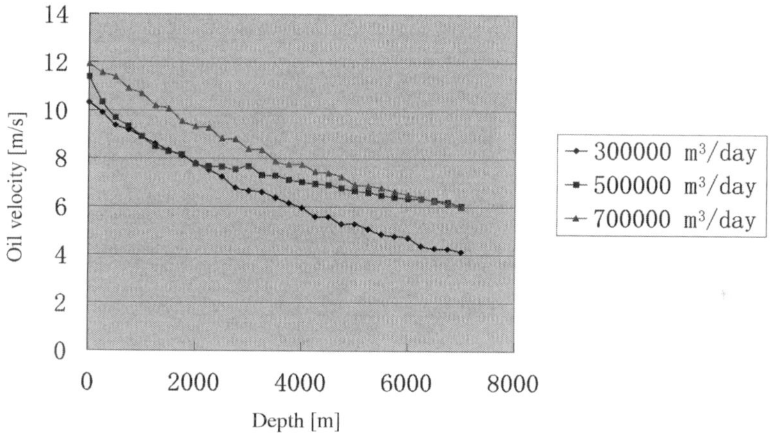

Figure 8.64 Oil velocity distribution.

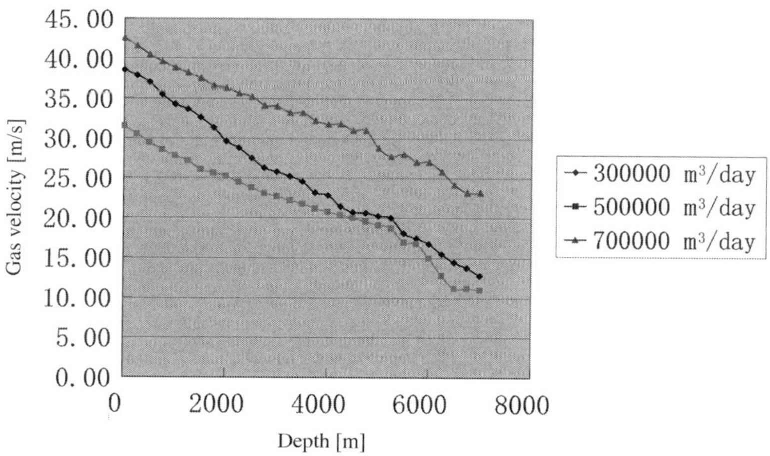

Figure 8.65 Gas velocity distribution.

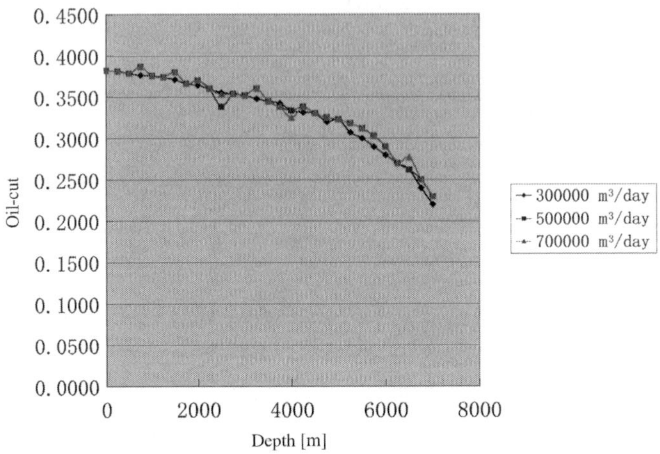

Figure 8.66 Oil-cut distribution.

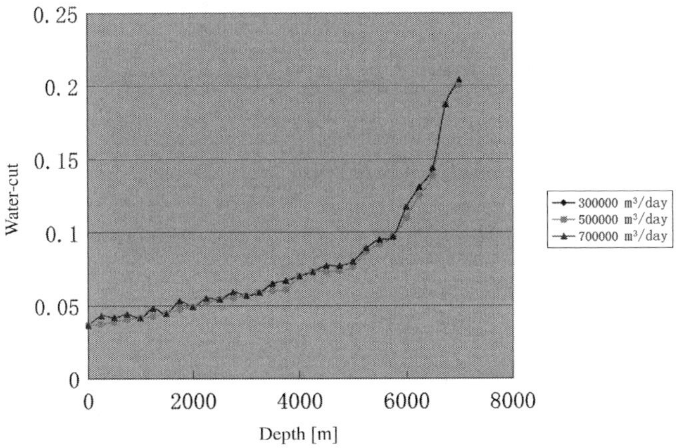

Figure 8.67 Water-cut distribution.

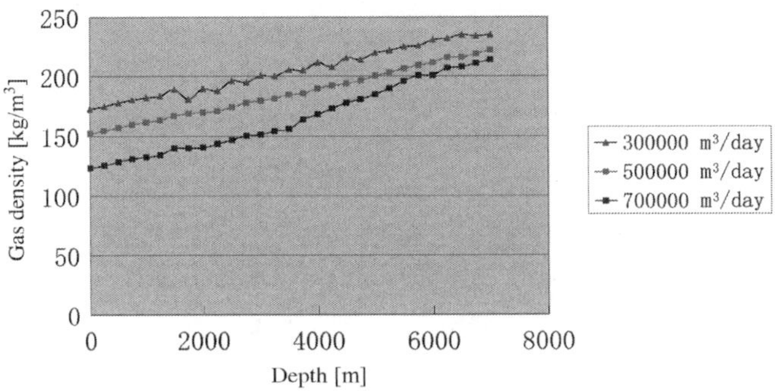

Figure 8.68 Gas density distribution.

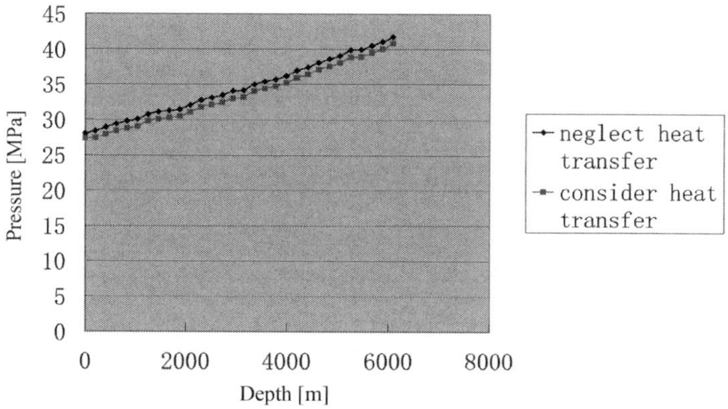

Figure 8.69 Pressure distribution.

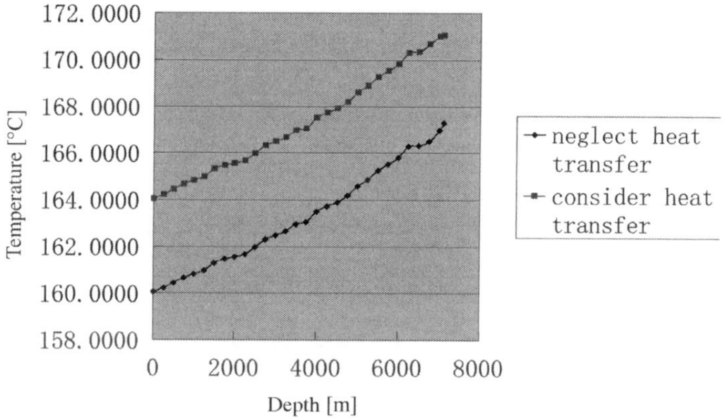

Figure 8.70 Temperature distribution.

the gas-cut weakly changes and the porosity, which can cause changes in the pressure and which has an effect on the gravity drop, is almost unchanged. Further, the gas velocity flow increases with an increase in the output resulting in a rise in friction causing the temperature to increase.

The values including and not including the heat transfer
From Figures 8.69 to 8.77, we can see that there is a strong influence on the oil-cut, water-cut and temperature and a weak effect on the pressure and water velocity. At the same time, the heat transfer has little effect on the oil velocity, gas velocity and gas density. (*f*) and (*g*) show that the water hold-up decreases and that of oil increases, respectively, with the heat transfer. The gas hole-up is a nonlinear distribution but the range is small (from 0.56 to 0.62).

8.7.6 **Comparison analysis**

Xu *et al.* (2011) developed a single phase model and adopted that same well as in this paper for their sample. Therefore, the pressure and temperature prediction performances were compared with the present model. From Figures 8.78 and 8.79, we find that the predicted values are different in the wellhead and consistent at the well-bottom.

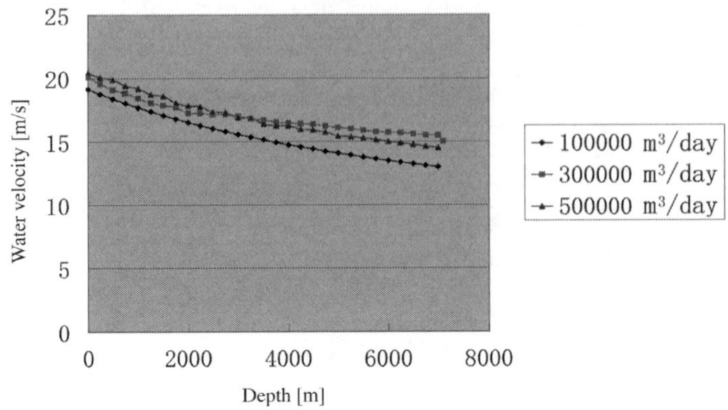

Figure 8.71 Water velocity distribution.

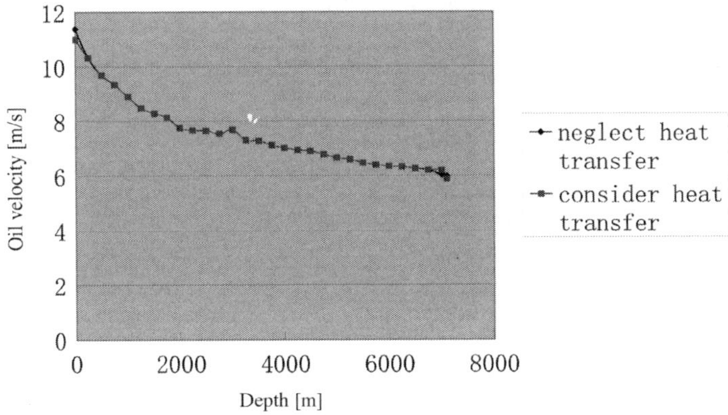

Figure 8.72 Oil velocity distribution.

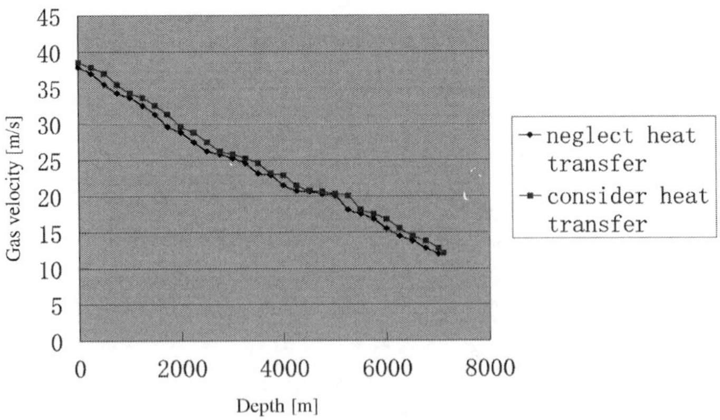

Figure 8.73 Gas velocity distribution.

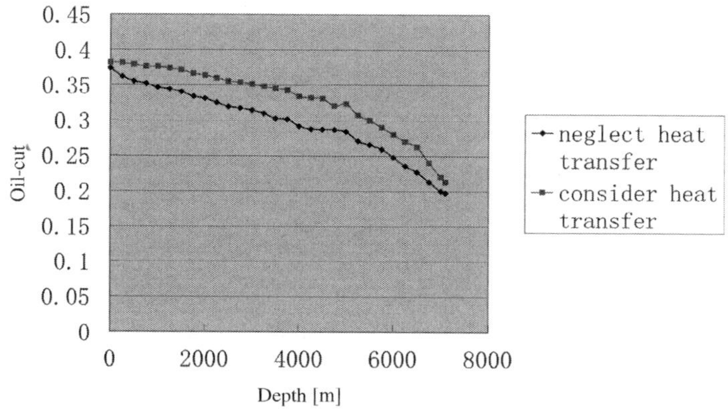

Figure 8.74 Oil-cut distribution.

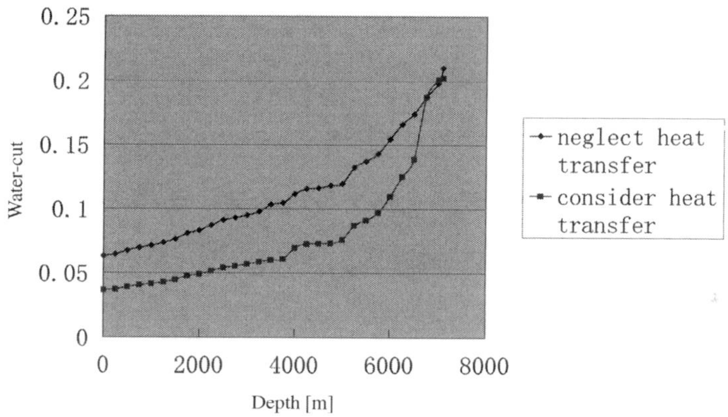

Figure 8.75 Water-cut distribution.

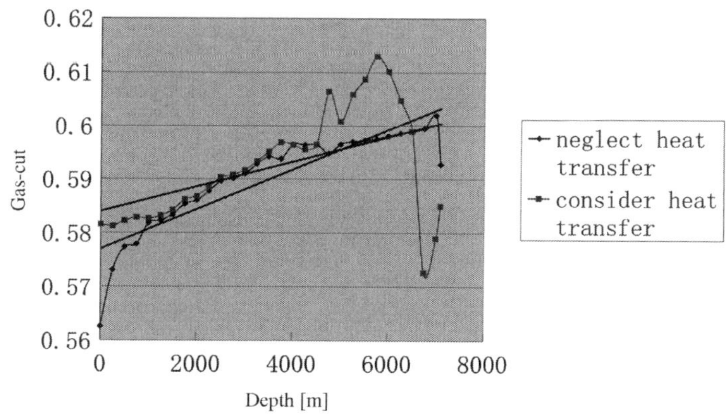

Figure 8.76 Gas-cut distribution.

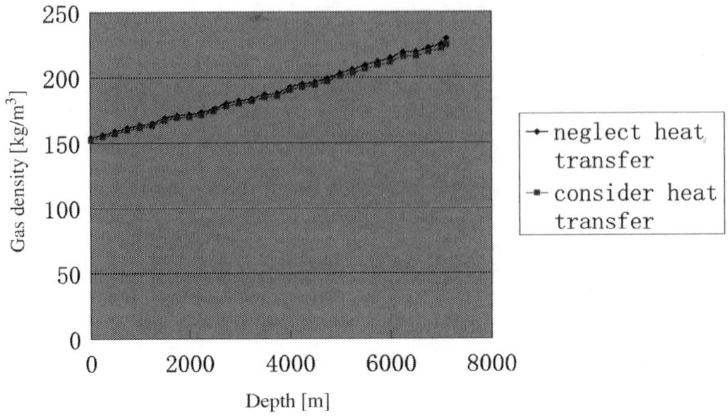

Figure 8.77 Gas density distribution.

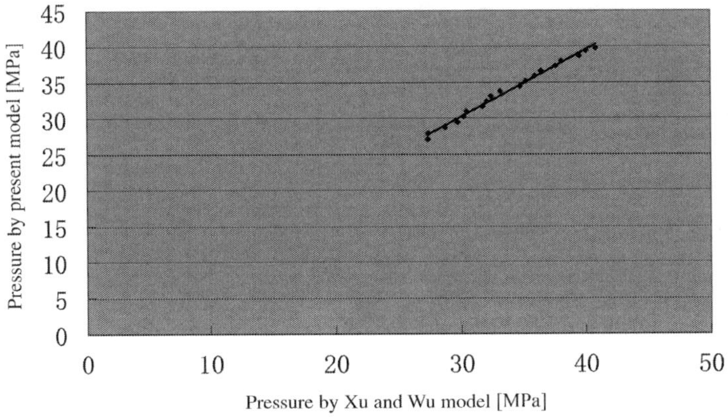

Figure 8.78 Pressure distribution.

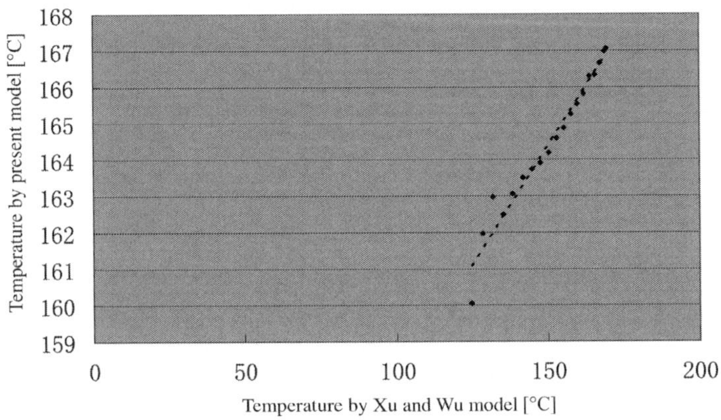

Figure 8.79 Temperature distribution.

CHAPTER 9

Predicting the pressure and temperature in shut-in

9.1 INTRODUCTION

Well control problems have always been difficult, as serious consequences arise if they not properly considered. In shut-in procedures, ascertaining the downhole status of the gas is essential to not only ensure effective well control measures, but also to have accurate information about pressure and temperature distributions. In particular, it is important to select the proper wellhead assembly as the maintaining maximum wellhead pressure is necessary to ensure safe shut-in processes. However, for high temperature-high pressure gas wells, it is often difficult to operate thermometers and pressure gauges. Thus, an effective and feasible method is one which predicts the distribution through the application of theory analysis technology. Predicting accurate temperature and pressure profiles in flowing wells can greatly improve the design of production facilities in petroleum and gas engineering.

One of the earliest works on predicting temperature profiles in a flowing well was presented by Kirkpatrick (1959). He presented a simple flowing-temperature-gradient chart that could be used to predict gas-lift valve temperatures at the injection depth. Much of the classic work in this area was developed by Ramey (1962), who presented approximate methods for predicting the temperature of either a single-phase incompressible liquid or a single-phase ideal gas flow in injection and production wells. Satter (1965) later improved Ramey's method by considering phase changes that occur within steam-injection projects. Shiu and Beggs (1980) simplified Ramey's method by correlating for a specific coefficient in Ramey's equation. Hagoort (2004) assessed Ramey's classic method for the calculation of temperatures in injection and production wells and showed that Ramey's method was an excellent approximation, except for an early transient period in which the calculated temperatures were significantly overestimated. They presented a simple graphical correlation to estimate the length of this early transient period. Willhite (1967) presented a detailed analysis of the overall heat-transfer mechanism in an injection well, and Coulter and Bardon (1979) developed a method for predicting temperatures in gas transmission lines. Hagoort (2005) presented a simple and physically transparent analytical solution for the prediction of wellbore temperatures in gas production wells. In these research papers, however, models were only built to predict temperature profiles, but not pressure profiles.

For research into the pressure of single phase flows, Rzasa and Katz (1945) presented an average temperature and average compressibility method, but the method tended to result in larger calculation errors. Sukkar and Corner (1955) in their studies only assumed that gas temperature was a constant and easily replaced using an average gas temperature, but the implementation was difficult to compute. In Cullender and Smith (1956), pressure, temperature and compressibility were all regarded as variables with variations in well depth. This method has been widely adopted by engineering personnel. However, such models only predicted pressure profiles but not temperature profiles.

For shut-in procedures in well control operations, there is some relevant research for the prediction of temperature and pressure distributions. Li and Zhuang (2002) analysed the pressure build-up principle and its recording time after shut-in, and proposed a method for acquiring the shut-in pressure for reservoirs with different permeability. Mingchang *et al.* (2005) studied well control problems according to the geological characteristics of carbonate rocks, and introduced a method for calculating the highest shut-in pressure allowed by a gas well. Beirute (1991) introduced a numerical simulation method for the down hole temperature distribution when oil wells were shut for days. However, from all this research, methods were found only for the prediction of pressure or temperature, but none which considered both. Wu *et al.* (2011; 2012) presented a coupled differential equations system model concerning the measurement of pressure and temperature simultaneously in HTHP wells according to mass, momentum, and energy balances.

During shut-in procedures, water or oil may enter the well along with gas, forming a gas-liquid two-phase flow. Transient flow models include a few flow patterns, such as slug flow, annular flow, churn flow and bubble flow (Barnea, 1987; Beirute, 1991). Detailed mechanisms for the transition of the different flow patterns for a well-bore flow under transient conditions are not well understood. Due to the complex nature of the gas-liquid flow in pipes, attempts have been made to develop predictive techniques through empirical or semi-empirical methods (such methods have dominated practical design procedures). Since the late 1980s, the trend has shifted towards a more fundamental modeling approach, also referred to as the mechanistic approach. Different researchers, such as Beggs and Brill (1973), Chierici *et al.* (1974), Kaichiro and Ishii (1984), Sharma *et al.* (1986), Andreussi and Persen (1987), Ansari *et al.* (1994), Grolman and Fortuin (1997), Ouyang and Aziz (1999; 2001), and others have proposed mechanistic models to describe the gas-liquid two-phase flow in vertical or inclined pipes. A two-phase flow in horizontal and near horizontal gas-liquid flows has also been studied (Hart *et al.*, 1989; Taitel and Dukler, 2004; Ouyang and Aziz, 2000; Taitel *et al.*, 1995; Xiao *et al.*, 1990). For multi-phase flows in pipes and wells, there are also some models (Brill, 1987; Griffith, 1984; Petalas and Aziz, 1996; Ouyang *et al.*, 1998). For further research into the shut-in condition, refer to Ding *et al.* (2012): Xu *et al.* (2013; 2014), Yang and Xu (2008).

9.2 PPT-SPDW

Considering the differential equation model for P and T, we set the following assumptions:

1. There is a steady gas flow of one dimension in the tubing, and all featured parameters are homochromatic at any transverse cross section.
2. The heat transfer in one dimension is steady from the tubing to the second interface, but the heat transfer in one dimension is unsteady from the second interface to the surrounding formation.
3. The heat losses from the tubing and the surrounding earth are radial.
4. The vertical formation temperature is considered linearly distributed based on known geothermal gradients.

9.2.1 **Physical model**

When a gas production well is shut, the pressure within the tubing is less than that in the formation, and as a result, the gas in the formation keeps moving into the tubing at the bottom. Through this tubing, gas flows from the bottom to the top with a mass flow rate. With the gas entering the tubing, the pressure inside increases gradually. After a settling time t_s, when the gas pressure at the bottom balances with the pressure in the corresponding formation, the gas stops entering, and the status in the wellhole tends to be stable. During the shut-in process, the state of the gas in the wellhole changes from steady flowing to transient flowing and finally settles at a comparatively static state.

9.2.2 **The coupled system model**

Consider the flow system depicted in Fig. 7.1, a straight cylindrical flow tube with an inclination angle θ, a constant cross-sectional flow area A, a hydraulic diameter d, and a total length Z. Through this tubing gas flows from the bottom to the top with a mass flow rate W. The distance co-ordinate in the flow direction along the tubing is denoted z.

Pressure model. The Cullender-Smith method (1956), developed for dry-gas wells, is generally applied to calculate wellhole pressure. Because it takes into account both variations in temperature and compressibility factor with depth, it is a more accurate method than the average Z_g and T, method used by the Railroad Commission and others. This method is therefore chosen as the basic model for this section.

The energy equation of steady flowing gas is:

$$\int_{P_{ws}}^{P_{ts}} \frac{Z_g T/P}{(1+(1.324\times 10^{-18} f(q_{sc}TZ_g)^2)/(d^5P^2)} dP = \int_0^L 0.03415\gamma_g \, dl \cos\theta \tag{9.1}$$

also is:

$$\int_{P_{ws}}^{P_{ts}} \frac{P/(Z_gT)}{(P/(Z_gT))^2 + (1.324\times 10^{-18} fq_{sc}^2)/d^5} dP = \int_0^L 0.03415\gamma_g \, dl \cos\theta. \tag{9.2}$$

Let:

$$F^2 = (1.324\times 10^{-18} fq_{sc}^2)/d^5,$$

then:

$$\int_{P_{ws}}^{P_{ts}} \frac{P/(Z_gT)}{(P/(Z_gT))^2 + F^2} dP = \int_0^L 0.03415\gamma_g \, dl \cos\theta. \tag{9.3}$$

Temperature model. Ramey (1962) and Willhite and Dietrich (1967) discussed the radial transfer of heat between the fluid and the earth in detail. The radial heat transfer from the fluid to the cement/earth can be described by:

$$dq = \frac{2\pi r_{to} U_{to}(T - T_h)}{w} dz. \tag{9.4}$$

The radial heat transfer from the cement/earth interface to the surrounding earth is:

$$dq = 2\pi K_e(T_h - T_e) w f(t_D) dz. \tag{9.5}$$

Obviously, the heat transfer from the fluid to the cement/earth is equal to that from the cement/earth interface to the surrounding earth, therefore we have:

$$T_h = \frac{K_eT_e + r_{to}U_{to}Tf(t_D)}{K_e + r_{to}U_{to}f(t_D)}. \tag{9.6}$$

Combining Equations (9.4), (9.5) and (9.6) gives the equation for the radial heat transfer between the fluid and the surrounding earth:

$$dq = \frac{2\pi r_{to}U_{to}K_e}{w(r_{to}U_{to}f(t_D) + K_e)}(T - T_e)dz. \tag{9.7}$$

Moreover,

$$dq = wC_P dT.$$

Combining the above two equations, we have:

$$\frac{dT}{dz} = a(T - T_e), \tag{9.8}$$

where:

$$a = \frac{2\pi r_{to}U_{to}K_e}{w(r_{to}U_{to}f(t_D) + K_e)}.$$

Heat transmission in the stratum. According to the assumption that the heat transfer from the second interface to the stratum is unsteady, the heat transmission model in the stratum is concluded as follows:

$$\frac{\partial T_e}{\partial t} = \frac{\lambda_e}{C_{P_e}\rho_e}\left(\frac{\partial^2 T_e}{\partial r^2} + \frac{1}{r}\frac{\partial T_e}{\partial r}\right). \tag{9.9}$$

Initial condition:

$$T(0) = T_0, \quad T_e(0) = T_e^0, \qquad 0 \le z \le h,$$

where T_0 and T_e^0 represents the gas temperature in the wellhole and stratum temperature in steady production state before shut-in procedures.

Boundary conditions:

$$\frac{\partial T_e}{\partial r} = 0, \quad \text{if } r \to \infty,$$

$$dQ = -2\pi r_{\text{cem}}\lambda_{\text{cem}}\left.\frac{\partial T_e}{\partial r}dz\right|_{r=r_{\text{cem}}}.$$

It follows from the dimensionless variables $r_D = \frac{r}{r_{\text{cem}}}$ and $t_D = \frac{\lambda_{et}}{r_{\text{cem}}^2}$. That, Equation (9.9) is changed into:

$$\frac{\partial T_e}{\partial t_D} = \frac{\lambda_e}{C_{P_e}\rho_e}\left(\frac{\partial^2 T_e}{\partial r_D^2} + \frac{1}{r_D}\frac{\partial T_e}{\partial r_D}\right). \tag{9.10}$$

The boundary conditions are converted into:

$$\left.\frac{\partial T_e}{\partial r_D}\right|_{r_D=1} = -\frac{dQ}{dz}(2\pi\lambda_e)^{-1}, \quad \left.\frac{\partial T_e}{\partial r_D}\right|_{r_D\to\infty} = 0. \tag{9.11}$$

Summing up the above, we have the following simultaneous equations models of pressure and temperature in shut-in process:

$$\begin{cases} \displaystyle\int_{P_{ws}}^{P_{ts}} \frac{P/(Z_g T)}{(P/(Z_g T))^2 + F^2}dP = \int_0^L 0.03415\gamma_g \, dl \cos\theta \\ \displaystyle\frac{dT}{dz} = a(T - T_e) \\ \displaystyle\frac{\partial T_e}{\partial t_D} = \frac{\lambda_e}{C_{P_e}\rho_e}\left(\frac{\partial^2 T_e}{\partial r_D^2} + \frac{1}{r_D}\frac{\partial T_e}{\partial r_D}\right) \\ \displaystyle\left.\frac{\partial T_e}{\partial r_D}\right|_{r_D=1} = -\frac{dQ}{dz}(2\pi\lambda_e)^{-1}, \quad \left.\frac{\partial T_e}{\partial r_D}\right|_{r_D\to\infty} = 0 \\ T(0) = T_0, \quad T_e(0) = T_e^0, \qquad 0 \le z \le h \end{cases}. \tag{9.12}$$

9.2.3 Solution model

Before solving the model, we have to deal with some parameters. The details are as follows:

(i) Calculating the gas condensing parameter Z_g:
If ($P < 35$ MPa)

$$Z_g = 1 + \left(0.31506 - \frac{1.0467}{T_{pr}} - \frac{0.5783}{T_{pr}^3}\right)\rho_{pr} + \left(0.053 - \frac{0.6123}{T_{pr}}\right)\rho_{pr}^2 + 0.6815\frac{\rho_{pr}^2}{T_{pr}^3},$$

where, $\rho_{pr} = 0.27\frac{P_{pr}}{T_{pr}}$, $T_{pr} = \frac{T}{T_{pc}}$, $P_{pr} = \frac{P}{P_{pc}}$.

Else

$$Z_g = (90.7x - 242x^2 + 42.4x^3)y^{(1.18+2.82x)} - (14.76x - 9.76x^2 + 4.58t^3)y + \frac{1 + y + y^2 + y^3}{(1-y)^3},$$

where,

$$F(y) = -0.06125P_{pr}xe^{-1.2(1-x)^2} + (90.7x - 242.2x^2 + 42.4x^3)y^{(2.18+2.82x)} + \frac{y + y^2 + y^3 - y^4}{(1-y^3)} - (14.76x - 9.76x^2 + 4.58x^3)y^2 = 0$$

$$x = \frac{1}{T_{pr}}.$$

(ii) The dimensionless time function. It can be calculated by:

$$\begin{cases} f(t_D) = 1.1281\sqrt{t_D}(1 - 0.3\sqrt{t_D}) & (t_D \le 1.5) \\ f(t_D) = \left(1 + \dfrac{0.6}{t_D}\right)[0.4063 + 0.5\ln(t_D)] & (t_D > 1.5) \end{cases}$$

$$t_D = \frac{t\alpha}{r_{wb}^2}.$$

(iii) The heat transfer coefficient U_{to} from different positions of the axis of the wellbore to the second surface:

$$\frac{1}{U_{to}} = r_{ti}\frac{1}{\lambda_{\text{ins}}}\ln\left(\frac{r_{ci}}{r_{to}}\right) + \frac{1}{h_c + h_r} + r_{ti}\frac{1}{\lambda_{\text{cem}}}\ln\left(\frac{r_{\text{cem}}}{r_{co}}\right). \tag{9.13}$$

On the right of Equation (9.13), are the thermal resistances for the insulated tubing, the hohlraum and the cement sheath, respectively. λ_{ins} and λ_{cem} are the heat conductivity of the heat insulating material and the cement sheath, and h_c and h_r are the coefficients of the convection heat transfer and the radiation heat transfer.

To simplify the calculation, we divide the wells into several short segments of the same length. The length of a segment varies depending on variations in wall thickness, hole diameter and well geometry. The model begins with a calculation at the bottom of the pipe. The steady production state values are taken as the initial values in the shut-in process.

Step 1. Obtain each point's inclination:

$$\theta_j = \theta_{j-1} + \frac{\theta_k - \theta_{k-1}}{\Delta s_k}\Delta s_j,$$

where j represents segment point of calculation, Δs_k represents measurement depth of inclination angle θ_k and θ_{k-1}, Δs_j step length of calculation.

Step 2. Give the initial values, T_0, T_e^0 and P_0:

Step 3. Let $T = T_k$, then we get the T_e at time t by solving the following equation:

$$\begin{cases} \dfrac{\partial T_e}{\partial t_D} = \dfrac{\lambda_e}{C_{P_e \rho_e}} \dfrac{\partial^2 T_e}{\partial r_D^2} + \dfrac{\lambda_e}{r_D C_{P_e \rho_e}} \dfrac{\partial T_e}{\partial r_D} \\ \left. \dfrac{\partial T_e}{\partial r_D} \right|_{r_D=1} = -a(T - T_e)(2\pi\lambda_e)^{-1} \\ \left. \dfrac{\partial T_e}{\partial r_D} \right|_{r_D \to \infty} = 0 \\ T_e(0) = T_e^0 \end{cases} . \tag{9.14}$$

Let T_{ej}^i be the temperature at the time i and radial j at the depth z, $i = 1, 2, \ldots, t_s$, $j = 1, 2, \ldots, N$, where t_s and N denote the last node of time and radial, respectively. T_e^0 is the stratum temperature in the initial state. We apply the finite difference method to discretize Equation (9.14):

$$\frac{T_{e,j}^{i+1} - T_{e,j}^i}{\tau} - \frac{\lambda_e}{C_{P_e \rho_e}} \frac{T_{e,j+1}^{i+1} - 2T_{e,j}^{i+1} + T_{e,j-1}^{i+1}}{\zeta^2} - \frac{\lambda_e}{r_D C_{P_e \rho_e}} \frac{T_{e,j+1}^{i+1} - T_{e,j}^{i+1}}{\zeta} = 0, \tag{9.15}$$

where τ is the interval of time and ζ is the interval of radial, respectively. It can be transformed into the standard form as follows:

$$-\left(\frac{\tau\lambda_e}{C_{P_e \rho_e}} + \frac{\tau\zeta\lambda_e}{r_D C_{P_e \rho_e}} \right) T_{e,j+1}^{i+1} + \left(\frac{2\tau\lambda_e}{C_{P_e \rho_e}} + \frac{\tau\zeta\lambda_e}{r_D C_{P_e \rho_e}} \right) T_{e,j}^{i+1} - \frac{\tau\lambda_e}{C_{P_e \rho_e}} T_{e,j-1}^{i+1} = \zeta^2 T_{e,j}^i. \tag{9.16}$$

Then the difference method is used to discretize the boundary condition. For $r_D = 1$, we have:

$$\left. \frac{\partial T_e}{\partial r_D} \right|_{r_D=1} - \left. \frac{aT}{2\pi\lambda_e} \right|_{r_D=1} = -\frac{aT_k}{2\pi\lambda_e}.$$

It follows that:

$$T_{e,2}^{i+1} = 1 - \left(1 + \frac{a\zeta}{2\pi\lambda_e} \right) T_{e,1}^{i+1} = -\frac{aT_k}{2\pi\lambda_e}. \tag{9.17}$$

For $r_D = N$, we have:

$$T_{e,N}^{i+1} - T_{e,N-1}^{i+1} = 0. \tag{9.18}$$

Combining equations (9.18), (9.17) and (9.16), we can compute the symbolic solution of the temperature T_e of the stratum. In this step, we will get the discrete distribution of T_e as the following matrix:

$$A = \begin{bmatrix} T_{e,1}^1 & T_{e,1}^2 & \cdots & T_{e,1}^i & \cdots & T_{e,1}^{t_s-1} & T_{e,1}^{t_s} \\ T_{e,2}^1 & T_{e,2}^2 & \cdots & T_{e,2}^i & \cdots & T_{e,2}^{t_s-1} & T_{e,2}^{t_s} \\ \cdots & \cdots & \cdots & \cdots & \cdots & \cdots & \cdots \\ T_{e,j}^1 & T_{e,j}^2 & \cdots & T_{e,j}^i & \cdots & T_{e,j}^{t_s-1} & T_{e,j}^{t_s} \\ \cdots & \cdots & \cdots & \cdots & \cdots & \cdots & \cdots \\ T_{e,N-1}^1 & T_{e,N-1}^2 & \cdots & T_{e,N-1}^i & \cdots & T_{e,N-1}^{t_s-1} & T_{e,N-1}^{t_s} \\ T_{e,N}^1 & T_{e,N}^2 & \cdots & T_{e,N}^i & \cdots & T_{e,N}^{t_s-1} & T_{e,N}^{t_s} \end{bmatrix},$$

where i represents the time and j represents the radial.

Step 4. Take the T_e at $r_D = 1$ into the system 9.2 and calculate the temperature T at time t and at depth z. In this step, we will get the discrete distribution of T as the following matrix:

$$A = \begin{bmatrix} T_1^1 & T_1^2 & \cdots & T_1^i & \cdots & T_1^{t_s-1} & T_1^{t_s} \\ T_2^1 & T_2^2 & \cdots & T_2^i & \cdots & T_2^{t_s-1} & T_2^{t_s} \\ \cdots & \cdots & \cdots & \cdots & \cdots & \cdots & \cdots \\ T_k^1 & T_k^2 & \cdots & T_k^i & \cdots & T_k^{t_s-1} & T_k^{t_s} \\ \cdots & \cdots & \cdots & \cdots & \cdots & \cdots & \cdots \\ T_{h-1}^1 & T_{h-1}^2 & \cdots & T_{h-1}^i & \cdots & T_{h-1}^{t_s-1} & T_{h-1}^{t_s} \\ T_h^1 & T_h^2 & \cdots & T_h^i & \cdots & T_h^{t_s-1} & T_h^{t_s} \end{bmatrix},$$

where i represents the time and k represents the depth.

Step 5. Calculate the pressure P at time t and at depth z through the system 9.12. In this step, we will get the discrete distribution of as the following matrix:

$$A = \begin{bmatrix} P_1^1 & P_1^2 & \cdots & P_1^i & \cdots & P_1^{t_s-1} & P_1^{t_s} \\ P_2^1 & P_2^2 & \cdots & P_2^i & \cdots & P_2^{t_s-1} & P_2^{t_s} \\ \cdots & \cdots & \cdots & \cdots & \cdots & \cdots & \cdots \\ P_k^1 & P_k^2 & \cdots & P_k^i & \cdots & P_k^{t_s-1} & P_k^{t_s} \\ \cdots & \cdots & \cdots & \cdots & \cdots & \cdots & \cdots \\ P_{h-1}^1 & P_{h-1}^2 & \cdots & P_{h-1}^i & \cdots & P_{h-1}^{t_s-1} & P_{h-1}^{t_s} \\ P_h^1 & P_h^2 & \cdots & P_h^i & \cdots & P_h^{t_s-1} & P_h^{t_s} \end{bmatrix}.$$

Step 6. Repeat Step 3 to Step 5 until P_{t_s}, T_{t_s}, and T_{e,t_s} are calculated.

Step 7. Calculate t_s by $\sum_{j=1}^{h} |T_j^{t_s} - T_j^{t_s-1}| \leq 100$, then take the value into the matrix to calculate the temperature and pressure.

9.2.4 Numerical simulation

As described previously, the algorithm begins with a pipe calculation at the bottom of the pipe. The calculations are performed for each successive "segment" of the pipe from the bottom to the top.

9.2.4.1 *Parameters*

The parameters for X well in Sichuan province, China are presented as follows:

Depth of the well = 7100 m
Ground temperature = 16°C
Ground thermal conductivity parameter = 2.06 W/m · K
Geothermal gradient = 2.18°C/100 m
Length of one segment = 10 m
Time of one segment = 10 minutes

Furthermore, the parameters for the pipes, the inclined well, the inclination, the azimuth and the vertical depth are in Tables 8.1, 8.2 and 8.3, respectively. The values of steady production state are taken as the initial values in the shut-in process.

9.2.4.2 *Results analysis*

Temperature profiles are shown in Figures 9.1, 9.2 and 9.3.

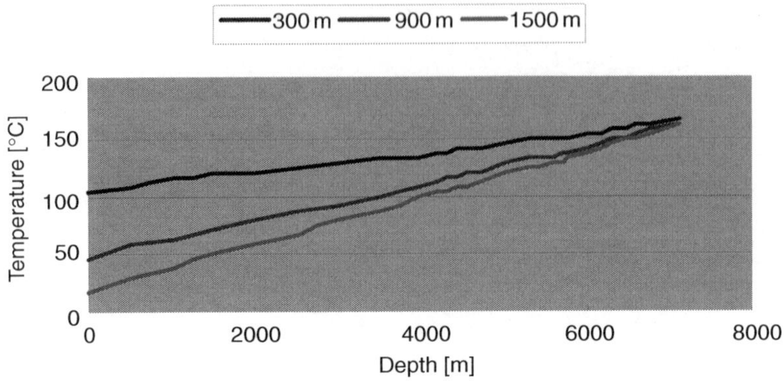

Figure 9.1 Temperature distribution in shut-in procedures.

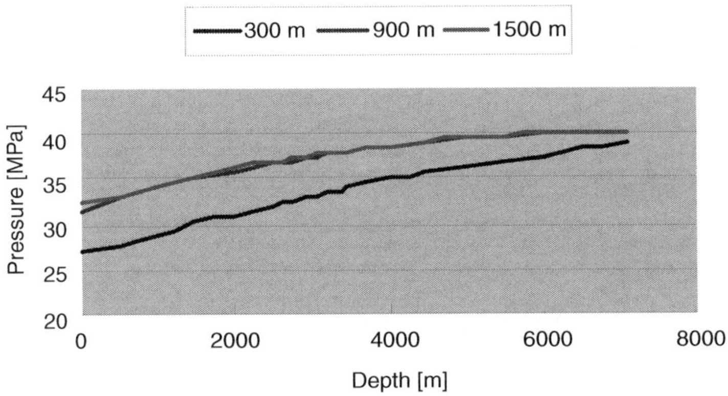

Figure 9.2 Gas temperature at well head.

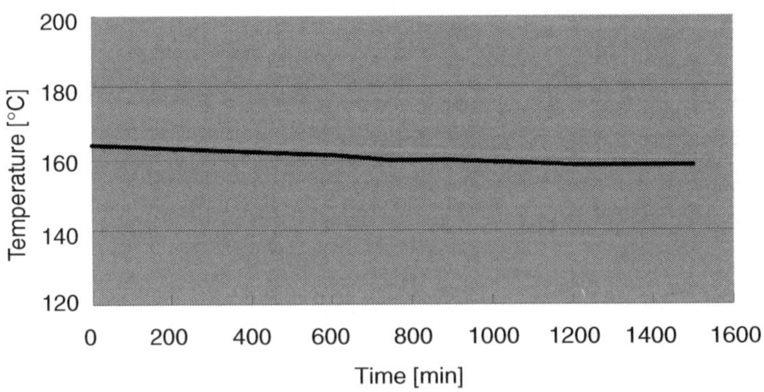

Figure 9.3 Gas temperature at the bottom.

From Figure 9.1, it can be seen that when the time is fixed, the temperature increases continuously from the top to the bottom of the pipe, and the speed is at progressively decreasing rate as the depth increases. When the depth is fixed, the temperature decreases progressively as time increases. The reason is that after shut-in, as the time increases, the heat transfers from the gas in the well-bore to the surrounding earth through the cement/earth interface, and the gas temperature

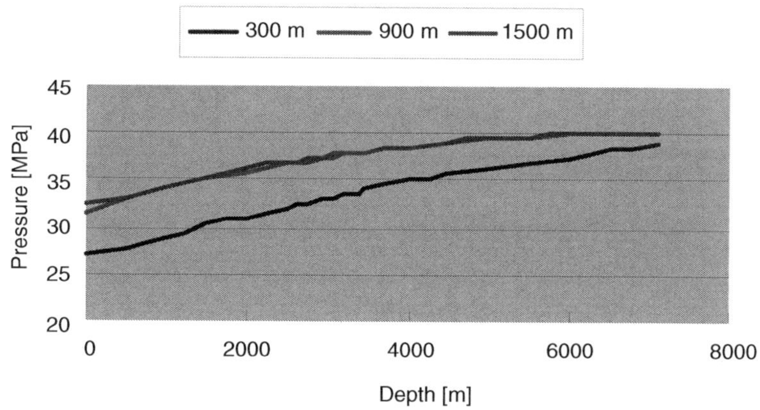

Figure 9.4 Pressure distribution in shut-in procedures.

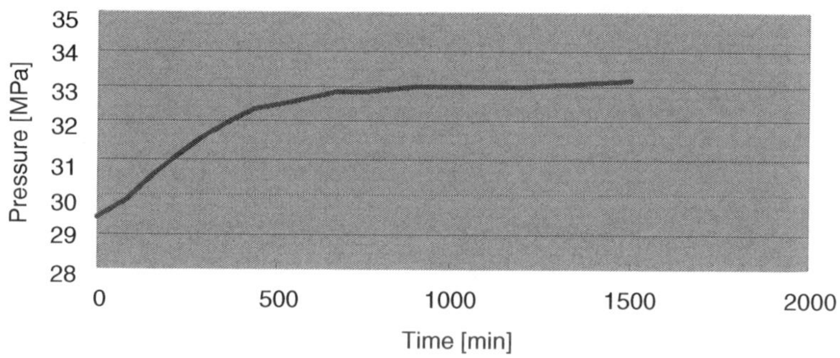

Figure 9.5 Gas pressure at well head.

falls gradually. At stability, the gas temperature approaches the formation temperature at each depth. The gas temperature variation trend at the well head is shown in Figure 9.2. As can be seen, the temperature drops quickly in the early stages, but stabilizes over time, until it approximately equals the ground temperature. The temperature at the bottom is relatively stable, as shown in Figure 9.3, and the temperature shows little variation as the time increases.

Figures 9.4, 9.5 and 9.6 show the pressure profiles.

The pressure distribution along with the well at different times is shown in Figure 9.4. From Figure 9.4, it can be seen that when the time is fixed, the pressure increases continuously with depth but at a decreasing rate. When the depth is fixed, the pressure drops as time increases, and stabilizes after a certain time. In addition, the gas pressure variation trend at the well head can be seen from Figure 9.5. The pressure changes quickly in the early stages, but stabilizes over time, and the general variation range in the pressure is small. The pressure at the bottom changes only slightly, as shown in Figure 9.6, and stays at a relatively stable level.

After settling, the gas flow in the wellhole achieves a relatively static state, so when the temperature and pressure of the gas only relate to depth, they are approximately equal to the linear functions for depth.

A coupled differential equations system model for pressure and temperature shut-in procedures for high temperature and high pressure deviated wells is built. In the model, which is based on the Cullender-Smith method, we focus on the heat transmission in the stratum, so as to take the time element into account. The basic data from the *X* well (high temperature and high pressure

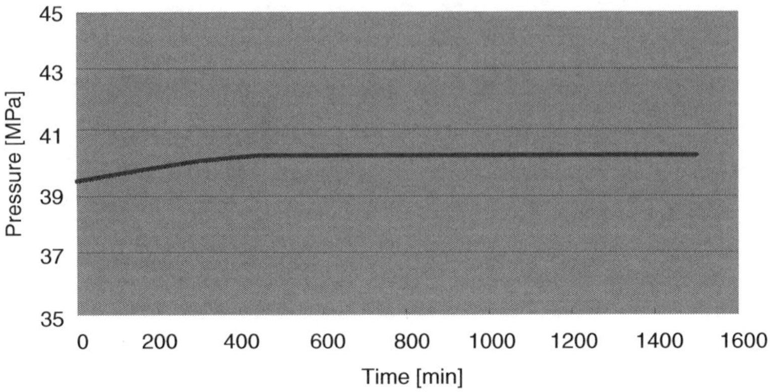

Figure 9.6 Gas pressure at the bottom.

gas well), 7100 m deep in China, are used for the case history calculations, and gas pressure and temperature curves along the depth of the well are plotted at different times. Our findings indicate that after shut-in the gas state in the wellhole changes from a steady flow to a transient flow, and then finally reaches a comparatively static state. During shut-in, the temperature and pressure change with time, until they are in accordance with the static gas column. The results allow us to analyse the well features in shut-in procedures, and better deal with well control problems.

9.3 PPTVD-TFSP

Considering the differential equation model for P, T, v and ρ, we set the following assumptions:

1. The gas phase and liquid phase are treated as a pseudo-single phase.
2. There is a one dimensional two-phase flow in the tubing, and all featured parameters are homochromatic at any transverse cross section.
3. The two phases have the same pressure, temperature and velocity at any transverse cross section.
4. There is no mass transfer between the two phases.
5. The vertical formation temperature is considered linearly distributed based on known geothermal gradients.

When a gas producing well is shut, the pressure within the tubing is less than that in the formation. As a result, the gas packer water in the formation continues to move into the tubing at the bottom. Through this tubing, the gas-water flows from the bottom to the top at a mass flow rate. With the gas-water entering the tubing, the pressure inside increases gradually. After settling, the pressure at the bottom balances with the pressure in the corresponding formation, the flow stops entering, and the status in the wellhole tends toward stability. During the shut-in process, the gas-water state in the wellhole changes from a steady flow to a transient flow, finally reaching a comparatively static state. This transformation can happen for both gas flows and liquid flows.

9.3.1 **The coupled system model**

Consider the flow system depicted in Figure 9.1, a straight cylindrical flow tube with an inclination angle θ, a constant cross-sectional flow area A, a hydraulic diameter d, and a total length Z. Through this tubing, fluid flows from the bottom to the top with a mass flow rate w. The distance coordinate in the flow direction along the tubing is denoted by z. Mass, momentum, and energy balances, along with the pressure, temperature, velocity and density, relative to the transient flow, as well as the stated equation are used to generate the constitutive equations.

Mass balance: Taking the bottom of well as the origin of the coordinate axis and vertical up as a positive direction, let dz and dt denote the differential depth and differential time respectively. It follows from mass balance law that:

$$\rho vAdt - \left[\rho vA + \frac{\partial}{\partial z}(\rho vA)\right]dt = \frac{\partial}{\partial t}(\rho Adz)dt \tag{9.19}$$

which is equivalent to

$$\frac{\partial \rho}{\partial z} + v\frac{\partial \rho}{\partial z} + \rho\frac{\partial v}{\partial z} = 0. \tag{9.20}$$

Momentum balance: During the time dt, the momentum flowing into dz is $\rho v^2 A$ and momentum flowing out from dz is $\rho v^2 A + \frac{\partial}{\partial z}(\rho v^2 A)dz$. The forces on dz includes pressure on the cross section, the gravity and friction. In addition, the increment momentum during time dt is $\frac{\partial}{\partial t}(\rho vAdz)$. It follows from Momentum Theorem that:

$$\frac{\partial(\rho v)}{\partial t} + \frac{\partial P}{\partial z} + \frac{\partial(\rho v^2)}{\partial z} + \rho g\cos\theta + \frac{f\rho v^2}{2d} = 0 \tag{9.21}$$

which is equivalent to

$$\rho\frac{\partial v}{\partial t} + v\frac{\partial \rho}{\partial t} + \frac{\partial P}{\partial z} + v^2\frac{\partial \rho}{\partial z} + 2\rho v\frac{\partial v}{\partial z} + \rho g\cos\theta + \frac{f\rho v^2}{2d} = 0. \tag{9.22}$$

Energy balance: The types of energy considered in this paper include inner energy, pressure energy, kinetic energy and potential energy, where the inner energy and pressure energy are collectively referred to as enthalpy. The energy flowing into dz includes enthalpy $H(z)$, kinetic energy $\frac{1}{2}mv^2A$, potential energy $mgAz\cos\theta$. Similarly, the energy flowing out of dz includes enthalpy $H(z) + \frac{\partial H}{z}$, kinetic energy $\frac{1}{2}\left[\rho v^3 + \frac{\partial}{z}(\rho v^3)dz\right]A$, potential energy $\rho vg(z+dz)\cos\theta$.

The heat between the fluid and the second interface should not be ignored. The differential element dz was shown in Figure 9.2. The radial heat transfer between the gas and the second interface tube is $\frac{2\pi r_{to}U_{to}K_e}{w(r_{to}U_{to}f(t_D)+K_e)}(T - T_e)dz$. The heat generated by the friction with tubing is $\frac{f\rho v^3}{2d}Adz$. In addition, the variable quantity of the energy during time dt is $\left[\frac{\partial \rho C_p T}{\partial t}\right]Adz$, where C_p is heat capacity.

As well known, for general enthalpy, the following equation holds:

$$\frac{\partial H}{\partial z} = C_p\frac{\partial T}{\partial z} - C_J C_p\frac{\partial P}{\partial z}. \tag{9.23}$$

From the above, it follows from Energy Conservation Law that:

$$\frac{\partial}{\partial t}(\rho C_p T)A + C_p\frac{\partial T}{\partial z} - C_J C_p\frac{\partial P}{\partial z} + \frac{\partial}{\partial z}\left[\frac{1}{2}\rho v^3 A + \rho gvAz\cos\theta\right] + a(T - T_e) + \frac{f\rho v^3 A}{2d} = 0, \tag{9.24}$$

where

$$a = \frac{2\pi r_{to}U_{to}K_e}{w(r_{to}U_{to}f(t_D) + K_e)}(T - T_e)dz.$$

Heat transmission in the stratum: According to the assumption that the heat transfer from the second interface to the stratum is unsteady and the heat transmission model in the stratum is concluded as follows,

$$\frac{\partial T_e}{\partial t}=\frac{\lambda_e}{C_p\rho_e}\left(\frac{\partial^2 T_e}{\partial r^2}+\frac{1}{r}\frac{\partial T_e}{\partial r}\right). \tag{9.25}$$

Initial condition:

$$T(0)=T_0,\quad T_e(0)=T_e^0,\qquad 0\le z\le h, \tag{9.26}$$

where T_0 and T_e^0 represents the gas temperature in the wellhole and stratum temperature in steady production state before shut-in procedures.

Boundary conditions:

$$\left.\frac{\partial T_e}{\partial r}\right|_{r\to\infty}=0,\quad dQ=-2\pi r_{\text{cem}}\lambda_{\text{cem}}\left.\frac{\partial T_e}{\partial r}dz\right|_{r=r_{\text{cem}}}$$

It follows from the dimensionless variables $r_D=r/r_{\text{cem}}$ and $t_D=\lambda_e t/r_{\text{cem}}^2$. That, Equation (9.25) is changed into:

$$\frac{\partial T_e}{\partial t_D}=\frac{\lambda_e}{C_p\rho_e}\left(\frac{\partial^2 T_e}{\partial r_D^2}+\frac{1}{r_D}\frac{\partial T_e}{\partial r_D}\right). \tag{9.27}$$

The boundary conditions are converted into:

$$\left.\frac{\partial T_e}{\partial r_D}\right|_{r_D=1}=-\frac{dQ}{dz}(2\pi\lambda_e)^{-1},\quad \left.\frac{\partial T_e}{\partial r_D}\right|_{r_D\to\infty}=0. \tag{9.28}$$

The kinetic energy makes sense only when the gas volume is far greater than the liquid volume, for gas well, so we can apply state equation of ideal gas, that is:

$$\rho=\frac{MP}{RZ_gT}, \tag{9.29}$$

where Z_g is condensing factor. By combining Equations (9.20), (9.22), (9.24) and (9.26)–(9.29), we have the following simultaneous equations models of pressure, temperature, velocity and density in shut-in process:

$$\begin{cases}\dfrac{\partial v}{\partial z}=-\dfrac{1}{\rho}\left(\dfrac{\partial\rho}{\partial t}+v\dfrac{\partial\rho}{\partial z}\right)\\[2mm] \dfrac{\partial P}{\partial z}=-\rho\dfrac{\partial v}{\partial t}-v\dfrac{\partial\rho}{\partial t}-v^2\dfrac{\partial\rho}{\partial z}-\dfrac{2\rho v^2}{2d}-\rho g\cos\theta\\[2mm] \dfrac{\partial T}{\partial z}=-\dfrac{1}{C_p}\left[\dfrac{\partial}{\partial t}(\rho C_pT)A-C_JC_p\dfrac{\partial P}{\partial z}+\dfrac{\partial}{\partial z}\left[\dfrac{1}{2}\rho v^3A+\rho gvAz\cos\theta\right]+a(T-T_e)+\dfrac{f\rho v^3A}{2d}\right]\\[2mm] \rho=\dfrac{MP}{RZ_gT}\\[2mm] \dfrac{\partial T_e}{\partial t_D}=\dfrac{\lambda_e}{C_p\rho_e}\left(\dfrac{\partial^2T_e}{\partial r_D^2}+\dfrac{1}{r_D}\dfrac{\partial T_e}{\partial r_D}\right)\\[2mm] \left.\dfrac{\partial T_e}{\partial r_D}\right|_{r_D=1}=-\dfrac{dQ}{dz}(2\pi\lambda_e)^{-1},\quad \left.\dfrac{\partial T_e}{\partial r_D}\right|_{r_D\to\infty}=0\\[2mm] P(z,0)=P_0(z),\quad T(z,0)=T_0(z),\quad v(z,0)=v_0(z)\\ P(0,t)=\bar{P}_0(t),\quad T(0,t)=\bar{T}_0(t),\quad v(0,t)=\bar{v}_0(t)\end{cases}. \tag{9.30}$$

9.3.2 Model solution

Before solving the model, we have to deal with some parameters. The details are as follows:

1. Gas condensing parameter Z_g:
   If ($P < 35$ MPa)

$$Z_g = 1 + \left(0.31506 - \frac{1.0467}{T_{pr}} - \frac{0.5783}{T_{pr}^3}\right)\rho_{pr} + \left(0.053 - \frac{0.6123}{T_{pr}}\right)\rho_{pr}^2 + 0.6815\frac{\rho_{pr}^2}{T_{pr}^3},$$

   where, $\rho_{pr} = 0.27\frac{P_{pr}}{T_{pr}}$, $T_{pr} = \frac{T}{T_{pc}}$, $P_{pr} = \frac{P}{P_{pc}}$.
   Else:

$$Z_g = (90.7x - 242x^2 + 42.4x^3)y^{(1.18+2.82x)} - (14.76x - 9.76x^2 + 4.58t^3)y + \frac{1+y+y^2+y^3}{(1-y)^3},$$

   where,

$$\begin{aligned} F(y) &= -0.06125P_{pr}xe^{-1.2(1-x)^2} + (90.7x - 242.2x^2 + 42.4x^3)y^{(2.18+2.82x)} \\ &\quad + \frac{y+y^2+y^3-y^4}{(1-y^3)} - (14.76x - 9.76x^2 + 4.58x^3)y^2 \\ &= 0 \end{aligned}$$

$$x = \frac{1}{T_{pr}}.$$

2. The dimensionless time function. It can be calculated by:

$$\begin{cases} f(t_D) = 1.1281\sqrt{t_D}(1 - 0.3\sqrt{t_D}) & (t_D \le 1.5) \\ f(t_D) = \left(1 + \dfrac{0.6}{t_D}\right)[0.4063 + 0.5\ln(t_D)] & (t_D > 1.5) \end{cases}$$

$$t_D = \frac{t\alpha}{r_{wb}^2}.$$

3. The heat transfer coefficient U_{to} from different positions of the axis of the wellbore to the second surface:

$$\frac{1}{U_{to}} = r_{ti}\frac{1}{\lambda_{\text{ins}}}\ln\left(\frac{r_{ci}}{r_{to}}\right) + \frac{1}{h_c + h_r} + r_{ti}\frac{1}{\lambda_{\text{cem}}}\ln\left(\frac{r_{\text{cem}}}{r_{co}}\right).$$

4. Friction factor f:

$$\frac{1}{\sqrt{f}} = 1.14 - 2\lg\left(\frac{0.00001524}{r_{ti}} + \frac{21.25}{Re^{0.9}}\right).$$

5. Dryness fraction of gas λ:

$$\lambda = \frac{1.205\gamma_g}{\frac{1000\gamma_l}{GWR} + 1.205\gamma_g},$$

   where *GWR* is a gas-liquid ratio, γ_g and γ_l are of relative density of the gas and liquid respectively.
6. Density of mixture:

$$\rho = \frac{M_t}{V_t},$$

   where $M_t = 1000\gamma_l + 1.205 \cdot GWR \cdot \gamma_l$, $V_t = 1 + 0.0003458\frac{ZT}{P} \cdot GWR$.
7. Velocity of mixture:

$$v = \frac{q_t}{A},$$

   where $q_t = \left(1 + 0.0003458\frac{ZT}{P}\right)Q_{gsc}/GWR$, A is the area of pipe, Q_{gsc} is the quantity of gas.

8. Gas-liquid mixture heat ratio C_p:

$$C_p = \lambda C_{pg} + (1-\lambda)C_{pl},$$

where $C_{pg} = 1697.5107P^{0.0661}T^{0.0776}$, $C_{pl} = 4.2\,\text{kJ/(kg}\cdot{}^\circ\text{C)}$.

9. Gas-liquid mixture Joule-Thomson parameter C_J:

$$C_J = -\frac{\lambda C_{pg}}{C_p}C_{Jl} + \frac{1-\lambda}{\rho_l C_p}.$$

To simplify the calculation, we divide the well into several short segments of the same length. The length of a segment varies depending on variations in well thickness, hole diameter, the fluid density inside and outside the pipe and well geometry. The model begins with a calculation at the bottom of the pipe and at the initial time. The steady production state values are taken as the initial values in the shut-in process. Based on the afore-mentioned discussion, we use the finite difference method based on the Taylor series expansion to solve the model.

Step 1. Obtain each point's inclination:

$$\theta_j = \theta_{j-1} + \frac{\theta_k - \theta_{k-1}}{\Delta s_k}\Delta s_j,$$

where j represents the segment point of calculation, Δs_k represents the measurement depth of inclination angle θ_k and θ_{k-1}, Δs_j the step length of calculation.

Step 2. Give the initial values, T_0, T_e^0 and P_0.

Step 3. Let $T = T_k$, then we get the T_e at time t by solving the following equation:

$$\begin{cases} \dfrac{\partial T_e}{\partial t_D} = \dfrac{\lambda_e}{C_{P_e\rho_e}}\dfrac{\partial^2 T_e}{\partial r_D^2} + \dfrac{\lambda_e}{r_D C_{P_e\rho_e}}\dfrac{\partial T_e}{\partial r_D} \\ \left.\dfrac{\partial T_e}{\partial r_D}\right|_{r_D=1} = -a(T-T_e)(2\pi\lambda_e)^{-1} \\ \left.\dfrac{\partial T_e}{\partial r_D}\right|_{r_D\to\infty} = 0 \\ T_e(0) = T_e^0 \end{cases} \quad (9.31)$$

Let T_{ej}^i be the temperature at the time i and radial j at the depth z, $i = 1, 2, \ldots, t_s$, $j = 1, 2, \ldots, N$, where t_s and N denote the last node of time and radial, respectively. T_e^0 is the stratum temperature in the initial state. We apply the finite different method to discretize Equation (9.31) as follows:

$$\frac{T_{e,j}^{i+1} - T_{e,j}^i}{\tau} - \frac{\lambda_e}{C_{P_e\rho_e}}\frac{T_{e,j+1}^{i+1} - 2T_{e,j}^{i+1} + T_{e,j-1}^{i+1}}{\zeta^2} - \frac{\lambda_e}{r_D C_{P_e\rho_e}}\frac{T_{e,j+1}^{i+1} - T_{e,j}^{i+1}}{\zeta} = 0, \quad (9.32)$$

where τ is the interval of time and ζ is the interval of radial, respectively. It can be transformed into the standard form as follows:

$$-\left(\frac{\tau\lambda_e}{C_{P_e\rho_e}} + \frac{\tau\zeta\lambda_e}{r_D C_{P_e\rho_e}}\right)T_{e,j+1}^{i+1} + \left(\frac{2\tau\lambda_e}{C_{P_e\rho_e}} + \frac{\tau\zeta\lambda_e}{r_D C_{P_e\rho_e}}\right)T_{e,j}^{i+1} - \frac{\tau\lambda_e}{C_{P_e\rho_e}}T_{e,j-1}^{i+1} = \zeta^2 T_{e,j}^i, \quad (9.33)$$

where τ is the interval of time and ζ is the interval of radial, respectively. It can be transformed into the standard form as follows:

$$-\left(\frac{\tau\lambda_e}{C_{P_e\rho_e}} + \frac{\tau\zeta\lambda_e}{r_D C_{P_e\rho_e}}\right)T_{e,j+1}^{i+1} + \left(\frac{2\tau\lambda_e}{C_{P_e\rho_e}} + \frac{\tau\zeta\lambda_e}{r_D C_{P_e\rho_e}}\right)T_{e,j}^{i+1} - \frac{\tau\lambda_e}{C_{P_e\rho_e}}T_{e,j-1}^{i+1} = \zeta^2 T_{e,j}^i. \quad (9.34)$$

Then the difference method is used to discretize the boundary condition. For $r_D = 1$, we have:

$$\left.\frac{\partial T_e}{\partial r_D}\right|_{r_D=1} - \left.\frac{aT}{2\pi\lambda_e}\right|_{r_D=1} = -\frac{aT_k}{2\pi\lambda_e}.$$

It follows that

$$T_{e,2}^{i+1} = 1 - \left(1 + \frac{a\zeta}{2\pi\lambda_e}\right)T_{e,1}^{i+1} = -\frac{aT_k}{2\pi\lambda_e}. \tag{9.35}$$

For $r_D = N$, we have:

$$T_{e,N}^{i+1} - T_{e,N-1}^{i+1} = 0. \tag{9.36}$$

Combining Equations (9.36), (9.35) and (9.34), we can compute the symbolic solution of the temperature T_e of the stratum. In this step, we will get the discrete distribution of T_e as the following matrix:

$$A = \begin{bmatrix} T_{e,1}^1 & T_{e,1}^2 & \cdots & T_{e,1}^i & \cdots & T_{e,1}^{t_s-1} & T_{e,1}^{t_s} \\ T_{e,2}^1 & T_{e,2}^2 & \cdots & T_{e,2}^i & \cdots & T_{e,2}^{t_s-1} & T_{e,2}^{t_s} \\ \cdots & \cdots & \cdots & \cdots & \cdots & \cdots & \cdots \\ T_{e,j}^1 & T_{e,j}^2 & \cdots & T_{e,j}^i & \cdots & T_{e,j}^{t_s-1} & T_{e,j}^{t_s} \\ \cdots & \cdots & \cdots & \cdots & \cdots & \cdots & \cdots \\ T_{e,N-1}^1 & T_{e,N-1}^2 & \cdots & T_{e,N-1}^i & \cdots & T_{e,N-1}^{t_s-1} & T_{e,N-1}^{t_s} \\ T_{e,N}^1 & T_{e,N}^2 & \cdots & T_{e,N}^i & \cdots & T_{e,N}^{t_s-1} & T_{e,N}^{t_s} \end{bmatrix},$$

where i represents the time and j represents the radial.

Step 4. Calculate the temperature, pressure, velocity and density of two-phase flow in different time and depth in the wellhole.

Step 5. Repeat from Step 3 to Step 4 until the time step is reached.

9.3.3 Numerical simulation

As described previously, the algorithm begins with a calculation at the bottom of the pipe at initial time. The calculations are performed for each successive segment of the pipe up to the surface as well as the time up to the testing time end point.

In this simulation, we study a pipe in X well, which is located in China. The relevant parameters are as follows:

Depth of the well = 7100 m
Critical pressure = 4.968 MPa
Gas specific weight = 0.6434 kg/m^3
Ground thermal conductivity parameter = 2.06 W/m·K
Ground temperature = 16°C
Geothermal gradient = 2.18°C/100 m
Porosity = 0.2
Roughness of the inner surface of the well = 0.000015
Comprehensive compression coefficient = 0.03
Length of one segment = 10 m
Time of one segment = 10 minutes

The parameters for the pipes, the inclined well, the inclination, the azimuth and the vertical depth are in Tables 8.1, 8.2 and 8.3, respectively. The values of steady production state are taken as the initial values for the shut-in process.

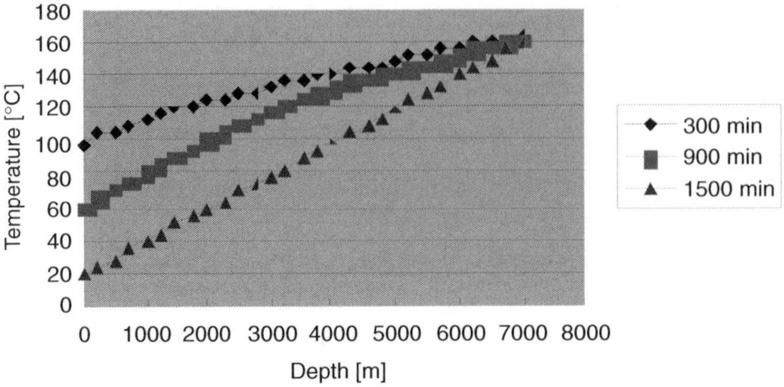

Figure 9.7 Temperature distribution.

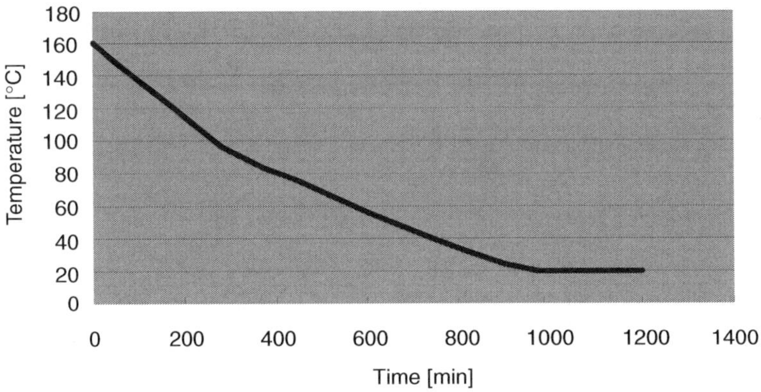

Figure 9.8 Fluid temperature at well head.

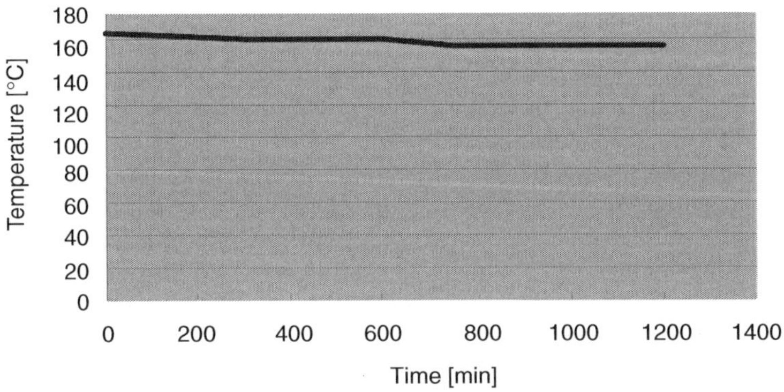

Figure 9.9 Fluid temperature at the bottom.

9.3.3.1 *Results analysis*

The temperature profiles are shown in Figures 9.7, 9.8, and 9.9.

From Figure 9.7, it can be seen that when the time is fixed, the temperature increases at a decreasing rate with depth. When the depth is fixed, the temperature decreases progressively with time. The reason for this is that after shut-in, as time increases, heat transfers from the fluid in the well-bore to the surrounding earth through the cement/earth interface, so the fluid temperature falls

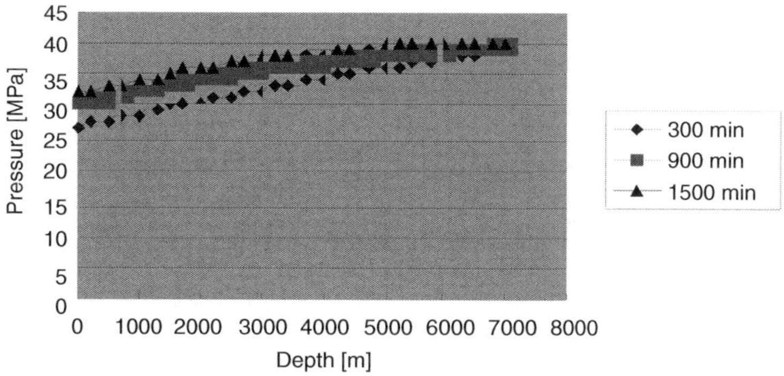

Figure 9.10 Pressure distribution.

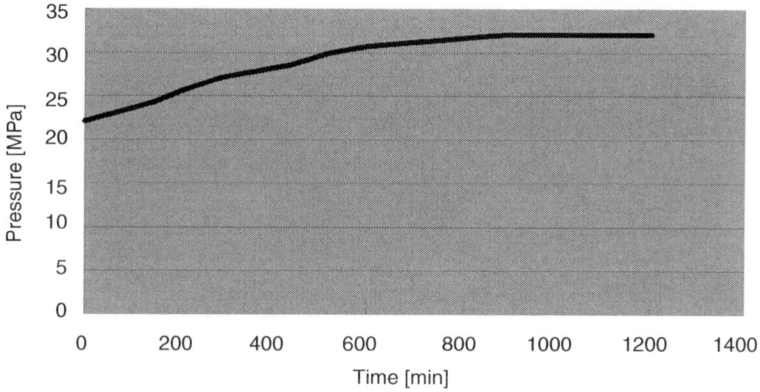

Figure 9.11 Fluid pressure at well head.

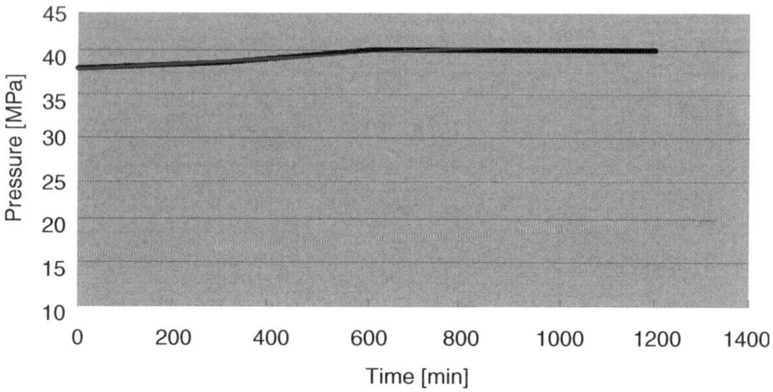

Figure 9.12 Fluid pressure at the bottom.

gradually. On reaching a stable state, the fluid temperature approaches the formation temperature at each depth. The variations in the fluid temperature at the well are shown in Figure 9.8, where it can be seen that the temperature drops quickly in the early stages, but stabilizes over time, until it approximately equals the ground temperature. The temperature at the bottom is relatively stable, as shown in Figure 9.9, and the temperature shows little variation over time.

Figures 9.10, 9.11 and 9.12 show the pressure profiles. From Figure 9.10, it can be seen that when the time is fixed, the pressure increases with depth. When the depth is fixed, the pressure

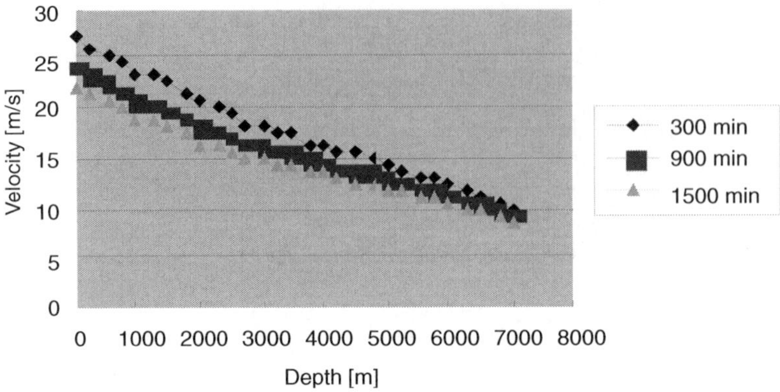

Figure 9.13 Velocity distribution.

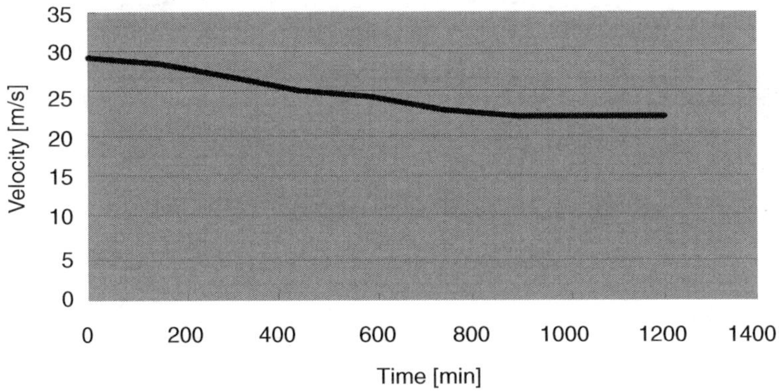

Figure 9.14 Fluid velocity at well head.

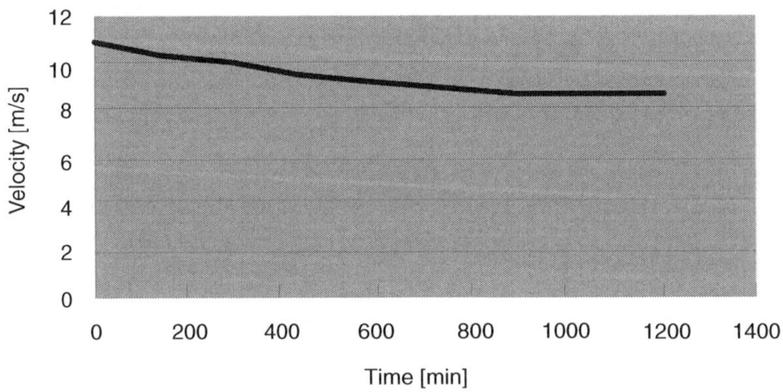

Figure 9.15 Fluid velocity at the bottom.

drops with time, and stabilizes after a certain time. The variation trends in the fluid pressure at the well head are shown in Figure 9.11, where it can be seen that the pressure changes quickly in the early stages, but stabilizes over time. Overall, however, the general pressure variation is small with the pressure at the bottom changing only slightly, as shown in Figure 9.12.

Figures 9.13, 9.14, 9.15 show the velocity profiles. From Figure 9.13, it can be seen that when the time is fixed, the velocity decreases with increasing depth, and when the depth is fixed, the

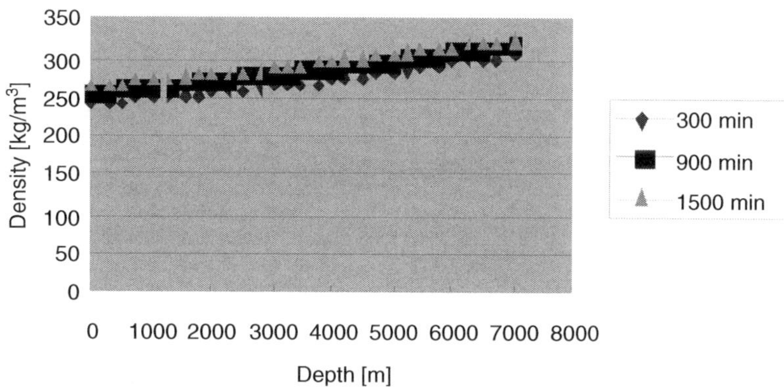

Figure 9.16 Density distribution.

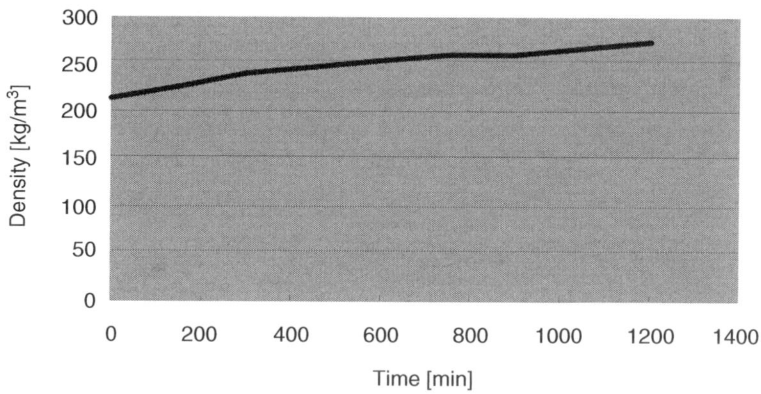

Figure 9.17 Fluid density at well head.

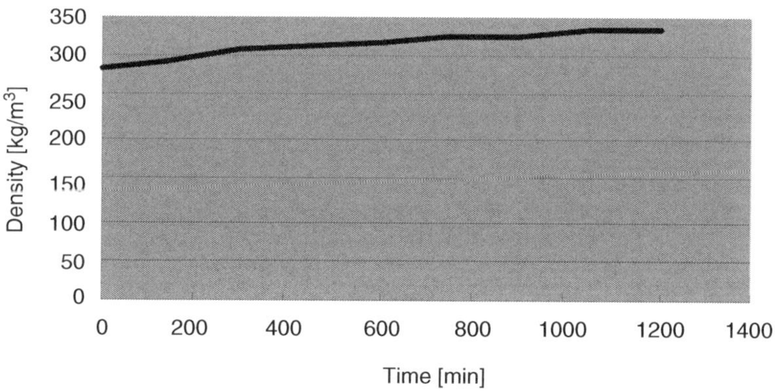

Figure 9.18 Fluid density at the bottom.

velocity decreases with time. With time, the velocity varies within a small range both at the well head and the bottom, which can be seen in Figures 9.14 and 9.15.

The density profiles are shown in Figures 9.16, 9.17 and 9.18. From Figure 9.16, it can be seen that when the time is fixed, the density increases slightly with depth. When the depth is fixed, the density increases with time. Furthermore, from Figures 9.17 and 9.18, the density can be seen to increase in a narrow range with time both at the well head and at the bottom. However, the general pressure variation range is small.

After settling, the two-phase fluid flowing in the wellhole reaches a relatively static state, where the temperature, pressure, velocity and density of the gas-liquid flow relate only to the depth, and are approximately equal to linear functions for depth.

CHAPTER 10

Software design and development

The software includes a calculation program and a database. The calculation process allows for basic data input, calculation, and the production of results. The database allows for data input, data save and delete functions for the tubing and casing.

10.1 CALCULATION PROGRAM

This section describes the main interface and a number of sub-interfaces, completes the calculations and produces the results. As shown in Figure 10.1, it is possible to calculate all the conditions or part of the conditions. When calculating all the conditions, the calculation is a one-time calculation of the completion string for the amount of computation required; a division calculation calculates separate user-specified conditions.

10.1.1 **All conditions calculation**

This module consists of a basic data input interface and a results interface, namely the complete input and output functions. By clicking the "File" menu option in the "New" button, and enter the calculation module for all operating conditions is entered. First, there is a "Basic information" window, where the desired interface for the calculation of the well data is entered, as shown in Figure 10.2.

After the input, by clicking "Next", the "Basic data" window appears. In this interface, the basic data is input, such as the inclination, the formation, the casing, and the tubing data. All data are given in a tabular form, so that the user is able to input data to the table, or import "EXCEL"

Figure 10.1 Interface of software.

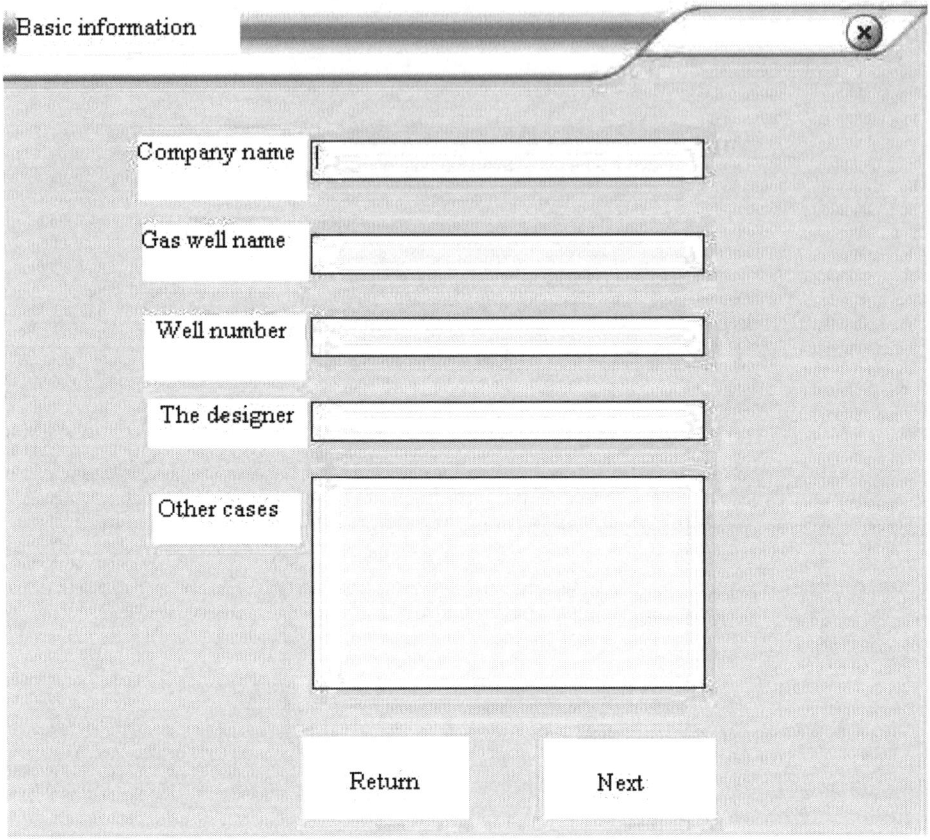

Figure 10.2 Basic information window.

data directly. If adopting direct import, users need to pay attention that the "EXCEL" files meet the software requirements. For detailed specifications refer to the template "EXCEL" file.

In the well inclination data window, clicking "Import data", imports the data from "EXCEL" into the table window. By clicking "Save data", the data in the table is saved as an EXCEL document. The well inclination interface is shown in Figure 10.3.

In the formation data window, in addition to the "Import data" and "Save data" buttons, there is a "Help" button, which displays the physical meaning of each parameter of the data in the formation. The formation data interface is shown in Figure 10.4.

In the casing structure data window, if the box is double-clicked, the relevant database content appears. Here, the relevant data based on the actual situation can be selected. The casing structure window is shown in Figure 10.5.

The pipeline structure data interface: in this interface, the pipeline structure data can be manually entered, modified and saved. Inclination data can also be imported directly into "EXCEL", as shown in Figure 10.6.

Because the required data interface input data needs to be calculated, it is necessary that the user input all related data. If some data is not entered, when the "Creation" button is clicked, a message will pop up prompting for the input data.

After completion of basic data input, clicking the "Creation" button opens up the "Placing tubular" window. In this interface, the relevant data needs to be entered. The specific physical meaning of the required parameters can be found with the "Help" button. Some parameters may

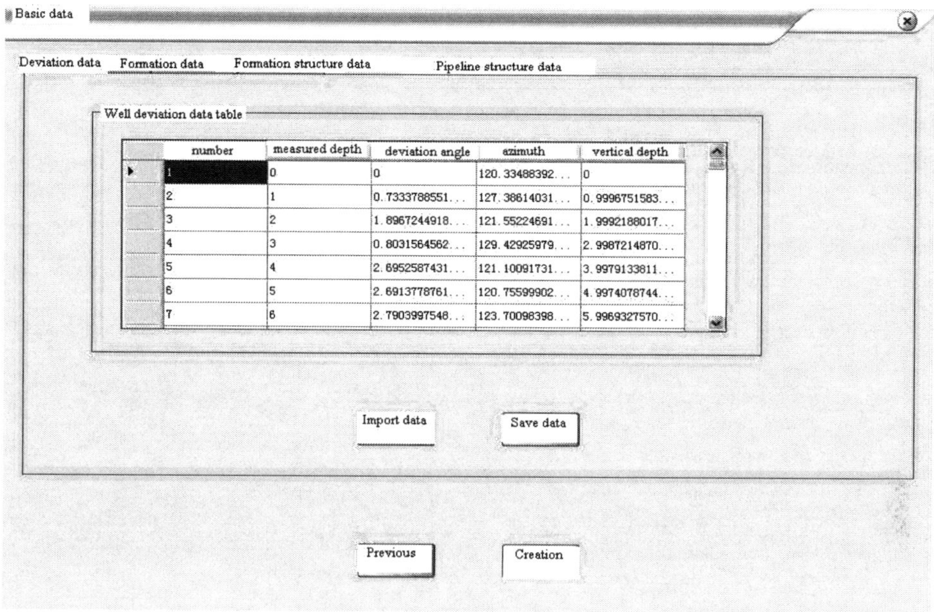

Figure 10.3 Basic data tablet.

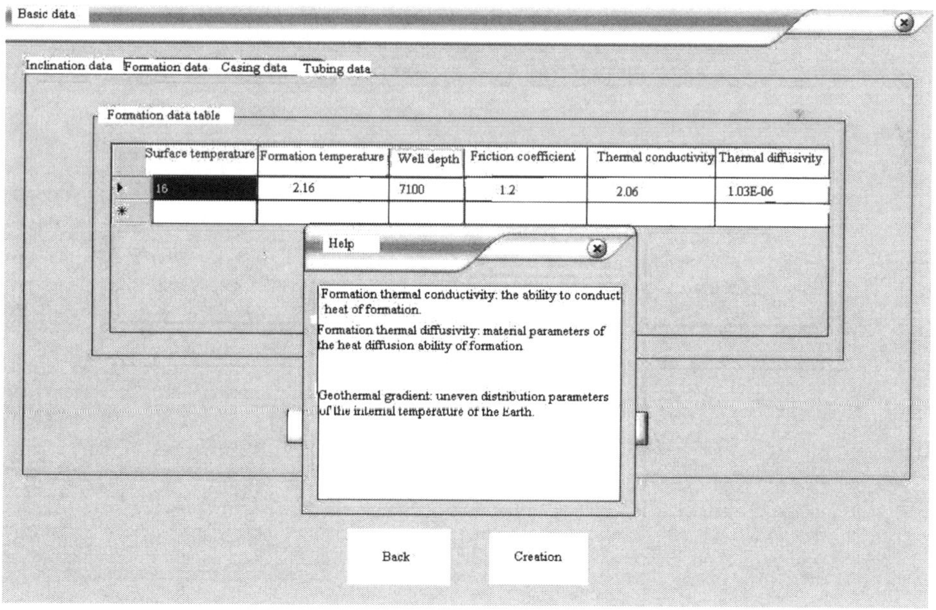

Figure 10.4 Basic data table for the formation.

have a variety of units, so it is necessary to pay attention to ensure the correct unit is selected. The window interface is shown in Figure 10.7.

To complete the data entry for the placement of the tubular, clicking "Next" gives entry to the packer setting window. In this window the relevant data for the perforating conditions are entered.

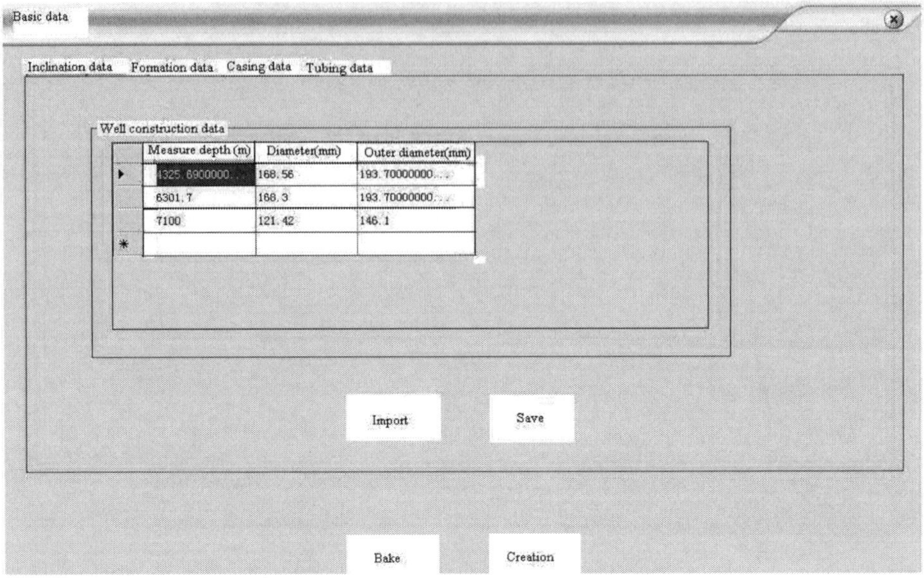

Figure 10.5 Basic data table for well construction.

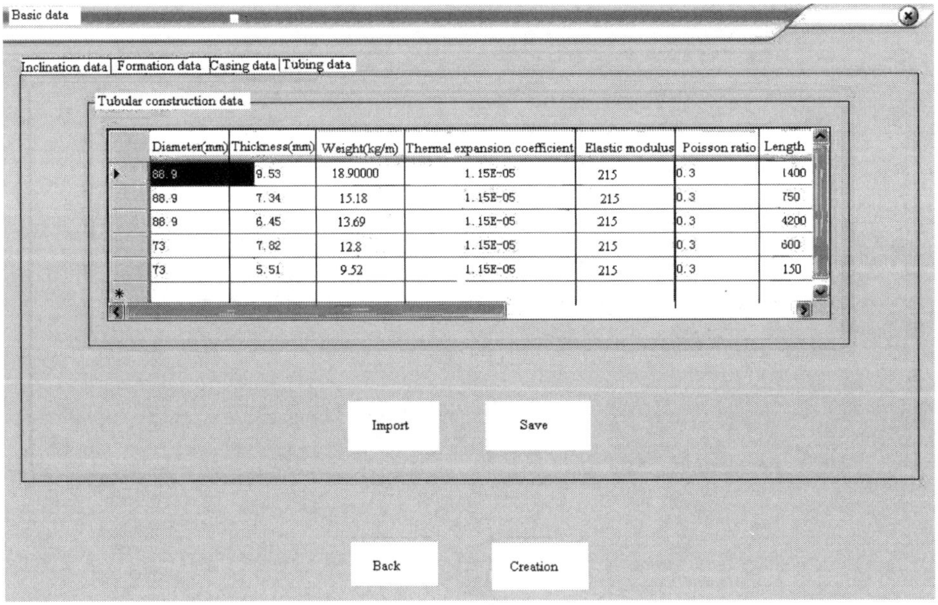

Figure 10.6 Basic data for tubular construction.

The specific physical meaning of the required parameters can be found with the "Help" button. Some parameters may have a variety of units, thus it is necessary to ensure the correct unit is selected. The packer setting window interface is shown in Figure 10.8.

To complete the data entry for the seated conditions, clicking "Next" leads to the perforation window are entered. The specific physical meaning of the required parameters can be found with

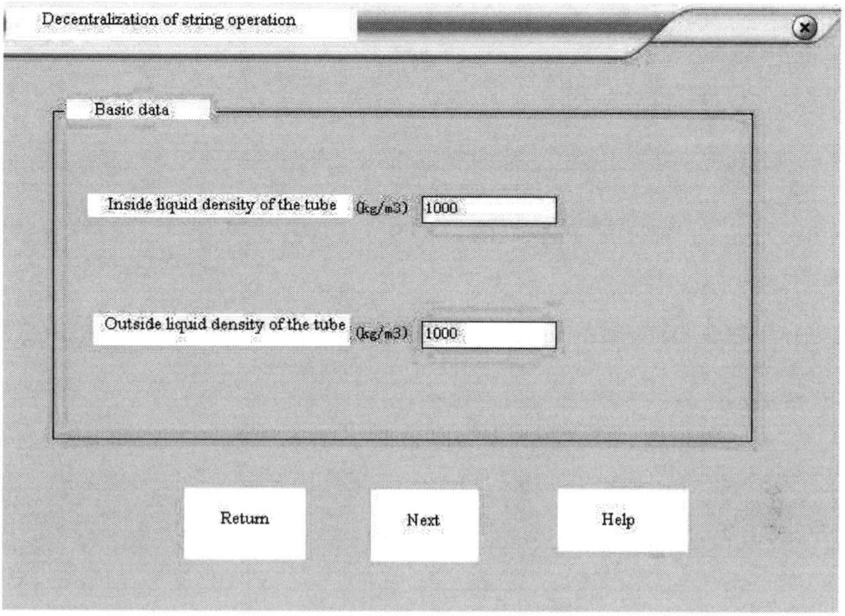

Figure 10.7 Basic data for the string decentralization.

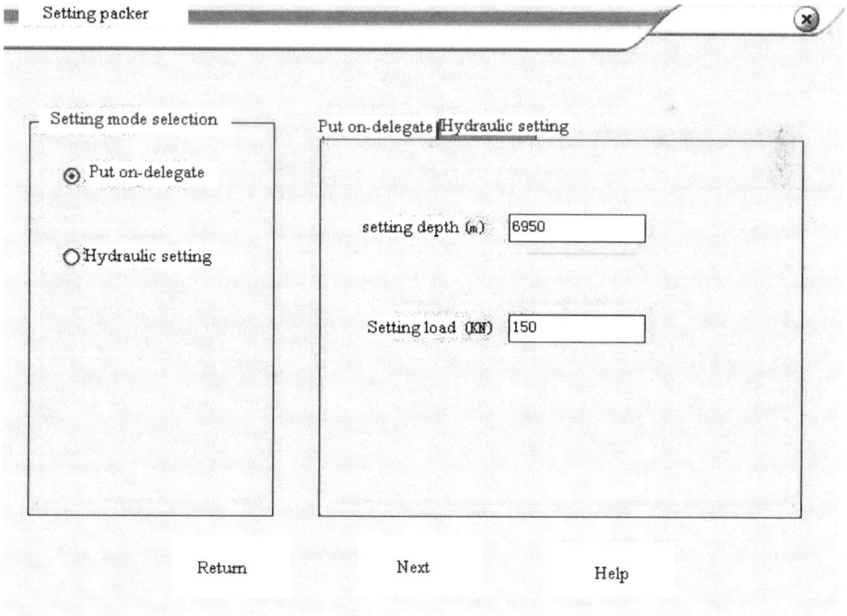

Figure 10.8 Basic packer setting data.

the "Help" button. Some parameters may have a variety of units, so it is necessary to pay attention to the selection of the correct unit. The perforation window interface is shown in Figure 10.9.

To complete the data input for the perforation conditions, clicking "Next" leads to the injection window. In this window the relevant data is input. The specific physical meaning of the required

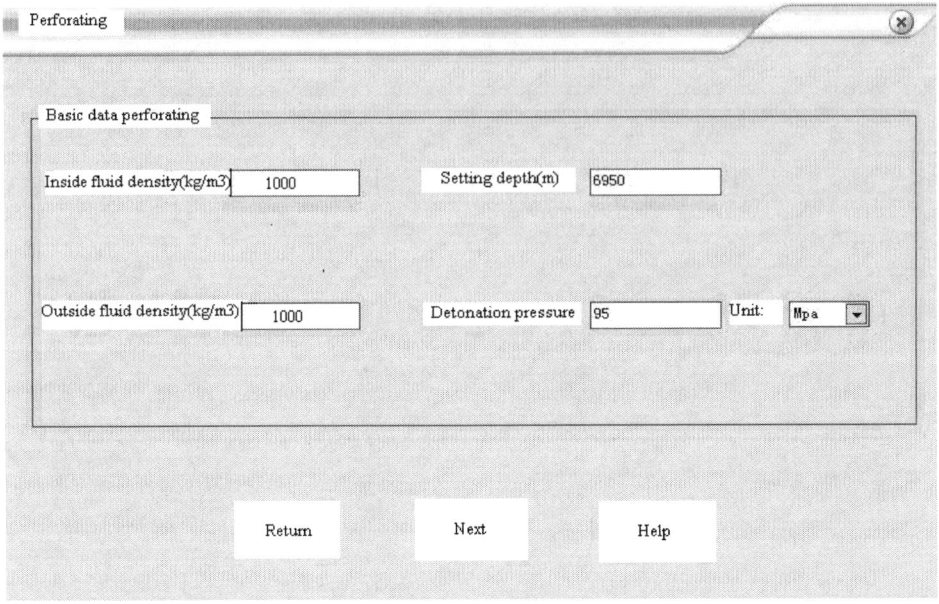

Figure 10.9 Basic perforating data.

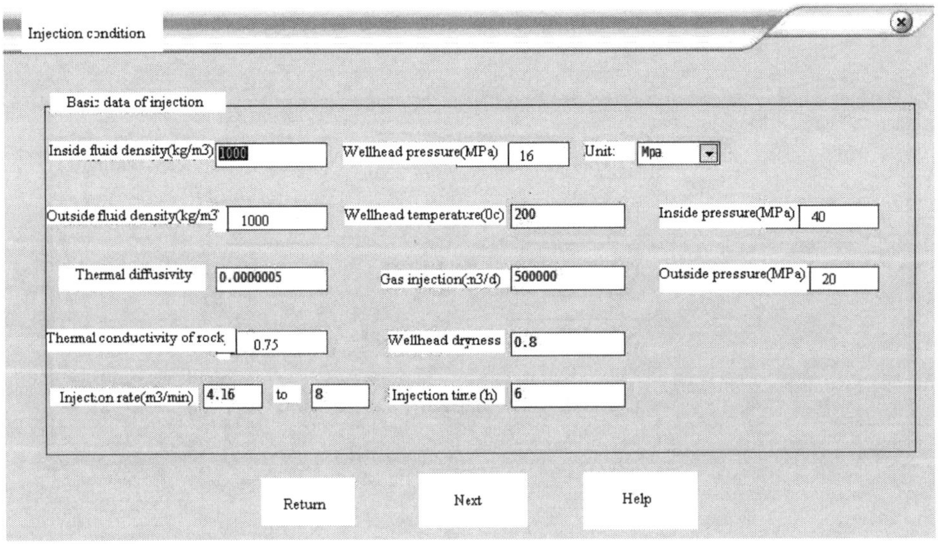

Figure 10.10 Basic injection data.

parameters can be found with the "Help" button. As some parameters may have a variety of units, it is necessary to pay attention to the selection of the correct unit. The injection window interface is shown in Figure 10.10.

After completing the data input for the injection conditions, clicking "Next" leads to where the production conditions are input. The specific physical meaning of the required parameters can be found with the "Help" button. Some parameters may have a variety of units, so it is important that the correct unit is selected. The production window interface is shown in Figure 10.11.

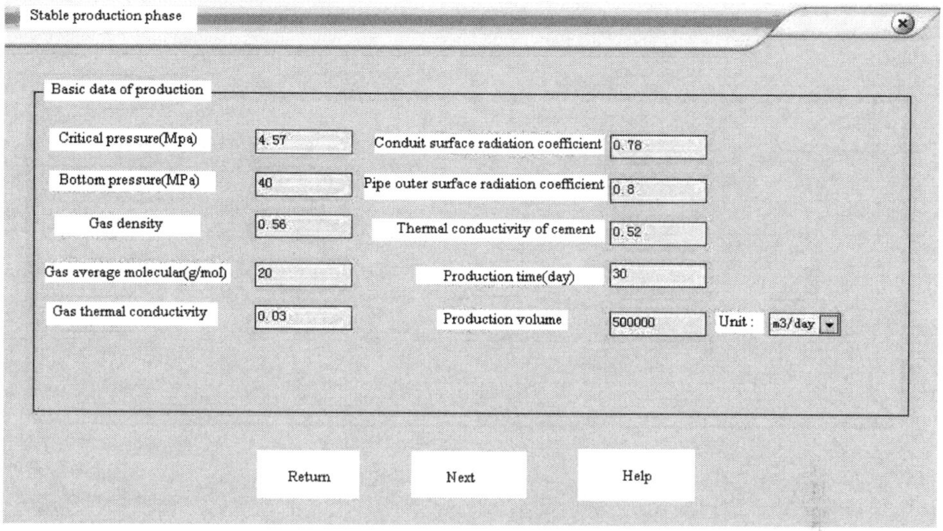

Figure 10.11 Basic production data.

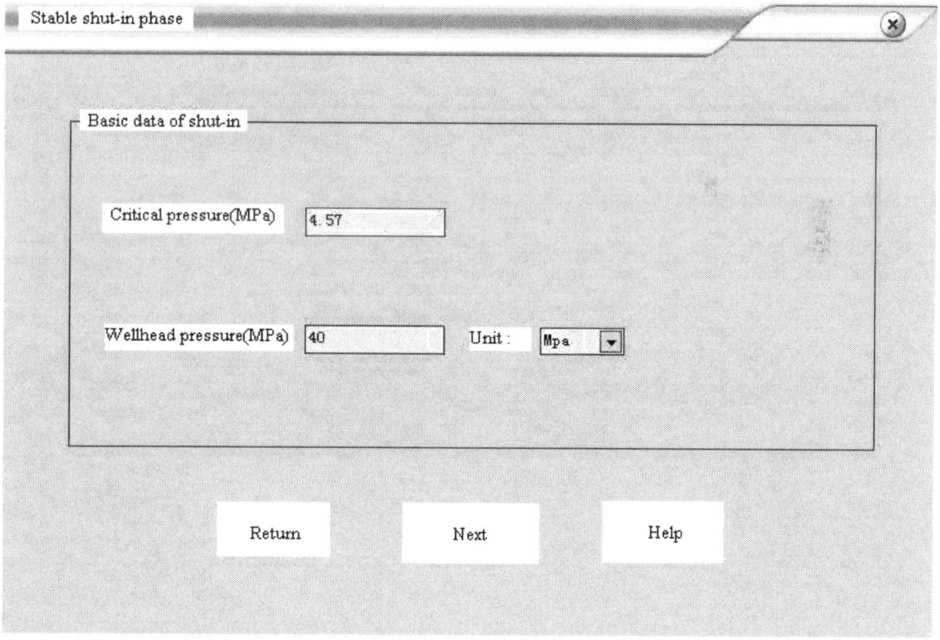

Figure 10.12 Basic shut-in data.

After data input for the production conditions is completed, clicking "Next" brings up the shut-in window, where the relevant data is input. The specific physical meaning of the required parameters can be found with the "Help" button. As some parameters may have a variety of units, it is important to ensure that the correct units are selected. The shut-in window interface is shown in Figure 10.12.

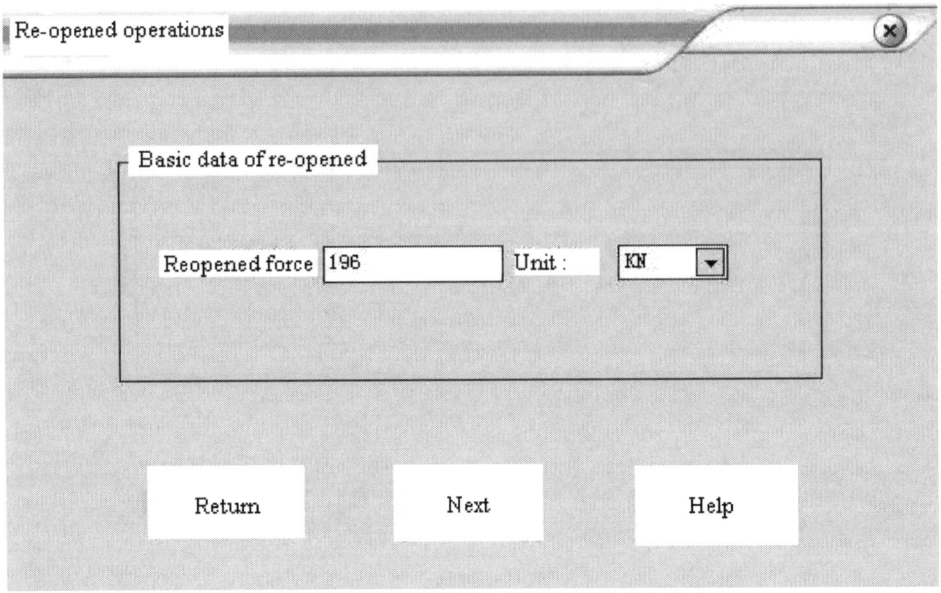

Figure 10.13 Basic re-opened data.

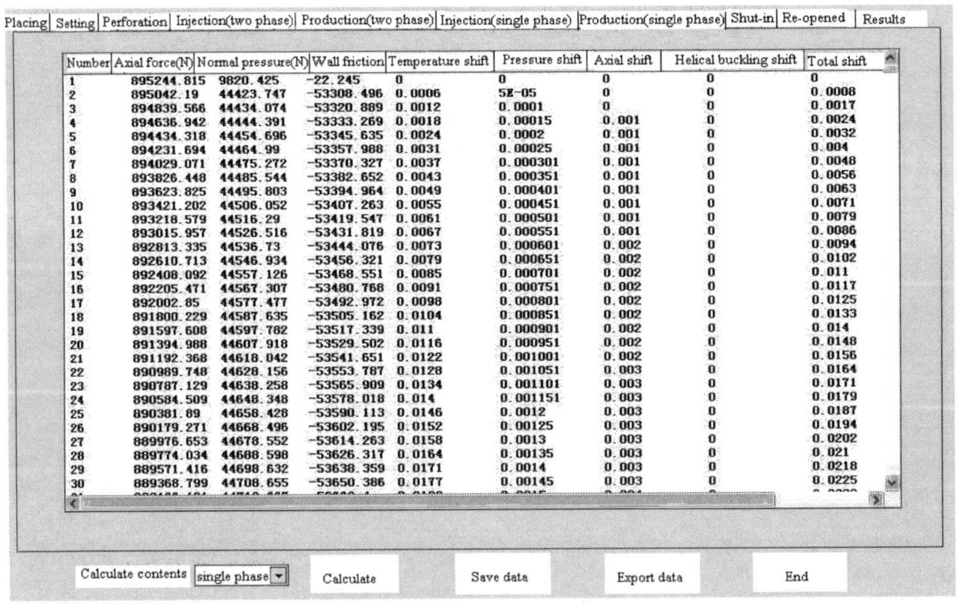

Figure 10.14 Placing tubular calculation results.

After completing the shut-in condition data input, clicking "Next" opens the window for the input of relevant data for the re-opened condition. The specific physical meaning of the required parameters can be found with the "Help" button. As some parameters may have a variety of units, it is important to ensure that the correct units are selected. The re-opened window interface is shown in Figure 10.13.

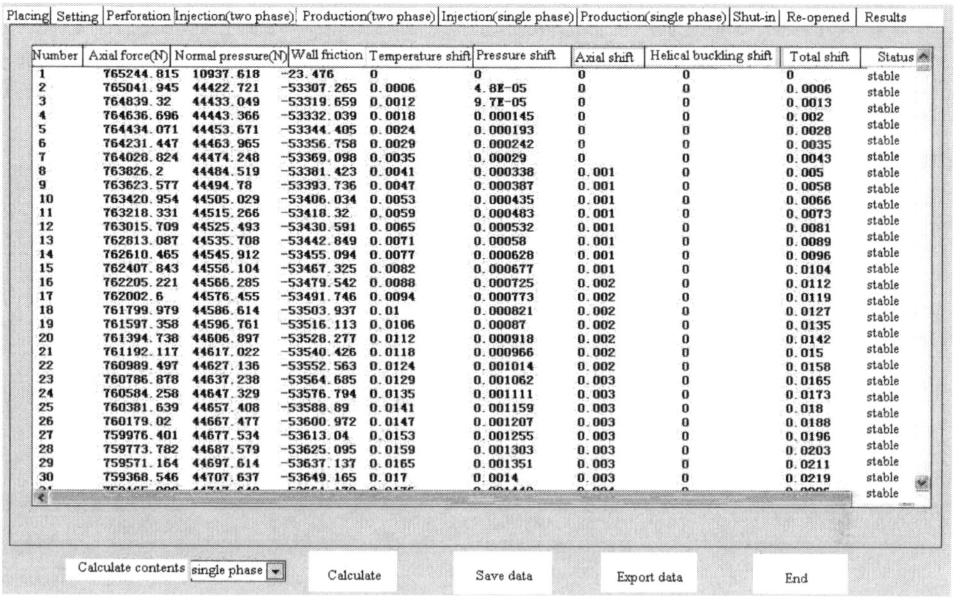

Number	Axial force(N)	Normal pressure(N)	Wall friction	Temperature shift	Pressure shift	Axial shift	Helical buckling shift	Total shift	Status
1	765244.815	10937.618	-23.476	0	0	0	0	0	stable
2	765041.945	44422.721	-53307.265	0.0006	4.8E-05	0	0	0.0006	stable
3	764839.32	44433.049	-53319.659	0.0012	9.7E-05	0	0	0.0013	stable
4	764636.696	44443.366	-53332.039	0.0018	0.000145	0	0	0.002	stable
5	764434.071	44453.671	-53344.405	0.0024	0.000193	0	0	0.0028	stable
6	764231.447	44463.965	-53356.758	0.0029	0.000242	0	0	0.0035	stable
7	764028.824	44474.248	-53369.098	0.0035	0.00029	0	0	0.0043	stable
8	763826.2	44484.519	-53381.423	0.0041	0.000338	0.001	0	0.005	stable
9	763623.577	44494.78	-53393.736	0.0047	0.000387	0.001	0	0.0058	stable
10	763420.954	44505.029	-53406.034	0.0053	0.000435	0.001	0	0.0066	stable
11	763218.331	44515.266	-53418.32	0.0059	0.000483	0.001	0	0.0073	stable
12	763015.709	44525.493	-53430.591	0.0065	0.000532	0.001	0	0.0081	stable
13	762813.087	44535.708	-53442.849	0.0071	0.00058	0.001	0	0.0089	stable
14	762610.465	44545.912	-53455.094	0.0077	0.000628	0.001	0	0.0096	stable
15	762407.843	44556.104	-53467.325	0.0082	0.000677	0.001	0	0.0104	stable
16	762205.221	44566.285	-53479.542	0.0088	0.000725	0.002	0	0.0112	stable
17	762002.6	44576.455	-53491.746	0.0094	0.000773	0.002	0	0.0119	stable
18	761799.979	44586.614	-53503.937	0.01	0.000821	0.002	0	0.0127	stable
19	761597.358	44596.761	-53516.113	0.0106	0.00087	0.002	0	0.0135	stable
20	761394.738	44606.897	-53528.277	0.0112	0.000918	0.002	0	0.0142	stable
21	761192.117	44617.022	-53540.426	0.0118	0.000966	0.002	0	0.015	stable
22	760989.497	44627.136	-53552.563	0.0124	0.001014	0.002	0	0.0158	stable
23	760786.878	44637.238	-53564.685	0.0129	0.001062	0.003	0	0.0165	stable
24	760584.258	44647.329	-53576.794	0.0135	0.001111	0.003	0	0.0173	stable
25	760381.639	44657.408	-53588.89	0.0141	0.001159	0.003	0	0.018	stable
26	760179.02	44667.477	-53600.972	0.0147	0.001207	0.003	0	0.0188	stable
27	759976.401	44677.534	-53613.04	0.0153	0.001255	0.003	0	0.0196	stable
28	759773.782	44687.579	-53625.095	0.0159	0.001303	0.003	0	0.0203	stable
29	759571.164	44697.614	-53637.137	0.0165	0.001351	0.003	0	0.0211	stable
30	759368.546	44707.637	-53649.165	0.017	0.0014	0.003	0	0.0219	stable
									stable

Figure 10.15 Packer setting calculation results.

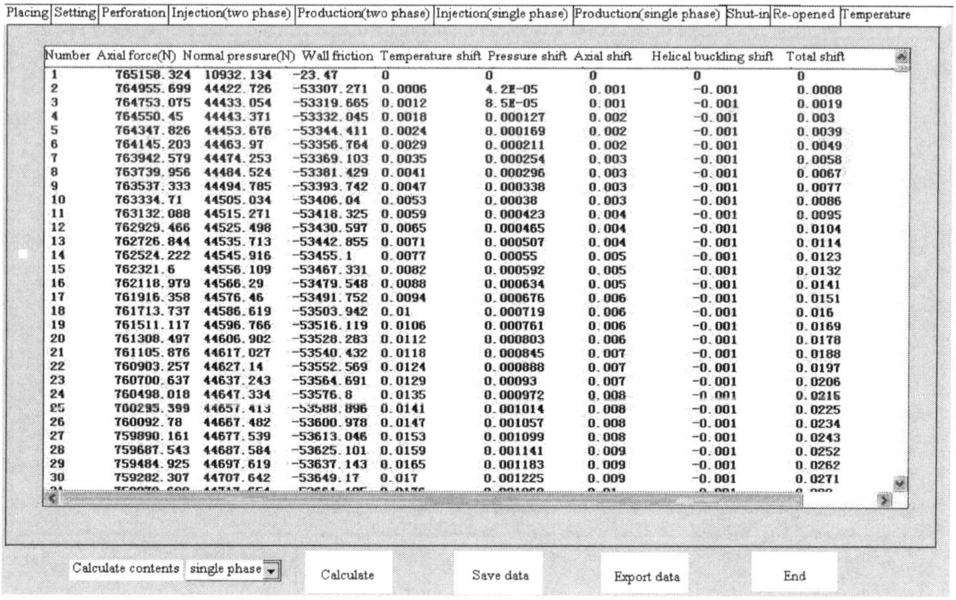

Number	Axial force(N)	Normal pressure(N)	Wall friction	Temperature shift	Pressure shift	Axial shift	Helical buckling shift	Total shift
1	765158.324	10932.134	-23.47	0	0	0	0	0
2	764955.699	44422.726	-53307.271	0.0006	4.2E-05	0.001	-0.001	0.0008
3	764753.075	44433.054	-53319.665	0.0012	8.5E-05	0.001	-0.001	0.0019
4	764550.45	44443.371	-53332.045	0.0018	0.000127	0.002	-0.001	0.003
5	764347.826	44453.676	-53344.411	0.0024	0.000169	0.002	-0.001	0.0039
6	764145.203	44463.97	-53356.764	0.0029	0.000211	0.002	-0.001	0.0049
7	763942.579	44474.253	-53369.103	0.0035	0.000254	0.003	-0.001	0.0058
8	763739.956	44484.524	-53381.429	0.0041	0.000296	0.003	-0.001	0.0067
9	763537.333	44494.785	-53393.742	0.0047	0.000338	0.003	-0.001	0.0077
10	763334.71	44505.034	-53406.04	0.0053	0.00038	0.003	-0.001	0.0086
11	763132.088	44515.271	-53418.325	0.0059	0.000423	0.004	-0.001	0.0095
12	762929.466	44525.498	-53430.597	0.0065	0.000465	0.004	-0.001	0.0104
13	762726.844	44535.713	-53442.855	0.0071	0.000507	0.004	-0.001	0.0114
14	762524.222	44545.916	-53455.1	0.0077	0.00055	0.005	-0.001	0.0123
15	762321.6	44556.109	-53467.331	0.0082	0.000592	0.005	-0.001	0.0132
16	762118.979	44566.29	-53479.548	0.0088	0.000634	0.005	-0.001	0.0141
17	761916.358	44576.46	-53491.752	0.0094	0.000676	0.006	-0.001	0.0151
18	761713.737	44586.619	-53503.942	0.01	0.000719	0.006	-0.001	0.016
19	761511.117	44596.766	-53516.119	0.0106	0.000761	0.006	-0.001	0.0169
20	761308.497	44606.902	-53528.283	0.0112	0.000803	0.006	-0.001	0.0178
21	761105.876	44617.027	-53540.432	0.0118	0.000845	0.007	-0.001	0.0188
22	760903.257	44627.14	-53552.569	0.0124	0.000888	0.007	-0.001	0.0197
23	760700.637	44637.243	-53564.691	0.0129	0.00093	0.007	-0.001	0.0206
24	760498.018	44647.334	-53576.8	0.0135	0.000972	0.008	-0.001	0.0215
25	760295.399	44657.413	-53588.896	0.0141	0.001014	0.008	-0.001	0.0225
26	760092.78	44667.482	-53600.978	0.0147	0.001057	0.008	-0.001	0.0234
27	759890.161	44677.539	-53613.046	0.0153	0.001099	0.008	-0.001	0.0243
28	759687.543	44687.584	-53625.101	0.0159	0.001141	0.009	-0.001	0.0252
29	759484.925	44697.619	-53637.143	0.0165	0.001183	0.009	-0.001	0.0262
30	759282.307	44707.642	-53649.17	0.017	0.001225	0.009	-0.001	0.0271

Figure 10.16 Perforation calculation results.

By selecting "single-phase flow calculation", and clicking the "Calculate" button, the software completes the calculation of the data for all the conditions required, and a list of each operating mode is displayed. By clicking on this list, different data for the different conditions is displayed as shown in Figures 10.14 to 10.21.

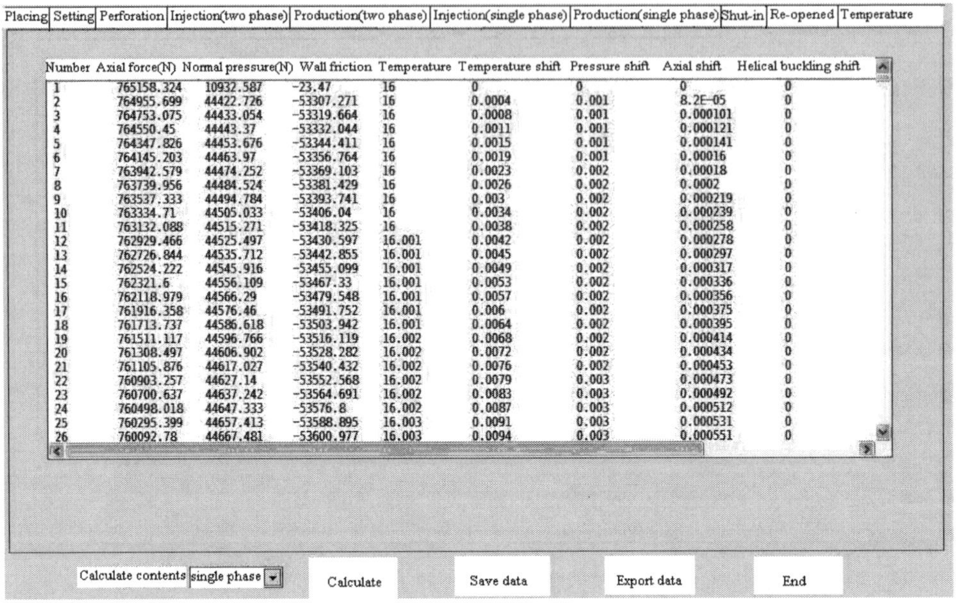

Figure 10.17 Injection (single phase) calculation results.

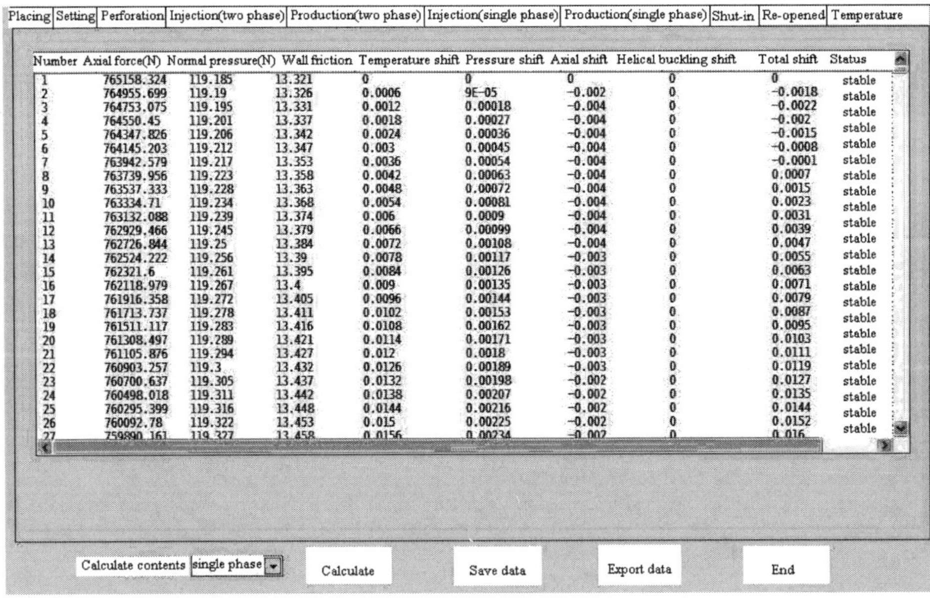

Figure 10.18 Production (single phase) calculation results.

By selecting the "two-phase flow calculation", and clicking the "calculate button", the two-phase flow data is calculated, as shown in Figures 10.22 and 10.23.

Calculation results and the associated graphs are exported by first clicking "Save Data", and then clicking on "Export Report". The export report window is as shown in Figure 10.24.

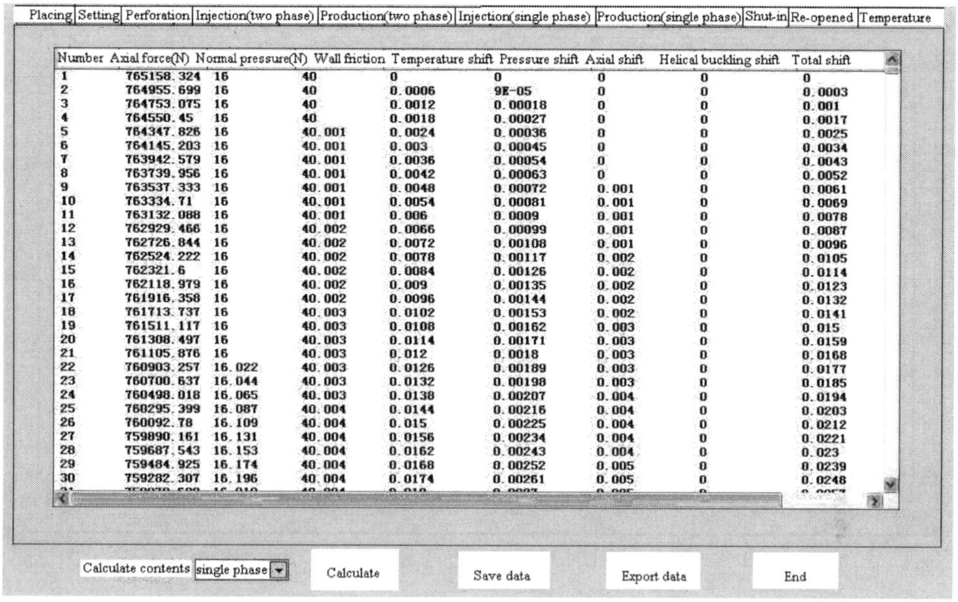

Number	Axial force(N)	Normal pressure(N)	Wall friction	Temperature shift	Pressure shift	Axial shift	Helical buckling shift	Total shift
1	765158.324	16	40	0	0	0	0	0
2	764955.699	16	40	0.0006	9E-05	0	0	0.0003
3	764753.075	16	40	0.0012	0.00018	0	0	0.001
4	764550.45	16	40	0.0018	0.00027	0	0	0.0017
5	764347.826	16	40.001	0.0024	0.00036	0	0	0.0025
6	764145.203	16	40.001	0.003	0.00045	0	0	0.0034
7	763942.579	16	40.001	0.0036	0.00054	0	0	0.0043
8	763739.956	16	40.001	0.0042	0.00063	0	0	0.0052
9	763537.333	16	40.001	0.0048	0.00072	0.001	0	0.0061
10	763334.71	16	40.001	0.0054	0.00081	0.001	0	0.0069
11	763132.088	16	40.001	0.006	0.0009	0.001	0	0.0078
12	762929.466	16	40.002	0.0066	0.00099	0.001	0	0.0087
13	762726.844	16	40.002	0.0072	0.00108	0.001	0	0.0096
14	762524.222	16	40.002	0.0078	0.00117	0.002	0	0.0105
15	762321.6	16	40.002	0.0084	0.00126	0.002	0	0.0114
16	762118.979	16	40.002	0.009	0.00135	0.002	0	0.0123
17	761916.358	16	40.002	0.0096	0.00144	0.002	0	0.0132
18	761713.737	16	40.003	0.0102	0.00153	0.002	0	0.0141
19	761511.117	16	40.003	0.0108	0.00162	0.003	0	0.015
20	761308.497	16	40.003	0.0114	0.00171	0.003	0	0.0159
21	761105.876	16	40.003	0.012	0.0018	0.003	0	0.0168
22	760903.257	16.022	40.003	0.0126	0.00189	0.003	0	0.0177
23	760700.637	16.044	40.003	0.0132	0.00198	0.003	0	0.0185
24	760498.018	16.065	40.003	0.0138	0.00207	0.004	0	0.0194
25	760295.399	16.087	40.004	0.0144	0.00216	0.004	0	0.0203
26	760092.78	16.109	40.004	0.015	0.00225	0.004	0	0.0212
27	759890.161	16.131	40.004	0.0156	0.00234	0.004	0	0.0221
28	759687.543	16.153	40.004	0.0162	0.00243	0.004	0	0.023
29	759484.925	16.174	40.004	0.0168	0.00252	0.005	0	0.0239
30	759282.307	16.196	40.004	0.0174	0.00261	0.005	0	0.0248

Figure 10.19 Shut-in calculation results.

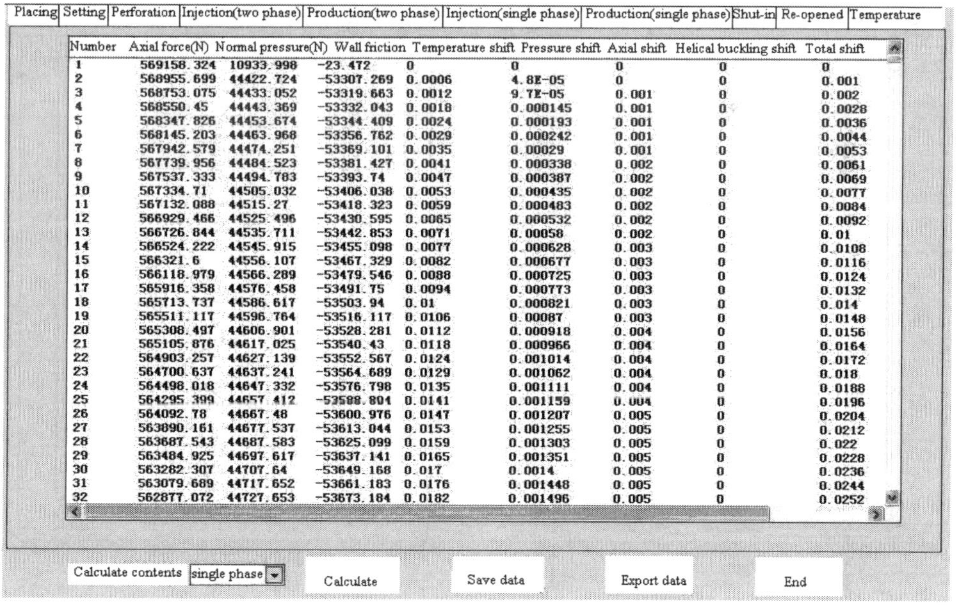

Number	Axial force(N)	Normal pressure(N)	Wall friction	Temperature shift	Pressure shift	Axial shift	Helical buckling shift	Total shift
1	569158.324	10933.998	-23.472	0	0	0	0	0
2	568955.699	44422.724	-53307.269	0.0006	4.8E-05	0	0	0.001
3	568753.075	44433.052	-53319.663	0.0012	9.7E-05	0.001	0	0.002
4	568550.45	44443.369	-53332.043	0.0018	0.000145	0.001	0	0.0028
5	568347.826	44453.674	-53344.409	0.0024	0.000193	0.001	0	0.0036
6	568145.203	44463.968	-53356.762	0.0029	0.000242	0.001	0	0.0044
7	567942.579	44474.251	-53369.101	0.0035	0.00029	0.001	0	0.0053
8	567739.956	44484.523	-53381.427	0.0041	0.000338	0.002	0	0.0061
9	567537.333	44494.783	-53393.74	0.0047	0.000387	0.002	0	0.0069
10	567334.71	44505.032	-53406.038	0.0053	0.000435	0.002	0	0.0077
11	567132.088	44515.27	-53418.323	0.0059	0.000483	0.002	0	0.0084
12	566929.466	44525.496	-53430.595	0.0065	0.000532	0.002	0	0.0092
13	566726.844	44535.711	-53442.853	0.0071	0.00058	0.002	0	0.01
14	566524.222	44545.915	-53455.098	0.0077	0.000628	0.003	0	0.0108
15	566321.6	44556.107	-53467.329	0.0082	0.000677	0.003	0	0.0116
16	566118.979	44566.289	-53479.546	0.0088	0.000725	0.003	0	0.0124
17	565916.358	44576.458	-53491.75	0.0094	0.000773	0.003	0	0.0132
18	565713.737	44586.617	-53503.94	0.01	0.000821	0.003	0	0.014
19	565511.117	44596.764	-53516.117	0.0106	0.00087	0.003	0	0.0148
20	565308.497	44606.901	-53528.281	0.0112	0.000918	0.004	0	0.0156
21	565105.876	44617.025	-53540.43	0.0118	0.000966	0.004	0	0.0164
22	564903.257	44627.139	-53552.567	0.0124	0.001014	0.004	0	0.0172
23	564700.637	44637.241	-53564.689	0.0129	0.001062	0.004	0	0.018
24	564498.018	44647.332	-53576.798	0.0135	0.001111	0.004	0	0.0188
25	564295.399	44657.412	-53588.804	0.0141	0.001159	0.004	0	0.0196
26	564092.78	44667.48	-53600.976	0.0147	0.001207	0.005	0	0.0204
27	563890.161	44677.537	-53613.044	0.0153	0.001255	0.005	0	0.0212
28	563687.543	44687.583	-53625.099	0.0159	0.001303	0.005	0	0.022
29	563484.925	44697.617	-53637.141	0.0165	0.001351	0.005	0	0.0228
30	563282.307	44707.64	-53649.168	0.017	0.0014	0.005	0	0.0236
31	563079.689	44717.652	-53661.183	0.0176	0.001448	0.005	0	0.0244
32	562877.072	44727.653	-53673.184	0.0182	0.001496	0.005	0	0.0252

Figure 10.20 Re-opened calculation results.

By clicking on "∗∗ condition", the corresponding conditions for the numerical results are displayed; and by clicking on "∗∗ Chart", the corresponding conditions for the curve are displayed. Save the file by clicking the "Save" button, which allows for a choice of formats, including PDF, WORD, EXCEL and others.

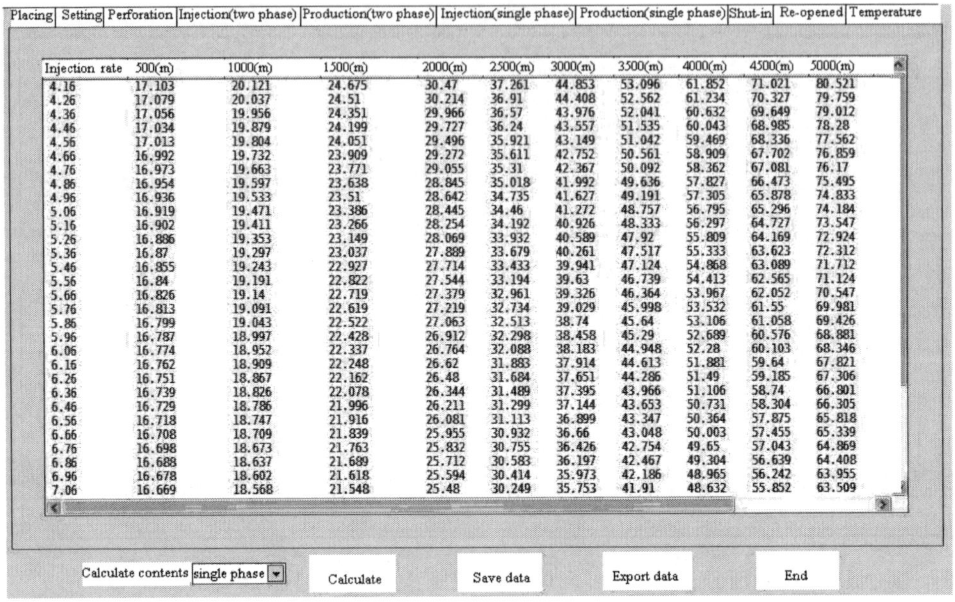

Injection rate	500(m)	1000(m)	1500(m)	2000(m)	2500(m)	3000(m)	3500(m)	4000(m)	4500(m)	5000(m)
4.16	17.103	20.121	24.675	30.47	37.261	44.853	53.096	61.852	71.021	80.521
4.26	17.079	20.037	24.51	30.214	36.91	44.408	52.562	61.234	70.327	79.759
4.36	17.056	19.956	24.351	29.966	36.57	43.976	52.041	60.632	69.649	79.012
4.46	17.034	19.879	24.199	29.727	36.24	43.557	51.535	60.043	68.985	78.28
4.56	17.013	19.804	24.051	29.496	35.921	43.149	51.042	59.469	68.336	77.562
4.66	16.992	19.732	23.909	29.272	35.611	42.752	50.561	58.909	67.702	76.859
4.76	16.973	19.663	23.771	29.055	35.31	42.367	50.092	58.362	67.081	76.17
4.86	16.954	19.597	23.638	28.845	35.018	41.992	49.636	57.827	66.473	75.495
4.96	16.936	19.533	23.51	28.642	34.735	41.627	49.191	57.305	65.878	74.833
5.06	16.919	19.471	23.386	28.445	34.46	41.272	48.757	56.795	65.296	74.184
5.16	16.902	19.411	23.266	28.254	34.192	40.926	48.333	56.297	64.727	73.547
5.26	16.886	19.353	23.149	28.069	33.932	40.589	47.92	55.809	64.169	72.924
5.36	16.87	19.297	23.037	27.889	33.679	40.261	47.517	55.333	63.623	72.312
5.46	16.855	19.243	22.927	27.714	33.433	39.941	47.124	54.868	63.089	71.712
5.56	16.84	19.191	22.822	27.544	33.194	39.63	46.739	54.413	62.565	71.124
5.66	16.826	19.14	22.719	27.379	32.961	39.326	46.364	53.967	62.052	70.547
5.76	16.813	19.091	22.619	27.219	32.734	39.029	45.998	53.532	61.55	69.981
5.86	16.799	19.043	22.522	27.063	32.513	38.74	45.64	53.106	61.058	69.426
5.96	16.787	18.997	22.428	26.912	32.298	38.458	45.29	52.689	60.576	68.881
6.06	16.774	18.952	22.337	26.764	32.088	38.183	44.948	52.28	60.103	68.346
6.16	16.762	18.909	22.248	26.62	31.883	37.914	44.613	51.881	59.64	67.821
6.26	16.751	18.867	22.162	26.48	31.684	37.651	44.286	51.49	59.185	67.306
6.36	16.739	18.826	22.078	26.344	31.489	37.395	43.966	51.106	58.74	66.801
6.46	16.729	18.786	21.996	26.211	31.299	37.144	43.653	50.731	58.304	66.305
6.56	16.718	18.747	21.916	26.081	31.113	36.899	43.347	50.364	57.875	65.818
6.66	16.708	18.709	21.839	25.955	30.932	36.66	43.048	50.003	57.455	65.339
6.76	16.698	18.673	21.763	25.832	30.755	36.426	42.754	49.65	57.043	64.869
6.86	16.688	18.637	21.689	25.712	30.583	36.197	42.467	49.304	56.639	64.408
6.96	16.678	18.602	21.618	25.594	30.414	35.973	42.186	48.965	56.242	63.955
7.06	16.669	18.568	21.548	25.48	30.249	35.753	41.91	48.632	55.852	63.509

Figure 10.21 Temperature variation results with a change in injection rates.

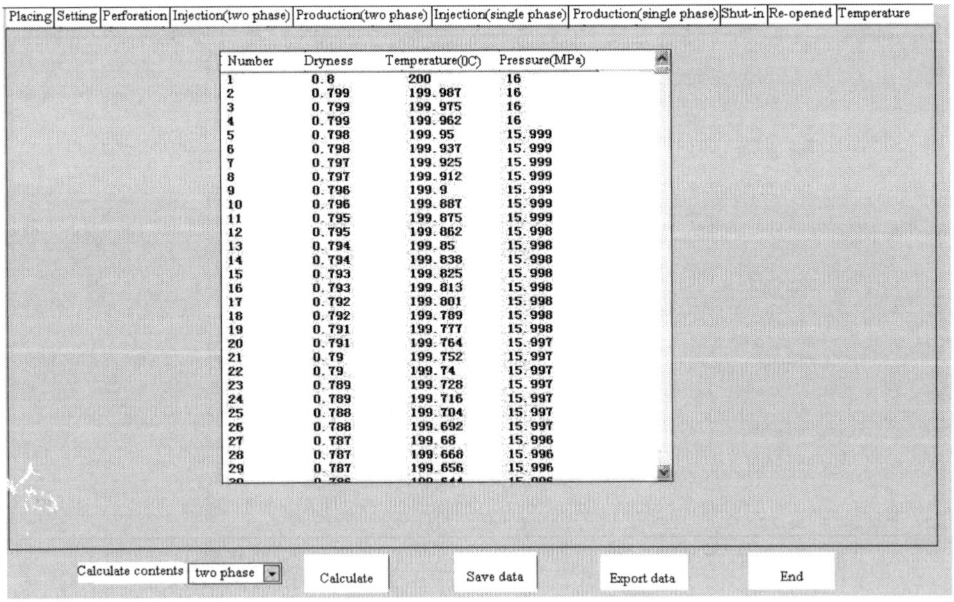

Number	Dryness	Temperature(0C)	Pressure(MPa)
1	0.8	200	16
2	0.799	199.987	16
3	0.799	199.975	16
4	0.799	199.962	16
5	0.798	199.95	15.999
6	0.798	199.937	15.999
7	0.797	199.925	15.999
8	0.797	199.912	15.999
9	0.796	199.9	15.999
10	0.796	199.887	15.999
11	0.795	199.875	15.999
12	0.795	199.862	15.998
13	0.794	199.85	15.998
14	0.794	199.838	15.998
15	0.793	199.825	15.998
16	0.793	199.813	15.998
17	0.792	199.801	15.998
18	0.792	199.789	15.998
19	0.791	199.777	15.998
20	0.791	199.764	15.997
21	0.79	199.752	15.997
22	0.79	199.74	15.997
23	0.789	199.728	15.997
24	0.789	199.716	15.997
25	0.788	199.704	15.997
26	0.788	199.692	15.997
27	0.787	199.68	15.996
28	0.787	199.668	15.996
29	0.787	199.656	15.996

Figure 10.22 Two phase results for injection.

Clicking on "** Chart" shows the corresponding curve for the parameters, including the axial force curve, the normal pressure curve, the temperature distribution, the temperature displacement curve, the axial force displacement curve, the helical buckling displacement curve, and the total displacement curve. The following lists some of the conditions of the curve.

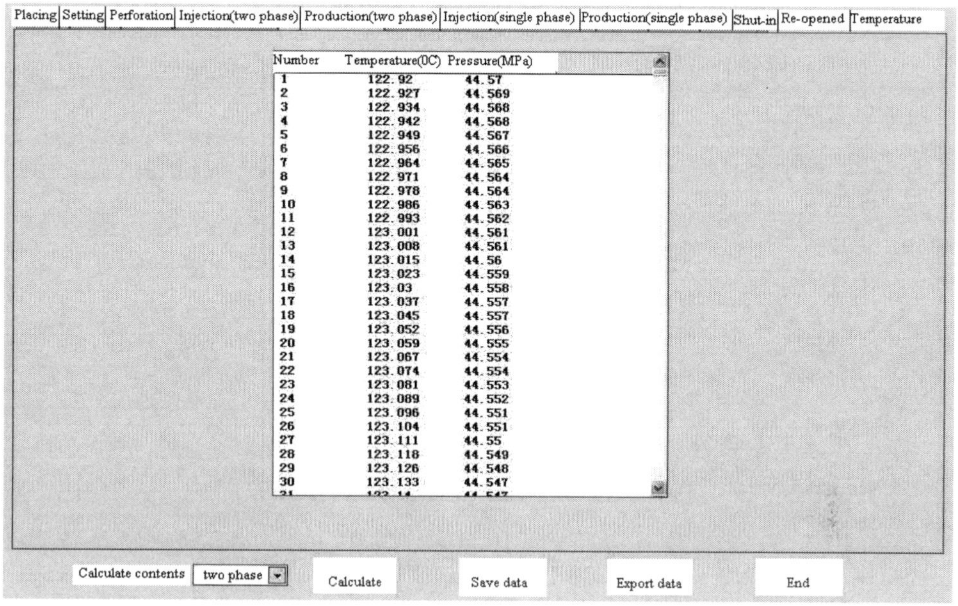

Figure 10.23 Two phase results for production.

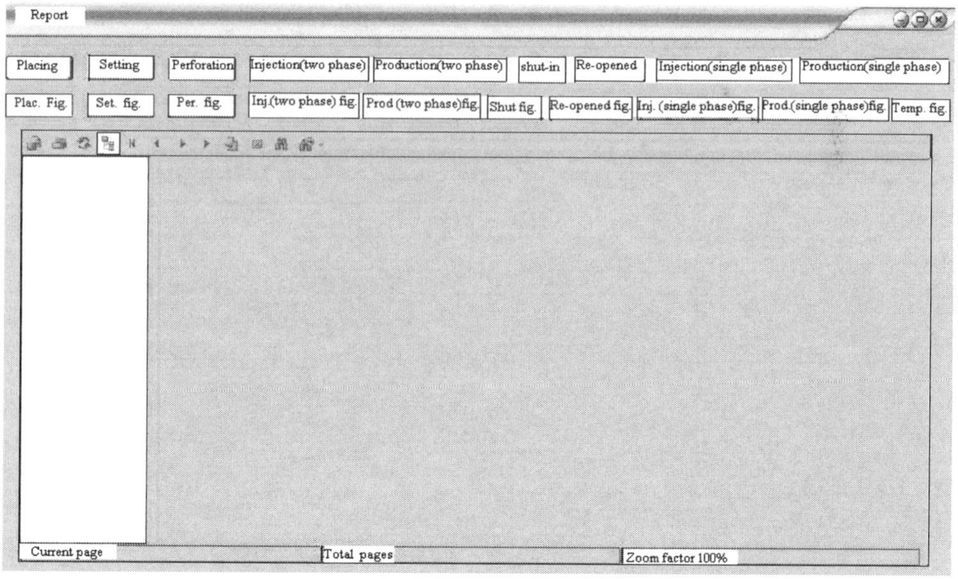

Figure 10.24 Export report window.

The following figures are calculated from a China Petroleum and Chemical Corp HTHP deep well. The axial force distribution curve for the placement of the tubular is shown in Figure 10.25. The axial force increases from the bottom to the wellhead, and is a basic linear distribution.

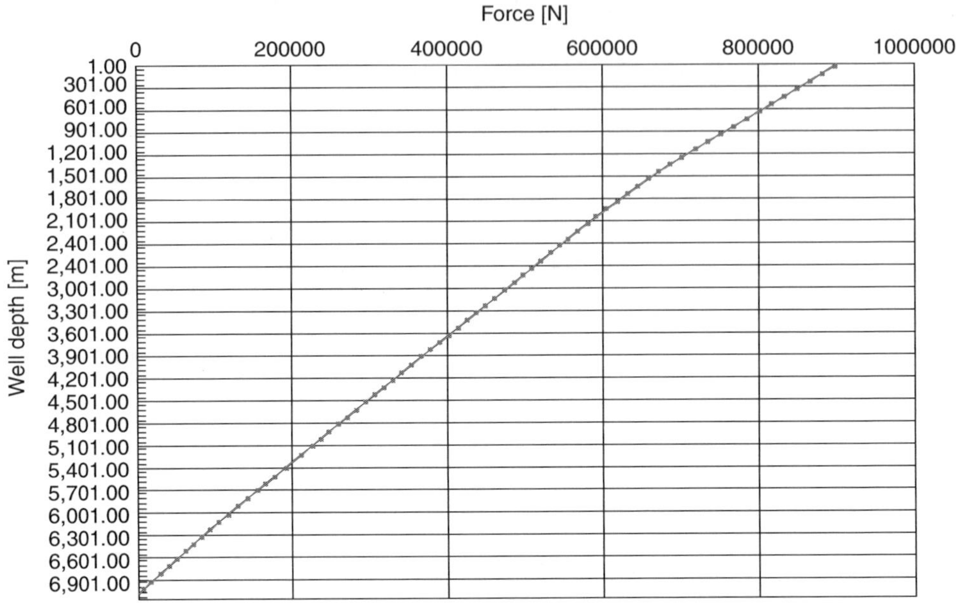

Figure 10.25 Axial force distribution for the placement of the tubular.

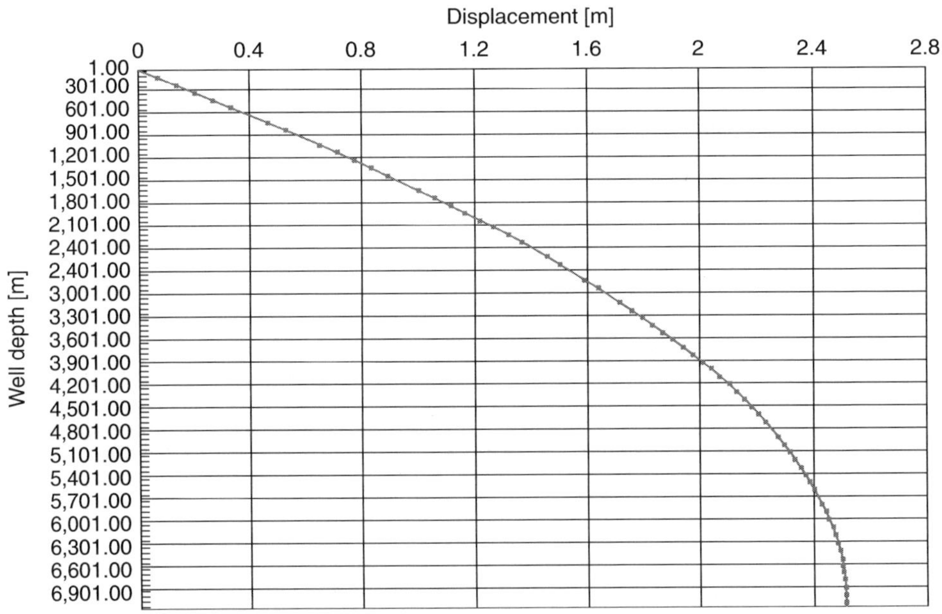

Figure 10.26 Temperature displacement distribution for the placement of the tubular.

The temperature displacement curve for the placement of the tubular is shown in Figure 10.26. The displacement decreases from the bottom to the wellhead, with the maximum at the well bottom.

The axial displacement curve for the placement of the tubular is shown in Figure 10.27. The displacement decreases from the bottom to the wellhead, with the maximum at the well bottom.

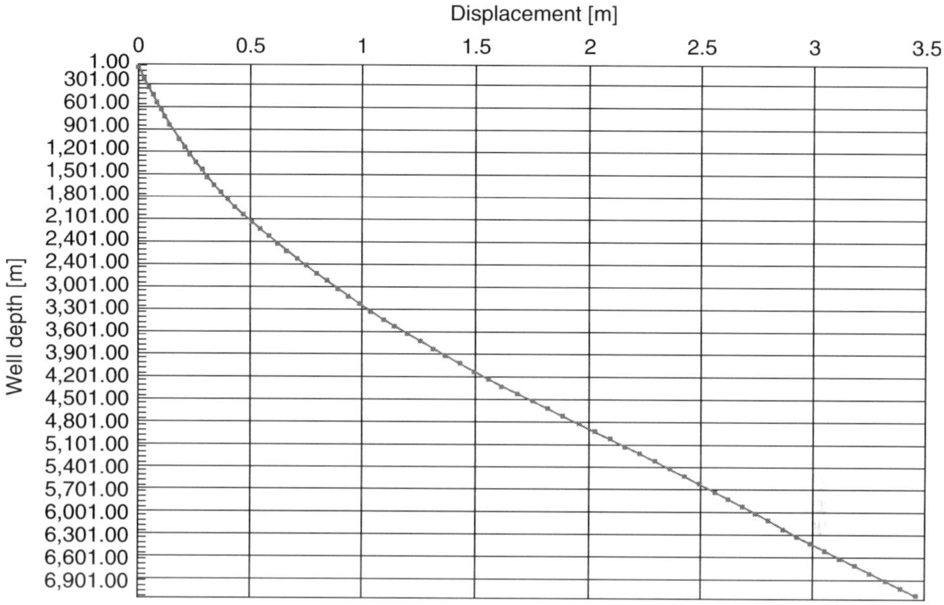

Figure 10.27 Axial displacement distribution for the placement of the tubular.

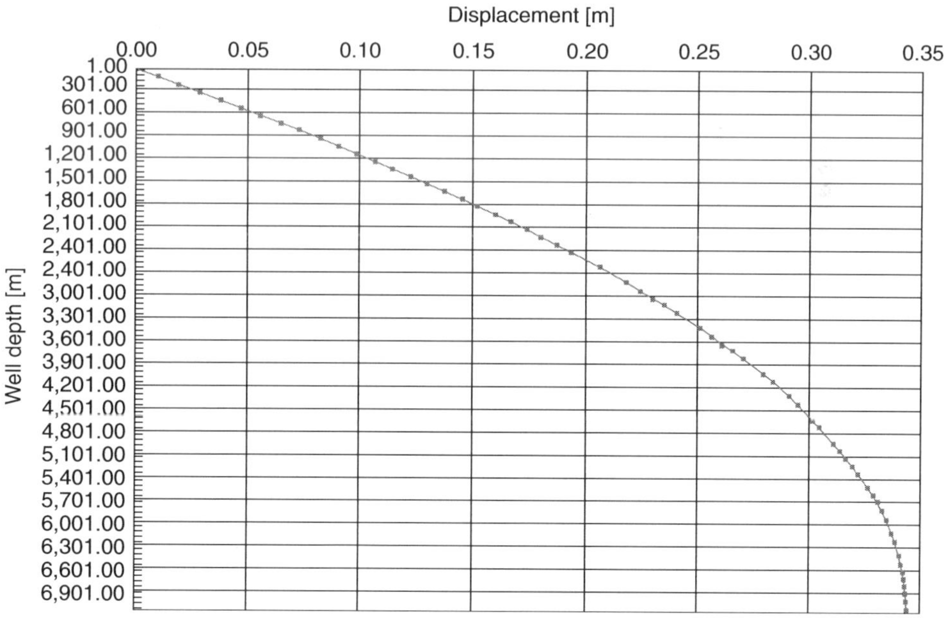

Figure 10.28 Pressure displacement distribution for the placement of the tubular.

The pressure displacement curve for the placement of the tubular is shown in Figure 10.28. The displacement is very small, and decreases from the bottom to the wellhead, with the maximum at the well bottom.

The helical buckling displacement curve for the placement of the tubular is shown in Figure 10.29. As a result of helical buckling, the string is shorter. The column at the bottom is shortened at the maximum, and decreases from the well bottom to the wellhead.

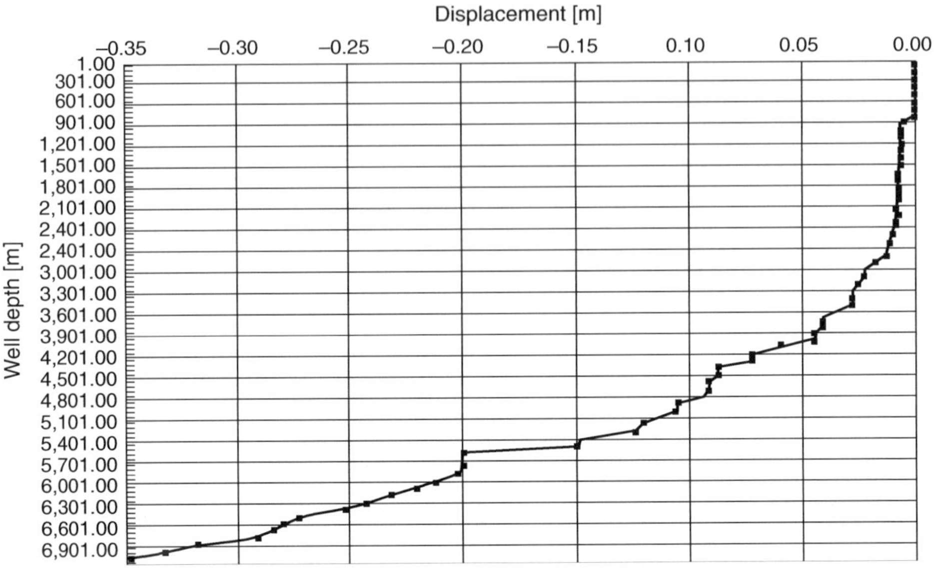

Figure 10.29 Helical buckling displacement distribution for the placement of the tubular.

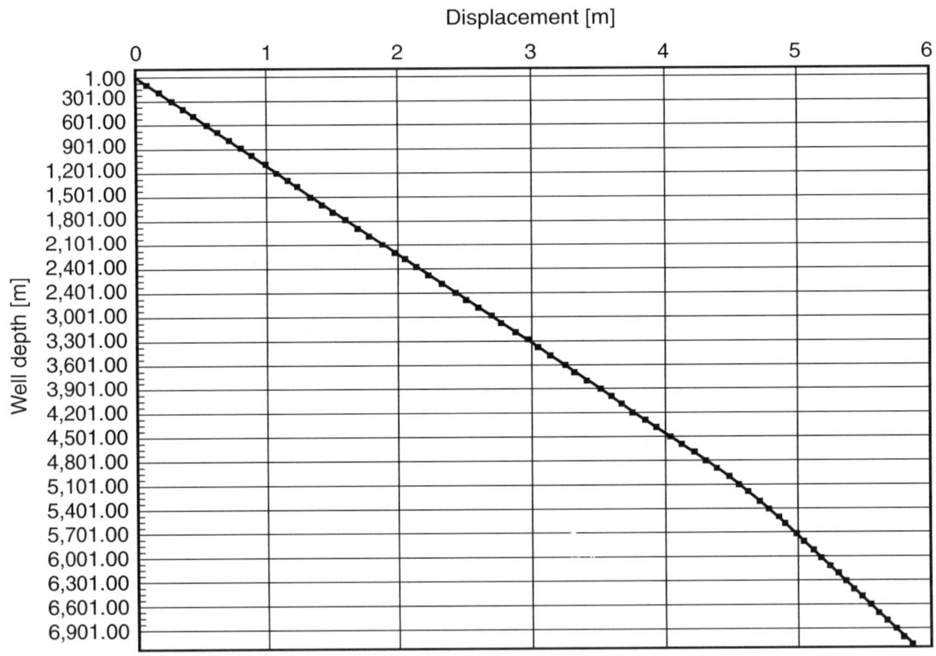

Figure 10.30 Total displacement distribution for the placement of the tubular.

The total displacement curve for the placement of the tubular is shown in Figure 10.30. The total displacement is the sum of all the above deformations, with the largest deformation occurring at the bottom, and the deformation gradually increasing from the top to the bottom.

The axial force distribution curve for the perforation is shown in Figure 10.31. The axial force increases from the bottom to the wellhead, and is a basic linear distribution.

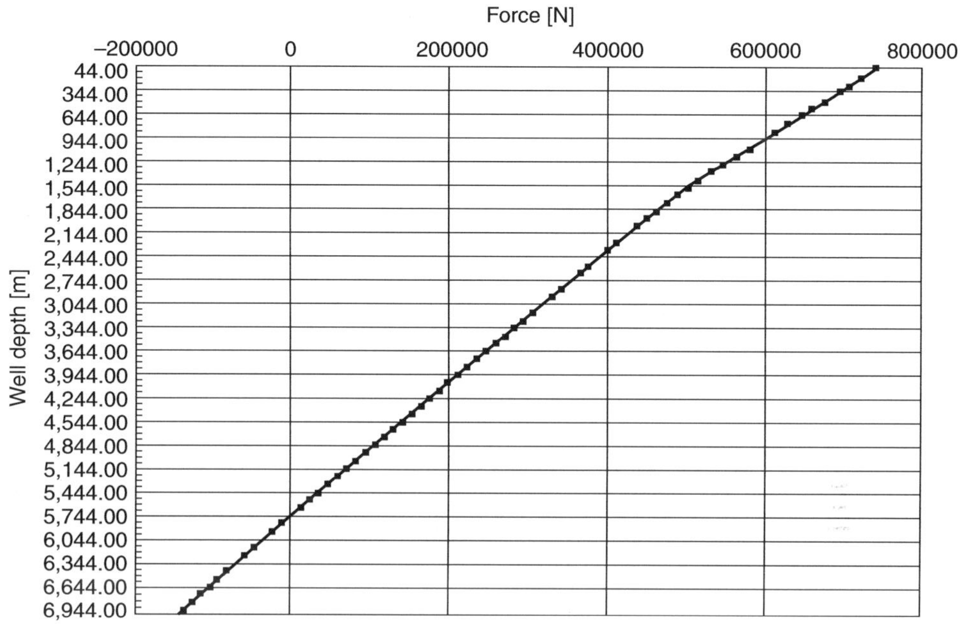

Figure 10.31 Axial force distribution for the perforation.

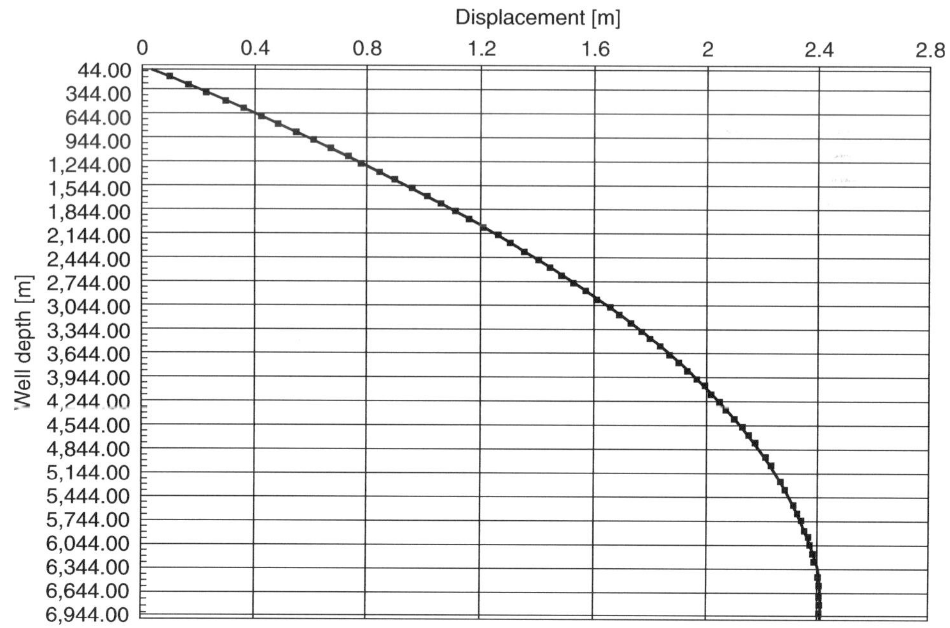

Figure 10.32 Temperature displacement distribution for the perforation.

The temperature displacement curve for the perforation is shown in Figure 10.32. The displacement decreases from the bottom to the wellhead, with the maximum at the well bottom.

The axial displacement curve for the perforation is shown in Figure 10.33. The displacement decreases from the bottom to the wellhead, with the maximum at the well bottom.

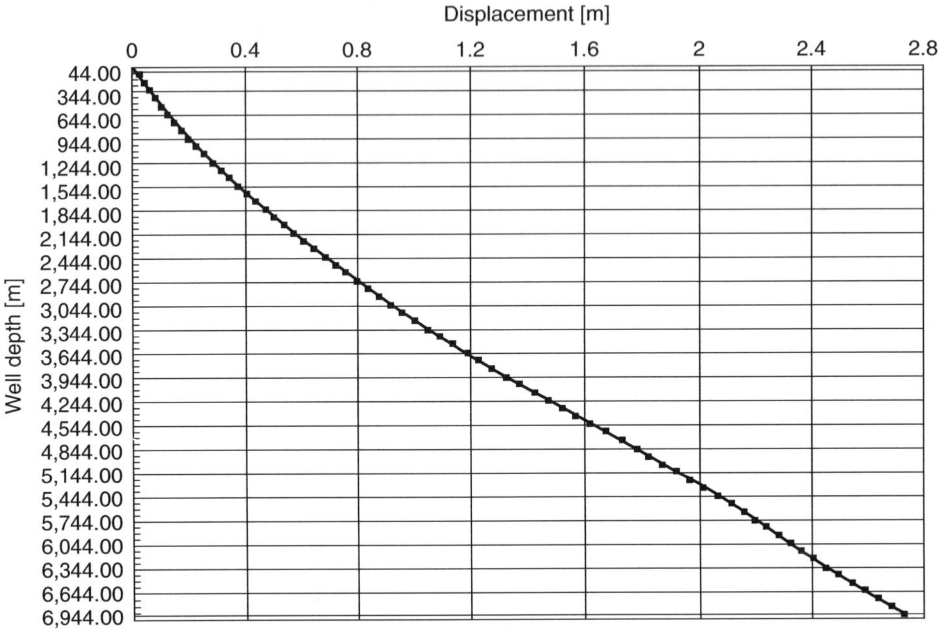

Figure 10.33 Axial displacement distribution for the perforation.

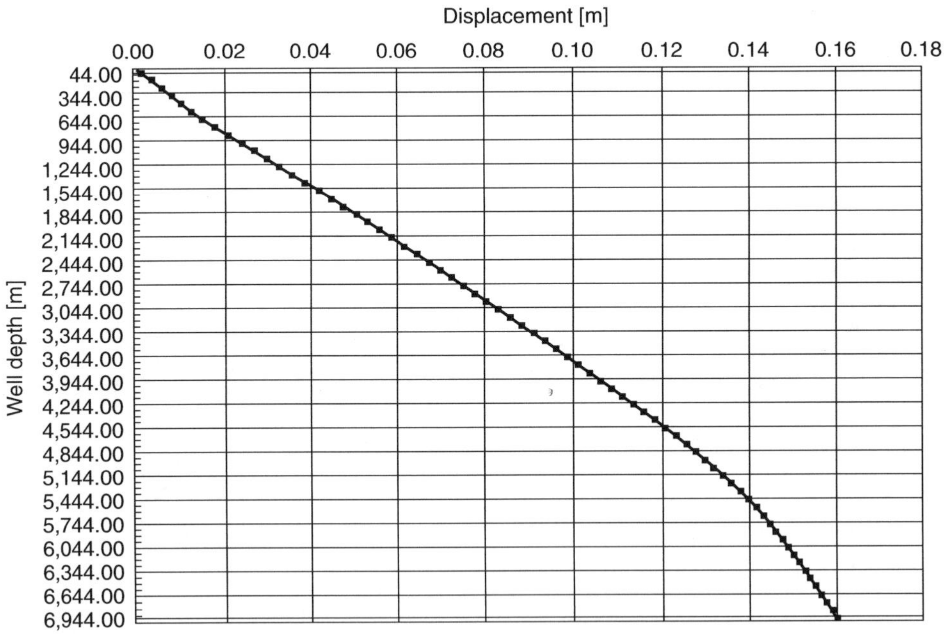

Figure 10.34 Pressure displacement distribution for the perforation.

The pressure displacement curve for the perforation is shown in Figure 10.34. The displacement is very small, and decreases from the bottom to the wellhead, with the maximum at the well bottom.

The helical buckling displacement curve for the perforation is shown in Figure 10.35. As a result of helical buckling, the string is shorter. The column at the bottom is shortened at the maximum, and decreases from the well bottom to the wellhead.

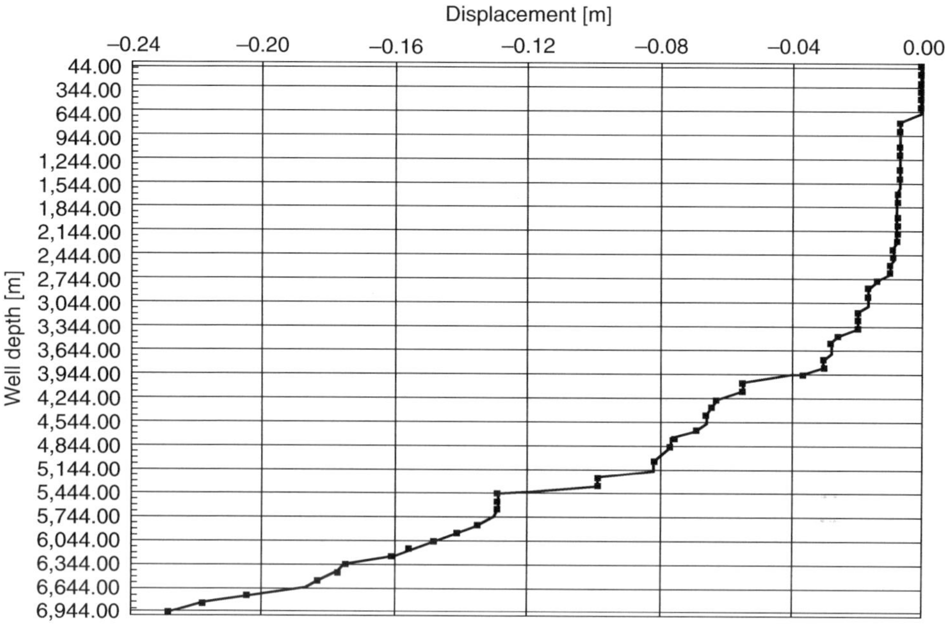

Figure 10.35 Helical buckling displacement distribution for the perforation.

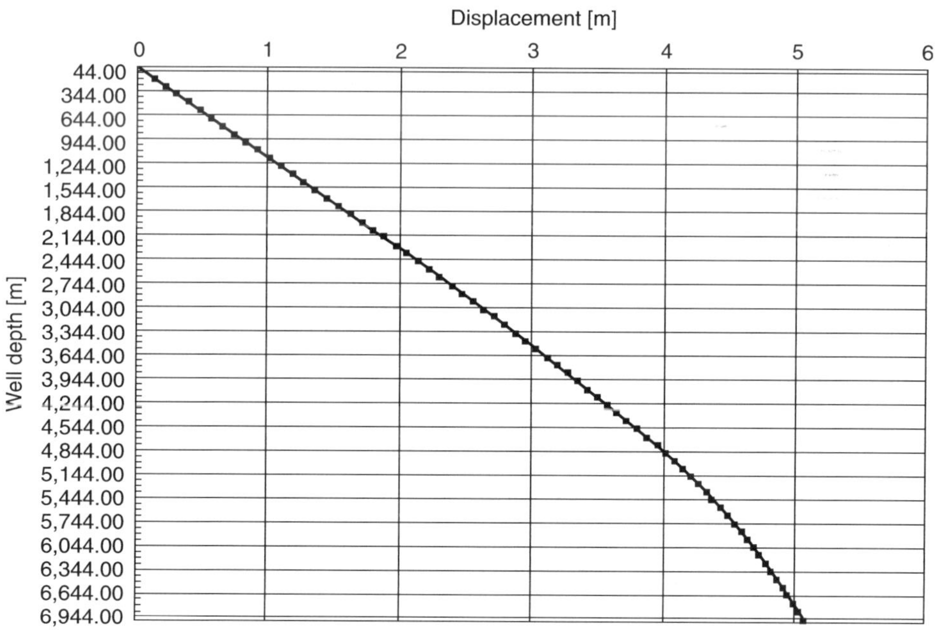

Figure 10.36 Total displacement distribution for the perforation.

The total displacement curve for the perforation is shown in Figure 10.36. The total displacement is the algebraic sum of all the above deformations, with the largest deformation occurring at the bottom, and the deformation gradually increasing from the top to the bottom.

When the original dryness is 0.8, the wellhead temperature is 200°C, the wellhead pressure is 16 MPa, so the dryness curve is as shown in Figure 10.37.

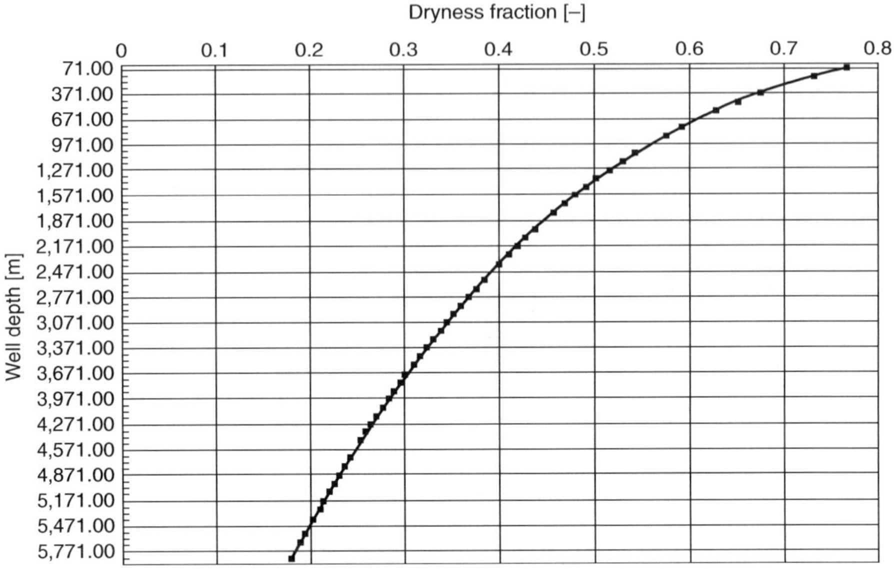

Figure 10.37 Injection (two phase) dryness distribution.

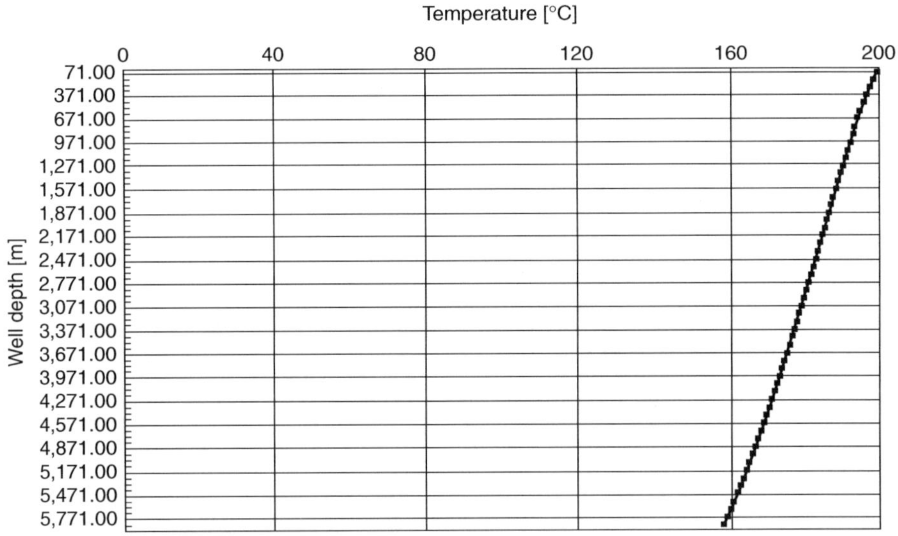

Figure 10.38 Injection (two phase) temperature distribution.

When the original dryness is 0.8, the wellhead temperature is 200°C, the wellhead pressure is 16 MPa, so the temperature curve is as shown in Figure 10.38.

When the original dryness is 0.8, the wellhead temperature is 200°C, the wellhead pressure is 16 MPa, so the pressure curve is as shown in Figure 10.39.

When the original dryness is 0.6, the wellhead temperature is 300°C, the wellhead pressure is 20 MPa, so the dryness curve is as shown in Figure 10.40.

When the original dryness is 0.6, the wellhead temperature is 300°C, the wellhead pressure is 20 MPa, so the temperature curve is as shown in Figure 10.41.

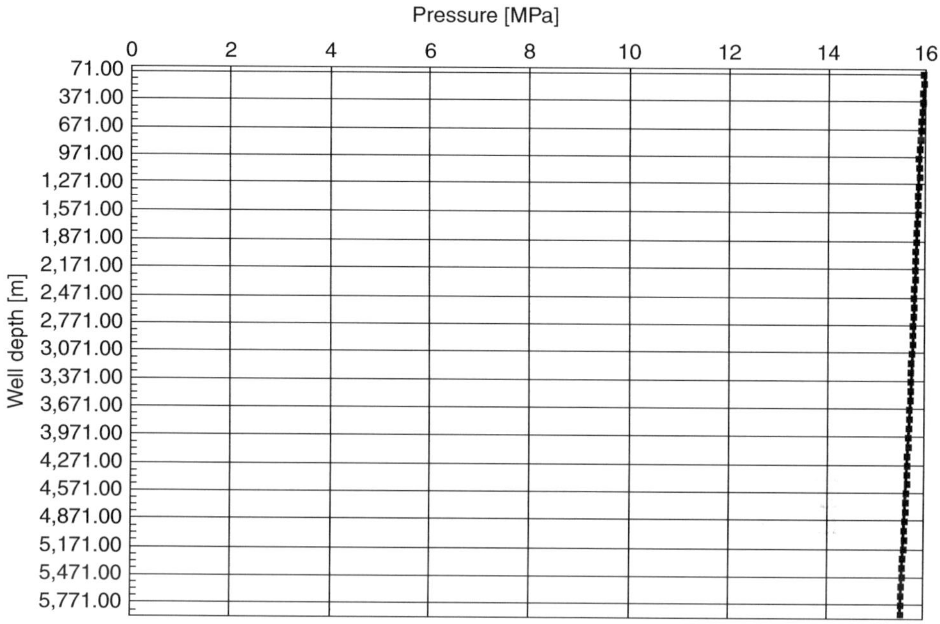

Figure 10.39 Injection (two phase) pressure distribution.

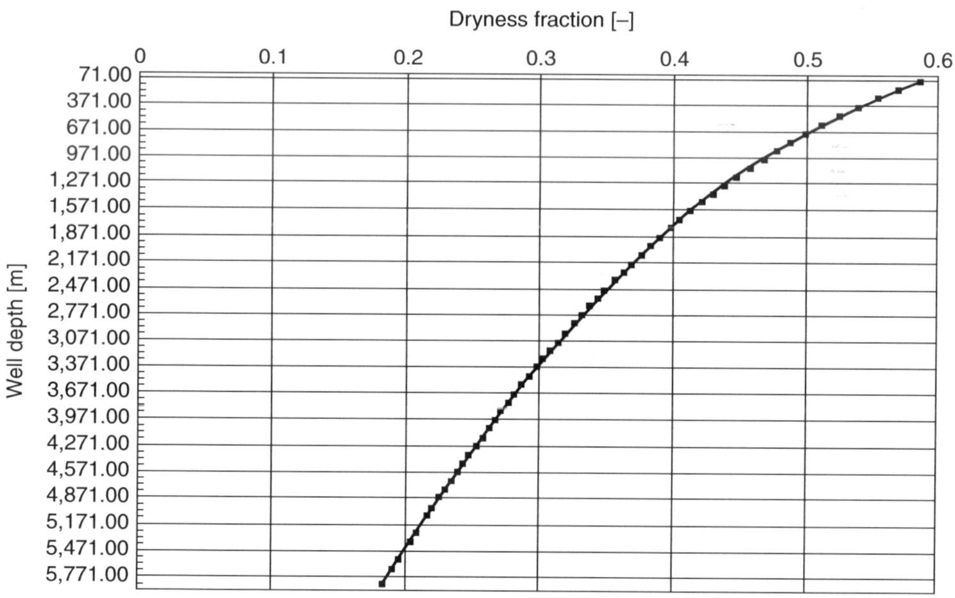

Figure 10.40 Dryness distribution for the injection (two phase).

When the original dryness is 0.6, the wellhead temperature is 300°C, the wellhead pressure is 20 MPa, so the pressure curve is as shown in Figure 10.42.

The temperature change curve with an injection velocity at 500 m depth is as shown in Figure 10.43.

The temperature change curve with an injection velocity at 1000 m depth is as shown in Figure 10.44.

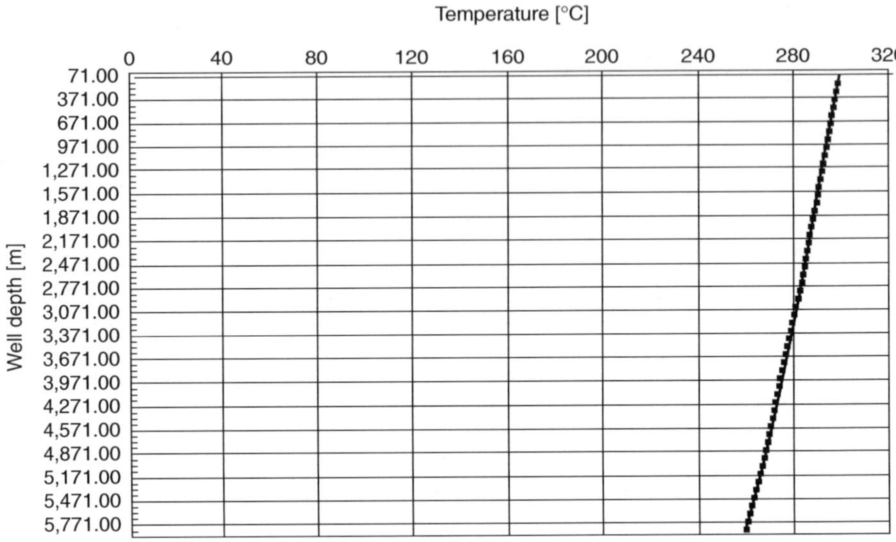

Figure 10.41 Temperature distribution for the injection (two phase).

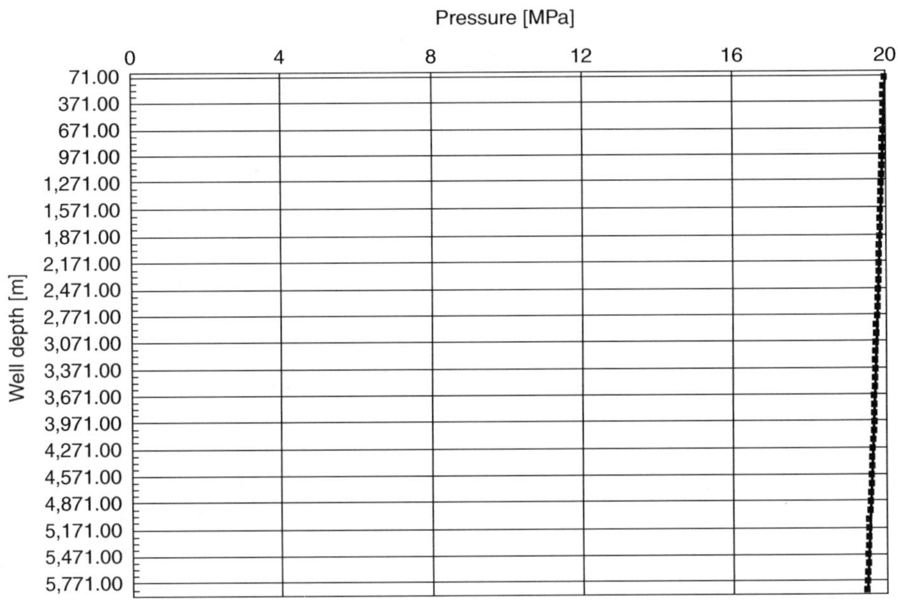

Figure 10.42 Pressure distribution for the injection (two phase).

The temperature change curve with an injection velocity at 1500 m depth is as shown in Figure 10.45.

The temperature change curve with an injection velocity at 6000 m depth is as shown in Figure 10.46.

The temperature change curve with an injection velocity at 6500 m depth is as shown in Figure 10.47 depth.

When the bottom hole pressure is 30 MPa, and the gas production is 400000 m^3/day, the temperature profile is as shown in Figure 10.48.

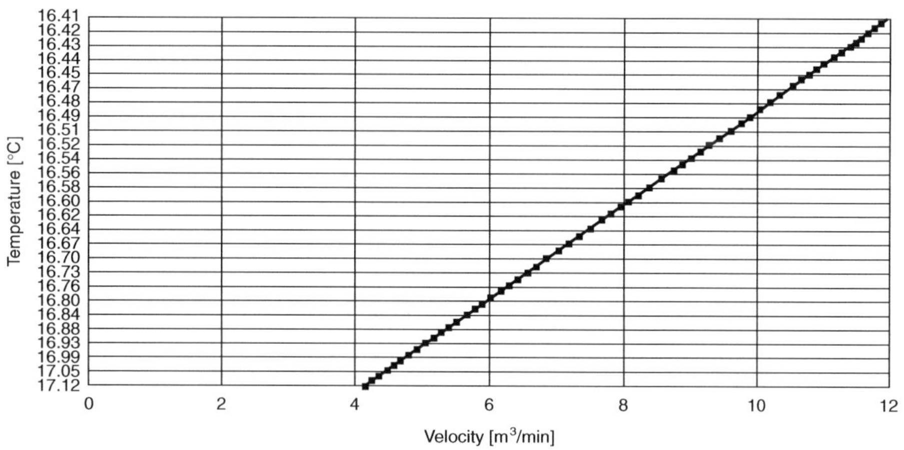

Figure 10.43 Temperature change with the injection velocity (500 m).

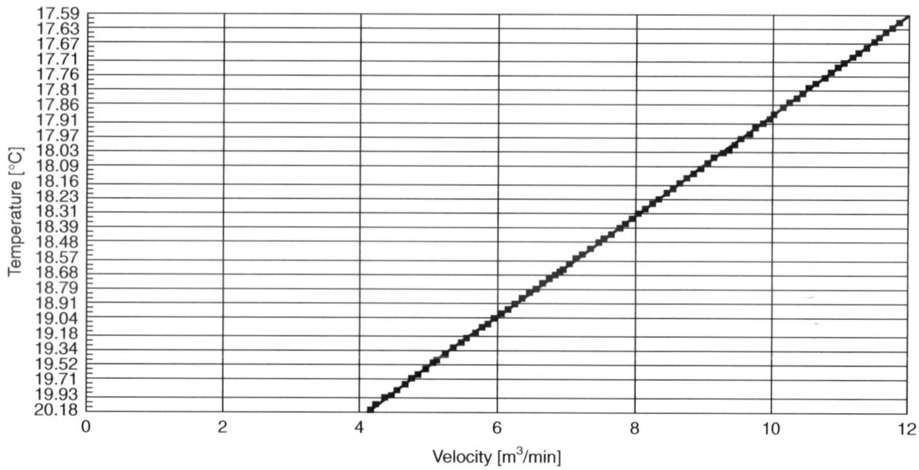

Figure 10.44 Temperature change with injection velocity at 1000 m.

When the bottom hole pressure is 30 MPa, and the gas production is 400000 m^3/day for the production condition, the pressure profile is as shown in Figure 10.49.

When the bottom hole pressure is 40 MPa, and the gas production is 500000 m^3/day for the production condition, the temperature profile is as shown in Figure 10.50.

When the bottom hole pressure is 40 MPa, and the gas production is 500000 m^3/day for the production condition, the pressure profile is as shown in Figure 10.51.

10.1.2 **Calculation according to conditions**

In this section, we show how it is possible to select for a specific condition without having to compute all conditions. In the calculation, click "basic data" in the main interface, and enter the required basic data.

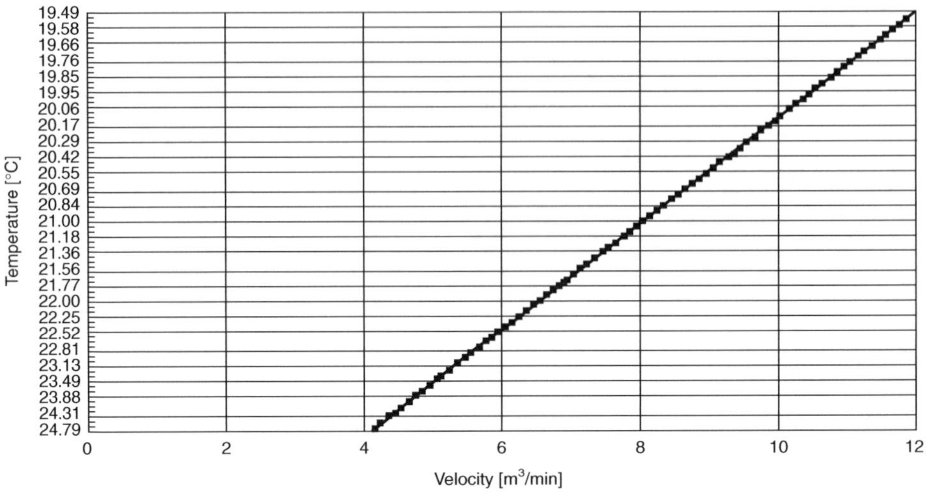

Figure 10.45 Temperature change with injection velocity at 1500 m.

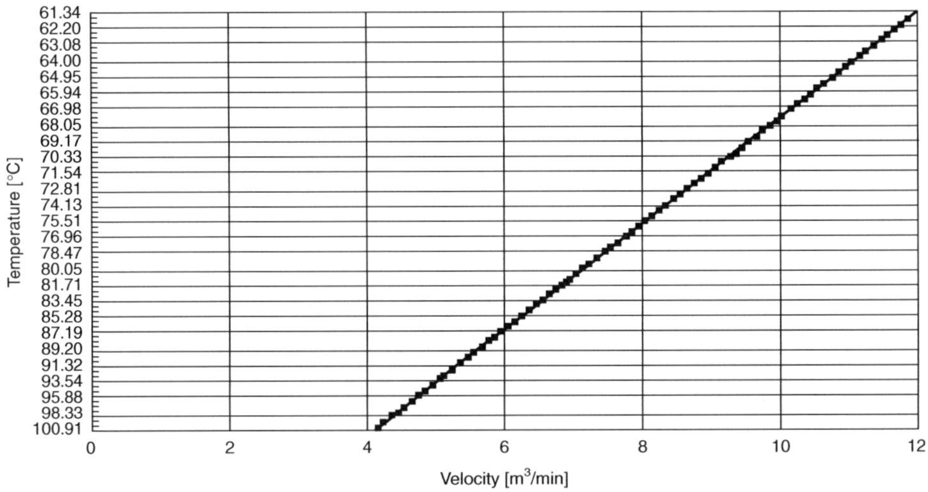

Figure 10.46 Temperature change with injection velocity at 6000 m.

1. **The basic data**

   The Basic data includes well inclination data, formation data, well construction data, pipeline structure placement data, placing data and the packer setting data. This part is the same as the previous calculation of the interface for all conditions. The detailed interface may reference the interface shown earlier.

2. **Select the interface conditions**

   Upon completion of basic data input, "Condition selection" and a pop-up condition selection screen come up where the desired interface for the calculation of the specific operating conditions can be selected. This is shown in Figure 10.52.

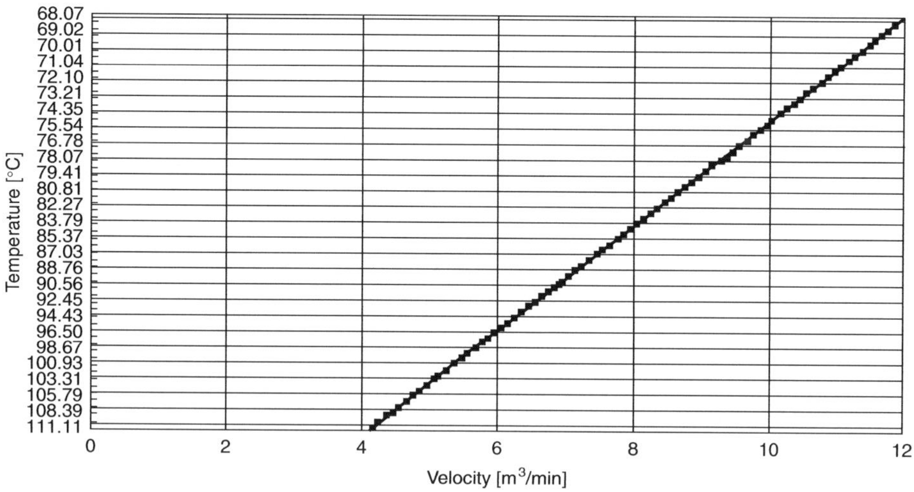

Figure 10.47 Temperature change with injection velocity at 6500 m.

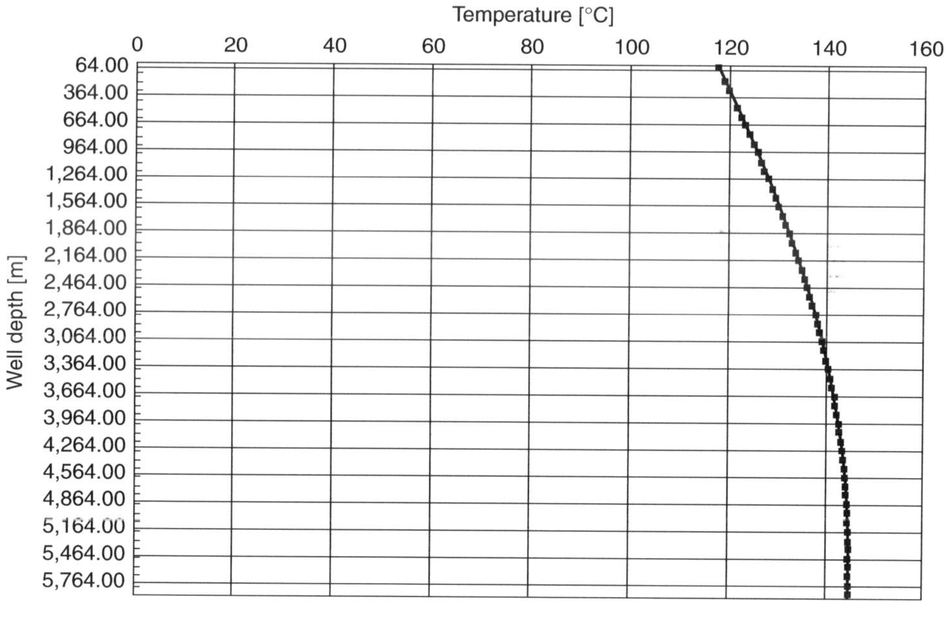

Figure 10.48 Temperature distribution for the production (two phase).

3. **The Results interface**

   Select a condition and click "Next" to move to the screen for the calculation of the interface. The data input interface required for each condition is the same as for the front interface of all calculated conditions. For this, refer to the previous figure.

   Complete the necessary data entry and click "Next" to enter the results interface. Click "Calculate" to calculate the required amount of computation, which is displayed in a list.

   If an output of the results is required, first click "Save" to save the results to the database, and click "Export Report" in the output interface.

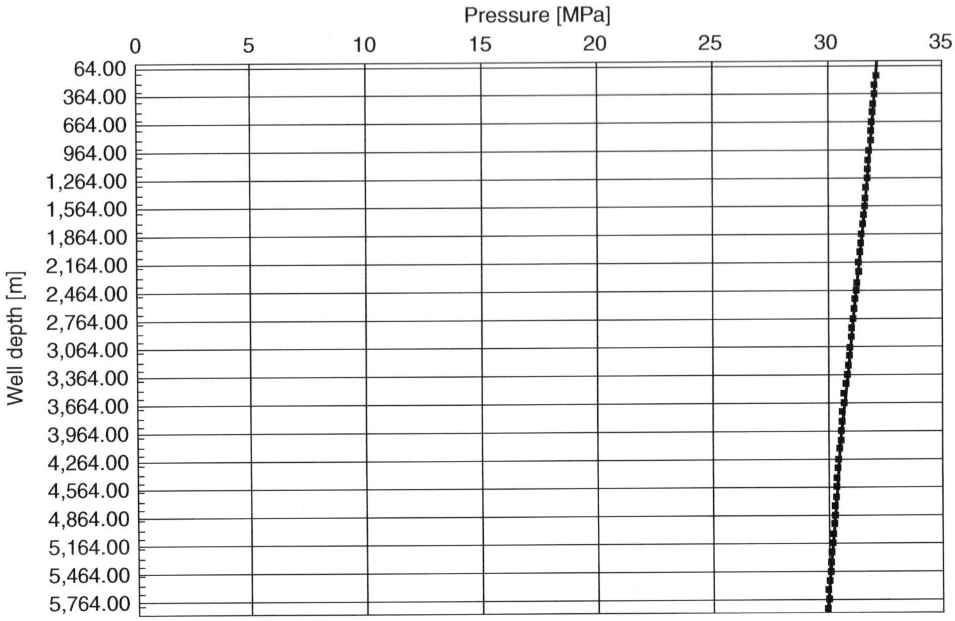

Figure 10.49 Pressure distribution for the production (two phase).

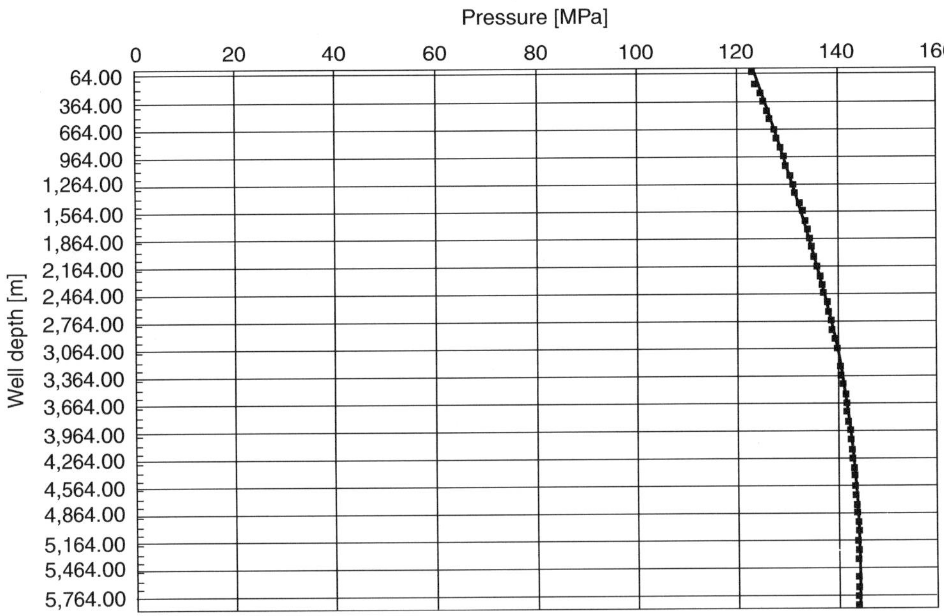

Figure 10.50 Pressure distribution for the production (two phase).

10.2 THE DATABASE

This section, describes the database creation and the maintenance of the tubing and casing. Data for the tubing and the casing can be added, modified and deleted. At the same time as changes are

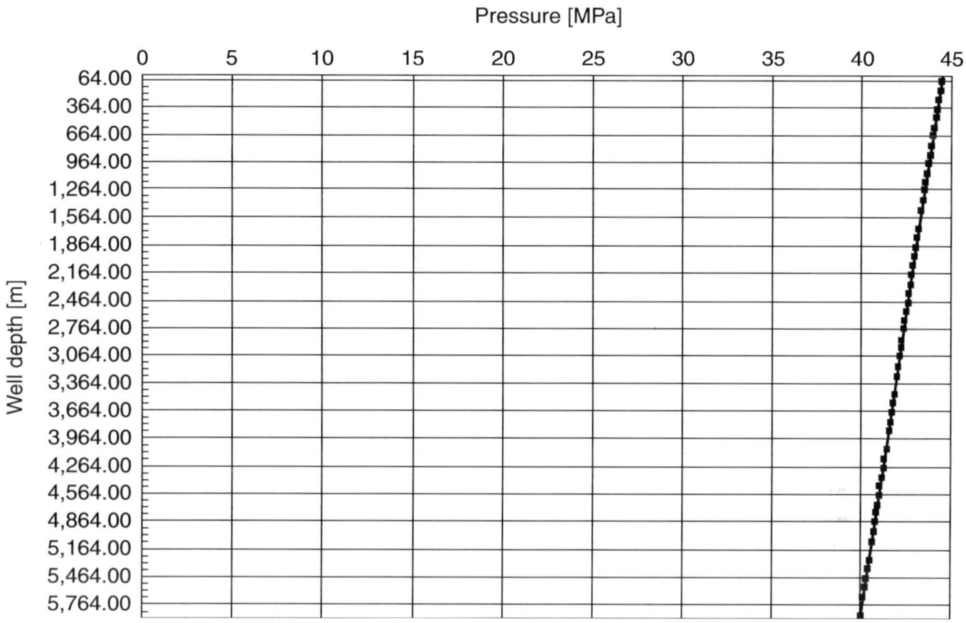

Figure 10.51 Pressure distribution for the production condition (two phase).

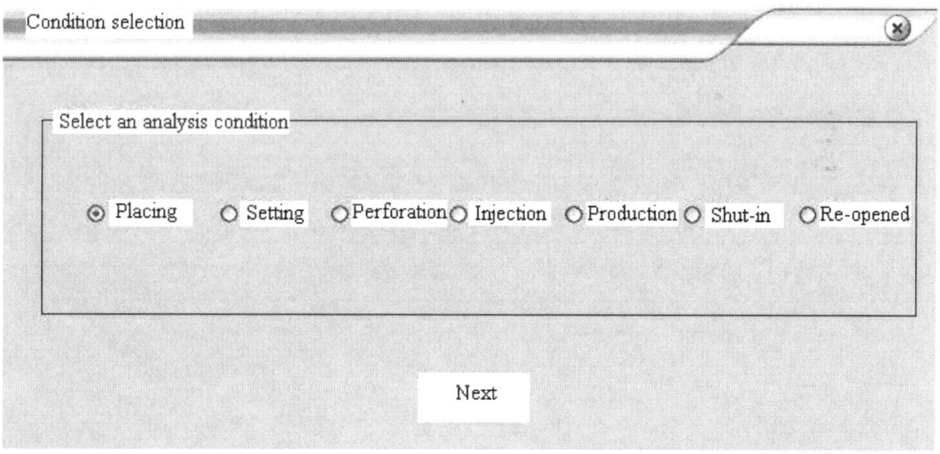

Figure 10.52 Selecting the interface conditions.

being made, the results for the latest changes in the tubing and casing data on the input interface can be seen using the drop-down list.

The database management interface enables the establishment of a pipeline database, which allows for both maintenance and removal. If more data needs to be added, it only needs to be input to the saved list. If there is a need to delete data, delete the corresponding data in the list and save. If it is necessary to modify the data, modify the data in the list and save.

The casing database management interface enables the establishment of a database on casing, maintenance and removal. If more data needs to be added, it only needs to be input to the list and saved. If there is a need to delete data, delete the corresponding data in the list and save. If it is necessary to modify the data, modify the data in the list and save.

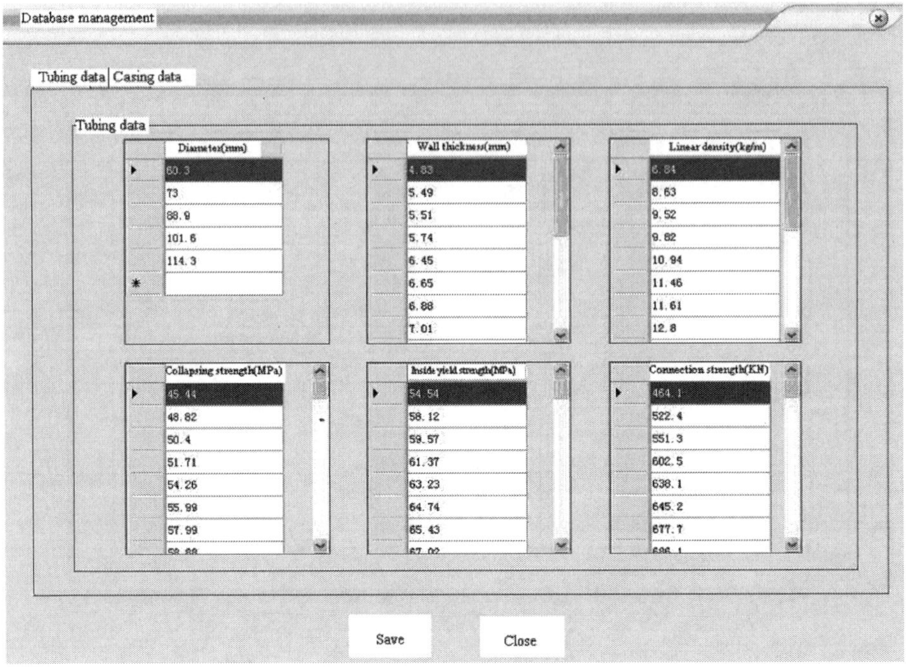

Figure 10.53 Database Management interface.

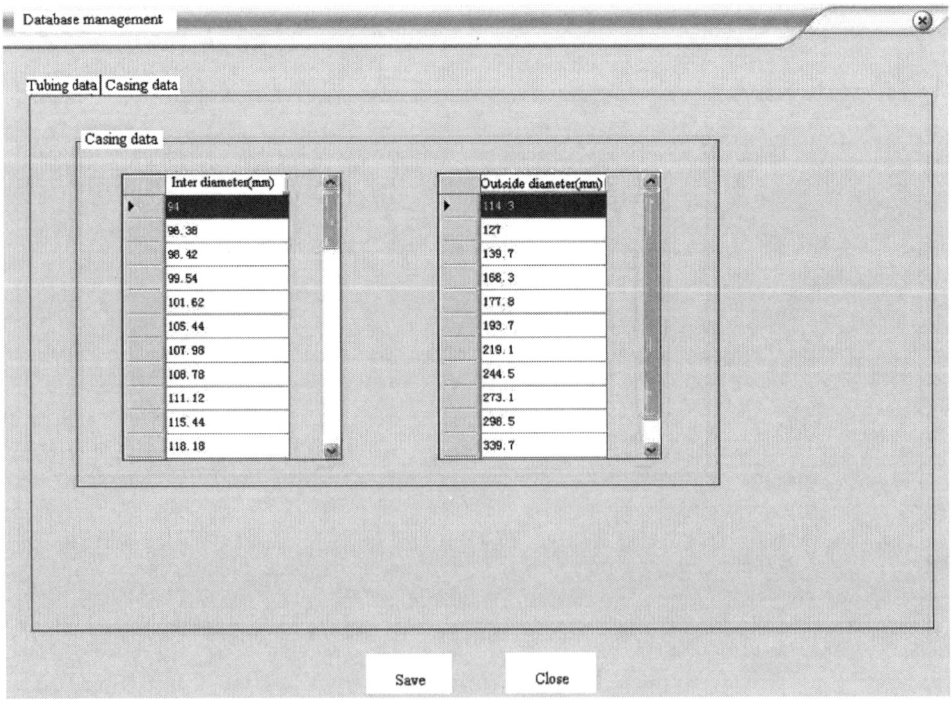

Figure 10.54 Casing database management interface.

References

Abbasbandy, S.: A numerical solution of Blasius equation by adomians decomposition method and comparison with homotopy perturbation method. *Chaos Soliton. Fract.* 31:1 (2007), pp. 257–260.

Açikgöz, M., Franca, F. & Lahey, R.Y. Jr: An experimental study of three-phase flow regimes. *Int. J. Multiphas. Flow* 18:3 (1992), pp. 327–336.

Ah, S.: A comprehensive wellbore steam/water flow model for steam injection and geothermal applications. *Soc. Petrol. Eng. J.* 21:5 (1981), pp. 527–534.

Akin, S.: Mathematical modeling of steam-assisted gravity drainage. *Comput. Geosci.* 32:2 (2006), pp. 240–246.

Akin, S. & Bagci, S.: A laboratory study of single-well steam-assisted gravity drainage process. *J. Petrol. Sci. Eng.* 32:1 (2001), pp. 23–33.

Alekseevskići, D., Gamkrelidze, R., Lychagin, V. & Vinogradov, A.: *Geometry I: Basic ideas and concepts of differential geometry*, vol. 28 of the Encyclopedia of Mathematical Sciences (ed R.V. Gamkrelidze. Springer, Berlin and New York, 1991.

Alves, I.N., Alhanati, F.J.S. & Shoham, O.: A unified model for predicting flowing temperature distribution, in wellbores and pipelines. *SPE Prod. Eng.* 7:4 (1992), pp. 363–367.

Anderson, J.D. Jr: *Computational fluid dynamics*. McGraw-Hill, New York, 1995.

Andreussi, P. & Persen, L.N.: Stratified gas-liquid flow in downwardly inclined pipes. *Int. J. Multiphas. Flow* 13:4 (1987), pp. 565–575.

Angeli, P. & Hewitt, G.F.: Pressure gradient in horizontal liquid–liquid flows. *Int. J. Multiphas. Flow* 24:7 (1999), pp. 1183–1203.

Ansari, A.M., Sylvester, N.D., Sarica, C., Shoham, O. & Brill, J.P.: A comprehensive mechanistic model for upward two-phase flow in wellbores. *SPE Prod. Facil.* 9:2 (1994), pp. 143–151.

Aziz, K., Govier, G.W. & Forarasi, M.: Pressure drop in wells producing oil and gas. *J. Can. Petrol. Technol.* 11:3 (1972), pp. 38–42.

Babadagli, T., Sahin, S., Kalfa, U., Celebioglu, D., Karabakal, U. & Topguder, N.N.: Evaluation of steam injection potential and improving ongoing CO_2 injection of the Bati Raman field, Turkey. *J. Petrol. Sci. Eng.* 68:1 (2009), pp. 107–117.

Bahonar, M., Azaiez, J. & Chen, Z.: Two issues in wellbore heat flow modelling along with the prediction of casing temperature in the steam injection wells. *J. Can. Petrol. Technol.* 50:1 (2011), pp. 43–63.

Barnea, D.: A unified model for predicting flow-pattern transitions for the whole range of pipe inclinations. *Int. J. Multiphas. Flow* 13:1 (1987), pp. 1–12.

Beggs, D.H. & Brill, J.P.: A study of two-phase flow in inclined pipes. *J. Petrol. Technol.* 25:5 (1973), pp. 607–617.

Beirute, R.M.: A circulating and shut-in well-temperature-profile simulator. *J. Petrol. Technol.* 43:9 (1991), pp. 1140–1146.

Birkhoff, G. & Rota, G.-C.: *Ordinary differential equations*. 3rd edn, Wiley, New York, 1978.

Bonizzi, M. & Issa, R.I.: On the simulation of three-phase slug flow in nearly horizontal pipes using the multi-fluid model. *Int. J. Multiphas. Flow* 29:1 (2003), pp. 1719–1747.

Boukadida, T. & LeRoux, A.Y.: A new version of the two-dimensional Lax-Friedrichs scheme. *Math. Comput.* 63:208 (1994), pp. 541–554.

Brauner, N.: The prediction of dispersed flows boundaries in liquid–liquid and gas–liquid systems. *Int. J. Multiphas. Flow* 27:5 (2001), pp. 885–910.

Brauner, N., Moalem Maron, D. & Rovinsky, J.: A two-fluid model for stratified flows with curved interfaces. *Int. J. Multiphas. Flow* 24:6 (1998), pp. 975–1004.

Brauner, N. & Ullmann, A.: Modeling of phase inversion phenomenon in two-phase pipe flows. *Int. J. Multiphas. Flow* 28:7 (2002), pp. 1177–1204.

Brill, J.: Multiphase flow in wells. *J. Petrol. Technol.* 39:1 (1987), pp. 15–21.

Buttazzo, G., Giaquinta, M. & Hildebrandt, S.: *One-dimensional variational problems: an introduction*, vol. 15 of the Oxford Lecture Series in Mathematics. Oxford University Press, 1999.

Cazarez, O., Montoya, D., Vital, A.G. & Bannwart, A.C.: Modeling of three-phase heavy oil–water–gas bubbly flow in upward vertical pipes. *Int. J. Multiphas. Flow* 36:6 (2010), pp. 439–448.

Cazarez-Candia, O. & Vásquez-Cruz, M.A.: Prediction of pressure, temperature, and velocity distribution of two-phase flow in oil wells. *J. Petrol. Sci. Eng.* 46:3 (2005), pp. 195–208.

Chaudhry, A.U.: *Oil well testing handbook*. Gulf Professional Publishing, 2004.

Cheatham, J.B. Jr: Helical postbuckling configuration of a weightless column under the action of m axial load. *Old SPE J.* 24:4 (1984), pp. 467–472.

Chen, K. & Xu, J.: A method for calculating pipe deformation in gas wells. *World J. Modell. Simul.* 4:3 (2008), pp. 188–195.

Chen, Y.-C. & Adnan, S.: Buckling of pipe and tubing constrained inside inclined wells. *Offshore Technology Conference*, Houston, Texas, USA, 1993.

Chierici, G.L., Ciucci, G.M. & Sclocchi, G.: Two-phase vertical flow in oil wells-prediction of pressure drop. *J. Petrol. Technol.* 26:8 (1974), pp. 927–938.

Coulter, D.M. & Bardon, M.F.: Revised equation improves flowing gas temperature prediction. *Oil Gas J.* 9 (1979), pp. 107–108.

Cullender, M.H. & Smith, R.V.: Practical solution of gas-flow equations for wells and pipelines with large temperature gradients. *Trans. AIME* 207 (1956), pp. 281.

Darcy, H.: *Les fontaines publiques de la ville de Dijon*. Paris: Dalmont, 1856.

Dawson, R.: Drill pipe buckling in inclined holes. *J. Petrol. Technol.* 36:10 (1984), pp. 1734–1738.

Deberne, N., Leone, J.F., Duque, A. & Lallemand, A.: A model for calculation of steam injector performance. *Int. J. Multiphas. Flow* 25:5 (1999), pp. 841–855.

Dekker, K. & Verwer, J.G.: *Stability of Runge-Kutta methods for stiff nonlinear differential equations*. Vol. 2 of CWI Monographs, VNorth-Holland. Amsterdam, 1984.

Ding, P. & Yan, X.Z.: Force analysis of high pressure water injection string, *Petrol. Drill. Tech.* 36:5 (2005), p. 23.

Ding, C., Xu, J. & Liu, Y.: Prediction (P, T) distribution in shut-in procedure for deviated wells and sensitive analysis. Sichuan University Research Report, Chengdu, PR China, 2012.

Dou, Y.H. & Zhang, F.X.: Mechanical analysis of well testing down-hole string in deep well with HTHP and its application. *Drill. Product. Technol.* 30:5 (2007), pp. 17–20.

Durrant, A.J. & Thambynayagam, R.K.M.: Wellbore heat transmission and pressure drop for steam/water injection and geothermal production: a simple solution technique. *SPE Reserv. Eng.* 1:2 (1986), pp. 148–162.

Farouq Ali, S.M.: A comprehensive wellbore steam/water flow model for steam injection. *JPT* 5 (1981), pp. 527–534.

Fayers, F.J.: Some theoretical results concerning the displacement of a viscous oil by a hot fluid in a porous medium. *J. Fluid Mech.* 13 (1962), pp. 65–76.

Gao, D.-L. & Gao, B.-K.: A method for calculating tubing behavior in hpht wells. *J. Petrol. Sci. Eng.* 41:1 (2004), pp. 183–188.

Ghorai, S., Suri, V. & Nigam, K.D.P.: Numerical modeling of three-phase stratified flow in pipes. *Chem. Eng. Sci.* 60:23 (2005), pp. 6637–6648.

Gould, T.: Vertical two-phase steam-water flow in geothermal wells. *J. Petrol. Technol.* 26:8 (1974), pp. 833–842.

Gourlay, A.R.: A multistep formulation of the optimized lax-wendroff method for nonlinear hyperbolic systems in two space variables. *Math. Comput.* 22:104 (1968), pp. 715–719.

Griffith, P.: Multiphase flow in pipes. *J. Petrol. Technol.* 36:3 (1984), pp. 361–367.

Grolman, E. & Fortuin, J.M.G.: Gas-liquid flow in slightly inclined pipes. *Chem. Eng. Sci.* 52:24 (1997), pp. 4461–4471.

Hagedorn, A. & Brown, K.: Experimental study of pressure gradients occurring during continuous two-phase flow in small-diameter vertical conduits. *J. Petrol. Technol.* 17:4 (1965), pp. 475–484.

Hagoort, J.: Ramey's wellbore heat transmission revisited. *SPE J.* 9:4 (2004), pp. 465–474.

Hagoort, J.: Prediction of wellbore temperatures in gas production wells. *J. Petrol. Sci. Eng.* 49:1 (2005), pp. 22–36.

Hagoort, J.: An analytical model for predicting the productivity of perforated wells. *J. Petrol. Sci. Eng.* 56:4 (2007), pp. 199–218.

Hammerlindl, D.J.: Movement, forces, and stresses associated with combination tubing strings sealed in packers. *J. Petrol. Technol.* 29:2 (1977), pp. 195–208.

Hammerlindl, D.J.: Packer-to-tubing forces for intermediate packers. *J. Petrol. Technol.* 32:3 (1980), pp. 515–527.

Harlow, F.H. & Welch, J.E.: Numerical calculation of time-dependent viscous incompressible flow of fluid with free surface. *Phys. Fluids* 8:12 (1965), pp. 2182–2189.

Hart, J., Hamersma, P.J. & Fortuin, J.M.H.: Correlations predicting frictional pressure drop and liquid holdup during horizontal gas-liquid pipe flow with a small liquid holdup. *Int. J. Multiphas. Flow* 15:6 (1989), pp. 947–964.

Hasan, A.R. & Kabir, C.S.: Heat transfer during two-phase flow in wellbores; Part I–Formation temperature. In: *SPE Annual Technical Conference and Exhibition*, Dallas, TX, USA, 1991, SPE paper 22866.

Hasan, A.R., Kabir, C.S. & Sarica, C.: *Fluid flow and heat transfer in wellbores*. Society of Petroleum Engineers, 2002.

Hasan, A.R., Kabir, C.S. & Sayarpour, M.: Simplified two-phase flow modeling in wellbores. *J. Petrol. Sci. Eng.* 72:1 (2010), pp. 42–49.

Hasan, A.R., Kabir, C.S. & Wang, X.: Wellbore two-phase flow and heat transfer during transient testing. *SPE Journal* 3:2 (1998), pp. 174–180.

He, X. & Kyllingstad, A.: Helical buckling and lock-up conditions for coiled tubing in curved wells. *SPE Drill. Completion* 10:1 (1995), pp. 10–15.

Hemeida, A.M.: Program calculates pressure gradient in two-phase flow. *Oil Gas J.* 85:10 (1987), pp. 36–38.

Holst, P.H. & Flock, D.L.: Wellbore behaviour during saturated steam injection. *J. Can. Petrol. Technol.* 5:4 (1966), pp. 184–193.

Howell, E.P., Seth, M.S. & Perkins, T.K.: Temperature calculations for wells which are completed through permafrost. In: *Fall Meeting of the Society of Petroleum Engineers of AIME*, San Antonio, TX, USA, 1972.

Huang, N.C. & Pattillo, P.D.: Helical buckling of a tube in an inclined wellbore. *Int. J. Nonlinear Mech.* 35:5 (2000), pp. 911–923.

Hurlburt, E.T. & Hanratty, T.J.: Prediction of the transition from stratified to slug and plug flow for long pipes. *Int. J. Multiphas. Flow* 28:5 (2002), pp. 707–729.

Kabir, C.S., Hasan, A.R., Jordan, D.L. & Wang, X.: A wellbore/reservoir simulator for testing gas wells in high-temperature reservoirs. *SPE Formation Eval.*11:2 (1996), pp. 128–134.

Kaichiro, M. & Ishii, M.: Flow regime transition criteria for upward two-phase flow in vertical tubes. *Int. J. Heat Mass Tran.* 27:5 (1984), pp. 723–737.

Kirkpatrick, C.: Advances in gas-lift technology. *Drill. Prod. Prac.* 14:3 (1959), pp. 24–60.

Lahey, R.T Jr, Açikgöz, M. & Franca, F.: Global volumetric phase fractions in horizontal three-phase flows. *AIChE J.* 38:7 (1992), pp. 1049–1058.

Latsa, M., Assimacopoulos, D., Stamou, A. & Markatos, N.: Two-phase modeling of batch sedimentation. *Appl. Math. Model.* 23:12 (1999), pp. 881–897.

Lauwerier, H.A.: The transport of heat in an oil layer caused by the injection of hot fluid. *Appl. Sci. Res.* 5:2 (1955), pp. 145–150.

Lesem, L.B., Greytok, F., Marotta, F. & McKetta, J. Jr: A method of calculating the distribution of temperature in flowing gas wells. *Trans. Am. Inst. Min. Metall. Pet. Eng.* 210 (1957), pp. 169–176.

Li, X.-F. & Zhuang, X.-Q.: Analysis on the pressure build-up and its recording time after shut-in. *Acta Petrol. Sin.* 23:5 (2002), pp. 110–112.

Li, Z.: Casing cementing with half warm-up for thermal recovery wells. *J. Petrol. Sci. Eng.* 61:2 (2008), pp. 94–98.

Liao, X.-W. & Feng, J.-L.: Pressure-temperature coupling calculation of transient wellbore heat transfer in deep geopressured gas reservoir. *Petrol. Explor. Devel.* 32:1 (2005), pp. 67–69.

Lin, Q., Zhang, L., Lin, Y.-P. & Xie, N.-X.: A new method using wellhead measurement to approximate unsteady-state gas-water two-phase flow in wellbore to calculate inflow performance. *J. Can. Petrol. Technol.* 47:10 (2008), pp. 14–19.

Liu, X., Jungang, L., Qianya, Z., Jinlai, F., Yingli, L. & Jingxin, S.: The analysis and prediction of scale accumulation for water-injection pipelines in the Daqing oilfield. *J. Petrol. Sci. Eng.* 66:3 (2009), pp. 161–164.

Liu, Y., Guo, F. & Xu, J.: Analyzing packer's deformation of tubular for unsetting process in HTHP wells under variable (T, P) fields. *Open Petrol. Eng. J.* 5 (2012), pp. 109–117.

Liu, Y., Xu, J., Wang, S. & Qi, B.: Analyzing axial stress and deformation of tubular for steam injection process in deviated wells based on the varied (T, P) fields. *Sci. World J.* (2013), PMC3791814.

Livescu, S., Durlofsky, L.J. & Aziz, K.: A semianalytical thermal multiphase wellbore-flow model for use in reservoir simulation. *SPE J.* 15:3 (2010), pp. 794–804.

Lubinski, A., Althouse, W.S. & Logan, J.L.: Helical buckling of tubing sealed in packers. *J. Petrol. Technol.* 14:6 (1962), pp. 655–670.

McAdams, W.H.: *Heat transmission*. McGraw-Hill, 1954.

Markatos, N.C. & Kirkcaldy, D.: Analysis and computation of three-dimensional, transient flow and combustion through granulated propellants. *Int. J. Heat Mass Tran.* 26:7 (1983), pp. 1037–1053.

Markatos, N.C. & Pericleous, K.A.: Laminar and turbulent natural convection in an enclosed cavity. *Int. J. Heat Mass Tran.* 27:5 (1984), pp. 755–772.

Markatos, N.C. & Singhal, A.K.: Numerical analysis of one-dimensional, two-phase flow in a vertical cylindrical passage. *Adv. Eng. Softw.* 4:3 (1982), pp. 99–106.

Marx, J.W. & Langenheim, R.H.: Reservoir heating by hot fluid injection. *Trans. AIME* 216 (1959), pp. 312–315.

McCann, R.C. & Suryanarayana, P.V.R.: Experimental study of curvature and frictional effects on buckling. In: *Offshore Technology Conference*, Houston, TX, USA, 1994.

Meriam, J.L. & Kraige, L.G.: *Engineering mechanics: dynamics*, vol. 2. Wiley, 2012.

Mingchang, Z., Zili, G., Shuquang, Z. & Jianhua, M.: Highest shut-in pressure allowed by gas well and some well-control problems in drilling site. *Natur. Gas Indus.* 25:11 (2005), p. 45.

Miska, S. & Cunha, J.C.: An analysis of helical buckling of tubulars subjected to axial and torsional loading in inclined wellbores. In: *SPE Production Operations Symposium*, Oklahoma City, OK, USA, 1995.

Mitchell, R.F.: New concepts for helical buckling. *SPE Drill. Eng.* 3:3 (1988), pp. 303–310.

Mitchell, R.F.: Comprehensive analysis of buckling with friction. *SPE Drill. Completion* 11:3 (1996), pp. 178–184.

Mitchell, R.F.: Effects of well deviation on helical buckling. *SPE Drill. Completion* 12:1 (1997), pp. 63–70.

Mitchell, R.F.: The effect of friction on initial buckling of tubing and flowlines. In: *IADC/SPE Drilling Conference*, 2006.

Moss, J.T. & White, P.D.: How to calculate temperature profiles in a water injection well. *Oil Gas J.* 57:11 (1959), pp. 174.

Mukherjee, H. & Brill, J.P.: Pressure drop correlations for inclined two-phase flow. *J. Energy Resour. Technol.* 107:4 (1985), pp. 549–554.

Nädler, M. & Mewes, D.: Flow induced emulsification in the flow of two immiscible liquids in horizontal pipes. *Int. J. Multiphas. Flow* 23:1 (1997), pp. 55–68.

Nenes, A., Assimacopoulos, D., Markatos, N. & Mitsoulis, E.: Simulation of airlift pumps for deep water wells. *Can. J. Chem. Eng.* 74:4 (1996), pp. 448–456.

Oglesby, K.D.: *An experimental study on the effects of oil viscosity, mixture velocity and water fraction on horizontal oil-water flow*. MSc Thesis, University of Tulsa, Tusla, OK, USA, 1979.

Orkiszewski, J.: Predicting two-phase pressure drops in vertical pipe. *J. Petrol. Technol.* 19:6 (1967), pp. 829–838.

Ouyang, L.-B., Arbabi, S. & Aziz, K.: General wellbore flow model for horizontal, vertical, and slanted well completions. *SPE J.* 3:2 (1998), pp. 124–133.

Ouyang, L.-B. & Aziz, K.: A mechanistic model for gas-liquid flow in pipes with radial influx or outflux. In: *SPE Annual Technical Conference and Exhibition*, Houston, TX, USA, 1999.

Ouyang, L.-B. & Aziz, K.: A homogeneous model for gas–liquid flow in horizontal wells. *J. Petrol. Sci. Eng.* 27:3 (2000), pp. 119–128.

Ouyang, L.-B. & Aziz, K.: Transient gas–liquid two-phase flow in pipes with radial influx or efflux. *J. Petrol. Sci. Eng.* 30:3 (2001), pp. 167–179.

Pacheco, E.F. & Farouq Ali, S.M.: Wellbore heat losses and pressure drop in steam injection. *J. Petrol. Technol.* 24:2 (1972), pp. 139–144.

Pan, L., Jayanti, S. & Hewitt, G.F.: Flow patterns, phase inversion and pressure gradient in air-oil-water flow in a horizontal pipe. *Proceedings of the ICMF'95*, Kyoto, Japan, 1995.

Paslay, P.R. & Bogy, D.B.: The stability of a circular rod laterally constrained to be in contact with an inclined circular cylinder. *J. Appl. Mech.* 31 (1964), pp. 605–610.

Petalas, N. & Aziz, K.: Development and testing of a new mechanistic model for multiphase flow in pipes. In: *The 1996 ASME Fluids Engineering Division Summer Meeting*, Part 1(of 2), San Diego, CA, USA, 1996, pp. 153–159.

Poettman, F.H. & Carpenter, P.G.: The multiphase flow of gas, oil, and water through vertical flow strings with application to the design of gas-lift installations. *Drilling and production practice*, 1952, API, 1952, pp. 257–317.

Qui, W., Miska, S. & Volk, L.: Drill pipe/coiled tubing buckling analysis in a hole of constant curvature. In: *SPE Permian Basin Oil and Gas Recovery Conference*, Midland, TX, USA, 1998.

Ramazanov, A.Sh. & Nagimov, V.M.: Analytical model for the calculation of temperature distribution in the oil reservoir during unsteady fluid inflow. Electronic scientific journal *Oil and Gas Business* 1 (2007).

Ramey, H. Jr: Wellbore heat transmission. *J. Petrol. Technol.* 14:4 (1962), pp. 427–435.

Rzasa, M.J. & Katz, D.L.: Calculation of static pressure gradients in gas wells. *Trans. AIME* 160:2 (1945), pp. 100–113.

Sagar, R., Doty, D.R. & Schmidt, Z.: Predicting temperature profiles in a flowing well. *SPE Prod. Eng.* 6:4 (1991), pp. 441–448.

Satter, A.: Heat losses during flow of steam down a wellbore. *J. Petrol. Technol.* 17:7 (1965), pp. 845–851.

Sharma, Y., Scoggins, M.W. Jr, Shoham, O. & Brill, J.P.: Simulation of transient two-phase flow in pipelines. *J. Energy Resour. Technol.* 108:3 (1986), pp. 202–206.

Shi, H.: *A study of oil-water flows in large diameter horizontal pipelines*. PhD Thesis, Ohio University, Athens, OH, USA, 2001.

Shi, H., Cai, J.-C. & Jepson, W.P.: The effect of surfactants on flow characteristics in oil/water flows in large diameter horizontal pipelines. In: *BHR Group Conference Series Publication*, vol. 35. Professional Engineering Publishing, Bury St. Edmunds, UK, 1999, 181–200.

Shi, H., Holmes, J.A., Diaz, L.R., Durlofsky, L.J. & Aziz, K.: Drift-flux parameters for three-phase steady-state flow in wellbores. In: *SPE Annual Technical Conference and Exhibition*, Houston, TX, USA, 2004.

Shiu, K. & Beggs, H.: Predicting temperatures in flowing oil wells. *J. Energy Resour. Technol.* 102 (1980), pp. 2–11.

Su, Z. & Gudmundsson, J.S: Pressure drop in perforated pipes: experiments and analysis. In: *SPE Asia Pacific Oil and Gas Conference*, Melbourne, VIC, Australia, 1994, Paper SPE 28800, pp. 563–574.

Sukkar, Y. & Cornell, D.: Direct calculation of bottom-hole pressures in natural gas wells. *Trans. AIME* 204 (1955), pp. 43–48.

Taitel, Y., Barnea, D. & Brill, J.P.: Stratified three phase flow in pipes. *Int. J. Multiphas. Flow* 21:1 (1995), pp. 53–60.

Taitel, Y. & Dukler, A.E.: A model for predicting flow regime transitions in horizontal and near horizontal gas-liquid flow. *AIChE J.* 22:1 (2004), pp. 47–55.

Taitel, Y., Shoham, O. & Brill, J.P.: Simplified transient solution and simulation of two-phase flow in pipelines. *Chem. Eng. Sci.* 44:6 (1989), pp. 1353–1359.

Tao, Z., Xu, J., Wu, Z., Wang, X. &. Song, J.: Predicting on distribution of temperature, pressure, velocity, and density of gas liquid two-phase transient flow in high temperature-high pressure wells. *Petrol. Sci. Technol.* 31:2 (2013), pp. 148–163.

Tiyao, Z., Linsong, C., Chunbai, H.E., Zhanxi, P. & Fengjun, Z.: Calculation model of on-way parameters and heating radius in a superheated steam injection wellbore. *Petrol. Explor. Devel.* 37:1 (2010), pp. 83–88.

Tortike, W.: Saturated-steam-property functional correlations for fully implicit thermal reservoir simulation. *SPE Reserv. Eng.* 4:4 (1989), pp. 471–474.

Trallero, J.L.: *Oil-water flow patterns in horizontal pipes*. PhD Thesis, University of Tulsa, Tulsa, OK, 1995.

Trallero, J.L., Sarica, C. & Brill, J.P.: A study of oil-water flow patterns in horizontal pipes. *SPE Prod. Facil.* 12:3 (1997), pp. 165–172.

Vielma, M., Atmaca, S., Sarica, C. & Zhang, H.-Q.: Characterization of oil/water flows in horizontal pipes. In: *SPE Annual Technical Conference and Exhibition*, Anaheim, CA, USA, 2007.

Wang, J., Macfadyen, R., & Zhang, Y.: Well bore heat loss-options and challenges for steam injector of thermal EOR project in Oman. In: *SPE EOR Conference at Oil & Gas West Asia*, Muscat, Oman, 2010, SPE 129224.

Wang, L.: A new algorithm for solving classical Blasius equation. *Appl. Math. Comput.* 157:1 (2004), pp. 1–9.

Wang, Y. & Dusseault, M.B.: A coupled conductive–convective thermo-poroelastic solution and implications for wellbore stability. *J. Petrol. Sci. Eng.* 38:3 (2003), pp. 187–198.

White, F.M.: *Fluid mechanics*. WCB/McGraw-Hill, New York, 1999.

Willhite, G.P.: Over-all heat transfer coefficients in steam and hot water injection wells. *J. Petrol. Technol.* 19:5 (1967), pp. 607–615.

Willhite, G.P. & Dietrich, W.K.: Design criteria for completion of steam injection wells. *J. Petrol. Technol.* 19:1 (1967), pp. 15–21.

Wolcott, J., Schenewerk, P., Berzins, T. & Karim, F.: A parametric investigation of the cyclic CO_2 injection process. *J. Petrol. Sci. Eng.* 14:1 (1995), pp. 35–44.

Woods, G.S., Spedding, P.L., Watterson, J.K. & Raghunathan, R.S.: Three-phase oil/water/air vertical flow. *Chem. Eng. Res. Des.* 76:5 (1998), pp. 571–584.
Wu, J. & Juvkam-Wold, H.C.: Helical buckling of pipes in extended reach and horizontal wells – Part 2: Frictional drag analysis. *J. Energy Resour. Technol.* 115:3 (1993), pp. 190–195.
Wu, J. & Juvkam-Wold, H.C.: Buckling and lockup of tubulars in inclined wellbores. *J. Energy Resour. Technol.* 117:3 (1995), pp. 208–213.
Wu, Z., Xu, J., Wang, S., Qi, B., Chen, K., Li, X. & Zhao, X.: Prediction of the dryness fraction of gas, pressure, and temperature in high temperature high pressure injection wells. *Petrol. Sci. Technol.* 30:7 (2012), pp. 720–736.
Wu, Z., Xu, J., Wang, X., Chen, K., Li, X. & Zhao, X.: Predicting temperature and pressure in hightemperature–high-pressure gas wells. *Petrol. Sci. Technol.* 29:2 (2011), pp. 132–148.
Xiao, J.J., Shonham, O. & Brill, J.P.: A comprehensive mechanistic model for two-phase flow in pipelines. In: *SPE Annual Technical Conference and Exhibition*, New Orleans, TX, 1990.
Xiao, Z.: The calculation of oil temperature in a well. *SPE* 17125, 1987, pp. 1–14.
Xu, J., Ding, C., Wang, S. & Qi, B.: The prediction of pressure, temperature, velocity, and density of two-phase flow in shut-in procedures for the HTHP gas wells. *Petrol. Sci. Technol.* 32:3 (2014), 335–344.
Xu, J., Ding, C., Wu, Z., Wang, S. & Qi, B.: Prediction pressure and temperature in shut-in procedures for HTHP deviated wells. *Petrol. Sci. Technol.* 31:21 (2013), pp. 2258–2271.
Xu, J., Liu, Y., Wang, S. & Qi, B.: Numerical modelling of steam quality in deviated wells with variable (T, P) fields. *Chem. Eng. Sci.* 84 (2012a), pp. 242–254.
Xu, J., Liu, Y., Wang, S. & Qi, B.: Prediction of pressure, temperature, holdup, velocity, and density distribution for steady-state bubbly gas three-phase flow in high-temperature–high-pressure (HTHP) wells. *Ind. Eng. Chem. Res.* 51:19 (2012b), pp. 6537–6562.
Xu, J., Liu, Y., Wu, Z., Wang, S. & Qi, B.: Predicting on distribution of pressure, temperature, velocity, density of three-phase bubbly flow in HTHP wells. *Petrol. Sci. Technol.* 32 (2014), pp. 1383–1391.
Xu, J., Luo, M., Hu, J., Wang, S., Qi, B. & Qiao, Z.: A direct Eulerian grp scheme for the prediction of gas-liquid two-phase flow in HTHP transient wells. *Abstr. Appl. Anal.* 2013 (2013a), Article ID 171732.
Xu, J., Luo, M., Wu, Z., Wang, S., Qi, B, & Qiao, Z.: Pressure and temperature prediction of transient flow in HTHP injection wells by Lax-Friedrichs method. *Petrol. Sci. Technol.* 31:9 (2013b), pp. 960–976.
Xu, J., Tao, Z., Wang, S. & Qiao, Z.: Predicting the distribution of pressure, temperature, velocity, and density of transient flow in HTHP production wells by the finite difference method. *Petrol. Sci. Technol.* 31:23 (2013c), pp. 2500–2508.
Xu, J. & Wu, Z.: A new mathematical model for the force analysis of tubular string in HTHP wells. *Appl. Comput. Math.* 11 (2012a), pp. 110–136.
Xu, J., Wu, Z., Wang, S. & Qi, B.: The prediction of distribution of temperature, pressure, density, and velocity in high-temperature–high-pressure gas wells. *Petrol. Sci. Technol.* 29:16 (2011), pp. 1705–1721.
Xu, J., Yao, L., Wu, Z., Wang, S. & Qi, B.: Predicting the dryness fraction of gas, pressure, and temperature for steam injection based on unsteady heat transmission. *Petrol. Sci. Technol.* 30:19 (2012c), pp. 1893–1906.
Yang, J. & Xu, J.: Modelling and simulation on shut-in wellbore and programming realization of natural gas well. *World J. Modell. Simul.* 4:3 (2008), pp. 172–181.
Yao, L., Xu, J., Wu, Z., Wang, S. & Qi, B.: Force analysis of the hydraulic packer in high temperature-high pressure deviated wells. *Sichuan University Research Report*, Chengdu, China, 2011.
Zhang, H.Q. & Sarica, C.: Unified modeling of gas/oil/water pipe flow-basic approaches and preliminary validation. In: *SPE Annual Technical Conference and Exhibition* 2005, Dallas, TX, USA, 2005, pp. 1–9.
Zhang, H.Q, Wang, Q., Sarica, C. & Brill, J.: Unified model for gas-liquid pipe flow via slug dynamics: Part 1: Model development. *J. Energy Resour. Technol.* 125:4 (2003), pp. 266–273.
Zhao, X. & Xu, J.: Numerical simulation of temperature and pressure distribution in producing gas well. *World J. Modell. Simul.* 4:2 (2008), pp. 94–103.

Appendix

Appendix 1: The process of Chapter 4

```
using System; using System.Collections.Generic;
using System.ComponentModel;
using System.Data;
using System.Drawing;
using System.Text;
using System.Windows.Forms;
using System.Data.OleDb;
using System.IO;
using System.Data.SqlClient;
namespace {
public partial class Xiaguanzhujieguo : Form
{
public static Xiaguanzhujieguo xiaguanzhujieguo = null;
// public Xuanzeshengchanwendu xuanzeshengchanwendu;
public Xiaguanzhubaobiao xiaguanzhubaobiao;
public Xiaguanzhujieguo()
{
InitializeComponent();
xiaguanzhujieguo = this;
} //
public double buchang;
public int rows3;
public int colums3;
public int n;
public double jingshen;
public double[] faibianhua = new double[100000];
public double[] kuosaibianhua = new double[100000];
public int[] jiange3 = new int[10000];
public double[] fai = new double[100000];
public double[] h = new double[100000];

//
public int rows1;
public int colums1;
public int[] jiange1 = new int[100000];
public double[] mizhong = new double[100000];
public double[] repengzhang = new double[100000];
public double[] tanxingmoliang = new double[100000];
public double[] bosongbi = new double[100000];
public double[] waijing = new double[100000];
public double[] neijing = new double[100000];
public double[] Ao = new double[100000];
public double[] Ai = new double[100000];
public double[] guanxingju = new double[100000];
public double[] qufujixian = new double[100000];
```

```
public double[] kangjiqiangdu = new double[10000];
public double[] neiqufuqiangdu = new double[10000];
public double[] lianjieqiangdu = new double[10000];

  //
public int rows2;
public int colums2;
public int[] jiange2 = new int[100000];
public double[] tongjing = new double[100000];
public double[] rc = new double[100000];
public double[] youxiajianxi = new double[100000];
public double[] Ah = new double[100000];

  //
public double zhouxiangkulun;
public double jingkouwendu;
public double ditiwendu;
public double dicengdaorexishu;
public double dicengrekuosanxishu;

  //
public double pin;
public double pout;

  //
public int zuofengfangshi;
public double jingkouyali;
public double linjieyali;
public double fenggeqi;

  //
public double Pshe;

  //
public double Pnei;
public double Pwai;
public double Q;
public double Kh;
public double time;
public double K;

  //
public double Tpc;
public double Ppc;
public double jingdiyali;
public double rg;
public double Mg;
public double Cci;
public double Cto;
public double qitidaorexishu;
public double Qgsc;
public double shuinihuandaorexishu;
public double shengchanshijian;
```

```
//
public double Tpc6;
public double Ppc6;
public double jingkouyalig;

//
public double shangtili;
public double jiefengyali;
public int shangtifangshi;

// Ftaoe
public void method(int x, int y, double buchang, double faibianhua, double kuosaibianhua,
double[] Ao, double[] Ai, double[] Ah, ref double[] Ftaoe, double[] qe, double[] fai,
double[] youxiaojianxi, double[] tanxingmoliang, double[] guanxingju, ref double[] fn,
ref double[] beita, ref string[] zhuangtai, ref double[] wuyinciyali,
ref double[] zhengyali, double zhouxiangkulun, ref double[] fve, ref double[] gama)
{
for (int i = y; i >= x; i–)
{
double t1 = Ftaoe[i] * ((faibianhua / 57.3) / ((y - x) * buchang));
double t2 = qe[i] * Math.Sin(fai[i] * Math.PI / 180);
double t3 = (t1 + t2) * (t1 + t2);
double t4 = Ftaoe[i] * Math.Sin(fai[i] * Math.PI / 180) * (kuosaibianhua / 57.3) / ((y - x) *
buchang);
double t5 = t4 * t4;
double t6 = t3 + t5;
fn[i] = Math.Pow(t6, 0.5);
double t7 = youxiaojianxi[i] / (tanxingmoliang[i] * guanxingju[i] * fn[i]);
double t8 = Math.Pow(t7, 0.5);
beita[i] = (Ftaoe[i] / 2) * t8;
if (beita[i] <= 1)
{
zhuangtai[i] = " ";
}
else if (beita[i] <= 1.469)
{
zhuangtai[i] = " ";
}
else
{
zhuangtai[i] = " ";
}
if (beita[i] <= 1)
{
wuyinciyali[i] = 1;
}
else if (beita[i] <= 1.449)
{
//wuyinciyali[i] = 1 + 4.0 * (beita[i] - 1) / 11 + 84 * (beita[i] - 1) * (beita[i] - 1) / 121;
wuyinciyali[i] = 8.0 * (beita[i] + 0.5) * (beita[i] + 0.5) / 9 - 1;
}
else
{
```

```
wuyinciyali[i] = beita[i] * beita[i];
}
zhengyali[i] = wuyinciyali[i] * fn[i];
fve[i] = (Ao[i] / Ah[i]) * zhouxiangkulun * zhengyali[i];
double temp1 = fn[i] / (tanxingmoliang[i] * guanxingju[i] * youxiaojianxi[i]);
double temp2 = Math.Pow(temp1, 0.25);
double temp3 = temp2 * temp2 * temp2;
gama[i] = (qe[i] * Math.Cos(fai[i] * Math.PI / 180)) / (2 * tanxingmoliang[i] * guanxingju[i] *
temp3);
Ftaoe[i - 1] = Ftaoe[i] + (qe[i] * Math.Cos(fai[i] * Math.PI / 180) + fve[i] - zhouxiangkulun *
zhengyali[i]) * buchang;
}
}
public void method1(int x, int y, double buchang, double faibianhua, double kuosaibianhua,
double[] Ao, double[] Ai, double[] Ah, ref double[] Ftaoe, double[] qe, double[] fai,
double[] youxiaojianxi, double[] tanxingmoliang, double[] guanxingju, ref double[] fn,
ref double[] beita, ref string[] zhuangtai, ref double[] wuyinciyali,
ref double[] zhengyali, double zhouxiangkulun, ref double[] fve, double K)
{
for (int i = y; i >= x; i–)
{
double t1 = Ftaoe[i] * ((faibianhua / 57.3) / ((y - x) * buchang));
double t2 = qe[i] * Math.Sin(fai[i] * Math.PI / 180);
double t3 = (t1 + t2) * (t1 + t2);
double t4 = Ftaoe[i] * Math.Sin(fai[i] * Math.PI / 180) * (kuosaibianhua / 57.3) / ((y - x) *
buchang);
double t5 = t4 * t4;
double t6 = t3 + t5;
fn[i] = Math.Pow(t6, 0.5);
double t7 = youxiaojianxi[i] / (tanxingmoliang[i] * guanxingju[i] * fn[i]);
double t8 = Math.Pow(t7, 0.5);
beita[i] = (Ftaoe[i] / 2) * t8;
if (beita[i] <= 1)
{
zhuangtai[i] = " ";
}
else if (beita[i] <= 1.469)
{
zhuangtai[i] = " ";
}
else
{
zhuangtai[i] = " ";
}
if (beita[i] <= 1)
{
wuyinciyali[i] = 1;
}
else if (beita[i] <= 1.449)
{
//wuyinciyali[i] = 1 + 4.0 * (beita[i] - 1) / 11 + 84 * (beita[i] - 1) * (beita[i] - 1) / 121;
wuyinciyali[i] = 8.0 * (beita[i] + 0.5) * (beita[i] + 0.5) / 9 - 1;
}
```

```
else
{
wuyinciyali[i] = beita[i] * beita[i];
}
zhengyali[i] = wuyinciyali[i] * fn[i];
fve[i] = (Ao[i] / Ah[i]) * zhouxiangkulun * zhengyali[i];

    Ftaoe[i - 1] = (Ftaoe[i] + (qe[i] * Math.Cos(fai[i] * Math.PI / 180) + fve[i] - zhouxiangkulun *
zhengyali[i]) * buchang) * K;
}
}
public void method2(int j, ref double[] z, ref double[] P5, ref double[] T5, double Ppc, double
Tpc)
{
double[] y = new double[10000];
double[] Tpr = new double[10000];
double[] Ppr = new double[10000];
double[] roupr = new double[10000];
double[] F = new double[10000];
double[] Fdao = new double[10000]; int j1 = 0;//
Ppr[j] = P5[j] / Ppc;
Tpr[j] = (T5[j] + 273) / Tpc;

    if (P5[j] < 35000000)
{
z[0] = 1;
do
{
j1++;
roupr[j1 - 1] = 0.27 * Ppr[j] / (z[j1 - 1] * Tpr[j]);
double t1 = 0.31506 - (1.0467 / Tpr[j]) - (0.5783 / Math.Pow(Tpr[j], 3));
double t2 = (0.5353 - 0.6123 / Tpr[j]) * Math.Pow(roupr[j1 - 1], 2);
z[j1] = 1 + t1 * roupr[j1 - 1] + t2 + 0.6815 * Math.Pow(roupr[j1 - 1], 2) / Math.Pow(Tpr[j], 3);
} while (Math.Abs(z[j1] - z[j1 - 1]) > 0.00001);

    z[j] = z[j1];

    }
else
{
y[1] = 0.001;

    double t = 1 / Tpr[j];
double t3 = -1.2 * Math.Pow((1 - t), 2);
double t4 = -0.06125 * Ppr[j] * t * Math.Exp(t3);
do
{
j1++;
    double t5 = y[j1] + Math.Pow(y[j1], 2) + Math.Pow(y[j1], 3) - Math.Pow(y[j1], 4);
double t6 = Math.Pow((1 - y[j1]), 3);
double t7 = (14.76 * t - 9.76 * Math.Pow(t, 2) + 4.58 * Math.Pow(t, 3)) * Math.Pow(y[j1], 2);
double t8 = 2.18 + 2.82 * t;
double t9 = (90.7 * t - 242.2 * Math.Pow(t, 2) + 42.4 * Math.Pow(t, 3)) * Math.Pow(y[j1], t8);
```

```
F[j1] = t4 + (t5 / t6) - t7 + t9;
double t10 = 1 + 4 * y[j1] + 4 * Math.Pow(y[j1], 2) - 4 * Math.Pow(y[j1], 3) + Math.Pow(y[j1], 4);
double t11 = Math.Pow((1 - y[j1]), 4);
y[j1];
double t13 = 1.18 + 2.82 * t;
double t14 = t8 * (90.7 * t - 242.2 * Math.Pow(t, 2) + 42.4 * Math.Pow(t, 3)) * Math.Pow(y[j1],
t13);
Fdao[j1] = t10 / t11 - t12 + t14;
y[j1 + 1] = y[j1] - F[j1] / Fdao[j1];
} while (Math.Abs(y[j1] - y[j1 + 1]) > 0.00001);

   z[j] = -t4 / y[j1 + 1];
}
}
public void method3(int j, ref double[] zguan, ref double[] P6, ref double[] wendu, double Ppcg,
double Tpcg)
{
double[] y = new double[10000];
double[] Tpr = new double[10000];
double[] Ppr = new double[10000];
double[] roupr = new double[10000];
double[] F = new double[10000];
double[] Fdao = new double[10000];
int j1 = 0;//
Ppr[j] = P6[j] / Ppcg;
Tpr[j] = (wendu[j] + 273) / Tpcg;

   if (P6[j] < 35000000)
{
zguan[0] = 1;
do
{
j1++;
roupr[j1 - 1] = 0.27 * Ppr[j] / (zguan[j1 - 1] * Tpr[j]);

   double t1 = 0.31506 - (1.0467 / Tpr[j]) - (0.5783 / Math.Pow(Tpr[j], 3));
double t2 = (0.5353 - 0.6123 / Tpr[j]) * Math.Pow(roupr[j1 - 1], 2);
zguan[j1] = 1 + t1 * roupr[j1 - 1] + t2 + 0.6815 * Math.Pow(roupr[j1 - 1], 2) / Math.Pow(Tpr[j], 3);
} while (Math.Abs(zguan[j1] - zguan[j1 - 1]) > 0.00001);
zguan[j] = zguan[j1];

   }
else
{
y[1] = 0.001;

   double t = 1 / Tpr[j];
double t3 = -1.2 * Math.Pow((1 - t), 2);
double t4 = -0.06125 * Ppr[j] * t * Math.Exp(t3);
do
{
j1++;
```

```
double t5 = y[j1] + Math.Pow(y[j1], 2) + Math.Pow(y[j1], 3) - Math.Pow(y[j1], 4);
double t6 = Math.Pow((1 - y[j1]), 3);
double t7 = (14.76 * t - 9.76 * Math.Pow(t, 2) + 4.58 * Math.Pow(t, 3)) * Math.Pow(y[j1], 2);
double t8 = 2.18 + 2.82 * t;
double t9 = (90.7 * t - 242.2 * Math.Pow(t, 2) + 42.4 * Math.Pow(t, 3)) * Math.Pow(y[j1], t8);
F[j1] = t4 + (t5 / t6) - t7 + t9;
double t10 = 1 + 4 * y[j1] + 4 * Math.Pow(y[j1], 2) - 4 * Math.Pow(y[j1], 3) + Math.Pow(y[j1], 4);
double t11 = Math.Pow((1 - y[j1]), 4);
double t12 = (29.52 * t - 19.52 * Math.Pow(t, 2) + 9.16 * Math.Pow(t, 3)) * y[j1];
double t13 = 1.18 + 2.82 * t;
double t14 = t8 * (90.7 * t - 242.2 * Math.Pow(t, 2) + 42.4 * Math.Pow(t, 3)) * Math.Pow(y[j1],
t13);
Fdao[j1] = t10 / t11 - t12 + t14;
y[j1 + 1] = y[j1] - F[j1] / Fdao[j1];
} while (Math.Abs(y[j1] - y[j1 + 1]) > 0.00001);

zguan[j] = -t4 / y[j1 + 1];

}
}
private void button2_Click(object sender, EventArgs e)
{
try
{
//
double[] fn = new double[10000];
double[] beita = new double[10000];
double[] wuyinciyali = new double[10000];
double[] zhengyali = new double[10000];
double[] fve = new double[10000];
string[] zhuangtai = new string[10000];
double[] guanbimozu = new double[10000];

// qe
double[] qe = new double[10000];
double[] diejia = new double[10000];
diejia[rows1 + 1] = 0;
for (int i1 = rows1; i1 >= 1; i1–)
{
for (int i2 = jiange1[i1 + 1]; i2 >= jiange1[i1]; i2–)
{
double t1 = mizhong[i2] * (jiange1[i1 + 1] - i2) * buchang + pin * 9.8 * Ai[i2] - pout * 9.8 *
Ao[i2];
diejia[i1] = mizhong[jiange1[i1 + 2]] * (jiange1[i1 + 2] - jiange1[i1 + 1]) * buchang + pin * 9.8 *
Ai[i1 + 1] - pout * 9.8 * Ao[i1 + 1] + diejia[i1 + 1];
qe[i2] = t1 + diejia[i1];
}
}

// qe_1
double[] qe_1 = new double[10000];
for (int i = 1; i <= n; i++)
{
```

```
qe_1[i] = mizhong[i] + pin * 9.8 * Ai[i] - pout * 9.8 * Ao[i];
}

  // Ftaoe_1
double[] Ftaoe_1 = new double[10000];
Ftaoe_1[n] = 0;

  for (int i = n; i >= 1; i–)
{
Ftaoe_1[i - 1] = Ftaoe_1[i] + qe_1[i] * Math.Cos(fai[i] * Math.PI / 180) * buchang;
}

  //
double[] Ftaoe = new double[10000];
Ftaoe[n] = (-pin * 9.8 * Ai[n] + pout * 9.8 * Ao[n]) * h[n];

  //
double[] gama = new double[10000];
for (int i = rows3 - 1; i >= 1; i–)
{
method(jiange3[i], jiange3[i + 1], buchang, faibianhua[i], kuosaibianhua[i], Ao,
Ai, Ah, ref Ftaoe, qe, fai, youxiajianxi, tanxingmoliang, guanxingju, ref fn,
ref beita, ref zhuangtai, ref wuyinciyali, ref zhengyali,
zhouxiangkulun, ref fve, ref gama);
}

  //
double K1 = 0;
for (int i = 1; i <= n; i++)
{
K1 = K1 + gama[i];
}
K1 = K1 / n;
if (K1 > 1)
{
K1 = 1 / K1;
}
if (K1 > 0.5)
{
K1 = 0.5;
}
for (int i = rows3 - 1; i >= 1; i–)
{
method1(jiange3[i], jiange3[i + 1], buchang, faibianhua[i], kuosaibianhua[i], Ao,
Ai, Ah, ref Ftaoe, qe, fai, youxiajianxi, tanxingmoliang, guanxingju, ref fn,
ref beita, ref zhuangtai, ref wuyinciyali, ref zhengyali,
zhouxiangkulun, ref fve, K1);
}

  //
double[] Po = new double[10000];
Po[n] = pout * 9.8 * h[n];
double[] Pi = new double[10000];
```

```
Pi[n] = pin * 9.8 * h[n];
double[] Up = new double[10000];
Up[n] = 0;
for (int i = n; i >= 1; i–)
{
Po[i - 1] = Po[i] - pout * 9.8 * Math.Cos(fai[i] * Math.PI / 180) * buchang;
Pi[i - 1] = Pi[i] - pin * 9.8 * Math.Cos(fai[i] * Math.PI / 180) * buchang;
double t1 = 2 * bosongbi[i] / (tanxingmoliang[i] * (Ao[i] - Ai[i]));
double t2 = Ao[i] * Po[i] - Ai[i] * Pi[i];
double t3 = t1 * t2 * buchang;
if (t3 < 0)
{
t3 = -t3;
}
Up[i - 1] = (Up[i] + t3);
}
for (int i = 1; i <= n + 1; i++)
{
Up[i] = Up[i] * 0.3;
}

   //
double[] wendu = new double[10000];
double[] Ut = new double[10000];
Ut[n] = 0;
for (int i = n; i >= 1; i–)
{
if (h[i] <= 20)
{
wendu[i] = jingkouwendu;
}
else
{
wendu[i] = (ditiwendu / 100) * (h[i] - 20) + jingkouwendu;
}
Ut[i - 1] = (Ut[i] + repengzhang[i] * (wendu[i] - jingkouwendu) * buchang);
}
for (int i = 1; i <= n + 1; i++)
{
Ut[i] = Ut[i] * 0.4;
}

   //
double[] Ftao = new double[10000];
double[] Uf = new double[10000];
Uf[n] = 0;
for (int i = n; i >= 1; i–)
{
Ftao[i] = Ftaoe[i] - Pi[i] * Ai[i] + Po[i] * Ao[i];
Uf[i - 1] = (Uf[i] + Ftao[i] * buchang / (tanxingmoliang[i] * (Ao[i] - Ai[i])));
guanbimozu[i] = zhouxiangkulun * fn[i];
}
for (int i = 1; i <= n + 1; i++)
```

```
{
Uf[i] = Uf[i] * 0.4;
}
//
double[] Um1 = new double[10000];
Um1[n] = 0;
for (int i = n; i >= 1; i–)
{
if (beita[i] <= 1)
{
Um1[i - 1] = Um1[i];
}
else if ((beita[i] <= 1.469))
{
double t1 = fn[i] / (tanxingmoliang[i] * guanxingju[i] * youxiajianxi[i]);
double t2 = Math.Pow(t1, 0.5);
double t3 = (8.0 / 11) * t2 * youxiajianxi[i] * youxiajianxi[i] * (beita[i] - 1);
Um1[i - 1] = Um1[i] - t3;
}
else
{
double t1 = fn[i] / (tanxingmoliang[i] * guanxingju[i] * youxiajianxi[i]);
double t2 = Math.Pow(t1, 0.5);
double t3 = 0.5 * t2 * youxiajianxi[i] * youxiajianxi[i] * beita[i];
Um1[i - 1] = Um1[i] - t3;
}
}
//
double[] Uzong = new double[10000];
for (int i = 1; i <= n; i++)
{
Uzong[i] = Up[i] + Ut[i] + Uf[i] + Um1[i];
}

    //

    //
double[] qulv = new double[10000];
for (int i = rows3 - 1; i >= 1; i–)
{
for (int i1 = jiange3[i + 1]; i1 >= jiange3[i]; i1–)
{
double t1 = (faibianhua[i] / 57.3) / ((jiange3[i + 1] - jiange3[i]) * buchang);
double t2 = t1 * t1;
double t3 = Math.Sin(fai[i1] * Math.PI / 180) * Math.Sin(fai[i1] * Math.PI / 180);
double t4 = (kuosaibianhua[i] / 57.3) / ((jiange3[i + 1] - jiange3[i]) * buchang);
double t5 = t4 * t4;
qulv[i1] = t2 + t3 * t5;
}
}

    //
double[] liju = new double[10000];
```

```
for (int i = 1; i <= n; i++)
{
if (beita[i] <= 1)
{
liju[i] = tanxingmoliang[i] * guanxingju[i] * qulv[i];
}
else if (beita[i] <= 1.469)
{
double t1 = tanxingmoliang[i] * guanxingju[i] * youxiajianxi[i] * fn[i];
double t2 = Math.Pow(t1, 0.5);
liju[i] = tanxingmoliang[i] * guanxingju[i] * qulv[i] + (16.0 / 11) * t2 * (beita[i] - 1);
}
else
{
double t3 = tanxingmoliang[i] * guanxingju[i] * youxiajianxi[i] * fn[i];
double t4 = Math.Pow(t3, 0.5);
liju[i] = tanxingmoliang[i] * guanxingju[i] * qulv[i] + t4 * beita[i];
}
}

  //
double[] zhouxiangyali = new double[10000];//
double[] neihuanyingli = new double[10000];//
double[] waihuanyingli = new double[10000];//
double[] neijingxiangyingli = new double[10000];//
double[] waijingxiangyingli = new double[10000];//
double[] neiwanquyingli = new double[10000];//
double[] waiwanquyingli = new double[10000];//
for (int i = 1; i <= n; i++)
{
zhouxiangyali[i] = -(Ftao[i] / (Ao[i] - Ai[i]));
neihuanyingli[i] = ((Ao[i] + Ai[i]) / (Ao[i] - Ai[i])) * Pi[i] - (2 * Ao[i] / (Ao[i] - Ai[i])) * Po[i];
waihuanyingli[i] = -((Ao[i] + Ai[i]) / (Ao[i] - Ai[i])) * Po[i] + (2 * Ai[i] / (Ao[i] - Ai[i])) * Pi[i];
neijingxiangyingli[i] = -Pi[i];
waijingxiangyingli[i] = -Po[i];
neiwanquyingli[i] = 4 * liju[i] * (neijing[i] / 2) / guanxingju[i];
waiwanquyingli[i] = 4 * liju[i] * (waijing[i] / 2) / guanxingju[i];
}

  //

  double[] qiangdujiaohe11 = new double[10000];
for (int i = 1; i <= n; i++)
{
double t1 = zhouxiangyali[i] + neiwanquyingli[i] - neijingxiangyingli[i];
double t2 = t1 * t1;
double t3 = zhouxiangyali[i] + neiwanquyingli[i] - neihuanyingli[i];
double t4 = t3 * t3;
double t5 = neijingxiangyingli[i] - neihuanyingli[i];
double t6 = t5 * t5;
double t7 = t2 + t4 + t6;
qiangdujiaohe11[i] = Math.Pow(t7, 0.5) / Math.Pow(2, 0.5);
}
```

```
double[] qiangdujiaohe21 = new double[10000];
for (int i = 1; i <= n; i++)
{
double t1 = zhouxiangyali[i] - neiwanquyingli[i] - neijingxiangyingli[i];
double t2 = t1 * t1;
double t3 = zhouxiangyali[i] - neiwanquyingli[i] - neihuanyingli[i];
double t4 = t3 * t3;
double t5 = neihuanyingli[i] - neijingxiangyingli[i];
double t6 = t5 * t5;
double t7 = t2 + t4 + t6;
qiangdujiaohe21[i] = Math.Pow(t7, 0.5) / Math.Pow(2, 0.5);
}
double[] qiangdujiaohe31 = new double[10000];
for (int i = 1; i <= n; i++)
{
double t1 = zhouxiangyali[i] + waiwanquyingli[i] - waijingxiangyingli[i];
double t2 = t1 * t1;
double t3 = zhouxiangyali[i] + waiwanquyingli[i] - waihuanyingli[i];
double t4 = t3 * t3;
double t5 = waijingxiangyingli[i] - waihuanyingli[i];
double t6 = t5 * t5;
double t7 = t2 + t4 + t6;
qiangdujiaohe31[i] = Math.Pow(t7, 0.5) / Math.Pow(2, 0.5);
}
double[] qiangdujiaohe41 = new double[10000];
for (int i = 1; i <= n; i++)
{
double t1 = zhouxiangyali[i] - waiwanquyingli[i] - waijingxiangyingli[i];
double t2 = t1 * t1;
double t3 = zhouxiangyali[i] - waiwanquyingli[i] - waihuanyingli[i];
double t4 = t3 * t3;
double t5 = waihuanyingli[i] - waijingxiangyingli[i];
double t6 = t5 * t5;
double t7 = t2 + t4 + t6;
qiangdujiaohe41[i] = Math.Pow(t7, 0.5) / Math.Pow(2, 0.5);
}
double[] qiandujiaohe = new double[10000];
double[] anquanxishu = new double[10000];
double[] anquan_kangji1 = new double[10000];
double[] anquan_neiqu1 = new double[10000];
double[] anquan_lianjie1 = new double[10000];
for (int i = 1; i <= n; i++)
{
qiandujiaohe[i] = Math.Max(Math.Max(qiangdujiaohe11[i], qiangdujiaohe21[i]),
Math.Max(qiangdujiaohe31[i],
qiangdujiaohe41[i]));
anquanxishu[i] = qufujixian[i] /
qiandujiaohe[i];
if (waihuanyingli[i] < 0)
{
anquan_kangji1[i] = -waihuanyingli[i] / kangjiqiangdu[i];
}
else
```

```
{
anquan_kangji1[i] = waihuanyingli[i] / kangjiqiangdu[i];
}
if (neihuanyingli[i] < 0)
{
anquan_neiqu1[i] = -neihuanyingli[i] / neiqufuqiangdu[i];
}
else
{
anquan_neiqu1[i] = neihuanyingli[i] / neiqufuqiangdu[i];
}
if (zhouxiangyali[i] < 0)
{
anquan_lianjie1[i] = -zhouxiangyali[i] / lianjieqiangdu[i];
}
else
{
anquan_lianjie1[i] = zhouxiangyali[i] / lianjieqiangdu[i];
}

    }

    //
int panduan = 0;
string[] ququzhuangtai = new string[10000];
for (int i = 0; i <= 9999; i++)
{
ququzhuangtai[i] = " ;
}
for (int i = 1; i <= n; i++)
{
if (panduan >= 10)
{
for (int jishu = i - panduan; jishu <= i - 1; jishu++)
{
ququzhuangtai[jishu] = " ";
}
}
if (beita[i] <= 1)
{
panduan = 0;
}
else
{
panduan = panduan + 1;
}
}

    //
listView1.Items.Clear();
for (int i = n; i >= 1; i–)
{
string[] subItems1 = new string[]
```

```
{
(n-i+1).ToString(),
Math.Round(Ftaoe_1[n - i + 1], 3).ToString(),
Math.Round(zhengyali[n - i + 1], 3).ToString(),
Math.Round(-guanbimozu[n - i + 1], 3).ToString(),
Math.Round(Ut[i], 4).ToString(),
Math.Round(Up[i], 6).ToString(),
Math.Round(Uf[i], 3).ToString(),
Math.Round(Um1[i], 3).ToString(),
Math.Round(Uzong[i], 4).ToString(),
ququzhuangtai[i]
};
listView1.Items.Insert(n - i, new ListViewItem(subItems1));
}
}
catch (Exception ex)
{
MessageBox.Show("ex.ToString(), ");
}

   }

   private void button1_Click(object sender, EventArgs e)
{
Xuanzegongkuang.xuanzegongkuang.Show();
this.Close();
}

   private void Xiaguanzhujieguo_Load(object sender, EventArgs e)
{
//mainform
buchang = Mainform.mainform.buchang;
rows3 = Mainform.mainform.rows3;
colums3 = Mainform.mainform.colums3;
n = Mainform.mainform.n;
jingshen = Mainform.mainform.jingshen;
faibianhua = Mainform.mainform.faibianhua;
kuosaibianhua = Mainform.mainform.kuosaibianhua;
jiange3 = Mainform.mainform.jiange3;
fai = Mainform.mainform.fai;
h = Mainform.mainform.h;

   rows1 = Mainform.mainform.rows1;
colums1 = Mainform.mainform.colums1;
jiange1 = Mainform.mainform.jiange1;
mizhong = Mainform.mainform.mizhong;
repengzhang = Mainform.mainform.repengzhang;
tanxingmoliang = Mainform.mainform.tanxingmoliang;
bosongbi = Mainform.mainform.bosongbi;
waijing = Mainform.mainform.waijing;
neijing = Mainform.mainform.neijing;
Ao = Mainform.mainform.Ao;
Ai = Mainform.mainform.Ai;
```

```
guanxingju = Mainform.mainform.guanxingju;

    rows2 = Mainform.mainform.rows2;
colums2 = Mainform.mainform.colums2;
jiange2 = Mainform.mainform.jiange2;
tongjing = Mainform.mainform.tongjing;
rc = Mainform.mainform.rc;
youxiajianxi = Mainform.mainform.youxiajianxi;
Ah = Mainform.mainform.Ah;
zhouxiangkulun = Mainform.mainform.zhouxiangkulun;
jingkouwendu = Mainform.mainform.jingkouwendu;
ditiwendu = Mainform.mainform.ditiwendu;
dicengdaorexishu = Mainform.mainform.dicengdaorexishu;
dicengrekuosanxishu = Mainform.mainform.dicengrekuosanxishu;
qufujixian = Mainform.mainform.qufujixian;
kangjiqiangdu = Mainform.mainform.kangjiqiangdu;
neiqufuqiangdu = Mainform.mainform.neiqufuqiangdu;
lianjieqiangdu = Mainform.mainform.lianjieqiangdu;

    pin = Mainform.mainform.pin;
pout = Mainform.mainform.pout;

    zuofengfangshi = Mainform.mainform.zuofengfangshi;
jingkouyali = Mainform.mainform.jingkouyali;
linjieyali = Mainform.mainform.linjieyali;
fenggeqi = Mainform.mainform.fenggeqi;

    Pshe = Mainform.mainform.Pshe;

    Pnei = Mainform.mainform.Pnei;
Pwai = Mainform.mainform.Pwai;
Q = Mainform.mainform.Q;
Kh = Mainform.mainform.Kh;
time = Mainform.mainform.time;
K = Mainform.mainform.K;

    Tpc = Mainform.mainform.Tpc;
Ppc = Mainform.mainform.Ppc;
jingdiyali = Mainform.mainform.jingdiyali;
rg = Mainform.mainform.rg;
Mg = Mainform.mainform.Mg;
Cci = Mainform.mainform.Cci;
Cto = Mainform.mainform.Cto;
qitidaorexishu = Mainform.mainform.qitidaorexishu;
Qgsc = Mainform.mainform.Qgsc;
shuinihuandaorexishu = Mainform.mainform.shuinihuandaorexishu;
shengchanshijian = Mainform.mainform.shengchanshijian;

    Tpc6 = Mainform.mainform.Tpc6;
Ppc6 = Mainform.mainform.Ppc6;
jingkouyalig = Mainform.mainform.jingkouyalig;
shangtili = Mainform.mainform.shangtili;
jiefengyali = Mainform.mainform.jiefengyali;
```

```
shangtifangshi = Mainform.mainform.shangtifangshi;
}

   private void button3_Click(object sender, EventArgs e)
{
string cons = "Persist Security Info = false; Data Source = .; Initial Catalog = jieguo; User ID =
sa; Password = sql";
//
System.Data.SqlClient.SqlTransaction st = null;
SqlConnection sc = null;
try
{
sc = new SqlConnection(cons);
//sc = new SqlConnection(strConn);
sc.Open();
//
st = sc.BeginTransaction();
//

   string que = "truncate table xiaguanzhu";
string que1 = "truncate table xiaguanzhu1";

   string charu;
string charu1;

   SqlCommand ss = new SqlCommand(que, sc, st);
SqlCommand ss1 = new SqlCommand(que1, sc, st);

   ss.ExecuteNonQuery();
ss1.ExecuteNonQuery();

   //
string[] b = new string[15];
string[] b1 = new string[15];
long iv = 100;//
int count = 0;
long iv1 = 1;
int count1 = 0;
//

   foreach (ListViewItem l in this.listView1.Items)
{
//MessageBox.Show("ITEM:" + l.Text.ToString());
if (count1 == iv1 || count1 == 0)
{
//

   //b[o]
for (int a = 0; a < l.SubItems.Count; a++)
{
//MessageBox.Show("SUBITEM:" + l.SubItems[a].Text.ToString());
b[a] = l.SubItems[a].Text.ToString();
}
```

```
//

    charu = "insert into xiaguanzhu values('" + b[0].ToString() + "'"
+ ",'" + b[1].ToString() + "'" + ",'" + b[2].ToString() + "','" +
b[3].ToString() + "','" + b[4].ToString() + "','" + b[5].ToString() +
"','" + b[6].ToString() +
"','" + b[7].ToString() +
"','" + b[8].ToString() +
"','" + b[9].ToString() +
"')";

    ss = new SqlCommand(charu, sc, st);
ss.ExecuteNonQuery();

    if (count1 == iv1)
{
count1 = 0;
}
}
count1++;
if (count == iv || count == 0)
{
//

    //b[o]
for (int a = 0; a < l.SubItems.Count; a++)
{
//MessageBox.Show("SUBITEM:" + l.SubItems[a].Text.ToString());
b1[a] = l.SubItems[a].Text.ToString();
}
//

    charu1 = "insert into xiaguanzhu1 values('" + b1[0].ToString() + "'"
+ ",'" + b1[1].ToString() + "'" + ",'" + b1[2].ToString() + "','" +
b1[3].ToString() + "','" + b1[4].ToString() + "','" + b1[5].ToString() +
"','" + b1[6].ToString() +
"','" + b1[7].ToString() +
"','" + b1[8].ToString() +
"','" + b1[9].ToString() +
"')";

    ss1 = new SqlCommand(charu1, sc, st);
ss1.ExecuteNonQuery();

    if (count == iv)
{
count = 0;
}
}
count++;
}
//
st.Commit();
```

```
MessageBox.Show("saved!");
}
catch (Exception ex)
{
//
if (st != null)
st.Rollback();
MessageBox.Show(ex.Message);
}

   finally
{
if (sc != null)
sc.Close();
sc = null;
}
}

   private void button4_Click(object sender, EventArgs e)
{
if (xiaguanzhubaobiao != null)
{
xiaguanzhubaobiao.Close();
xiaguanzhubaobiao = new Xiaguanzhubaobiao();
xiaguanzhubaobiao.MdiParent = Mainform.mainform;
xiaguanzhubaobiao.Show();
}
else
{
xiaguanzhubaobiao = new Xiaguanzhubaobiao();
xiaguanzhubaobiao.MdiParent = Mainform.mainform;
xiaguanzhubaobiao.Show();
}

   }
}
}
```

Appendix 2: The source code of Chapter 5

```
using System;
using System.Collections.Generic;
using System.ComponentModel;
using System.Data;
using System.Drawing;
using System.Text;
using System.Windows.Forms;
using System.Text.RegularExpressions;


   namespace
public partial class Zuofeng : Form
{
```

```
public static Zuofeng zuofeng = null;
public Shekong shekong;
public Zuofengb zuofengb;

   //
public int zuofengfangshi;
public double jingkouyali;
public double linjieyali;
public double fenggeqi;

   public Zuofeng()
{
InitializeComponent();
zuofeng = this;
}


   private void button1_Click(object sender, EventArgs e)
{
Xiaguanzhu.xiaguanzhu.Show();
this.Hide();
}


   private void button2_Click(object sender, EventArgs e)
{
try
{
if (!radioButton1.Checked && !radioButton2.Checked)
{
MessageBox.Show("");
}
else
{
if (radioButton1.Checked)
{
if (textBox2.Text == "" || textBox1.Text == "")
{
MessageBox.Show("请先输入上提下放座封数据");
}
else
{
fenggeqi = double.Parse(textBox1.Text);
jingkouyali = double.Parse(textBox2.Text) * 1000;
zuofengfangshi = 1;

   //mainform
Mainform.mainform.fenggeqi = fenggeqi;
Mainform.mainform.jingkouyali = jingkouyali;
Mainform.mainform.zuofengfangshi = zuofengfangshi;

   //
if (shekong != null)
```

```
{
shekong.Show();
this.Hide();
}
else
{
shekong = new Shekong();
shekong.MdiParent = Mainform.mainform;
shekong.Show();
this.Hide();
}

    }


    }
else if (radioButton2.Checked)
{
if (textBox16.Text == "" || textBox17.Text == "")
{
MessageBox.Show(" ");
}
else
{
fenggeqi = double.Parse(textBox16.Text);
linjieyali = double.Parse(textBox17.Text) * 1000000;
zuofengfangshi = 2;

    //mainform
Mainform.mainform.fenggeqi = fenggeqi;
Mainform.mainform.linjieyali = linjieyali;
Mainform.mainform.zuofengfangshi = zuofengfangshi;

    //
if (shekong != null)
{
shekong.Show();
this.Hide();
}
else
{
shekong = new Shekong();
shekong.MdiParent = Mainform.mainform;
shekong.Show();
this.Hide();
}
}


    }
}
}
catch (Exception ex)
```

```
{
MessageBox.Show("+ ex.ToString(), ");
}


    }

    private void textBox1_Validating(object sender, CancelEventArgs e)
{
string patten = ("^[0-9]*(.[0-9]+)?$");
Regex r = new Regex(patten);
if (!r.IsMatch(textBox1.Text))
{
errorProvider1.SetError(textBox1, " ");
e.Cancel = true;
}
else
{
errorProvider1.SetError(textBox1, "");
}
}

    private void textBox2_Validating(object sender, CancelEventArgs e)
{
string patten = ("^[0-9]*(.[0-9]+)?$");
Regex r = new Regex(patten);
if (!r.IsMatch(textBox2.Text))
{
errorProvider1.SetError(textBox2, " ");
e.Cancel = true;
}
else
{
errorProvider1.SetError(textBox2, "");
}
}


    private void textBox16_Validating(object sender, CancelEventArgs e)
{
string patten = ("^[0-9]*(.[0-9]+)?$");
Regex r = new Regex(patten);
if (!r.IsMatch(textBox16.Text))
{
errorProvider1.SetError(textBox16, " ");
e.Cancel = true;
}
else
{
errorProvider1.SetError(textBox16, "");
}
}
private void textBox17_Validating(object sender, CancelEventArgs e)
```

```
{
string patten = ("^[0-9]*(.[0-9]+)?$");
Regex r = new Regex(patten);
if (!r.IsMatch(textBox17.Text))
{
errorProvider1.SetError(textBox17, " ");
e.Cancel = true;
}
else
{
errorProvider1.SetError(textBox17, "");
}
}

   private void button3_Click(object sender, EventArgs e)
{

   if (zuofengb != null)
{
zuofengb.Close();
zuofengb = new Zuofengb();
zuofengb.MdiParent = Mainform.mainform;
zuofengb.Show();
}
else
{
zuofengb = new Zuofengb();
zuofengb.MdiParent = Mainform.mainform;
zuofengb.Show();
}
}
}
}
```

Appendix 3: The source code of Chapter 6

```
   using System; using System.Collections.Generic;
   using System.ComponentModel;
   using System.Data;
   using System.Drawing;
   using System.Text;
   using System.Windows.Forms;
   using System.Data.OleDb;
   using System.IO;
   using CrystalDecisions.CrystalReports.Engine;
   using CrystalDecisions.Shared;
   using System.Data.SqlClient;


   namespace{
public partial class Zhurubaobiao : Form
{
```

```
public Zhurubaobiao()
{
InitializeComponent();
}

   private void button1_Click(object sender, EventArgs e)
{
string cons = "Persist Security Info = false; Data Source = .; Initial Catalog = jieguo; User ID =
sa; Password = sql";
SqlConnection sc = new SqlConnection(cons);
string strsel = "select * from zhuru";
SqlDataAdapter ada = new SqlDataAdapter(strsel, sc);
DataSet1 ds = new DataSet1();
ada.Fill(ds, "zhuru");
CrystalReport4 rpt = new CrystalReport4();
rpt.SetDataSource(ds);
crystalReportViewer1.ReportSource = rpt;
}

   private void button2_Click(object sender, EventArgs e)
{
string cons = "Persist Security Info = false; Data Source = .; Initial Catalog = jieguo; User ID =
sa; Password = sql";
SqlConnection sc = new SqlConnection(cons);
string strsel = "select * from zhuru1";
SqlDataAdapter ada = new SqlDataAdapter(strsel, sc);
DataSet1 ds = new DataSet1();
ada.Fill(ds, "zhuru1");
CrystalReport11 rpt = new CrystalReport11();
rpt.SetDataSource(ds);
crystalReportViewer1.ReportSource = rpt;
}
 private void button3_Click(object sender, EventArgs e)
{
try
{
string cons = "Persist Security Info = false; Data Source = .; Initial Catalog = jieguo; User ID =
sa; Password = sql";
SqlConnection sc = new SqlConnection(cons);
string strsel = "select * from zhurudanxiang";
SqlDataAdapter ada = new SqlDataAdapter(strsel, sc);
DataSet1 ds = new DataSet1();
ada.Fill(ds, "zhurudanxiang");
CrystalReport16 rpt = new CrystalReport16();
rpt.SetDataSource(ds);
crystalReportViewer1.ReportSource = rpt;
}
catch (Exception ex)
{
MessageBox.Show("+ ex.ToString(), ");
}
}
```

```
   private void button4_Click(object sender, EventArgs e)
{
try
{
string cons = "Persist Security Info = false; Data Source = .; Initial Catalog = jieguo; User ID =
sa; Password = sql";
SqlConnection sc = new SqlConnection(cons);
string strsel = "select * from zhurudanxiang1";
SqlDataAdapter ada = new SqlDataAdapter(strsel, sc);
DataSet1 ds = new DataSet1();
ada.Fill(ds, "zhurudanxiang1");
CrystalReport18 rpt = new CrystalReport18();
rpt.SetDataSource(ds);
crystalReportViewer1.ReportSource = rpt;
}
catch (Exception ex)
{
MessageBox.Show("+ ex.ToString(), ");
}
}

   private void button19_Click(object sender, EventArgs e)
{
try
{
string cons = "Persist Security Info = false; Data Source = .; Initial Catalog = jieguo; User ID =
sa; Password = sql";
SqlConnection sc = new SqlConnection(cons);
string strsel = "select * from wendusudu";
SqlDataAdapter ada = new SqlDataAdapter(strsel, sc);
DataSet1 ds = new DataSet1();
ada.Fill(ds, "wendusudu");
CrystalReport20 rpt = new CrystalReport20();
rpt.SetDataSource(ds);
crystalReportViewer1.ReportSource = rpt;
}
catch (Exception ex)
{
MessageBox.Show("+ ex.ToString(), ");
}
}
}
}
```

Appendix 4: The source code of Chapter 7

```
using System; using System.Collections.Generic;
   using System.ComponentModel;
   using System.Data;
   using System.Drawing;
   using System.Text;
   using System.Windows.Forms;
   using System.Text.RegularExpressions;
```

```
namespace{
public partial class Shengchan : Form
{
public static Shengchan shengchan = null;
public Guanjing guanjing;
public Shengchanb shengchanb;

//
public double Tpc;
public double Ppc;
public double jingdiyali;
public double rg;
public double Mg;
public double Cci;
public double Cto;
public double qitidaorexishu;
public double Qgsc;
public double shuinihuandaorexishu;
public double shengchanshijian;


public Shengchan()
{
InitializeComponent();
shengchan = this;
}

private void button1_Click(object sender, EventArgs e)
{
Zhuru.zhuru.Show();
this.Hide();
}

private void button2_Click(object sender, EventArgs e)
{
try
{
if (textBox2.Text == "" || textBox3.Text == "" || textBox4.Text == "" ||
textBox5.Text == "" || textBox6.Text == "" || textBox7.Text == "" ||
textBox8.Text == "" || textBox9.Text == "" || textBox10.Text == ""
|| textBox11.Text == "")
{
MessageBox.Show(" ");
}

else
{
if (comboBox1.SelectedIndex == 0)
{
Qgsc = double.Parse(textBox9.Text);
}
else
{
```

```
Qgsc = double.Parse(textBox9.Text) * 10000;
}

   Tpc = 647.3;
Ppc = double.Parse(textBox2.Text) * 1000000;
jingdiyali = double.Parse(textBox3.Text) * 1000000;
rg = double.Parse(textBox4.Text);
Mg = double.Parse(textBox5.Text);
qitidaorexishu = double.Parse(textBox6.Text);
Cci = double.Parse(textBox7.Text);
Cto = double.Parse(textBox8.Text);
shuinihuandaorexishu = double.Parse(textBox10.Text);
shengchanshijian = double.Parse(textBox11.Text) * 86400;

   //mainform
Mainform.mainform.Tpc = Tpc;
Mainform.mainform.Ppc = Ppc;
Mainform.mainform.jingdiyali = jingdiyali;
Mainform.mainform.rg = rg;
Mainform.mainform.Mg = Mg;
Mainform.mainform.qitidaorexishu = qitidaorexishu;
Mainform.mainform.Cci = Cci;
Mainform.mainform.Cto = Cto;
Mainform.mainform.Qgsc = Qgsc;
Mainform.mainform.shuinihuandaorexishu = shuinihuandaorexishu;
Mainform.mainform.shengchanshijian = shengchanshijian;

   //
if (guanjing != null)
{
guanjing.Show();
this.Hide();
}
else
{
guanjing = new Guanjing();
guanjing.MdiParent = Mainform.mainform;
guanjing.Show();
this.Hide();
}

   }
}
catch (Exception ex)
{
MessageBox.Show("+ ex.ToString(), ");
}

   }

   private void textBox2_Validating(object sender, CancelEventArgs e)
{
string patten = ("^[0-9]*(.[0-9]+)?$");
```

```
Regex r = new Regex(patten);
if (!r.IsMatch(textBox2.Text))
{
errorProvider1.SetError(textBox2, " ");
e.Cancel = true;
}
else
{
errorProvider1.SetError(textBox2, "");
}
}

   private void textBox3_Validating(object sender, CancelEventArgs e)
{
string patten = ("^[0-9]*(.[0-9]+)?$");
Regex r = new Regex(patten);
if (!r.IsMatch(textBox3.Text))
{
errorProvider1.SetError(textBox3, " ");
e.Cancel = true;
}
else
{
errorProvider1.SetError(textBox3, "");
}
}

   private void textBox4_Validating(object sender, CancelEventArgs e)
{
string patten = ("^[0-9]*(.[0-9]+)?$");
Regex r = new Regex(patten);
if (!r.IsMatch(textBox4.Text))
{
errorProvider1.SetError(textBox4, " ");
e.Cancel = true;
}
else
{
errorProvider1.SetError(textBox4, "");
}
}

   private void textBox5_Validating(object sender, CancelEventArgs e)
{
string patten = ("^[0-9]*(.[0-9]+)?$");
Regex r = new Regex(patten);
if (!r.IsMatch(textBox5.Text))
{
errorProvider1.SetError(textBox5, " ");
e.Cancel = true;
}
else
{
```

```
errorProvider1.SetError(textBox5, "");
}
}

   private void textBox6_Validating(object sender, CancelEventArgs e)
{
string patten = ("^[0-9]*(.[0-9]+)?$");
Regex r = new Regex(patten);
if (!r.IsMatch(textBox6.Text))
{
errorProvider1.SetError(textBox6, " ");
e.Cancel = true;
}
else
{
errorProvider1.SetError(textBox6, "");
}
}

   private void textBox7_Validating(object sender, CancelEventArgs e)
{
string patten = ("^[0-9]*(.[0-9]+)?$");
Regex r = new Regex(patten);
if (!r.IsMatch(textBox7.Text))
{
errorProvider1.SetError(textBox7, " ");
e.Cancel = true;
}
else
{
errorProvider1.SetError(textBox7, "");
}
}

   private void textBox8_Validating(object sender, CancelEventArgs e)
{
string patten = ("^[0-9]*(.[0-9]+)?$");
Regex r = new Regex(patten);
if (!r.IsMatch(textBox8.Text))
{
errorProvider1.SetError(textBox8, " ");
e.Cancel = true;
}
else
{
errorProvider1.SetError(textBox8, "");
}
}

   private void textBox9_Validating(object sender, CancelEventArgs e)
{
string patten = ("^[0-9]*(.[0-9]+)?$");
Regex r = new Regex(patten);
```

```
if (!r.IsMatch(textBox9.Text))
{ errorProvider1.SetError(textBox9, " ");
e.Cancel = true;
}
else
{
errorProvider1.SetError(textBox9, "");
}
}

   private void textBox10_Validating(object sender, CancelEventArgs e)
{
string patten = ("^[0-9]*(.[0-9]+)?$");
Regex r = new Regex(patten);
if (!r.IsMatch(textBox10.Text))
{
errorProvider1.SetError(textBox10, " ");
e.Cancel = true;
}
else
{
errorProvider1.SetError(textBox10, "");
}
}

   private void textBox11_Validating(object sender, CancelEventArgs e)
{
string patten = ("^[0-9]*(.[0-9]+)?$");
Regex r = new Regex(patten);
if (!r.IsMatch(textBox11.Text))
{
errorProvider1.SetError(textBox11, " ");
e.Cancel = true;
}
else
{
errorProvider1.SetError(textBox11, "");
}
}

   private void button3_Click(object sender, EventArgs e)
{

   if (shengchanb != null)
{
shengchanb.Close();
shengchanb = new Shengchanb();
shengchanb.MdiParent = Mainform.mainform;
shengchanb.Show();
}
else
{
shengchanb = new Shengchanb();
```

```
shengchanb.MdiParent = Mainform.mainform;
shengchanb.Show();
}
}

   private void Shengchan_Load(object sender, EventArgs e)
{
comboBox1.SelectedIndex = 0;
}

   private void textBox9_TextChanged(object sender, EventArgs e)
{

   }
}
}
```

Subject index

Note: page references in **bold** indicate diagrams and tables.

Multiphysics Modeling

Series Editors: Jochen Bundschuh & Mario César Suárez Arriaga

ISSN:1877-0274

Publisher: CRC/Balkema, Taylor & Francis

1. Numerical Modeling of Coupled Phenomena in Science and Engineering: Practical Use and Examples
   Editors: M.C. Suárez Arriaga, J. Bundschuh & F.J. Domínguez-Mota
   2009
   ISBN: 978-0-415-47628-72.

2. Introduction to the Numerical Modeling of Groundwater and Geothermal Systems: Fundamentals of Mass, Energy and Solute Transport in Poroelastic Rocks
   J. Bundschuh & M.C. Suárez Arriaga
   2010
   ISBN: 978-0-415-40167-83.

3. Drilling and Completion in Petroleum Engineering: Theory and Numerical Applications
   Editors: Xinpu Shen, Mao Bai & William Standifird
   2011
   ISBN: 978-0-415-66527-8

4. Computational Modeling of Shallow Geothermal Systems
   Rafid Al-Khoury
   2011
   ISBN: 978-0-415-59627-5

5. Geochemical Modeling of Groundwater, Vadose and Geothermal Systems
   Editors: J. Bundschuh & M. Zilberbrand
   2011
   ISBN: 978-0-415-668101-1

6. Mathematical and Numerical Modeling in Porous Media: Applications in Geosciences
   Editors: Martín A. Díaz Viera, Pratap N. Sahay, Manuel Coronado & Arturo Ortiz Tapia
   2012
   ISBN: 978-0-415-66537-7